Louise Gillette
(619)438-1216

Dictionary of Botany

D0059094

The Wordsworth
Dictionary of Botany

–

George Usher

Wordsworth Reference

First published as *A Dictionary of Botany* by Constable & Co Ltd,
London, 1966.

This edition published 1996 by Wordsworth Editions Ltd.
Cumberland House, Crib Street, Ware, Hertfordshire SG12 9ET.

Copyright © George Usher 1966.

All rights reserved. This publication may not be
reproduced, stored in a retrieval system,
or transmitted, in any form or by any means, electronic,
mechanical, photocopying, recording or otherwise,
without the prior permission of the publishers.

ISBN 1-85326-374-5

Printed and bound in Great Britain by Mackays of Chatham PLC.

AUTHOR'S PREFACE

THIS work has been compiled to meet the needs of university students and grammar school pupils. It has been difficult to decide the limits of botany as a subject. As the scope of each science is widening and they are becoming integrated, I have endeavoured to take the broadest view. In an attempt to achieve this, terms which are used in soil-science, statistics and biochemistry have been incorporated as well as those particular to botany. At the risk of being anachronistic, obsolete terms have been included. Definitions of all the phyla, classes, orders and families have been given, but descriptions of genera and species have been omitted.

The problems of plant taxonomy are so immense that to include the various differences described by different authors would involve writing several volumes. A line had to be drawn somewhere, and the following system has therefore been adopted. The flowering plants have been described according to Engler and Prantl's "*Die Natürlichen Pflanzenfamilien*" (EP), and Hutchinson's "*Families of Flowering Plants*" (H), referring to Bentham and Hooker's "*Genera plantarum*" (BH). If two or more authors describe an order etc., Engler and Prantl's description has been adopted. The gymnosperms have been described according to Engler and Prantl, whilst the descriptions of all the other plants, except the fungi, are in accordance with G. M. Smith's "*Cryptogamic Botany*". Those of the fungi follow E. A. Bessey's "*Morphology and Taxonomy of Fungi*".

In the text the sign = means that the terms are synonymous, e.g. SWARM SPORE = ZOOSPORE, but See "————" means that the term is referred to in the text indicated, e.g. CROSS BREEDING See "*Cross*".

A dictionary is hardly an original work, but rather a collection from the works of other writers. To the authors of the very many works I have consulted, and sometimes quoted, I offer my sincere thanks, emphasizing that any misunderstanding in the definitions is my responsibility entirely. Inevitably in a book of this nature, mistakes will have crept in, and I would be grateful if readers would draw my attention to them so that corrections can be made in a revised edition should it be called for.

GEORGE USHER

August, 1965.

A

A-. A class of flowers with freely exposed nectar.

Å = Ångström unit.

A-LAYER (HORIZON). The top layer of a soil usually divided into A_{00}—leaf-litter, A_0—raw humus, A_1—dark-coloured humus layer, A_2—the next layer which may or may not be leached.

A-SPORE = ALPHA-SPORE.

AB. A flower with partially concealed nectar.

Ab-. Latin prefix meaning 'from'.

ABAXIAL. The surface facing away from the axis.

ABBREVIATED. Shortened rather suddenly.

ABERRANT. Showing some unusual difference in structure, having characteristics not strictly according to type.

ABERRATION. A peculiarity of an individual plant, usually due to some special environmental factor, and not capable of transmission to the offspring.

ABHYMENIAL. Opposite the spore-producing surface.

ABIOGENESIS. The concept that life can arise from dead matter. The theory was generally supported until it was finally repudiated by Pasteur. (The theory of Spontaneous Generation.)

ABJECTION. The separation of a spore from its stalk, forcibly by the fungus.

ABJUNCTION. The separation of a spore from its stalk by means of a septum.

ABOÖSPORE. An oöspore produced without sexual fusion.

ABORTION. A state of being incompletely developed, or producing such a state.

ABRUPT. As if cut off horizontally; truncate.

ABRUPTLY PINNATE. Said of a compound leaf without a terminal leaflet.

ABSOLUTE GROWTH RATE. The rate of increase in size of a growing plant (or part of it) in a given time, under specified conditions.

ABSCISS LAYER = ABSCISSION LAYER.

ABSCISS-PHELLOID. The unsuberised cells of an abscission layer in the bark.

ABSCISSION. (1) The shedding of leaves, foliage branches, floral parts, fruits, or bark, by the laying-down of an abscission layer. *i.e.* in contrast to decaying or falling off. The abscission of leaves is probably retarded by the migration of auxin from the leaf-blade.
(2) The liberation of a fungus spore by the disappearance of an adjoining layer, or wall.

ABSCISSION LAYER. (See *abscission*). A layer of parenchyma bounded on both sides by cork. Found at the base of leaf-petioles etc. of woody dicotyledons and gymnosperms. Disintegration of the parenchyma causes the organ to be separated from the parent. A similar tissue in bark causes it to scale.

ABSOLUTE TRANSPIRATION. The rate of water loss from a plant determined experimentally.

ABSORPTION RATIO. The ratio of the final internal concentration of a particular ion in plant tissue (i) to the final external concentration (e). That is i/e.

ABSTRICTION. A general term for the release of a spore from its stalk, especially by constriction.

ACANACEOUS. Prickly.

ACANTHA. A prickle or spine.

ACANTHACEAE. A family of the Tubiflorae (EP, BH), Personales (H). Found in the tropics and sub-tropics. Mostly shrubs or herbs with a wide range of habit. The leaves are usually opposite and decussate, and entire, borne on tumid nodes. There are no stipules. The inflorescence is commonly a dichotomous cyme, often short in the leaf-axil. Racemes occur, and solitary flowers are common. Coloured, large bracts and bractioles are usually present. The flowers have stamens and ovaries. The ovary stands above the other floral parts, of the irregular flower, usually with a nectiferous disc below the ovary. The four or five parts of the calyx are fused, as are the petals, into two lips. Four or two (rarely five) stamens are fused on the petals, and one to three staminoids are frequently present. Anthers often have one lobe smaller than the other. The two fused loculi of the ovary lie in the plane of the floral axis. The axile placenta is usually elastic, with two to many anatropous ovules in each loculus. The fruit is a bi-locular capsule. There is usually no endosperm. The family contains many garden plants.

ACANTHINAE. A sub-order of the Tubiflorae, and including such genera as *Linum* (flax).

ACANTHO-. A Greek prefix meaning 'thorny'.

ACANTHOCARPOUS. Spiny-fruited.

ACANTHOCLADOUS. Spiny-branched.

ACANTHOPHYSIS. A thick, sterile hair with short-pointed outgrowths on the surface.

ACARODOMATIUM. A hollow protective structure formed by some mite-harbouring plants. This may be a symbiosis.

ACAROPHILY, ACROPHYTISM. A symbiosis between plants and mites.

ACAYALLAGIC. See *Caryallagic*.

ACAUDATE. Without a tail.

ACAULINE, ACAULOSE, ACAULESCENT. Stemless, or nearly so.

ACAULIS. Latin, meaning 'without visible stem'.

ACCELERATOR. Any substance which increases the efficiency of an enzyme.

ACCESSORY BUD. A bud additional to the normal axillary bud.

ACCESSORY CELL. A cell associated with the guard cell of a stoma, but differing in structure from it and from the ordinary cells of the epidermis.

ACCESSORY CHARACTER. A characteristic of a species which are not essential to its diagnosis, but is sometimes used to differentiate races.

ACCESSORY CHROMOSOME = SEX CHROMOSOME.

ACCESSORY FRUCTIFICATION. An asexual reproductive structure of lower plants.

ACCESSORY MULTIPLICATION. Any asexual reproductive process.

ACCESSORY SPORE. (1) An asexual spore.

(2) A conidium differing from the normal type for the species.

ACCOMMODATION. The capacity of a plant to adapt itself to changes in the environment, providing that the changes occur gradually.

ACCRESCENT. Enlarged and persistent. Especially of a calyx which enlarges as the fruit ripens.

ACCULEATE. Having narrow spines.

ACCUMBENT. (1) Lying against.

(2) Of an embryo having the edges of the cotyledons against the radicle.

ACENTRIC. A piece of a chromosome or chromatid without the centromere.

ACEPHALOUS. Lacking a head. Said of a style without a well-developed stigma.

ACERACEAE. (EP, BH, H) A family of the Sapindales. Spineless trees or shrubs. The leaves are simple, palmately-lobed or pinnate, lack stipules, and are opposite. There are more than two sepals. There are few ovules, and a disk is present. The ovary is bilocular, developing into a two-winged indehiscent samara. Confined to the temperate areas of the Northern Hemisphere. *e.g.* Maple, Sycamore, Sugar Maple.

ACEROSE, ACEROUS. Needle-shaped.

ACERVATE. Heaped: growth in heaps or groups.

ACERVULUS. A compact mass of hyphae bearing conidiophores and conidia. Formed by some parasitic ascomycetes. The corresponding structure of saprophytic ascomycetes is called a *sporodochium.*

ACETABULIFORM. Saucer-like.

ACETALDEHYDE. CH_3CHO. Probably an intermediate in the aerobic and anaerobic respiration of carbohydrates. Retards or inhibits germination and growth of some fungi.

ACETYLATION. The production of an acetyl radical from pyruvate during photosynthesis. The enzyme concerned is Co-enzyme A. The complete reaction is represented by:—

(1) $CH_3. CO. COO^- + CoA + 2H_2O - 2H$
pyruvate

$$\longrightarrow$$

$CH_3. COO. CoA + HCO_3^- + H^+$
acetyl CoA

(2) Acetyl CoA + oxalacetate $\longrightarrow$ citrate + CoA

N-ACETYL-2-GLUCOSAMINE. A glucose derivative, and the unit which repeated in the chitin molecule:—

ACETYL CoA. (Acetyl Coenzyme A) Activated acetic acid, with an energy-rich sulphur bond. One of the end products of the oxidative decarboxylation of pyruvic acid, produced near the beginning of the Citric Acid (Kreb's) Cycle. See DPN and DPNH.

ACETYL-S-CoA. An intermediate produced by Coenzyme A when it detaches carbon dioxide from the αketo-acid during respiration.

-ACEUS. A Latin suffix meaning 'like'.

ACHARIACEAE (H, BH). A family of the Parietales. Herbs or small shrubs containing three genera and three species which are confined to South Africa. The flowers are regular and monoecious. The calyx has 3-5 free sepals, the petals are also 3-5 and fused. The 3-5 stamens are fused to the petals. The superior ovary consists of 3-5 fused carpels, each of which contains numerous ovules on parietal placentas. The fruits are capsules, and the seeds are endospermous.

ACHATOCARPACEAE (H). A family of the Bixales. There are no stipules and the unisexual flowers lack petals. The embryo is ring-shaped.

ACHENE, ACHAENOCARP, AKENE. A dry indehiscent, one-seeded fruit, formed from a single carpel and with the seed distinct from the fruit wall.

ACHLAMYDEOUS. Of a flower, lacking a perianth.

ACHLAMYDOSPOREAE (BH). The sixth series of the Monochlamydeae. The ovary is unilocular and contains 1-3 ovules which are not apparent until after fertilization. The seeds are endospermous, but lack a testa and are adnate to the receptacle or pericarp.

ACHLYOGETONACEAE. A family of the Chytridiales found in fresh-water algae and some nematodes.

ACHROMATIC. Without colour.

ACHROMATIC SPINDLE (See 'Spindle'). In cell-division, a system of apparent fibres which are not stained by basic dyes. They join the poles of the nucleus, and diverge towards the centre of the cell.

ACHROMATIN. The nuclear sap, which does not stain with basic dyes.

ACHROOS. Colourless.

ACICULAR. Needle-shaped; stiff, slender and pointed.

ACICULATE. Having fine scratches on the surface.

ACID DYE. Dyes having an active acid radical, combining with metals, and staining particularly cytoplasm.

ACID FERMENTATION. (See *Fermentation*).

ACID HUMUS. A humus formed in a soil lacking calcium, and other bases.

ACIDOPHILE. Of cells (or cell contents) stained by acid dyes.

ACID PLANT. A plant having acid sap (with pH less than 7) due to the production of ammonium salts of organic acids, *e.g.* malic or oxalic acids. *e.g.* Wood Sorrel, and Begonia.

ACIES. The edge.

ACINACEOUS. With numerous pips.

ACINACIFORM. Curved, scimitar-shaped.

ACME. The period of maximum vigour of an individual, race, or species.

ACONITRASE. An enzyme involved in the Citric Acid Cycle (Kreb's Cycle). It dehydrates citrate forming *cis*-aconitrate, which is then transformed to D-isocitrate, by the addition of a water molecule.

ACOTYLEDONES = CRYPTOGAMAE.

ACOTYLEDONOUS. Without cotyledons (said of the embryo of a higher plant).

ACQUIRED CHARACTERISTICS. The characters of an individual developed due to environmental conditions. Adherents to the theory of acquired characters believe that these characters can be inherited.

ACQUIRED VARIATION. A structural anomaly which becomes evident as the individual develops.

ACRANDROUS. Having antheridia at the top of the stem, as in some mosses.

ACRASIALES. An order of the Myxothallophyta, Mycetozoa, or the single order of the Acrasieae, depending on the authority. Sometimes united with the Labyrinthuales as a separate subclass. They are saprophytic, usually on dung. The swarm-spores lack flagella, giving rise to myxamoebae as the vegetative phase. These aggregate to form a pseudoplasmodium of individual cells during spore-formation. The naked fruiting-body is stalked or sessile, kept together by slime.

ACRASIEAE. A class of the Myxothallophyta. (For characters see *Acrasiales*.)

ACRO-. Greek prefix meaning 'apical'.

ACROCARPOUS. Borne at the end of a branch.

ACRODROMOUS. Of the veins of leaves, which having run parallel to the edge of the leaf, fuse at the tip.

ACROGAMAE. A division of the Angiospermae in which fertilization takes place through the micropyle (the usual way). A method of classification proposed by Treub.

ACROGENS. The ferns and mosses.

ACROGENOUS. Formed at the apex.

ACROGYNAE. A sub-order of the Jungermanniales. The leafy gametophytes generally have the leaves in three vertical rows. The apical cell of the branches of the gametophyte develops into an archegonium, so that the sporophyte is always terminal.

ACROGYNY. Said of the gametophyte of some Bryophyta, when growth stops on the formation of the archegonia. The apical cell usually forming the last archegonium.

ACRONYCHIUS (Latin). Meaning 'curved like a claw'.

ACROPETAL. Produced in succession from the base upwards, so that the oldest members are at the base, and the youngest at the top.

ACROPETALOUS INFLORESCENCE = 'RACEME'.

ACROPHYTIUM. An alpine plant-formation.

ACROPLEUROGENOUS. Borne at the tips and along the sides.

ACROSCOPIC. On the side towards the apex.

ACROSPIRE = PLUMULE.

ACROSPORE. A spore borne at the tip of a hypha.

ACROSTICHOID SORUS. A fern sorus consisting of several fused sori.

ACROTONIC. Of orchids having the tip of the anther next to the rostellum.

ACTINIDIACEAE (H). A family of the Theales. The simple leaves are alternate. The bracts are not spurred or modified to form pitchers, and the plants are not epiphytic. The petals are imbricate, and the pedicel has no bracts. There are numerous stamens. In the bud the anthers are inflexed.

ACTINODROMOUS. Having the main veins of a leaf radiating from the tip of the petiole.

ACTINOMORPHIC. Of flowers etc. that are radially symmetrical.

ACTINOMYCETACEAE. A family of the Actinomycetales, producing a mycelium, arthrospores, but not producing aerial conidia.

ACTINOMYCETALES. A group of organisms intermediate between the bacteria and fungi. There is a fine mycelium, not exceeding $1\cdot5\ \mu$ in diameter, and are Gram positive. Reproduce by fission, arthrospores, and/or conidia. No endospores are produced, and sexual reproduction is not known. Some cause diseases of plants and animals. Some, *e.g. Streptomyces* produce valuable antibiotics.

ACTINOMYCOSIS. A disease of man or animals caused by an actinomycete *e.g.* Lumpy Jaw of Cattle.

ACTINOSTELE. A protostele which is stellate in cross-section, with the protoxylem at the tips of the star. A primitive type of stele found in the early Pteridophytes.

ACTION SPECTRUM. Relates the amount of photosynthetic activity for an equal amount of light to the wavelength of the light.

ACTIUM. A rocky sea-shore formation.

ACTIVATION BY LIGHT. The changing of a complex chemical into a different one by light energy, during photosynthesis.

ACTIVATOR = 'ACCELERATOR'.

ACTIVE TRANSPORT. The accumulation of materials in a cell, against the concentration gradient, *i.e.* involving the use of energy.

ACULEATE. Bearing prickles, or covered with needle-like outgrowths.

ACULEOLATE. Having somewhat spine-like processes.

ACUMINATE. Narrowing to a point.

ACUTANGULAR. Said of a stem which has several longitudinal sharp edges.

ACUTE. Having a sharp, and rather abrupt point: said usually of a leaf-tip.

ACUTIFOLIUS. Latin, meaning 'with an acute leaf'.

ACYCLIC. Of floral parts not in whorls, but in spirals.

AD-. (1) Latin prefix meaning 'to'.
(2) -AD, a suffix denoting an ecad.

ADAPTABLE. Able to originate an ecad.

ADAPTATION. Adjustments of an organism to their environment. These adaptations may be to individuals or groups, and may be inherited, or not inherited.

ADAXIAL. The surface, usually of a leaf, facing the stem. Next to the axis.

ADAPTED RACE = PHYSIOLOGIC RACE.

ADAPTIVE ENZYME. An enzyme produced, especially by bacteria in response to growing on an unusual substrate, *i.e.* the enzyme is not normally present, but develops soon after growth on the new substrate has begun, so that it can be utilized.

ADAPTOR COMPOUND. A carrier compound, *e.g.* RNA, responsible for the transfer of a substance to the site of reaction.

ADCRUSTATION. The deposition of a substance on a cell-wall, in contrast to being laid down between the existing molecules. This latter process is called encrusting.

ADDITIVE FIXATIVE. A fixative that combines chemically with the protein of the material being preserved.

ADELPHOGAMY. Union between a vegetative mother-cell and one of its daughter cells.

ADELPHOUS. Said of an androecium with the filaments of the stamens partly or wholly united.

ADEN-. Greek prefix meaning 'a gland'.

ADENINE. 6-aminopurine. Combines with the sugar ribose to form the nucleoside adenosine. This occurs in RNA, DNA, ATP, ADP, and AMP. Adenine is found in beet-juice, and tea-leaves. It also promotes bacterial growth.

ADENOID. Gland-like.

ADENOPHORE. A stalk supporting a gland.

ADENOSE. Having glands; gland-like.

ADENOPHYLLOUS. Having glandular leaves.

ADENOSINE. A nucleoside derived from the combination of adenine with a ribose sugar, and giving rise to the nucleotide adenylic acid, one of the nucleic acids.

ADENOSINE DIPHOSPHATE (ADP). The diphosphate ester of adenosine. Closely linked with adenosine triphosphate in the transfer of energy through 'high energy' bonds during the respiration of carbohydrates. The energy is released during the production of ester phosphates by the ADP, which acts as a phosphorylating agent.

ADENOSINE MONOPHOSPHATE. A nucleotide, the monophosphate derivative of adenosine. = Adenylic acid.

ADENOSINE TRIPHOSPHATE (ATP). The triphosphate ester of adenosine. Found in mitochondria. Two of the phosphate radicals have 'high energy' links of the pyrophosphate type (P-O-P). These are changed to 'low energy' links during the formation of ester phosphates by the ATP. Thus energy is released during phosphorylation. ATP also acts as a phosphate donor, converting glucose to glucose-6-phosphate. This is necessary before glucose can further be converted to a disaccharide or polysaccharide. Glucose + ATP → Glucose-6-phosphate + ADP.

ADENYLIC ACID = ADENOSINE MONOPHOSPHATE. A constituent of DPN.

ADERMIN. A vitamin necessary for the growth of lactic acid bacteria, certain yeasts, and fungi.

ADESMY. An abnormal condition in a flower, when parts that are normally united are separate.

ADHESION. The fusion of members of distinct whorls of a flower, *e.g.* when the calyx grows up with the ovary wall. In the wider sense the union of dissimilar organs.

ADIABATIC. Incapable of translocation.

ADNASCENS (Latin). Meaning 'growing upon something'.

ADNATE. Referred to various parts of fungi. Of gills and tubes, widely joined to the stipe; of pellicle, scales etc. tightly attached to the surface.

ADNATION. (1) One organ attached to another by its whole length.
(2) The fusion of the vascular bundles of floral parts.
(3) Sometimes used as an equivalent to *adhesion*.

ADNEXED. Of the gills and tubes of Basidiomycetes. Touching, but not attached to the stipe.

ADOXACEAE (EP, BH, H). A family of the Rubiales (EP, BH), Saxifragales (H). Contains one genus *Adoxa*. Found in North Temperate regions. The plant has a creeping rhizome which branches monopodially and bears flower-shoots with a few radical leaves and a pair of opposite cauline leaves. The flowers are in small heads of cauline flowers (usually 5). The outer whorl of the perianth usually has three parts, the inner of four parts on the terminal flower, and five on the lateral flowers. The inner whorl is fused at the base to form a short, broad tube. The whole flower is green and inconspicuous. The stamens alternate with the petals. The ovary is of 3-5 (rarely 2) fused carpels, which are semi-inferior, and contain one pendulous ovule each. The fruit is a drupe with several stones, and the seeds are endospermous.

ADP = 'ADENOSINE DIPHOSPHATE'.

ADPRESSED = APPRESSED.

ADSORPTION. A surface phenomenon whereby substances in solution migrate to a surface (phase boundary) building up to a higher concentration than in the rest of the solution. *E.g.* activated charcoal will remove Gentian violet out of solution. See *Freundlich's isotherm*.

ADSPERSED. Widely distributed; scattered.

ADUNCATE. Hooked.

ADVENTITIOUS. (1) Of organs or tissues developing in an abnormal position, *e.g.* roots developing from stems.
(2) Invading from other formations.

ADVENTIVE. A plant that has not become permanent in a particular area.

ADVERSE. Facing the main axis.

ADYNAMANDRY. Self-sterility.

AEGICERATACEAE (H). A family of the Myrsinales. Characterized by the absence of endosperm. The seed and embryo are long, and germination takes place within the fruit. The cotyledons are fused to form a tube around the plumule. The anthers are introrse, and the loculi of the fruit are transversely septate. The fruit is a capsule. The family is confined to tropical Asia, Malaya, and tropical and sub-tropical Australia.

AECIDIOSPORE = AECIOSPORE.

AECIDIUM. A cup-shaped aecium with the free edge of the periderm toothed. May be synonymous with aecium.

AECIOSPORE, AECIDIOSPORE. A spore formed in a aecium.

AECISPORE = AECIOSPORE.

AECIUM. A globular, cup-shaped, tubular or irregular fungus fruit-body which bursts through the epidermis of the host. Typically it consists of surrounding sterile fungus membrane—the periderm, enclosing the fertile portion which produces chains of aeciospores. Typical of the Rust fungi.

AENIUS. Latin for 'brass-coloured'.

AEQUI-. Latin prefix meaning 'equal'.

AEQUALIS. Similar in size

AEQUI-HYMENIIFEROUS. Having the hymenium developed equally over each gill surface.

AERATING ROOT. A root of some plants growing in soft mud. They contain aerenchyma, and stand above the surface of the ground acting as ventilating organs.

AERATION TISSUE = AERENCHYMA.

AERENCHYMA. A tissue of unthickened cells surrounding large airspaces. The walls formed by the cells are perforated so that the air spaces are continuous throughout the organ. This type of tissue is usually associated with hydrophytes.

AEREOLAE. Cracks developing in the surface of a crustaceous lichen due to growth strains. These cracks aid in aeration.

AERIAL PLANT = EPIPHYTE.

AERIAL ROOT. A root, usually arising adventitiously from a stem. Usually climbing organs, but may contain chlorophyll, or form a parasitic sucker, thus aiding in nutrition.

AEROBE, AEROBIONT. A plant needing elemental oxygen for respiration.

AEROBIC RESPIRATION. Enzymatic destruction of a substrate to release energy, involving elemental oxygen, evolving carbon dioxide and water.

AEROBIOSIS. Existence in the presence of oxygen.

AEROPHYTE = EPIPHYTE.

AEROTAXIS. The movement of a motile plant or gamete, in relation to the oxygen concentration in the water.

AEROTROPISM. A growth curvature in relation to the oxygen concentration.

AERUGINOSE, AERUGINOUS. Bluish-green.

AESTIVAL, AESTIVALIS (Latin). Occurring in the summer, or characteristic of summer.

AESTIVATION. The arrangement of the structure in a bud, usually of flower parts, especially sepals and petals.

AETHALLIUM, AETHALIUM. The fruit-body of some myxomycetes. It is sessile or slightly raised involving most of all of the plasmodium, and consists of a number of sporangia more or less confluent, and incompletely individualised.

AETIOLOGY = ETIOLOGY.

AEXTOXICACEAE (H). A family of the Celastrales. The leaves and inflorescence are covered with scales. There are no stipules, and the flowers are dioecious in axillary racemes, in a bud completely enclosed by the bracteole. An annular or glandular disk is present, but does not enclose the carpels or ovary.

AFFINITY. Likeness, especially in relationship.

AFFIXED. Inserted upon.

AFTER RIPENING. (1) A period through which some seeds must pass after ripening, before they will germinate.
(2) The chemical and physical changes which go on inside a seed or other dormant plant-structure, and lead to the development of conditions when growth can commence.
(3) A similar period required by the spores of some fungi.

AGAMANDROECIOUS. Having male and neuter flowers in the same inflorescence.

AGAMIC, AGAMOUS. Asexual.

AGAMOGENESIS. Asexual reproduction.

AGAMOGYNOECIOUS. Having female and neuter flowers in the same inflorescence.

AGAMOHERMAPHRODITE. Having hermaphrodite and neuter flowers in the same inflorescence.

AGAMOSPERMY. The production of seeds without the fusion of gametes.

AGAMOSPECIES = APOMITIC SPECIES. See *Apomixis*.

AGAMOTROPIC. Said of a flower which does not shut after having once opened.

AGAR-AGAR. A carbohydrate free of nitrogen produced by various sea-weeds and used as a mucilage on which micro-organisms are grown.

AGARICACEAE. A family of the Agaricales which has the basidia on radiating gills. The Mushrooms and Toadstools. Mostly saprophytes, but a few parasites.

AGARICALES. An order of the Basidiomycetae. The fruit-body is generally macroscopic, and the basidia are exposed, at least at maturity. The hymenium is one layer thick and is borne on gills, or pores. The basidium is club-shaped, simple, bearing apical sterigmata. They usually carry 4 basidiospores, but there may be 2 or 8.

AGARICOLOUS. Living on mushrooms or toadstools.

AGAVALES (H). An order of the Corolliferae. Tree-like perennials with a thick woody rhizome and the leaves clustered at the bottom or top of the stem. The leaves are thick and fleshy or fibrous and sometimes prickly. The flowers are often small in much-branched pannicles, and are bracteate. They are mostly actinomorphic and bisexual to dioecious with the perianth lobes dry or fleshy. There are 6 stamens with the bilocular anthers opening inwards or at the sides. The ovary is superior or inferior with 3 or 1 loculae. The ovules are axile or attached centrally. The fruit is a capsule or berry, and the seeds are endospermous. There are many species in Australia and in the tropics and sub-tropics, often in dry regions.

AGAVACEAE (H). A family of the Agavales. The perianth is not dry, and the segments are usually united into a tube. The leaves are often tufted at the top of the stem. The family is confined to the tropics and sub-tropics with a few genera in Australia, and includes New Zealand Hemp, and Bombay Aloe fibre.

AGDESTIDACEAE (H). A family of the Chenopodiales containing one genus. The anthers open by longitudinal slits, and the ovary of 2 or more locular is half inferior winged by the persistent dry calyx-segments. Found in tropical and central South America.

AGE AND AREA, THEORY OF. That the areas of distribution of the members of a group of allied species are correlated with their ages as species in any given region.

AGGLOMERATE. Crowded, or heaped into a cluster.

AGGLOMERATE SOIL. A soil with the grains uncoated, or with a loose friable coat. The non-mineral material forms aggregates in the spaces between the mineral particles. The particles are not stuck together, or only slightly so.

AGGLUTINATE. Stuck together.

AGGREGATE. Closely packed together.

AGGREGATE FRUIT. A collection of small simple fruits derived from a flower with several free carpels, *e.g.* blackberry.

AGGREGATE RAY. A group of closely placed, narrow vascular rays.

AGGREGATE SPECIES. A group of closely related species called by a single name.

AGGREGATION. The coming together of plants into groups.

AGRAD. A cultivated plant.

AGRESTAL, AGRESTIS (Latin). Growing in cultivated land, but not itself cultivated; *e.g.* a weed.

AGRIUM. A formation on cultivated land.

AGROCERIC ACID. $C_{16}H_{44}O_3$. A constituent of humus.

11

AGRONOMY. The study of the cultivation of field crops, and related topics.

AGROSTEROL. $C_{26}H_{44}O$. A constituent of humus.

AGROSTOLOGY. The study of grasses.

AGYNOUS. Said of an abnormal flower in which the gynecium is not developed.

AGYRIALES = ATICHIACEAE.

AIPHYTIUM. An ultimate formation.

AIR CAVITY, AIR SPACE. (1) A large intercellular space in a leaf into which a stoma opens.
(2) A cavity in the upper surface of the thallus of some liverworts, opening externally by an air-pore and containing chains of photosynthetic cells.
(3) A large intercellular space in which air is stored in some water-plants.

AIR PLANT = EPIPHYTE.

AIR PORE. See *stoma, air cavity*.

AIR SPACE. See *air cavity*.

AIR SPORA. Living cells, free-floating in the air, *e.g.* spores, pollen.

AITIOGENIC, AITIOGENOUS. Said of a reaction by a plant to an external stimulus, generally a movement of some kind.

AITIONASTIC. Said of the curvature of a plant-member in response to a diffuse stimulus.

AIZOACEAE (EP) = FICOIDEAE (BH) = FICOIDACEAE (H). A family of the Centrospermae, of 20 genera and 650 species, found chiefly in South Africa, Central and South America, Asia and Australia. These are xerophytic herbs, or undershrubs with opposite or alternate exstipulate leaves which are often fleshy. The flowers are in cymes of hermaphrodite regular flowers. The four or five perianth-lobes may be free or fused, and there are 3, 5, or many anthers. The ovary is of three or many fused carpels. It may be superior or inferior, of three loculae with many ovules in each. Dédoublement is common in the androecium, and then the other anthers are replaced by petaloid staminodis. Ovaries are usually superior with axile placentation. The fruit is usually a capsule. The seeds have curved embryos around the perisperm. *E.g.* Mesembryanthemum.

AKANIACEAE (EP, H) (Included in the Sapindaceae by BH). A family of the Sapindales with one genus *Akania* in East Australia. Tree with alternate imparipinnate leaves. The inflorescence is a panicle of regular, hermaphrodite flowers. The calyx and corolla are each of 5 free lobes, with the petals contorted. There is no disk, and usually 8 anthers, the 5 external ones being opposite the sepals. The superior ovary is trilocular with two anatropous, pendulous ovules in each. The fruit is a locular capsule. The seeds have a fleshy endosperm with a straight embryo.

AKARYOTE. The stage in the nuclear cycle of the Plasmodiophoraceae, before meiosis when little or no chromatin is visible in the nucleus. Sometimes applied to the same stage in the life-cycle of any lower plant.

AKENE = ACHENE.

AKINETE. A thick-walled, non-motile spore, containing oil or other food reserves, formed singly within a cell, with the spore wall indistinct from the cell-wall.

ALA (pl. ALAE). (1) One of the side-petals of a leguminous flower.
(2) A membranous outgrowth of a fruit, aiding in wind-dispersal.
(3) A narrow, leafy outgrowth down the stem of a decurrent leaf.

ALBASTRUM. Latin meaning 'flower bud'.

ALANGIACEAE (EP, H). A family of the Myrtiflorae (EP), Araliales (H), part of the Cornaceae (BH). Contains one genus *Alangium*, with 22 species found throughout the tropics. Trees or shrubs bearing cymes of flowers. The 4-10 sepals and petals are free, although the petals may be slightly coherent at the base, or absent. The stamens are usually the same number as the petals and are free. The inferior ovary has 1, 2, or 3 loculi, each with one pendulous ovule. The ovule has 2 integuments, and the fruit is a one-seeded drupe. The seeds are endospermous.

ALANINE. $CH_3CH(NH_2)$ COOH. An amino acid, occurring combined in proteins.

ALATE. Winged: applied to stems when decurrent leaves are present.

ALBIDUS. Latin meaning 'whitish'.

ALBINO. A more or less white plant. An abnormal form lacking pigment (or nearly so).

ALBUGINACEAE. A family of the Peronosporales containing the single parasitic genus *Albugo (Cystopus)*. The unbranched conidiophores are borne on the end of sympodially-branched hyphae, in a palisade-like layer under the epidermis of the host. The conidia are cut off at the apex of the conidiophore, in chains, with each conidium separated from its neighbour by a sterile disjunctor cell, which dissolves before dispersal. The conidia germinate forming zoospores.

ALBUMEN. An obsolete name for the endosperm.

ALBUMINS. A group of simple proteins which are soluble in water, and coagulated by heat.

ALBUMINOIDS. A general term for proteins.

ALBUMINOUS CELL. A cell, rich in contents, associated with the phloem in the stems and leaves of some Gymnosperms, and possibly conducting proteins.

ALBURNUM. An obsolete term for sap-wood.

ALBUS. Latin meaning 'white'.

ALCOHOL. A hydroxy-derivative of the paraffins. *E.g.* C_2H_6 is the paraffin ethane, and C_2H_5OH is the alcohol (ethyl alcohol). The alcohols occurring in plants include the monohydric alcohols, *e.g.* ethyl alcohol; the trihydric alcohols, with three OH groups, *e.g.* glycerol (occurring as its oxidized derivatives) and the polyhydric alcohols, *e.g.* sugars, with the general formula R.OH, where R is a fairly long chain, or cyclic.

$$\begin{array}{cc} H \quad H & CH_2OH \\ | \quad | & | \\ H—C—C—OH & CHOH \\ | \quad | & | \\ H \quad H & CH_2OH \\ \text{ethyl alcohol} & \text{glycerol} \end{array}$$

ALCOHOLIC FERMENTATION. The production of various alcohols during the anaerobic respiration of carbohydrates by various microorganisms. *E.g.* the production of ethyl alcohol by yeasts fermenting sugar found in malt, during brewing.

$$C_6H_{12}O_6 = 2C_2H_5OH + 2CO_2$$

$$\text{sugar} \qquad \text{ethyl} \qquad \text{carbon}$$
$$\text{alcohol} \qquad \text{dioxide}$$

ALDEHYDES. Oxidation products of primary alcohols, having the general

formula $C_nH_{2n}O$, but having a $-\overset{\displaystyle H}{\underset{\displaystyle \diagdown O}{C}}$ group. *E.g.* $H-\overset{\displaystyle H}{\underset{\displaystyle H}{C}}-\overset{\displaystyle H}{\underset{\displaystyle H}{C}}-OH$ is ethyl

alcohol, and $H-\overset{\displaystyle H}{\underset{\displaystyle H}{C}}-\overset{\displaystyle H}{\underset{\displaystyle O}{C}}\diagdown$ is the derived acetaldehyde.

ALDOHEXOSE. An aldose with six carbon atoms.

ALDOL. $CH_3.CH(OH).CH_2CHO$. The condensation product of two acetaldehyde molecules.

$$CH_3.CHO + CH_3.CHO = CH_3.CH(OH).CH_2.CHO$$

ALDOL CONDENSATION. A general term applied to the condensation of any two aldehyde molecules.

ALDOPENTOSE. An aldose with five carbon atoms, *e.g.* ribose.

ALDOSE. A monosaccharide sugar with an aldehyde (CHO) group in the first position, and a HCOH group in the second position. *E.g.*

$$
\begin{array}{l}
CHO \ldots\ldots\ldots 1 \\
H-C-OH \ldots\ldots 2 \\
H-C-OH \ldots\ldots 3 \\
H-C-OH \ldots\ldots 4 \\
CH_2OH \ldots. \ 5
\end{array}
$$

D-ribose (an aldopentose)

ALDOTETROSE. An aldose with four carbon atoms.

ALDOTRIOSE. An aldose with three carbon atoms.

ALEPIDOTE. Smooth; without scales or scurf.

ALEPPO GALL. A tumourous growth on various Oak species. Contains gallic acid, and gallitannic acid. Extract used for tanning leather.

ALEURIOSPORE. A terminal, thick-walled spore, formed from the end of a hypha, from which it is separated by a cross-wall.

ALEURONE GRAINS. Granules of protein found in cells. These cells often form a distinct layer—the aleurone layer, *e.g.* in the endosperm of Maize.

ALGAE. A division of the Thallophyta, containing all the holophytic members. All are aquatic, or sub-aquatic, varying from the unicellular types to the large seaweeds. There is never any vascular tissue, nor differentiation into stem, root, and leaf. The reproductive organs are essentially one-celled, and the gametes are mostly flagellate.

ALGANIN = ALGIN.

ALGICOLOUS. Living on algae.

ALGIN. The magnesium-calcium salt of alginic acid. Found in inner cell-wall of the Brown Algae. It is used commercially in the manufacture of confectionary, and to some extent in the manufacture of man-made fibres.

ALGOLOGY. The study of Algae.

ALIEN. A plant introduced by man, and which has become naturalized.

-ALIS. A Latin suffix meaning 'belonging to'.

ALISMACEAE = ALISMATACEAE. A family of the Helobiae (EP), Alismatales (H). BH include the Butomaceae (EP). Cosmopolitan, mainly in the northern hemisphere, with 11 genera and 75 species. They are perennial water, or marsh herbs, with rhizomes. The erect, radical leaves are floating, or submerged, and none of the family are saprophytes. The inflorescence is usually a much-branched raceme. The flowers are hermaphrodite or unisexual with the perianth of two whorls of three, calyx-like lobes. There are 6 to many extrorse anthers, or sometimes 3, and the superior ovary has 6 to many separate carpels, carpels each containing 1 (rarely 2 or more) anatropous ovules. The fruit is a group of achenes. The seeds lack endosperm, and the embryo is U-shaped. BH include the following characters:—variously-shaped leaves, introrse anthers, a large number of ovules scattered over the inner surface of the carpels, and a straight embryo.

ALISMATALES (H). An order of the Calyciferae. Herbs with a rhizome living in fresh or brackish water. A few are saprophytes. The leaves are radical, opposite, alternate, or clustered. The bisexual or unisexual flowers may be very small, and are subtended by a bract. The perianth is often divided into petals and sepals, but may be absent. There are 3 to many stamens. The superior gynecium is made up of separate carpels, or they may be fused at the base. The many to 1 ovules are basal, or at the inner angle. There is no endosperm.

ALKALOIDS. Cyclic vegetable-bases, with a heterocyclic ring, and containing nitrogen. They have a physiological effect on animals, *e.g.* morphine, and atropine. Sometimes used to include nitrogen bases that do not have a heterocyclic ring, *e.g.* caffeine.

ALIFORM. Wing-shaped.

ALLANTOID. Sausage-shaped.

ALLENTOSPHAERIACEAE. A family of the Sphaeriales. Saprophytes, or weak parasites on twigs etc. The stroma is not entirely of hyphae. The asci on long stalks are intermingled with sterile hairs that disappear soon after maturity. The whole forms a hymenium. The ascospores are yellowish to colourless, and are mostly allantoid. The conidia are long and slender.

ALLELE, ALLELOMORPH, ALLELOMORPHIC GENE. Different forms of a gene, having the same locus on homologous chromosomes, and are subject to Mendelian (alternative) inheritance.

ALLELISM. The relationship between alleles.

ALLIACEOUS. Onion-like, in smell or form.

ALLIANCE. A sub-class, consisting of a number of selected families of plants.

ALLITIC SOIL. A soil in which the clay fraction has a high proportion of aluminium.

ALLOCARPY. Fruiting after cross-fertilization.

ALLOCHRONIC SPECIATION. The production of new species during the passage of time, tending to form a gradation from one species to the next.

ALLOCHROUS (ALLCHROOUS). Changing from one colour to another.

ALLOGAMY. Cross-fertilization.

ALLOMETRIC COEFFICIENT. The ratio of relative growth rates. Expressed as d logY /d logX, where Y is the measure of the organ, or part of it, and X is the measurement of the whole organism, or organ.

ALLOPATRIC SPECIATION. The formation of new species over a length of time, by the geographical isolation of groups of the common ancestor, *e.g.* by a mountain range.

ALLOPHANOIDS. Hydrated aluminium silicates found in soils. They are gels having the ratio of silicon dioxide to aluminium oxide varying about 2.

ALLOPOLYPLOIDY. An artificial polyploid formed by the hybridization of different species, *i.e.* the individual has two or more sets of chromosomes, which have different chromosome compliments.

ALLOSOME. See *Sex chromosome*. Any chromosome other than a typical one.

ALLOSOMAL INHERITANCE. Inheritance of characters carried on an allosome.

ALLOSYNDESIS. (1) The pairing in a cross of two polyploids, or of chromosomes derived from opposite parents.
(2) Pairing in an allopolyploid between chromosomes derived from ultimate diploid ancestors, opposed to autosyndesis.

ALLOTETRAPLOID. An allopolyploid formed by the doubling of the chromosome number in a diploid hybrid: *i.e* a tetraploid with a diploid set of chromosomes from each parent.

ALLOTROPOUS FLOWER. A flower having the nectar freely-exposed and available to a wide variety of insects.

ALLOTYPE. One of the original types used to describe a new species.

ALLUVIAL SOIL. A soil derived from marine, esturine, or river deposits. A young soil, with an undeveloped profile. Usually has a high fertility.

ALPESTRIS. Latin meaning 'growing at a high altitude, but below the snow-line'.

ALPHA-KETOGLUTARATE OXIDASE. An enzyme in the Citric Acid Cycle (Kreb's Cycle), which oxidises α-ketoglutarate to succinic acid, by oxidative decarboxylation.

ALPHA-SPORE. A fertile spore of the imperfect state of the Diaporthaceae. This family of the fungi also produces beta-spores, which are generally filiform and sterile.

ALPINE. Growing at high levels, *i.e.* above the tree-line.

ALSAD. A grove plant.

ALSIUM. A grove formation.

ALSTROEMERIACEAE (H). A family of the Alstroemeriales. Found in Central and South America. The stems are herbaceous, and erect, or woody and climbing. The inflorescence is terminal, and frequently surrounded by a whorl of leaves. The ovary is interior, and the fruit a capsule.

ALSTROEMERIALES (H). An order of the Corolliferae, found mainly in the southern hemisphere. The rootstock is a rhizome with fibrous, or tuberous roots. The stems are erect or climbing with alternate, linear to ovate leaves. The flowers are showy and borne in terminal clusters, or racemes. There are 6 perianth segments which may be free or partly fused, and are usually alike, but may be dissimilar. The 6 stamens are usually free, but may be partly connate.

The ovary is superior or inferior either 3-locular with parietal placentation, or unilocular with parietal placentas. The fruit is a capsule or berry, and the seeds have endosperm.

ALTERNATE. Of leaves or branches that are arranged singly on the parent axis.

ALTERNATE HOST. One of the two unlike hosts of a heteroecious rust.

ALTERNATION OF GENERATIONS. The alternate production of independent haploid sexual, and diploid asexual, generations in a life-cycle. This may be further complicated in some Algae and Fungi. If the two generations look alike, the alternation is *homologous*, and if they do not look alike, the alternation is *antithetic*.

ALTERNATIVE HOST. One of the hosts of a plant pathogen, usually referred to the wild host(s) of a disease-causing organism of a cultivated plant.

ALTERNATIVE INHERITANCE. See *Allelomorph*.

ALTERNE. A sudden change in a plant species which is spread over a wide area. Such a change is brought about by variation in the soil or other environmental conditions.

ALVEOLE ALVEOLUS. (1) A small surface pit.
(2) A pore of a polyporous fungus.

ALVEOLATE. Pitted.

AMARANTACEAE (EP, BH). A family of the Centrospermae (EP), Curvembryae (BH), found in temperate and tropical regions, and contains 72 genera, and 700 species. They are generally herbs or shrubs with opposite or alternate, entire leaves, without stipules. The flowers are bisexual (rarely unisexual), and regular, borne in axillary cymes, or are solitary. The perianth segments are membranous (usually), free or fused, and 4-5 in number. There are 1-5 stamens opposite the members of the perianth. These may be free, or fused (a) to each other, (b) to the perianth, or (c) to the disk. The superior ovary consists of 2-3 fused carpels, which may be united to the perianth. There is one loculus containing many to 1, campyloptropous ovules. The fruit is a berry or not. The embryo is curved, and the seed contains endosperm.

AMARANTHACEAE (H). A family of the Chenopodiales (H). A widely dispersed family. There are no stipules and the flower has no petals. There are 3-5 calyx lobes with the stamens opposite them. The anthers open by longitudinal slits, and the filaments are joined at the base, often with a staminode between them. The ovary is unilocular, and the fruits are free, not united in a fleshy mass.

AMARYLLIDACEAE (EP, H). BH include the Velloziaceae. A family of the Liliiflorae (EP), Epigynae (BH), Amaryllidales (H). There are 90 genera, with 1050 species, mainly found in the tropics and sub-tropics. They are usually xerophytes, with bulbs or rhizomes. The inflorescence is usually on a scape, with a spathe. The flowers may be solitary, in a cyme, or umbel-like head. The hermaphrodite flowers are regular, or zygomorphic. The perianth has 6 segments in two whorls of 3. These may be free or united, and are petaloid. The anthers are in 2 whorls of 3, free, and usually introrse. Staminodes may be present. The gynecium is of 3 fused carpels and is inferior (rarely half-inferior or superior) with 3 loculi (rarely 1) with axile placentation, and numerous anatropous ovules. A corona present in some members, *e.g.* Daffodil. The fruit is a capsule or berry. The seeds are endospermous with a small straight embryo. Vegetative reproduction by bulbils is common.

17

AMARYLLIDALES (H). An order of the Corolliferae. Herbs with bulbs (rarely a rhizome). The leaves are radical, and usually linear. The flowers are showy in umbels (rarely solitary) on a leafless stalk, subtended by one or more papery bracts. There are generally 6 stamens. A corona may be present. The 3-carpelled ovary is superior or inferior, usually having 3 loculi with axile placentas. The fruit is a capsule or berry. *E.g.* Onion, Daffodil.

AMBIENT. Relating the environment or surroundings.

AMBIGUOUS. Of uncertain origin, or doubtful systematic position.

AMBILINEARITY. The inheritance of cytoplasmic particles (plastogenes or plasmagens) from either male or female parent.

AMBIPAROUS BUD. A bud containing both young vegetative leaves and young flowers.

AMEIOSIS. Having one division (instead of two) of the nucleus at meiosis, so that the chromosome number of the mother-cell is not reduced.

AMENTACEOUS. Bearing catkins.

AMENTIFERAE = AMENTACEAE. The catkin-bearing plants. This group is unacceptable taxonomically.

AMENTIFEROUS = AMENTACEOUS.

AMENTIFORM. Catkin-like.

AMENTUM. A catkin.

AMERISTIC. Unable to complete the development, due to lack of nourishment.

AMEROSPORE. A one-celled spore.

AMIDE. A derivative of an organic acid in which the hydroxyl (OH) group is replaced by an amino (NH_2) group. *E.g.* $CH_3.COOH$ is acetic acid, and $CH_3.CO.NH_2$ is acetamide.

AMIDE PLANT. A plant which accumulates amides, especially asparagine or glutamine, rather than ammonium salts. There is probably no clear-cut distinction between these plants and *acid plants (q.v.).*

AMINE. A derivative of ammonia in which one or more of the hydrogen atoms is replaced by alkyl group(s), *e.g.*

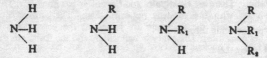

ammonia   Primary amine   secondary amine   tertiary amine

AMINE. A salt (ester) of an amino-acid.

AMINO ACID. A derivative of acetic acid in which one hydrogen of the alkyl group is replaced by an NH_2 group. *E.g.* $CH_3.COOH$ is acetic acid and $CH_2.NH_2.COOH$ is amino-acetic acid (glycine). The substitution of a second hydrogen by one of a number of side-chains gives rise to a series of amino acids. *E.g.* $CH.NH_2.CH_3.COOH$, is analine. Amino acids are synthesized by autotrophic plants and ultimately built up into proteins.

AMITOSIS. The division of the nucleus by a median constriction, without a normal mitosis taking place. The chromosomes do not become visible, and it is unlikely that two identical nuclei result.

AMMOCHTHAD, AMMODYTE. A plant living in sand.

AMMOCHTHADIUM. A sand-bank formation.

AMMONIA PLANT. A plant which forms ammonia and organic acids from amino acids.

AMMONIFICATION. The decomposition of nitrogenous, organic substances to produce ammonia.

AMMONIFIERS. Bacteria capable of liberating ammonia from proteins, or protein derivatives.

AMMONISATION = AMMONIFICATION. Applied especially to bacteria producing nitrogenous substances in the soil.

AMOEBIDIACEAE. A family of the Eccrinales, with the one genus *Amoebidium*.

AMOEBOID. Resembling an amoeba—having no cell-wall, or fixed form: creeping, and putting out pseudopodia.

AMORPH. A recessive mutant gene which has no obvious effect on the characteristic affected by the non-mutant allelomorph. Its lack of effect is the same as its absence.

AMP = ADENOSINE MONOPHOSPHATE.

AMPELIDACEAE (H) (BH) = VITACEAE (EP) (H).

AMPHIASTER. The two asters and the spindle connecting them. Produced during mitosis or meiosis.

AMPHICARPIC. Having two kinds of fruits.

AMPHICLINOUS PROGENY, AMPHICLINOUS HYBRID. A family resulting from a cross, and having some of the offspring resembling one parent, and some the other.

AMPHICRIBRAL BUNDLE. A vascular bundle in which the phloem surrounds a central strand of xylem.

AMPHIDIPLOID = ALLOTETRAPLOID.

AMPHIGASTRIUM. The lower leaves of a leafy liverwort. They are always inserted transversely.

AMPHIGENOUS. Growing all around.

AMPHIGYNOUS. Said of the antheridium of the Pythiaceae, which has the oögonial incept growing through it.

AMPHIKARYON. A nucleus with two haploid sets of chromosomes. A normal diploid nucleus.

AMPHILEPSIS. Inheritance of characteristics from both parents.

AMPHIMIXIS. The fusion of two distinct cells and nuclei, *e.g.* antheridium and archegonium, resulting in the formation of a new individual.

AMPHIPLOIC SIPHONOSTELE. See *Solenostele*.

AMPHISPHAERIACEAE. A family of the Sphaeriales, distinguished by the circular opening of the perithecium.

AMPHISPORE. A thick-walled uredospore, produced under dry conditions by some Uredinales, and germinating after a resting period.

AMPHITENE. The stage in meiosis when the spireme threads are united in pairs.

AMPHITHECIUM. (1) The outer layer of the embryo of the Bryophytes, derived early by the periclinal division of the cell.
(2) The thalline margin of a lichen apothecium.

AMPHITRICHOUS. Having a flagellum at each end of the cell.

AMPHITROPOUS. Of an ovule which bends both ways on the stalk.

AMPHIVASAL BUNDLE. A vascular bundle in which the xylem surrounds the central strand of phloem.

AMPHOLYTE = AMPHOTERIC ELECTROLYTE.

AMPHOTERIC. Having both acidic and basic properties.

AMPHOTERIC ELECTROLYTE. A substance that can exist as positively or negatively charged ions in solution, *e.g.* proteins and their derivatives. The pH of the solution determines the charge on the ions. The pH at which the number of positive and negative ions is equal, is called the isoelectric point.

AMPLECTANT, APLECTANS. Latin meaning 'covering, embracing'.

AMPLEXICAUL. Said of a sessile leaf with its base clasping the stem horizontally.

AMPLIATE. Enlarged.

AMPULLIFORM, AMPULLACEAL, APULLACEOUS. Flask-shaped.

AMYGDALIN, AMYGDALINE. A glucoside containing a cyanide radicle, giving the characteristic flavour to almonds.

AMYGDOLOID. Almond-shaped.

AMYLACEOUS. Starchy.

AMYLASE. An enzyme responsible for the break-down of starch; part of the diastase complex. β amylase attacks the amylose part of the starch molecule, breaking it down to maltose. α amylose attacks the remaining part of the molecule, and any amylose breaking it down to dextrins (5-10 glucose units), which may be further attacked by the β amylase. Only the 1:4 linkages in the amylose, and amylopectin are attacked by these two enzymes.

AMYLOID. (1) Of spores, turning blue with iodine solution, due to the presence of starch.
(2) Starch-like

AMYLOPECTIN. A polysaccharide made up of branched chains containing a large number of glucose units linked in the 1:4 or 1:6 position. A constituent of most starches.

The linkage forming amylopectin from glucose units

AMYLOPLAST. A colourless plastid modified for starch storage, *e.g.* in the potato.

AMYLOSE. A long straight-chained polysaccharide, made up of about 300 glucose units linked in the 1:4 position (see *amylopectin*). A constituent of starch, and is responsible for the formation of the typical blue colour of starches when treated with iodine solution.

AMYLOSTATOLITH. A starch grain acting as a statolith.

AMYLUM STARS. Star-shaped organs of vegetative reproduction growing on the lower nodes of the Charales. They are so-called because they are densely-packed with starch.

ANABIONT. An organism in which anabolic processes predominate over catabolic processes, *e.g.* green plants.

ANABOLISM. The formation of complex molecules from simple ones by living organisms. This process involves the storage of energy.

ANACARDIACEAE (EP, H). A family of the Sapindales. (BH) include Corynocarpaceae and Julianaceae. There are 60 genera with 500 species, found mainly in the tropics. They are trees or shrubs, with alternate or opposite leaves without stipules. There are resin-canals present, but the leaves are not dotted with glands. There are large numbers of flowers in panicles. The flowers have a disk, and are regular with the perianth segments in fives; they are hypogenous to epigenous. There are 10-5 stamens, usually, with an ovary of 3-1 (rarely 5) fused carpels. Only one ovule is fertile. It is anatropous, pendulous or adnate to the ovary, or on a basal funicle. The fruit is usually a drupe with a resinous mesocarp. The embryo is curved, and there is no endosperm. *E.g.* Mango, Cashew nut, Pistachio nut.

ANACHOROPTERIDACEAE. The single order of the Anachoropteridales.

ANACHOROPTERIDALES. An order of the Primofilicales. The leaves are pinnately divided with all the segments on the same plane. The sporangia are borne on the flattened edges of the pinnules.

ANACROGYNAE. A sub-order of the Jungermanniales, in which the apical cell does not develop into an achegonium. See *ACROGYNAE*. 'Anacrogyny' is the adjective describing this condition.

ANADROMIC HELICOID BRANCHING. A type of branching found in some fern leaves. The new branches arise from only one side of the main axis. This leads to the development of a helix. If the side bearing the branches faces away from the main axis of the leaf, the helix is anadromic.

ANAEROBIC RESPIRATION. The enzymatic breakdown of a substrate releasing energy, without using oxygen. The anaerobic respiration of a given weight of a substrate, yields less energy than the aerobic respiration of the same weight of the same substrate. Facultative anaerobes can use free oxygen, obligate anaerobes do not. See *alcoholic fermentation*.

ANAKINETIC. Leading to the restoration of energy, and the formation of reactive, energy-rich substances.

ANAKINETOMERES. Reactive, energy-rich molecules.

ANALINE. $CH_3.CH(NH_2).COOH$. An amino acid, possibly formed from pyruvic acid during respiration. This takes place by the removal of an oxygen atom (reduction) and the addition of an NH_2^+ radical (transamination).

ANALOGOUS. Of structures having the same appearance or function, but that are not necessarily homologous.

ANANDROUS. Without stamens.

ANAPHASE. The stage during mitosis or meiosis when the centromeres divide, so that each chromatid has its own centromere. The chromatids then migrate along the spindle to opposite poles. Anaphase ends when the chromosomes are formed into two groups at the opposite poles of the spindle.

ANAPHYSIS. A sterigma-like thread in the apothecium of some lichens.

ANASCISTIC. Said of tetrads that divide twice longitudinally during meiosis.

ANASTOMOSIS. (1) Joining by cross-connections to form a network.
(2) The fusion of two distinct branches from filaments in the Thallophyta.
(3) The fusion of vascular bundles, occurring mainly at the nodes.

ANATROPOUS. Of an ovule with the micropyle close to the underside of the funicle, and with the chalaza at the opposite end.

ANAUXITE. A kaoline-type of clay. $Al_2O_3.3SiO_2.2H_2O$.

ANCAD. A canyon plant.

ANCEPS. Latin meaning 'two-edged'.

ANCIPITOUS. Two-edged.

ANCISTROCLADACEAE (H). a family of the Ochnales. They are climbing shrubs with hooked branches. The calyx is enlarged and wing-like in the fruit. The petals are contorted. The ovary has one loculus and contains one ovary.

ANCYLISTALES = LAGENIDIALES.

ANCYLISTIDACEAE = LAGENIDIACEAE.

ANDREAEACEAE. The single family of the Andreaeales.

ANDREAEALES. An order of the Musci, or sometimes considered to be the single order of the sub-class Andreaeobrya. It contains the one genus *Andreaea*. This is a small dark plant with small opaque cells in the thallus. It is unique among the mosses as the stalked capsule opens by 4 longitudinal slits.

ANDRO-. A Greek prefix meaning 'male'.

ANDROCYTE. A sperm mother-cell. A pair of androcytes are formed by the division of the androgonial mother-cell, and are metamorphosed into atherozoids in the antheridium of the Bryophyta, and Pteridophyta.

ANDRODIOECIOUS. Of a species in which male and hermaphrodite individuals are found.

ANDROECIUM. (1) A collective term for the stamens of a flower.
(2) A group of antheridia of mosses.

ANDROGENESIS. Development from a male cell.

ANDROGONIAL CELL. The developing antheridium of the Bryophyta and Pteridophyta contains 4 primary androgonial cells, which divide repeatedly to form the cubical androgonial cells. These are called androcyte mother-cells after their last division. These divide diagonally to give the androcytes.

ANDROGYNOUS. (1) Having stamens and ovaries in distinct parts of the same inflorescence.
(2) Having male and female organs on, or in the same branch of the thallus.

ANDROMONOECIOUS. Of an individual plant, or of a species containing individuals which have male and hermaphrodite flowers on the same plant.

ANDROPHORE. An elongation of the receptacle of a flower between the corolla and the stamens.

ANDROSPORANGIUM. A sporangium producing androspores.

ANDROSPORE. A special form of zoospore produced by the Oedogoniales. It germinates to form the dwarf male filament.

ANDROTERMONE. A male-determining substance.

ANEMO-. A Greek prefix meaning 'wind'.

ANEMOCHOROUS. Said of plants whose fruits or seeds are dispersed by wind; especially plants which retain their seeds or fruit throughout the winter, ready for dispersal in the spring.

ANEMOPHILOUS. Having seeds, pollen spores etc. dispersed by wind.

ANEMOTROPISM. The active response of a plant to an air-current.

ANEUPLOID. A polyploid having some of the chromosomes of one set missing, so that the chromosome number is not an exact multiple of the haploid number, *i.e.* an unbalanced polyploid. See *euploid*.

ANEURIN. A vitamin, possibly promotes root-growth. A growth factor for some fungi and bacteria.

ANFRACTUOSUS. Latin for 'sinuous'.

ANGIOCARPOUS, ANGIOCARPIC. Of a fungus fruit-body which is closed, at least until the spores are mature.

ANGIOSPERMAE. A class (sub-division) of the Spermatophyta. Characterized by the ovules (female gametophytes) being completely enclosed in an ovary which forms part of the flower, and swells at maturity to form the fruit. The pollen-grains (male gametophytes) are produced in anthers. Typically the xylen has vessels.

ANGIOSPERMOPSIDA = ANGIOSPERMAE.

ÅNGSTRÖM UNIT. 10^{-8} cm. The wave-length of yellow light is approximately 5600 Å.

ANGUILLULIFORM. Worm-like in form.

ANGULAR DIVERGENCE. The angle between the lines of insertion of two adjacent leaves.

ANGUSTATE. Narrow.

ANGUSTI-. Latin prefix meaning 'narrow'.

ANGUSTIFOLIATE. Having narrow leaves.

ANISO-. Greek prefix meaning 'unequal'.

ANISOCOTYLY. Having cotyledons of unequal size in the seedling.

ANISOGAMY. Having male and female gametes of different size, but both are motile. See *heterogamy*.

ANISOGENOMATIC. Having a chromosome complement of unlike sets of chromosomes.

ANISOGENY. Having different inheritance in reciprocal crosses.

ANISOMERY. Of flowers having different numbers of flowers in successive whorls.

ANISOPHYLLY = HETEROPHYLLY.

ANISOPLANOGAMETES. Flagellate gametes of different size.

ANISOLEPIDIACEAE. A family of the Hypochytridiales. Its members are holocarpic, intrametrical, monocentric, and the thallus develops into a sporangium or resting spore.

ANISOTROPIC. Crystalline.

ANNEXED = ADNATE.

23

ANNOTINUS. Latin meaning 'applied to branches of the previous year's growth'.

ANNUAL. A plant completing its life-cycle in a single season: *i.e.* a seed germinates and the mature plant so produced dies, having produced seeds, within the season.

○**ANNUAL RINGS.** See *growth rings*.

ANNULAR. Ring-like.

ANNULAR THICKENING. The laying down of lignin in separate rings in the walls of xylem vessels and tracheids.

ANNULATE. (1) Ringed. Made up of ringed segments, or thickened in rings.
(2) Having a membranous ring on the stipe (of Agarics).

ANNULUS. (1) The ring of tissue around the inside of the top of a moss capsule. The cells below it dry out, causing the lid to fall off, and allowing the spores to be dispersed.
(2) An arc of cells around the sporangium of a fern. The outer walls of the cells are not thickened, so that when the cells dry out, tensions are set-up which burst open the sporangium.
(3) The ring-like remains of the veil left on the stipe after the expansion of the pileus of some Agarics.
(4) A ring of cells around a bulbil and takes part in the vegetative reproduction of some fungi.

ANODERM. Having no skin.

ANONACEAE (EP, H). (BH) include the Eupomatiaceae (EP). A family of the Ranales (EP) (BH), Anonales (H). Includes 80 genera and 820 species of chiefly tropical (especially Old World) aromatic trees, shrubs or climbers. The leaves are usually simple, in two rows, and have no stipules. The flowers are regular, and hermaphrodite (rarely unisexual), with 3 separate (rarely partly fused) sepals, and 6 free (rarely 4-3) petals in two whorls. The many hypogenous stamens are spirally arranged, and the anthers are often overtopped by the truncate enlarged connective. Rarely there are 4-3 stamens. The gynecium usually consists of numerous superior free carpels, but there may be few fused to form a unilocular ovary, with separate styles. There are 1 to many anatropous ovules which are basal or parietal. The fruit is dry or fleshy, usually an aggregate of berries. The seeds are often arillate, and have a ruminant, copious endosperm. The embryo is minute.
(BH) include no perianth, and perigynous flowers in their characteristics.

ANONALES (H). An order of the Lignosae. These are woody plants with alternate, simple leaves without stipules. The flowers are hermaphrodite and are hypgynous to perigynous. The carpels are free (rarely fused) with parietal placentas. Petals are usually present. The numerous stamens are free. The seeds have a ruminant endosperm and a minute embryo.

ANTAGONISM. (1) The inhibition of the growth of one organism by another. This may be due to the production of antibiotics, for competition for food etc.
(2) The ability of one toxic salt to reduce or remove the toxic effect of another.

ANTAGONISTIC SYMBIOSIS. The parasitism of one lichen on another.

ANTAPICAL PLATE. One of the plates found at the posterior end of the Peridinales.

ANTE-. Latin prefix meaning 'before'.

ANTECEDENT GENOM. The genom when it plays the principle role in determining inheritance.

ANTEPETALOUS, ANTIPETALOUS. Of floral parts situated opposite the petals.

ANTEPOSITION, ANTEPOSED. Situated opposite another plant member.

ANTERIOR. (1) Of lateral flowers, the part farthest away from the main axis *i.e.* facing the bract.
(2) The end of a motile organism which goes first during locomotion.

ANTESEPALOUS, ANTISEPALOUS. Inserted opposite the sepals.

ANTHER. The part of the flower of Angiosperms producing pollen (microspores), borne at the end of the stamens, and usually consisting of four sporangia.

ANTHERIDIAL CELL. Found in bryophytes and pteridophytes. A cell which develops into an atheridium.

ANTHERIDIAL CHAMBER. The cavity in a thallus containing an antheridium.

ANTHERIDIAL FILAMENT. A filament developing in the capitulum of the Charales. When mature each cell of the filament becomes an antheridium, and produces a single antherozoid.

ANTHERIDIOPHORE, ANTHERIDIAL RECEPTACLE. (1) A special male branch found in some of the Marchanitales.
(2) A special branch bearing one or more antheridia.

ANTHERIDIUM, ANTHERID. The male sex-organ producing antherozoids (microgametes), which are usually motile, but in some of the Phycomycetes, the antheridium contacts the oögonium (female sex-organ) into which it produces a fertilization tube, and through which the cell contents of the antheridium pass, resulting in the fusion of male and female nuclei.

ANTHEROZOID. A small motile male gamete with flagella, and produced in an atheridium.

ANTHESIS. (1) The opening of a flower-bud.
(2) The duration of life of a flower from the opening of the bud, to setting of the fruit.

ANTHO-. A Greek prefix meaning 'flower'.

ANTHOCARP. A fruit with a persistent perianth or other floral parts.

ANTHOCEROTACEAE. The single family of the Anthocerotales.

ANTHOCEROTIDAE. A sub-class of the Hepaticae, ANTHOCEROTAE, a class of the Bryophyta, ANTHOCEROTALES, a single class of the Anthocerotae, or an order of the Antheocerotidae. The Hornworts. The gametophyte is a simple thallus, with large chloroplasts, may be one per cell, and pyrenoids are present. The antheridia develop from hypodermal cells on the upper surface of the thallus, and lie in a roofed chamber. The archegonia are embedded in the thallus. The lower part of the capsule is meristematic and is continuously growing, having stomata and photosynthetic tissue. The sporogenous tissue is derived from the amphithecium of the embryo.

ANTHOCYANINS. A water-soluble glucoside (flavonoid) forming the red-brown-violet group of sap-pigments, especially in flowers. Leucoanthocyanins are colourless in alkaline, but are converted to anthocyanins in acid.

ANTHOCYANS. A general term to include the anthocyanins and anthocyanidines. The soluble pigments especially in fruits and flowers, mostly red and blue. They have the general form:—

ANTHOID. Appearing like a flower.

ANTHOLYSIS. Retrograde metamorphosis of a flower.

ANTHOPHORE. An elongation of the receptacle between the calyx and the corolla.

ANTHOPHYTA = ANGIOSPERMAE.

ANTHOTAXY. Arrangement of the flower.

ANTHOXANTHINS. See *Flavonic substances*.

ANTHOXANTHIN. A glycoside giving the yellow colours in cell-sap.

ANTHRACINY. The breakdown of organic material by fungi, followed by its further digestion in the alimentary tracts of insects, worms etc., giving a dark-coloured soil.

ANTHRACNOSE. A plant disease having characteristic limited lesions and hypoplasia.

ANTHRANILIC ACID. A possible precursor of tryptophane in some fungi. $C_6H_4(COOH).NH_2$. *o*-amino-benzoic acid.

ANTHROPOGENIC CLIMAX. A climax of vegetation formed under the influence of human activity.

ANTHROPOPHYTE. A plant introduced incidentally in the course of cultivation.

-ANTHUS. Greek suffix meaning '-flowered'.

ANTI-. Latin prefix meaning 'opposite'.

ANTIBIOSIS. Antagonism between two organisms resulting in one overcoming the other, or, at least, to the detriment of one of them.

ANTIBIOTIC. Any substance damaging to life, but especially a substance produced by micro-organisms damaging, or killing, other micro-organisms, or higher plants. The term is usually referred to substances of medical importance, *e.g.* Penicillin, Streptomycin etc.

ANTICAL. The upper surface of a thallus, stem or leaf.

ANTICLINAL. Of cell-walls, growing along the diameter of a plant axis.

ANTICOUS. Placed on the anterior-side of an organ.

ANTIDROMY, ANTIDROMOUS. Left- and right-hand twining in the same species of plant.

ANTIENZYME. A substance inhibiting the function of an enzyme.

ANTIMORPH. Having an effect opposite to that of the normal, non-mutant allelomorph.

ANTIPETALOUS. Opposite the petals.

ANTIPHYTE. A sporophyte.

ANTIPODALCELLS. Three cells in the embryo-sac, lying at the opposite end to the micropyle. They have a nucleus and cytoplasm, but no cell-wall.

ANTISEPALOUS. Opposite the sepals.

ANTITHETIC. The alternation of morphologically unlike generations, *i.e.* the generations are not homologous. See *Alternation of Generations*.

ANTITROPIC. Twisting clock-wise.

ANTITROPOUS = ORTHOTROPOUS.

ANTONIACEAE (H). A family of the Loganiales. Found in Malaya, Tropical America, and Tropical Africa. This is a family of trees and shrubs. The corolla lobes are valvate, with 4 or more (rarely 1) stamens inserted on the throat. An epicalyx may be present. The ovary usually has many ovules in each loculus. The seeds are winged at each end, and the fruit is a capsule which usually dehisces by splitting into 7.

ANTRORSE. Pointing upwards and inwards.

APATITE. A mineral in igneous rocks, giving rise to phosphorus salts in the soil. $Ca_{10}(PO_4)_6X_2$, where X_2 is F_2, Cl_2, CO_3, or $(OH)_2$.

APANDROUS. (1) Lacking, or having non-functional male organs.
(2) Forming oöspores when no antheridia are present.

APETALAE = INCOMPLETAE.

APETALOUS, APETALY. Without petals.

APHAPTOTROPISM. Not reacting to a contact stimulus.

APHELIOTROPIC = APHOTOTROPIC.

APHLEBIAE. Aborted pinnae found in some living and fossil ferns.

APHOTIC. Capable of growth in the absence of light.

APHOTOMETRIC. (1) Said of a leaf that does not respond to light.
(2) Said of motile organisms that always turn the same end towards the light.

APHOTOTACTIC. Of a motile organism that does not move in response to the light intensity.

APHOTOTROPIC. Growing away from the light.

APHYLLOPHORALES = POLYPORALES.

APHYLLOUS, APHYLLY. Lacking leaves.

APICAL CAP. Rings on the lateral walls, at the distal ends of the cells in the filament of some of the Oedogoniales.

APICAL CELL. (1) The single meristematic cell at the tips of the branches of the thallus of the more complex Algae.
(2) The meristematic cell at the tip of a bryophyte thallus. It forms new tissue by division, but may give rise to two apical cell, initiating the dichotomy of the thallus.
(3) The single meristematic cell at the tips of root and shoot branches of the Pteridophyta.

APICAL CELL THEORY. The theory that the tissues of higher plants develop from a single cell. It has been disproved by a study of the complex apices of the Gymnosperms.

APICAL GROWTH. The elongation of a filament of the Thallophyta by growth at the apex only.

27

APICAL MERISTEM. The group of actively dividing cells found at, or near the tip of a stem, root or sometimes, a leaf. It originates from a single cell in the Pteridophytes, and from a group of cells in the Spermatophyta. It brings about an increase in length by forming the primary plant-body.

APICAL PLACENTATION. Of ovule(s), when they are attached at the top of the ovary.

APICAL PLATES. A series of plates found on the top of the anterior half of the Peridinales.

APICULATE. Ending in a short, sharp point.

APICULUS. (1) A short point at the end of a spore.
(2) The projection by which a basidiospore is attached to the sterigma.
(3) A short point formed by the elongation of a vein.

APILEATE. Having no pileus.

APLANETISM. Having non-motile spores instead of gametes.

APLANOGAMETE. A non-flagellate gamete.

APLANOSPORE. A non-motile spore, with its wall free from the wall of the cell in which it was formed.

APLEROTIC. Of oöspores of Fungi which do not fill the oögonium.

APO-. Greek prefix meaning 'from'.

APOBASIDIOMYCETE. A Gasteromycete having apobasidia.

APOBASIDIUM. A basidium with terminal spores which are arranged symmetrically on the basidium.

APOCARPAE (BH). A series of the Monocotyledons having the perianth in one or two whorls, or absent, with a superior apocarpous ovary. The seeds have no endosperm.

APOCARPOUS. Of an ovary made-up of separate carpels.

APOCRENIC ACID. One of the constituents of humus.

APOCYACEAE (EP, BH, H). A family of the Contortae (EP), Gentianales (BH), Apocynales (H). This family has 180 genera with 1400 species, found mostly in the Tropics. They are usually twining shrubs, and mainly large lianes. They contain latex-vessels, and the bundles are collateral. The leaves are simple, opposite or alternate or in whorls of three, usually with close parallel lateral veins. Small interpetiolar stipules are rarely present. The inflorescence is typically a panicle, but is frequently modified. Bracts and bractioles are present. The flowers are hermaphrodite and regular, usually with 4-5 parts. The calyx consists of 5 fused, but deeply-lobed parts with the odd sepal posterior, often with glands at the base. The petals are five and fused to form a funnel or saucer, often hairy within, and sometimes asymmetric. The 5 anthers are free, alternate with the petals, and joined to them, usually forming a stylar head, and without a coronal appendage. The pollen is granular and is freely transported. A disk is usually present. The ovary is superior or half-inferior, consisting of two carpels, which are fused, or free (united by the style). There are 1-2 locular with α many to 2 ovules, which are not arranged in parietal pairs, in each loculus. The ovules are anatropous, and pendulous. The styles are usually simple, united to form a head, with the base free, and a ring of hairs below. The fruit is 2 follicles, a berry or capsule, or 2 indehiscent fruits. The seeds are usually flat often with a crown of hairs. The embryo is straight, and the endosperm may, or may not be present.

APOCYNALES (H). An order of the Lignosae. These plants are woody or herbaceous with opposite leaves, and are found mainly in the Tropics. There are no stipules. The petals are united, and a corona is often present. The stamens are the same number as the corolla-lobes. The pollen in granular or glutinate. The 2 carpels are often free, or become free in the fruit, with the styles free, but united above to form a common stigma (rarely free). The seeds usually have an endosperm, and a straight embryo. They are often winged or have long hairs.

APOCYTE, APOCYTIUM. A multinucleate mass of protoplasm with no cell-walls.

APOGAMY. Reproduction without the fusion of gametes, and usually without meiosis. The term may include any form of vegetative reproduction.

APOGENY. Sterility.

APOGEOTROPIC. Negatively geotropic.

APOHELIOTROPIC. Negatively heliotropic.

APOMIXIS = APOGAMY.

APONOGETONACEAE (EP, H). A family of the Helobieae (EP), Aponogeto-nales (H), and part of the Naiadeae (BH). There is 1 genus *Aponogeton* which has 25 species which are distributed throughout the Tropics, mainly in South Africa. These are fresh-water plants with a rhizome and floating leaves (some species have submerged leaves as well). The tissue between the veins breaks-up as the leaf grows. The hermaphrodite flowers are regular, and project above the surface in spikes. The perianth usually has 2 segments, but may be 1 or 3. There are usually 6 stamens in 2 whorls of 3 with 2-locular anthers. The superior gynecium consists of 3-6 free carpels, each containing 2 or many anatropous, erect, ovules. The fruit is leathery, with seeds containing a straight embryo and no endosperm.

APONOGETONALES (H). An order of the Calyciferae. These are freshwater or marine perennials, rooting from a rhizome. The leaves are oblong, or linear, forming a basal sheath. The flowers are hermaphrodite, or unisexual in spikes which may be simple, or forked, free or at first enclosed in a leaf-sheath. There are no bracts, and the perianth has 3-1 segments, or none. They may be bract-like. There are usually 6 stamens, but there may be more, or one. The anthers have 1 or 2 locular. The gynecium is apocarpous, or with 1 ovary, each containing 2 to many ovules. The seeds have no endosperm.

APOPETALOUS. (1) Having many petals.
(2) Sometimes used in exactly the opposite sense, *i.e.* having no petals.

APOPHYSIS. (1) The sterile tissue at the base of a moss capsule.
(2) The enlargement of the distal end of a scale of a pine cone.
(3) A swelling, or swollen hypha of a fungus.

APOPLASMODIAL. Of Acrasiales, having non-fusion of the myxamoebae.

APOPLAST. The complex made-up of the xylem cell-wall, and the cellulose wall and contents of the adjoining cell, through which water, salts etc. have to pass before reaching the other tissues of the plant.

APOROGAMY. The penetration of the ovule by the pollen-tube, by some other path than the micropyle.

APOSPORY. (1) The production of a diploid gametophyte from the sporophyte due to the absence of meiosis.
(2) The assumption of the reproductive function of the spores by unspecialized cells.

APOSTASIACEAE (H). A family of the Haemodorales, found from India and Ceylon, through Malaya to Tropical Australia. The flowers are actinomorphic or slightly zygomorphic and are in spikes or racemes. The stamens are 1 or 3 with filaments that are connate at the base and with the style. The trilocular ovary is inferior.

APOSTROPHE. The arrangement of the chloroplasts against the radial walls of the palisade layer in bright light.

APOTHECIUM. The fructification of some Ascomycetes which is flattened or cup-shaped with the asci on the upper surface in a palisade-like layer, usually mixed with sterile hyphae. It is *angiocarpic* if it is closed before maturity, and *gymnocarpic* if open.

APOTROPOUS. Said of an anatropous ovule with a ventral raphe.

APPENDAGE. Any external outgrowth of a plant which does not have any apparent essential function.

APPENDICLED. Having small appendages.

APPENDICULATE. (1) Bearing appendages.
(2) Having outgrowths at the throat of a corolla of a flower.
(3) Of mosses having a fringe of small pieces.
(4) Of agarics retaining the fragments of the veil.

APPLANATE. Flattened.

APPLIED. Lying against each other by a flat surface.

APPOSITION. The thickening of a cell-wall by the depositing of material on the surface, rather than between the microfibrils.

APPRESSED, ADPRESSED. Closely flattened; Compacted, but not joined.

APPRESSORIUM. The swelling at the tip of a germ-tube or hypha of some parasitic fungi. It is in close contact with the host, and puts out a fine tube through the host's cell or stoma.

APPROXIMATE. (1) Close together, but not joined.
(2) Of gills of agarics, near to, but not touching the stipe.

APTANDRACEAE (H). A family of the Olacales, found in tropical America, and tropical West Africa. There are no stipules. The stamens are united into a column, with disk-glands outside them. The fruit is a drupe, which is more or less enclosed by a large persistent calyx.

APTEROUS. Wingless.

APUD = 'In', used to avoid repetition, *e.g.* Clark apud Stras. in *Ann. Appl. Biol.*

AQUATIC. Living in water.

AQUEOUS TISSUE. A tissue of enlarged cells which store water.

AQUIFOLIACEAE (EP, H). A family of the Sapindales (EP) Celastrales (H). *E.g.* Holly. There are 5 genera and 300 species. They are shrubs and trees with leathery alternate leaves with small stipules, or absent, and a cymose inflorescence. The flowers are regular and unisexual, or bisexual and lack a disk. There are usually 4 sepals which are imbricate. Usually there are 4 petals which are imbricate (rarely contorted). There are 4 free stamens. The ovary has 4 fused, superior, loculi, each containing 1 or 2 pendulous, anatropous ovules. The fruit is a drupe. The seeds contain an endosperm with a straight embryo.

AQUILARIACEAE (H). A family of the Thymelaeales. The leaves are alternate and scattered, with the petals represented by scales. There are 10 or 5 ᵉtamens, and the style is short, or absent, with a large stigma. The ovary has more than 1 loculus, each containing a single pendulous (usually) ovule.

ARABANS. Polysaccharides derived from arabinose, *e.g.* hemicelluloses, mucilages, gums, pectin. See *Pentosans*.

ARABINOSE. A pentose sugar; an aldose.

$$
\begin{array}{c}
\text{CHO} \\
| \\
\text{H—C—OH} \\
| \\
\text{HO—C—H} \\
| \\
\text{HO—C—H} \\
| \\
\text{CH}_2\text{OH} \\
\text{L—Arabinose}
\end{array}
$$

ARABLE LAND. Cultivated land used for growing crops, *i.e.* not covered by natural vegetation, or permanently grass fields.

ARACEAE (EP, H). A family of the Spathiflorae (EP), Arales (H). Found in temperate and tropical regions, *e.g. Arum*, Coco biscuits, Elephant-foot yam, Indian Ipecacuanha. Usually terrestrial, rarely aquatic, with usually a conspicuous spathe and spadix, and many roots.

ARACHNOID. Covered with, or made up of fine hairs or fibres.

ARALES (H). An order of the Corolliferae, found in temperate and tropical regions. They are herbs, rarely climbing and woody, and very rarely aquatic. The leaves are radical or cauline, but if they are cauline, they are alternate. They are usually entire. Sometimes they are variously divided, and are often spear-shaped. The flowers are bisexual or unisexual, with a small perianth, or it may be absent. The stamens are hypogynous, and may be free or united. The ovary is superior, and the fruit usually a berry. There is an endosperm. The very small flowers are densely arranged on a spike (spadix), subtended by, or enclosed by, a large bract.

ARALIACEAE (EP, BH, H). A family of the Umbelliflorae (EP), Umbellales (BH), Aralilales (H), with 55 genera, and 700 species. They are found mainly in the tropics, especially Indo-Malaya, and tropical America, *e.g.* Ivy. Members of this family are usually trees or shrubs, some are palm-like and others are twiners. The leaves are usually alternate, rarely opposite or whorled: they are stipulate, with the mature leaves often more simple than the immature ones. There are resin canals. The flowers are usually in simple umbels, but may be in racemes, or heads. They are bisexual and regular, usually epigynous, with the parts in fives (sometimes 3 to many). The 5 free calyx lobes are usually small. There are usually 5 free petals, which are imbricate or valvate. They may be fused or absent. The 5 anthers are free (there may be 3 to many). The structure, and position of the ovary varies. Typically it is inferior, with 5 fused loculi, but there may be 1 to many fused parts. It may be half-inferior, or superior. Each locules contains 1 anatropous, pendulous ovule. The fruit is a drupe, with as many seeds as ovules. The embryo is small, and there is copious, sometimes ruminate, endosperm.

ARALIALES (H). An order of the Lignosae. The members are woody, rarely herbs. The leaves are simple or compound, and stipule may be present. The small flowers are often in umbels or heads. The petals are free or united, with the stamens alternating with the petals or corolla-lobes. The ovary is inferior, and the seeds have copious endosperm which is often ruminate.

ARANEOSE, ARANEOUS = ARACHNOID.

ARBUSCLE. (1) A dwarf tree, or a shrub looking like a tree.
(2) A much-branched haustorium, formed by some endophytic fungi. It is sometimes later digested by the host.

ARBUSCULES. Fine tufted hyphae formed by the fungi of a mycorrhizal association. They are usually found in the cells near the endodermis.

ARCHEGONIATAE. One of the main divisions of the plant kingdom. It includes the Bryophyta and Pteridophyta. The female sex-organ is an archegonium, and there is a regular alternation of generations.

ARCHEGONIUM. The flask-shaped container of the ovum (egg cell) of liverworts, mosses, ferns, and some gymnosperms. The swollen base (venter) contains the egg-cell, and is surmounted by the neck, with neck canal-cells.

ARCHESPORIAL CELLS. The inner cells of the developing sporangium of a pteridophyte. Most of these cells ultimately give rise to spore mother-cells.

ARCHESPORIUM. The cells of a sporangium which give rise to the archesporial cells.

ARCHICARP. The cell, hypha or coil of hyphae of Ascomycetes that develops into the fruit-body, or part of it.

ARCHICHLAMYDEAE (EP). A sub-class of the Angiospermae, having the sepals and petals separate from each other or the perianth is incomplete, or absent.

ARCHIGONIOPHORE. Special stalked vertical branches, of the Marchantiaceae, which bear the archegonia.

ARCHILICHENS. Lichens having bright-green gonidia.

ARCHIMYCETES. Fungi having a simple thallus, and reproducing by zoöspores. It includes the Plasmodiophorales, and Myxochytridiales.

ARCHOPLASM, ARCOPLASM, ARCHIPLASM. (1) In cell-division, the substance of the radiations surrounding the centrosome. It consists of cytoplasm and material from the nucleus.
(2) The modified cytoplasm of the Golgi apparatus.
(3) Idioplasm.

ARCUATE. Like an arc: bent like a bow.

ARDELLA. A small spot-like apothecium.

ARENACEOUS, ARENICOLOUS. From the Latin 'arenarius'. Showing best growth on sandy soil.

AREOLA. (1) A small pit.
(2) A thinner area in the siliceous deposit in the walls of some diatoms.
(3) A small space on the surface of a lichen, delimited by lines or cracks.

AREOLATE. Of the leaves of mosses, when they are divided into small areas. Also see *areola*.

AREOLATION. The net pattern formed by the boundaries of cells.

AREOLE. The area occupied by a group of spines or hairs on a cactus.

ARESCENT. Drying.

ARGENTATE. From the Latin 'argenteus'. Silvery in appearance.

AGILLICOLOUS. Living on clay soils.

ARGININE. δ-guanidine-αamino-valeric acid ($HN = C(NH_2) — NH.CH_2.(CH_2)_2.CH(NH_2).COOH$). One of the principle amino-acids making up plant protein.

ARIL. A fleshy, coloured covering on the seed. It arises as an upgrowth of the funicle or base of the ovule, and may be a tuft of hairs.

ARILLODE. A structure looking like an aril, but not formed from the placenta.

ARISTA = AWN.

ARISTOLOCHIACEAE (EP, BH, H). A family of the Aristolochiales. There are 5 genera and 300 species, found in the tropics and warm temperate regions, except Australia. They are mostly climbing shrubs, with a few herbs, having broad medullary rays. The well-developed leaves are usually simple, often cordate, and without stipules. The bisexual flowers are epigynous, regular, or zygomorphic. The perianth is usually of 3 fused petaloid lobes. There are 6-36 anthers which are free or united with the style. The ovary is inferior and of 4-6 loculi, each containing many anatropous, horizontal or pendulous, ovules. The fruit is a capsule. The embryo is small embedded in a rich endosperm.

ARISTOLOCHIALES (EP, BH, H). An order of the Archichlamydeae (EP, BH) Herbaceae (H). The members have a soft wood with broad medullary rays, or are parasitic or epiphytic herbs. The flowers are bisexual or unisexual, and hypgynous to epigynous. There are no petals, but the calyx is petaloid. There are many to few stamens. The ovary is usually inferior with 3-6 loculi. The placenation is axile, or parietal.

ARMATURE. The persistent woody scales at the base of the leaves and stems of Cyatheaceous ferns.

ARMED. Having thorns.

ARMILLA = ANNULUS.

ARMILLATE. Edged, fringed, frilled.

ARMOUR, ARMOR. A covering of old leaf-bases on the stems of cycads and some ferns.

AROIDEAE (BH) = ARACEAE (EP). A family of the Nadiflorae (BH).

ARRECT. Stiffly upright.

ARRHENOKARYON. A nucleus having two separate sets of haploid chromosomes.

ARRHIZAL, ARRHIZOUS. Not having roots.

ARTEFACT. A structure not usually present in a cell, but produced during investigation, *e.g.* during staining.

ARTHROGENOUS. (1) Of bacteria where the individual develops into a spore.
(2) Developed from pieces separated from the parent plant.

ARTHROPHYTES = SPHENOPSIDS. The Equisetales.

ARTHROSPORE. (1) A spore formed by the breaking of the hyphae, especially in the Actinomycetes.
(2) A spore formed by the segmentation of, and the separation from the parent cell.

ARTHROSTERIGMA. An individual sterigma of the lichens.

ARTICLE. A joint of a stem or fruit, which breaks at maturity.

ARTICULATAE = EQUISETINAE.

ARTICULATE. (1) Jointed.
(2) Breaking into distinct pieces without tearing, at maturity.

ARTICULATE LEAF. A leaf that is cut off by an absciss layer.

ARTICULATION. A joint in a stem or fruit, at which natural separation occurs.

ARTIFICIAL CHARACTER. A character chosen arbitrarily, without consideration of the natural relationships of the plant.

ARTIFICIAL CLASSIFICATION. A classification based at least in part on artificial characters.

ARTIFICIAL COMMUNITY. A plant community maintained by artificial means, *e.g.* a garden.

ARUNDINACEOUS. Reed-like.

ARUNDINULACEAE. A family of the Eccrinales, distinguished by there being two nuclei in the zygote.

ARVENSIS. Latin meaning 'of arable land'.

ASCENDING, ASCENDENT. (1) Of conidiophores that curve upwards.
(2) Of gills on a cone-like or unexpanded pileus.
(3) Becoming vertical, by upward curving.
(4) Of an ovule which arises obliquely from close to the base of the ovary.

ASCENDING AESTIVATION. Aestivation in which each petal overlaps the edge of the petal posterior to it.

-ASCENS. Latin suffix meaning 'tending towards'.

ASCIDIUM. A pitcher-shaped leaf, no part of a leaf.

ASCIGEROUS, ASCIFEROUS. Bearing asci. The stage in the life-cycle of an ascomycete when the asci are produced.

ASCIGEROUS CENTRUM. The special tissue of a pyrenomycete which develops into the asci and paraphyses.

ASCIIFORM. Hatchet-shaped.

ASCLEPIADACEAE (EP, BH, H). A family of the Contortae (EP), Gentianales (BH), Apocynales (H), found mainly in warmer regions, especially in South Africa, rare in temperate countries. The pollen is in waxy masses (pollinia). The filaments of the stamens are joined into a tube. The fruits are follicles which dehisce ventrally.

ASCOCARP. The fruiting-body of an ascomycete, usually developed from one fertilized ascogonium, and made up from asci, paraphysis (sterile hairs) and a denser surrounding periderm.

ASCOCONIDIOPHORE. An ascus-like conidiophore.

ASCOCONIDIUM. A conidium in an ascoconidiophore.

ASCOCORTICIACEAE. A family of the Taphrinales, having the hymenium on a membranous hypothecium of interwoven hyphae.

ASCOGENIC CELL. The cells bearing the asci in the Laboulbeniales. Homologous with the ascogenous hyphae of the other Euascomycetes.

ASCOGENOUS, ASCOGENIC. Ascus-producing, or ascus supporting.

ASCOGENOUS HYPHAE. The hyphae growing from the zygote of the Euascomycetes. Several ascogenous hyphae develop from each zygote, and each hypha bears several asci.

ASCOGONIUM. The female sex-organs of the Euascomycetes, which after fertilization, produce the ascogenous hyphae.

ASCOGONIUM. The cell, or group of cells of the Ascomycetes, fertilized by the antheridium.

ASCOHYMENIALES. The group of Ascomycetes in which the asci and paraphyses form a hymenium.

ASCOIDEACEAE. A family of the Saccharomycetales. They are saprophyets which develop a distinct mycelium. The ascus is elongated with many asci.

ASCOLICHENES. A sub-class of the Lichens which have an ascomycete as the fungus component.

ASCOLOCULARES. A group of Ascomycetes which produce asci, without paraphyses, in cavities in the stroma.

ASCOMA. A sporocarp having asci.

ASCOMYCETAE. A class of Fungi. The hyphae (when present) are septate. The reproductive organ is the ascus which may be club-shaped, cylindrical, round, or pear-shaped. The ascospores are produced inside the ascus, immediately after a reduction division. There are usually 8, but always a multiple of 2.

ASCOPHORE. (1) An ascus-producing hypha in an ascocarp.
(2) = APOTHECIUM.

ASCOSPORE. A spore produced as the result of sexual fusion and subsequent reduction division in the Ascomycetes.

ASCOSTOME. The pore at the top of an ascus.

ASCOSTROMA. A stroma, in or on which asci are formed.

ASCUS. A sac-like cell in the perfect stage of an Ascomycete, in which the ascospores are produced.

ASEPALOUS. Having no sepals.

ASEPTATE. Without cross-walls.

ASEXUAL. Lacking sex-organs or sex-spores.

ASEXUAL REPRODUCTION. Reproduction without nuclear fusion, *e.g.* spore formation or vegetative reproduction.

ASPARAGINE. One of the amino-acids in proteins. $COOH.CH_2.CH-(NH_2).CONH_2$. The amide of aspartic acid, and occurs widely in plants.

ASPECT. (1) The degree of exposure to sun, wind etc.
(2) The effect of seasonal changes on the appearance of vegetation.

ASPECT SOCIETY. A plant community which is dominated by a single species, or group of species, at a particular season.

ASPERATE. Roughened with projections, points or bristles.

ASPERGILLALES. An order of the Euascomycetes, having a closed ascocarp with the asci irregularly distributed in it. The periderm is made up of loosely, or compactly interwoven hyphae.

ASPERGILLACEAE. A family of the Aspergillales, *e.g. Aspergillus, Penicillium.* The mature ascocarp has a thin cortex, or may be more like a stroma with a thick cortex surrounding one or more masses of asci. The conidia are mostly catenulate.

ASPERGILLIFORM. Tufted, like a brush.

ASPERGILLIN. (1) The black water-soluble pigment in the spores of *Aspergillus niger*.
(2) The various antibiotics produced by *Aspergillus* spp.

ASPERGILLOSIS. Any disease of animals or man caused by *Aspergillus* spp.

ASPERIFOLIAE = BORAGINACEAE.

ASPERMOUS. Without seeds.

ASPERULATE. Finely asperate.

ASPOROGENIC. Not forming spores.

ASPOROGENOUS YEAST. A yeast which does not produce ascospores.

ASSIMILATE. Any substance produced in a plant during the manufacture of food.

ASSIMILATION. The absorption and utilization of simple food-substances.

ASSIMILATION NUMBER. The amount of photosynthesis taking place in unit mass of a leaf in an hour, and considering the weight of chlorophyll in the tissue. It is expressed as the ratio of weight of carbon-dioxide (in mg.) absorbed/time (in hours)/weight of chlorophyll (in mg.) in the tissue.

ASSIMILATORY QUOTIENT, ASSIMILATION QUOTIENT. The ratio of carbon-dioxide intake to oxygen output (CO_2/O_2). This ratio is usually near 1, agreeing with the equation:—

$$6CO_2 + 6H_2O = 6C_6H_{12}O_6 + 6O_2.$$

ASSIMILATIVE. Concerned with growth before reproduction: non-reproductive.

ASSOCIATION. The largest group of natural vegetation. A stable plant-community named after the dominant type of plant.

ASSOCIES. A developing association, which is therefore unstable.

ASSURGENT = ASCENDING.

ASTELIC. Without a stele.

ASTER. (1) A group of radiating cytoplasmic fibrils surrounding the centrosome immediately before, and during cell-division.
(2) The star-shaped arrangement of the chromosomes during metaphase.

ASTERACEAE = COMPOSITAE.

ASTERALES (BH, H). An order of the Gamopetalae (BH), Herbaceae (H), of world-wide distribution. They are herbaceous, to woody, but rarely trees. The leaves are alternate, opposite, or all radical. There are no stipules. The flowers are in heads, which are rarely compound, and the head is surrounded by an involucre of bracts. The heads may have all the flowers ligulate and bisexual, or have the outer flowers ligulate and female (rarely sterile) and the inner ones tubular and bisexual. The anthers are united into a tube, and the filaments are free and inserted on the corolla tube. The unilocular ovary is inferior, with a usually 2-lobed style, and containing 1 erect ovule. The fruit is an achene, usually crowned by a pappus of hairs which is a modified calyx.

ASTEROID. Star-shaped.

ASTEROPHYSIS. A stalked, stellate structure found in the hymenium of some fungi.

ASTEROXYLACEAE. A family of the Psilophytales, with the single genus *Asteroxylon*. It had a horizontal rhizome, which bore leafless underground branches, and erect branches with small leaves. The sporangia were dehiscent, with a jacket-layer several cells thick, and were borne on the tips of small leafless branches.

ASTICHOUS. Not arranged in rows.

ASTRAL RAY. One of the cytoplasmic fibrils that seem to play some part in the delimitation of new nucleus during cell-division.

ASTROCENTRE = CENTROSOME.

ASTROID. The star-shaped figure formed by looped chromosomes collected at the equator of the spindle during cell-division.

ASTROPHIOLATE. Lacking a strophiole.

ASTROSCLEREIDE. A star-shaped sclereide.

ASTROSPHERE = ATTRACTION SPHERE.

ASYMMETRICAL. Said of an organ or organism which has the two sides of a central plane unalike.

ASYNAPSIS. The failure of chromosomes to pair at pachytene, or the absence of chiasmata formation.

ATAVISM. Being like a remote ancestor, rather than like the parents.

ATELOMITIC. Said of a chromosome having the spindle fibre attached somewhere along its side.

ATER. The Latin for 'black'.

ATICHIACEAE. A family of the Myriangiales or Erysiphales. Sooty moulds on leaves, associated with scale-insects. The fungus is apparently completely external to the leaf, having no separate hyphae, but forming a gelatinous mass of anastomosing hyphae in a cushion, or stellate thallus. The cell-walls are gelatinous when wet, and horny when dry. Accessory spores (propagula) are produced in clusters, and the asci are scattered in the tissue of the thallus.

ATOMATE. Having a powdered surface.

ATP = ADENOSINE TRIPHOSPHATE.

ATRATE, ATRATOUS. Blackened, blackening.

ATRICOLOR. Inky.

ATRO-. A Latin prefix meaning 'black'.

ATROPHY. The reduction in size and utility of a tissue or organ.

ATROPURPUREUS. Latin meaning 'dark purple'.

ATROVIRENS. Latin meaning 'dark-green'.

ATROPOUS. See *Orthotropous*.

ATTACHMENT = SPINDLE ATTACHMENT.

ATTACHMENT CONSTRICTION = SPINDLE ATTACHMENT.

ATTACHMENT ORGAN. (1) An enlargement of the base of an algal thallus which attaches it to a solid object.
(2) A hooked hair etc. attaching a fruit to an animal, and hence aiding in dispersal.

ATTENUATE. (1) Narrowed.
(2) Of a pathogen having reduced virulence, or pathogenicity.

ATTRACTION PARTICLE = CENTRIOLE.

ATTRACTION SPHERE. The mass of protoplasm, with the striations radiating through it from the centrosomes during cell-division.

ATTRACTION SPINDLE. The terminal portions of the spindle formed during cell-division.

-ATUS. A Latin suffix meaning 'having' or 'provided with'.

ATYPICAL. Unusual, or not normal.

AULOPHYTE. A plant living non-parasitically in a cavity in another plant.

AUREUS. Latin for 'golden'.

AURICLES. Ear-shaped appendages, usually on leaves.

AURICULARIACEAE. A family of the Auriculariales. Parasitic or saprophytic on plants, usually on wood. The fructifications are often gelatinous. The basidia are not separated into hypobasidium and epibasidium. Clamp-connections are present in some genera, and absent in others. ·

AURICULARIALES. An order of the Heterobasidiae, or Eubasidii. The elongated basidium, or the epibasal portion of it is divided transversely into 4 cells, each bearing a single basidium and basidiospore. The nuclear division is stichobasal. The members are saprophytes or parasites, with an amorphous mycelium or a fructification which is usually gelatinous.

AUROFUSARIN. An orange-yellow pigment found in *Fusarium culmorum*.

AUROGLAUCIN. A golden-orange pigment of *Aspergillus* spp.

AUSTERE. Astringent, harsh.

AUSTRALIS. Latin for 'southern'.

AUSTROBAILEYACEAE (H). A family of the Laurales. The leaves are often covered with transparent dots, and are mostly opposite. The carpels are free, or only one. The style has two lobes, and the endosperm is not ruminate.

AUTCHTHONOUS. Native.

AUTECOLOGY. The ecology of an individual species.

AUTEUFORM. An autoecious rust having all spore stages.

AUTOALLOGAMY. Said of a species of flowering plants which has some members capable of self-pollination, and others capable of cross-pollination.

AUTOBASIDIOMYCETE. A basidiomycete with a non-septate basidium.

AUTOBASIDIUM. (1) A basidium which is aseptate.
(2) A basidium with spores at the side, and asymmetrical.

AUTOCARP. A fruit resulting from self-fertilization.

AUTODELIQUESCENT. Becoming liquid by the process of self-digestion.

AUTOECIOUS. Said of the Rusts producing the different spore forms on one host, *i.e.* completing the life-cycle on one host.

AUTOGAMY. (1) Self-fertilization.
(2) The fusion of sister cells.
(3) The fusion of nuclei in pairs within one cell of a female organ, without cell fusion having taken place.

AUTOICOUS. Having male and female inflorescences on the same plant.

AUTOLYSIS. The destruction of cell-contents by enzymes produced in the cell. This is due to the disordered metabolism of the cell, and may be caused by substances not normally present in the cell.

AUTOMIXIS. The mingling of chromatin derived from two sex cells produced by one individual plant.

AUTONOMISM. Movement due to an internal stimulus, *e.g.* protoplasmic flow in a cell.

AUTOPARASITE. A parasite living on another parasite.

AUTOPHYTE. A plant which synthesises complex food substances from simple compounds, *e.g.* green plants.

AUTOPOLYPLOID. A polyploid with identical sets of chromosomes, *i.e.* derived from the same species.

AUTOPOTAMOUS. Originating in fresh water.

AUTOSOME. A chromosome, other than a sex-chromosome. A normal chromosome, with its normal homologue.

AUTOSPORE. An aplanospore of algal cells, similar in shape and size to the parent cell.

AUTOSYNDESIS. The pairing of chromosomes from the same polyploid parents, or remote polyploid ancestors.

AUTOTETRAPLOID. A tetraploid with four similar sets of chromosomes.

AUTOTROPH. A plant that can use carbon dioxide as its sole source of carbon, or more generally a plant that can manufacture complex organic compounds from simple inorganic sources. AUTOTROPHIC BACTERIA can use carbon dioxide.

AUTOTROPISM. A tendency to grow in a straight line.

AUTOTROPIC REVERSAL. If during tropic growth, an organ grows beyond the direct line of the stimulus, and then reverses its growth to lie along the line of the stimulus, it has shown autotropic reversal.

AUTOXENOUS = AUTOECIOUS.

AUXANOGRAM. The differential growth of micro-organisms, especially yeasts, on Petri dishes containing media prepared to determine the nutrient requirements.

AUXANOMETER. A piece of apparatus used to measure the elongation of plant shoots. It incorporates a lever to exaggerate the actual growth.

AUXILIARY CELL. An accessory cell in the thallus of some Floridae from which the gonimoblast filaments grow, after the zygote nucleus has migrated into it from the carpogomium.

AUXIMONES. A general term for growth promoting substances of unknown composition. A little used term.

AUXINS. Plant growth-regulators; plant growth-substances; plant hormones. A substance produced in one tissue, migrating to effect the development of another tissue. The tissues producing the auxin are typically meristematic, and the effect of the auxin specific. They correlate the growth of the plant. Certain substances, *e.g.* gibberillin, produced by one plant may effect the growth of another plant if applied artificially. The best known is $C_{18}H_{32}O_5$ which causes elongation of individual cells in the growing tips of plants.

AUXITHALS. A little used term for growth substances.

AUXOCYTE. A cell in which meiosis has begun.

AUXOSPIREME. The spireme formed after syndesis in meiosis.

AUXOSPORE. A rejuvenescent spore produced under adverse environmental conditions by some diatoms. When produced by pennate diatoms, formation is always associated with sexual reproduction, by the enlargement of a zygote, or parthenogenetic gamete.
Those of the Centrales are formed singly inside a cell and may, or may not, be the result of sexual fusion.

AVAILABLE WATER. All the water in a soil that can be taken in by a plant at any given time.

AVERRHOACEAE (H). A family of the Rutales, found in tropical Asia and including the Bilmbi fruit and Carambola fruit. The leaves are usually alternate, and rarely spotted with glands. Stipules are normally absent. There is no disk, the petals are contorted, and there are many ovules in each loculus.

AVERSE. Turned back.

AVERSION. The inhibition of growth at the edges of neighbouring colonies of microorganisms especially in cultures of one species. See *antagonism, barrage*.

AWN. (1) A bristle on the flowering glumes of the Graminae.
(2) A long thread-like outgrowth on certain fruits.

AXENY. The resistance of a plant to a pathogen, without the presence of morphological barriers. 'Passive' resistance.

AXIAL. Related to the axis of a plant. A term used to describe a structure which is morphologically a shoot.

AXIAL CELL. The primary axial cell is a central cell in the early stages in the development of the archegonium of Bryophytes and Pteridophytes. It gives rise to the cover cell, and the lower central cell.

AXIAL FIELD. A longitudinal thickened strip running down the skeleton of some of the Pennales.

AXIAL GRADIENT. A physiological gradient along the axis of a plant, with the activity being higher at one end, gradually becoming less as it passes along the axis, and being least at the other end.

AXIL. The angle between the upper surface of a leaf, and the stem which bears it.

AXILE. (1) Attached to the central axis.
(2) Coinciding with the longitudinal axis.

AXILE CHLOROPLAST. A chloroplast lying along the axis of the cell.

AXILE PLACENTATION. Said of an ovary having the ovules attached to the tissues of the central axis.

AXILE STRAND. The conducting elements in the centre of the plant axis, giving off vascular branches to the leaves. = PROTOSTELE, but includes the central supporting strand occurring in some mosses.

AXILLANT. Subtending at an angle.

AXILLARY. Situated in an axil, referred especially to buds.

AXIS. The main part of the plant body, consisting of the stem and root. It has a central stele, cortex and epidermis, and bearing leaves, emergences, and/or hairs.

AXOSPERMOUS. Having axile placentation.

AZUREUS. The Latin for 'sky-blue'.

AZYGOTE. An individual developed without fertilization, from a haploid individual.

AZYGOSPORE. A zygospore formed parthenogenetically, and is a resting spore. The term is sometimes confined to zygospores which are formed from gametes.

B

B. A class of flowers which have their nectar fully concealed.

B[1]. A class of flowers similar to B but having the flowers in heads.

B HORIZON. The second main zone of a soil profile, where substances accumulate if the soil is leached.

B-SPORE = BETA SPORE. See *Alpha-spore*.

BACCA. A berry formed from an inferior ovary. It may mean a berry in general.

BACCATE. Like a berry: pulpy.

BACCIFEROUS. Bearing berries.

BACCIFORM. Shaped like a berry.

BACILLAR, BACILLIFORM. Rod-shaped.

BACILLAREAE. See *Bacillariophyceae*.

BACILLARIOPHYCEAE. A class of the Chrysophyta. The diatoms. They are unicellular or colonial, having a cell-wall which is highly ornamented, containing cilia, and in two halves, called valves. The plastids contain the brown pigment isofucoxanthin as well as chlorophyll. The food reserves are in the form of fats and volutin, never starch. The cells are non-flagellate; the gametes are isogametes, and asexual reproduction takes place by the normal cell division. Some members form auxospores or endospores, and others produce flagellated zooids which may be gametic in nature. This class usually forms part of the fresh-water or marine plankton.

BACILLUS. Any rod-shaped bacterium. Also used as a generic name.

BACK CAVITY. The widened opening between the inside faces of the two guard cells of a stoma.

BACK CROSS. The fertilization of gametes from a hybrid by gametes from one of the parents of the hybrid. The term may be applied to a genetically equivalent cross to this, but does not include a test cross (*q.v.*).

BACTERIA. Unicellular plants without chlorophyll. They reproduce by binary fission, and are probably related to the Fungi. Most are saprophytic or parasitic, but some are autotrophic. Some are pigmented, and others are important pathogens of plants and animals, including man.

BACTERICIDAL. Killing bacteria.

BACTERIOCHLOROPHYLL. Chlorophyll found in bacteria.

BACTERIODS. Irregular, enlarged forms of rod-shaped bacteria especially of the root-nodule forming species. These are ultimately absorbed by the cells of the root-nodule.

BACTERIOLOGY. The study of bacteria.

BACTERIOPHAGE. A virus which kills bacteria.

BACTERIOPURPURIN. A purple pigment found in certain bacteria, *e.g.* the sulphur bacteria.

BACTERIORHIZA. A symbiosis between a root and bacteria.

BACTERIOSTATIC. Said of a substance which inhibits the growth and/or multiplication of bacteria, but does not kill them.

BACTERIUM. A rod-shaped member of the Bacteria.

BACULIFORM. Rod-shaped (of fungus spores).

BADIOUS, BADIUS. Latin meaning 'chestnut brown'.

BALANCE. The adjustment of the genetic material of an individual to allow for the healthy growth and development of the individual. INTERCHRO-MOSOME BALANCE—The adjustment of whole chromosomes to give the healthy development of an organism. INTRACHROMOSOME BALANCE—The balance of genes within a chromosome. GENIC

BALANCE—Healthy growth and development produced by adjustment of the proportion of genes. POLYGENETIC BALANCE—The effect produced by the adjustment of the proportions of polygenes which have opposite effects. A POLYGENIC COMBINATION occurs if the polygenes occur on the same chromosome. The balance is INTERNAL, and shown by a homozygote if it is achieved within a single representative of a chromosome. A RELATIONAL BALANCE is shown by a heterozygote, as it is achieved by two homologous chromosomes acting together. If a new balance is formed by a change in the proportion of genes, as in a secondary polyploid, a SECONDARY BALANCE is formed, and is capable of competing with the original balance.

BALANCED LETHALS. See *Lethal genes*.

BALANCED MEDIUM, BALANCED SOLUTION. A culture (or solid medium) with the elements necessary for growth, in such proportions that the toxic effect of individual salts are mutually eliminated. They are used for the growth of algae, fungi, bacteria.

BALANITACEAE (H). A family of the Malpighiales. The leaves are alternate, compound or unifoliate, and have no stipules.

BALANOPHORACEAE (EP, BH, H). A family of the Achlamydosporae (BH), Santalales (EP, H), with 15 genera and 40 species, all but one of which is tropical. They are parasites on roots of trees, and are fleshy herbs without chlorophyll. The inflorescence is a spike or head with small unisexual flowers, and develops from the rhizome. The male flowers have 3-4 perianth segments which may be free or fused, with the same number of anthers as perianth segments, and are opposite them. The female flower usually has no perianth. The ovary is inferior with 1-2 or rarely 3 fused loculi. There is 1 ovule which has no integument, and is usually pendulous. The fruit is a nut, or drupe-like, and the seeds have endosperm.

BALANOPSIDACEAE (EP, H). The single family of the Balanopsidales (EP, H) (*q.v.*) = BALANOPSEAE (BH) and is the single family of the Unisexales (BH).

BALANOPSIDALES (EP, H). An order of the Archichlamydeae (EP), Lignosae (H). These are woody plants with simple alternate leaves without stipules. The flowers are dioecious, the male flowers are in catkins, and the female are solitary in an involucre of bracts. There are no sepals or petals. There are only a few stamens. The superior ovary has 2 fused loculi. It is imperfectly 2-locular each with 2 ascending ovules on parietal, sub-basal placentas. The fruit is a drupe, and the seeds have some endosperm, and a straight embryo.

BALLISTOSPORE. A fungus spore attached asymmetrically to the sterigma, and discharged violently.

BALSAMIFEROUS. Yielding balsam.

BALSAMINACEAE (EP, H). A family of the Sapindales (EP), Geraniales (H). BH include this family in the Geraniaceae. There are 2 genera with 430 species, found mainly in tropical Asia and Africa. They are herbs with translucent stems. The alternate leaves are pinnately nerved, are not peltate, and lack stipules. The flowers are bisexual and zygomorphic. There are 5 free sepals, with the 2 anterior ones small or aborted, and the posterior one spurred. The sepals are petaloid. There are 5 petals, and the lateral ones are joined in pairs. The stamens are free, but the anthers are joined, forming a cap over the ovary, the growth of which ultimately breaks the stamens at the base. The superior ovary has 5 fused carpels with 5 loculi each with many anatropous pendulous ovules. Each ovule has a dorsal raphe. The fruit is an explosive capsule, and the seeds have no endosperm.

BAND. A strand of thickened tissue in the thallus of a liverwort; it has a strengthening function.

BANGIALES. The single order of the Bangioideae.

BANGIOIDEAE. A sub-class of the Rhodophyceae, whose members have intercalary growth of the thallus, and the zygote divides directly to form carpospores.

BANNER = STANDARD.

BAR OF SANIO = SANIO'S BAR.

BARB. A hooked, or doubly-hooked hair.

BARBATE. Bearded; having one or more groups of hairs.

BARBEUIACEAE (H). A family of the Chenopodiales, with one genus found in Madagascar. There are no petals, and the anthers open by longitudinal slits. The ovary is 2 or more locular, and superior. The carpels are not joined around a central column. The fruits are free from each other, not united into a fleshy mass, and are capsular.

BARBEYACEAE (H). A family of the Urticales. The leaves are opposite and without stipules. The flower has a calyx, which is enlarged in the female flower, and venose in the fruit. There is one pendulous ovule, and the anthers are erect in the bud. The seed has a straight embryo, and there is no endosperm.

BARBULE. The inner row of peristome teeth in the capsules of some mosses.

BARK. Protective suberised (corky) tissue cut off by a secondary meristem (phellogen). The phellogen cuts off cells which become suberised (phelloderm), so that all the cells outside the phellogen ultimately die, forming the bark. If the phellogen is deep, cells of the cortex, and phloem may be involved. In a non-technical sense, the bark means all the tissues outside the cambium.

BARK BOUND. A condition when the bark does not split during the growth of the cells below, thus compressing these cells.

BARRAGE. The space between two mycelia across which they will not grow into contact with each other.

BARRAGE SEXUEL. The repulsion between sexually incompatible hyphae. Incompatible strains will grow away from each other.

BARREN. (1) Lacking pollen.
(2) Unable to produce seeds.

BASAL. Situated at the base. See *Basal placentation.*

BASAL BODY. (1) The part of the thallus of the Blastocladiaceae fixed to the substrate by rhizoids at the lower end.
(2) A deeply staining granule at the base of a flagellum. It may be the centrosome. See *Blepharoplast.*

BASAL CELL. (1) The lowest cell cut off from the archegonial initial of the Filicales and Hepaticae.
(2) The lower cell cut off from the antheridial initial of the Hepaticae. It remains sterile.
(3) One of the pair of cells formed by the initial division of a germinating spore of the Lycopodinae, in the early stages of the development of the gametophyte. It gives rise to a rudimentary rhizoidal cell.
(4) An attaching cell.
(5) The lower cell of a crozier.
(6) A uninucleate cell which may be the oögonium in the Uredinales.

BASAL CORPUSCLE = BASAL GRANULE = BASAL BODY.

BASAL PLACENTATION. Having the placenta at the base of the ovary.

BASAL WALLS. The walls which separate the hypobasal and epibasal halves of a fern embryo when this consists of 8 cells.

BASE. (1) The end of a plant organ nearest to the point of attachment of another organ.
(2) A substance which combines with an acid to form a salt, without the evolution of a gas.

BASE EXCHANGE. The inter-change of the cations of the soil-water with those of the soil-colloid complex, or root surface.

BASE-EXCHANGE CAPACITY. The amount of base held on a clay under specified conditions, *e.g.* of pH. The definition by various authors varies from this only in detail.

BASELLACEAE (EP, H). A family of the Centrospermae (EP), Chenopodiales (H), included in the Chenopodiacea by (BH). There are 4 genera, with 15 species found in tropical America, Asia, and Africa. They are perennials with rhizomes or tubers, giving rise to annual climbing shoots, which often have fleshy leaves. The flowers are in racemes or panicles, and are often conspicuous, each subtended by two bractioles. The 2 sepals are free, as are the 5 petals. The 5 free stamens are opposite the petals, having anthers opening by terminal pores or pore-like slits. The superior ovary consists of 3 fused carpels forming a single loculus, with a terminal style and 3 stigmas. There is a single basal campylotropous ovule. The seeds have copious endosperm, surrounded by a spirally twisted embryo.

BASE PAIR. One of the pairs of bases forming the cross-links in the DNA molecule. They are always paired as follows:—
Adenine and Thymine, Guanine and Cytosine or 5-methyl cytosine.

BASI-. A Greek and Latin suffix meaning 'base'.

BASIC DYE. A stain having an organic basic radicle which is active, and combines with an acidic radicle which is usually inorganic. They stain nucleo-protein.

BASIC NUMBER. The haploid number of chromosomes, represented by *x*, *i.e.* the number of chromosomes in a set. See *Chromosome Set*.

BASICHROMATIN. A form of chromatin containing a fairly high proportion of nucleic acid, and hence staining deeply with basic dyes.

BASIC TYPE. The chromosome complement characteristic of a species, and varying only in heterozygotes. See *Prime Type*.

BASIDIAL. Relating to a basidium.

BASIDIOBOLACEAE. A family of the Entomophthorales, containing the single genus *Basidiobolus*. It is separated from the other families because the cells are uninucleate, and the cell-wall gives a reaction for cellulose without primary treatment.

BASIDIOCARP. The basidia-producing fruit-body of the Basidiomycetes.

BASIDIOGENETIC. Produced on a basidium.

BASIDIOLICHENES. A tropical group of Lichens which have a Basidiomycete as the fungal component.

BASIDIOMYCETAE. A class of Fungi characterized by the production of a definite number (usually 4) spores externally on a 1 – 4-celled structure—the basidium. Meiosis takes place at the formation of these basidiospores, which germinate to form a haploid mycelium. Two of these mycelia fuse, usually with the formation of clamp-connections, giving rise to a diploid mycelium, which ultimately produces the fruiting-body.

BASIDIOPHORE. A fruit-body bearing mycelia.

BASIDIOSPORE. See *Basidiomycetae*.

BASIDIUM. (1) See *Basidiomycetae*.
(2) A sterigma in some moulds. Not a good use of the term.

BASIFIXED. Said of an anther which is attached by its base to the filament.

BASIFUGAL = ACROPETAL.

BASIGAMIC, BASIGAMOUS. Said of an embryo-sac in which the synerdidae and egg-nucleus lie at the base, and not at the end nearest the micropyle.

BASILAR. Near, pertaining to, or growing from, the base.

BASINERVED. Of a leaf having the main nerves running from the base.

BASIPETAL. Developing so that the oldest structures are at the apex, and the youngest at the base.

BASOPHILIC. Staining with a basic dye.

BASISCOPIC. Facing the base.

BASITONIC. Of orchids, where the base of the anther lies against the rostellum.

BASS-FIBRE, BASS-WOOD, BASTWOOD FIBRE. Sclereids and fibres formed in the secondary phloem of some trees, *e.g. Tilia* (Lime).

BAST. Sometimes used to mean 'phloem' and sometimes to mean 'bast-fibre'. A very vague term which, because of its vagueness, should not be used.

BATHYPHYTIUM. A lowland plant formation.

BATIDACEAE (H) = BATIDIACEAE (EP) = BATIDEAE (BH). The single family of the Batidales (EP) *(q.v.)*, or a family of the Chenopodiales (H), while BH places it in the Curvembryae. It contains one genus *Batis*, found in tropical America and Pacific islands.

BATIDALES. An order of the Archichlamydeae (EP). Coastal shrubs with opposite, fleshy, linear leaves and spikes of dioecious flowers. The male flowers are in the axile of 4-ranked bracts, with a cup-like perianth of clawed petals, with the claws united at the base. There are 4 stamens, with anthers opening by longitudinal slits. The female flower consists of a superior ovary of 2 fused carpels, forming 4 loculi, each with one anatropous ovule. The seed lacks endosperm, and the embryo is straight.

BAUERACEAE (H). A family of the Cunoniales (H). The leaves are trifoliate, opposite and sessile. If stipules are present, they are adnate to the base of the petiole. The flowers have petals, and a disk. There are 2 free, filiform styles, and the fruit is a subdidymous, compressed, truncate capsule.

BEADED. Used of the gill of an agaric which has a row of small droplets of liquid along its edge.

BEAKED. Having a long, pointed prolongation.

BEARDED. Having an awn, or bearing long hairs like a beard.

BEGONIACEAE (EP, BH, H). A family of the Parietales (EP), Passiflorales Cucurbitales (H). There are 4 genera. They are mostly perennial herbs with thick rhizomes or tubers. Some are climbers with adventitious roots. The leaves are radical, or alternate in two ranks. There are large stipules. The inflorescence is axillary, dichotomous, tending to be a bostryx. The first axis usually ends in male flowers, and the last, and sometimes the last but one, has female flowers. The male flower has a petaloid perianth of free parts. There are either 2 valvate petals, or 4 decussate ones. The numerous anthers are either free or joined, with the connective often elongated. The anthers have 2 loculi, which are straight. The perianth of the female flower has 2-5 segments. The inferior ovary usually has 2-3 fused carpels, with 2-3 loculi. There are many anatropous ovules on axile placentas. The styles are more or less free. The ovary is usually winged, resulting in a winged capsular fruit. The seed has no endosperm.

BEIDELLITE. A clay, $Al_2O_3.3SiO_2.O.4H_2O^{\pm}$.

BELT TRANSECT. A narrow strip of land taken as a sample of an area in studying the vegetation.

BENNETTITALES. An order of Gymnosperms of the Mesozoic. They bore bisporangiate strobili, having the megasporophylls and microsporophylls in a spiral inside a cluster of sterile leaves, *i.e.* rather like an angiosperm flower.

BENTHOS. Organisms living on, or near the bottom of a large mass of water.

BERBERIDACEAE (EP, BH, H). A family of the Ranales (EP, BH), Berberidales (H). BH include the Lardizabalaceae (EP). There are 12 genera and 200 species, found in North temperate regions and tropical mountains of South America. They are perennial small trees, shrubs, or herbs, the latter usually with a rhizome. All have pinnate or unifoliate leaves. The flowers are bisexual and regular; usually in racemes. Typically the perianth is in 4 whorls of 3 free segments, and the free anthers in 2 whorls of 3. The ovary is of 1 carpel, and superior. The 2 outer whorls of the perianth are true perianth, and the 2 inner ones are 'honey leaves' usually with nectaries at the base. The anthers are introrse, but usually open by 2 posterior valves. The valve, with the pollen on it moves upwards and turns around, so that the pollen faces the centre of the flower. The ovary contains 1 basal, or many ventral ovules. The fruit is dry, or a berry, the dry fruit opening in various ways. The embryo is straight, and there is a lot of endosperm.
(BH) include members with unisexual flowers, and extrorse anthers.

BERBERIDALES (H). An order of the Herbaceae. The members are herbs to shrubby, or climbers. The stems often have broad medullary rays. The flowers are hypogynous, bisexual or unisexual, and cyclic. The petals (which are rarely absent), are small. The stamens are mostly definite in number, free, and opposite the petals. The carpels are usually 1-3 in number and free. Rarely there are many carpels. The seeds have copious endosperm and the embryos are of variable size. The leaves are without stipules, alternate, simple to compound. The order is found mainly in North temperate regions.

BERBERIDEAE (BH) = BERBERIDACEAE (EP) + LARDIZABALACEAE (EP).

BERRY. A many-seeded fleshy fruit. All the parts of the ovary are fleshy, except the outer pericarp.

BETULACEAE (EP). A family of the Fagales (EP). BH include it in the Cupuliferae, and H gives a slightly different definition (see below). There are 6 genera and 105 species, including Beech, Hazel, and Alder. The family is found in North temperate regions, and tropical mountains. They are trees or shrubs with alternate undivided leaves. There are membranous deciduous stipules. The flowers are unisexual, the males in terminal catkins, and the female, which are wind-pollinated, in terminal heads. Typically there is a dichomomous cyme of 3 male flowers in the axil of each catkin leaf, although the central one may be absent. The male flowers are united to the bract, and may, or may not, have a perianth. There are 2-12 anthers. The female flower may have an epigynous perianth, or it is absent. 2 carpels are fused to form an inferior ovary, which is bilocular at the base. Each loculus has 1 pendulous anatropous ovule which has 1 integument. The fruit is a 1-seeded nut without an endosperm.

BETULACEAE (H). A family of the Fagales (H). The flowers are unisexual. The male has a small calyx with 4-2 stamens. The female flower has no calyx, so that the ovary is superior. The ovary is bilocular, each loculus having one ovule.

BETWEEN RACE. A race of plants which is intermediate in character between the typical species and one of its well-marked sub-species or varieties.

BI-. From the Latin 'bis' meaning twice. A prefix meaning 'twice' 'two-' 'having two'.

BIARTICULATE. Having two nodes or joints.

BIAS. A consistent error in observed quantity, due to inaccurate observation, calculation, or estimation.

BIATORINE. Referred to a lichen apothecium which is soft, waxy, and often brightly coloured.

BIBRACTEATE. Bearing two bracts.

BIBULOUS. Capable of absorbing water.

BICALCARATE. Two-spurred.

BICAPSULAR. Having a capsule consisting of two chambers.

BICARPELLARY, BICARPELATE. Said of a gynecium which consists of 2 carpels.

BICARPELLATAE (BH). The third series of the Gamopetalae. The ovary is usually superior, and consists of 2 (rarely 1 or 3) carpels. There are the same number of stamens as there are corolla lobes, or sometimes fewer, and they alternate with the corolla lobes.

BICILIATE = BIFLAGELLATE.

BICOLLATERAL BUNDLE. A vascular bundle with phloem inside and outside the xylem.

BICONIC. Looking like two cones placed base to base.

BICONJUGATE. Said of a compound leaf having two main ribs, each bearing a pair of leaflets.

BICRENATE. Having two rounded teeth.

BICUSPID, BICUSPIDATE. Having two short horn-like points.

BIDENTATE. Having two teeth.

BIENNIAL. A plant that develops vegetatively, and produces a food-storage organ in the first year. It flowers in the second year only, then dies.

BIFACIAL. Flattened, with the upper and lower surfaces differing in structure from each other.

BIFARIOUS. Having two rows of structures on either side of a central axis.

BIFID. Forked; having a fissure near the centre.

BIFLAGELLATAE. A group of the Phycomycetes with spores having two flagella.

BIFLAGELLATE. Having two flagella.

BIFOLIATE, BIFOLIOATE. Said of a compound leaf with two leaflets.

BIFURCATE. Forked, or twice forked.

BIGEMINATE. In two pairs.

BIGNONIACEAE (EP, BH, H). A family of the Tubiflorae (EP) Personales (BH), Bignoniales (H). It contains 100 genera with 800 species. They are found mostly in Brazil. The family consists of trees, shrubs, or climbers, rarely herbs. The leaves, which do not have stipules, are mostly compound, digitate or pinnate, sometimes with the terminal leaflet forming a tendril. Many are xerophytes with condensed stems. Bracts and bractioles are present. The flowers are bisexual, zygomorphic and hypogynous. They are usually bell-shaped, with five fused calyx lobes and five fused corolla lobes. The 4 stamens are free and attached to the petals (H includes 4 or 2 stamens). They are didynamous with the anther lobes usually one above the other. A posterior staminod is always present ((H) states that staminodes may or may not be present). The superior ovary has 2 fused carpels on a hypogenous disk with 2 (rarely 1) loculi ((H) states 4, 2 or 1 loculi), each containing many erect anatropous ovules on an axile placenta. The capsule is 7-septate, and the seeds are usually flattened with large membranous wings. There is no endosperm.

BIGNONIALES (H). An order of the Lignosae. They are trees, shrubs or climbers, rarely herbs, which may, or may not have tendrils. The leaves are alternate or opposite, mostly compound, and usually lacking stipules. The flowers are bisexual with the calyx lobes imbricate or valvate. The corolla lobes are fused, and it is usually zygomorphic. The lobes are imbricate, or rarely contorted, with 5, 4, or 2 stamens attached to them. Staminodes may be present. The superior ovary of 4-1 loculi is attached to a disk, and has axile or parietal placentation. The style is terminal. The fruit is a capsule or is indehiscent. Endosperm is absent, or if present, very scanty.

BIGUTTULATE. Containing two vacuoles or two oil-droplets.

BILABIATE. Having two lips.

BILABIATE DEHISCENCE. Opening by a transverse split across the top.

BILATERAL SYMMETRY. Can be halved in only one plane, so that the halves are mirror images of each other.

BILOCULAR. Having two chambers.

BIMODAL. Said of a frequency distribution which has two modes.

BINATE. In two parts: occurring in pairs.

BINOMIAL NOMENCLATURE. Naming by two Latin, or Latinized names, the first being the generic name, and the second the specific. *E.g. Ranunculus repens* and *Ranunculus bulbosa* belong to the same genus, but are different species. The name is usually followed by an abbreviation of the name of the person first naming the species, *e.g.* L. is the abbreviation for Linnaeus.

BINOMIAL SERIES. A series obtained by expanding the sum of two quantities to any power. $(a + b)^n$. *E.g.* if $n = 4$, then the series is $a^4 + 4a^3b + 6a^2b^2 + 4ab^3 + b^4$.

BINUCLEOLATE. Of an ascospore with two oil-droplets.

BIO-ASSAY. The evaluation of the effect of a substance on an organism under controlled conditions, *e.g.* the bio-assay of fungicides.

BIOBLAST = CHONDRIOSOME.

BIOCOENOSIS. The association of plants and animals, especially in relation to a given feeding area of the animals.

BIOCOENOSIUM. A community of both animals and plants.

BIOCOENOTIC. Referring to the inter-relationship between organisms in a community.

BIOGEN. A hypothetical protein molecule which is unstable, but has the essential properties of a living thing.

BIOGENESIS. The origin of living organisms from other living organisms. See *Abiogenesis.*

BIOGENETIC LAW. The hypothesis that, during its development, an individual repeats the evolution of its race in a shortened form.

BIOGENOUS. Living on another living organism, *i.e.* parasitic.

BIOLOGICAL BARRIER. The activity of any plant or animal which prevents an area being colonized by plants.

BIOLOGIC FORM = BIOLOGIC RACE = PHYSIOLOGIC RACE (FORM).

BIOLOGICAL CONTROL. The introduction, or preservation and encouragement of a natural enemy of a pest, disease, or disease-carrying organism, which may help in its control.

BIOLOGICAL SPECIALIZATION = PHYSIOLOGICAL SPECIATION.

BIOLOGICAL SPECTRUM. A table showing the percentage frequency of occurrence of the various plants and animals in a defined area.

BIOLOGICAL TYPE = LIFE FORM.

BIOLUMINESCENCE. The enzymatic production of light by living organisms, without their being previously illuminated, *e.g.* by some bacteria and fungi; wrongly called phosphorescence. The light is produced by the action of the enzyme luciferase in oxidizing the light-producing molecule luciferin.

BIOME. A large community of plants and animals, characterized by its particular type of dominant vegetation and its associated animals, *e.g.* tundra.

BIOMETER. A method of assessing the amount of life by measuring the respiration, *e.g.* in a soil sample.

BIOMETRY. The use of mathematics, especially statistics in the study of living organisms.

BION, BIONT. An individual plant, independent, and capable of a separate existence.

BIONOMIC. Relating to the environment; ecological.

BIOPHAGUS = BIOGENOUS.

BIOPHILOUS = BIOGENOUS.

BIOPHORE. A minute, hypothetical particle assumed to be capable of growth and reproduction.

BIOPLASM = PROTOPLASM.

BIOS FACTOR. A substance essential for the growth of a plant, especially yeasts, obtained from the environment. It is a mixture of aneurin, biotin, and other substances. Biotin = Vitamin H, and is probably identical to the co-enzyme R.

BIOSERIES. In evolution, a historical sequence formed by the changes in any one inheritable character.

BIOTA. The plants and animals of a given region.

BIOTIC. Related to life.

BIOTIC ADAPTATION. Changes in form or physiology, presumed to have arisen as a result of competition with other plants.

BIOTIC CLIMAX. A climax community maintained by living organisms, *e.g.* grassland prevented from becoming forest by the grazing of animals.

BIOTIC FACTOR. The influence of one living organism on another.

BIOTIN. Vitamin H. An essential growth substance for yeasts.

$$
\begin{array}{c}
\text{O} \\
\| \\
\text{C} \\
\diagup \quad \diagdown \\
\text{HN} \qquad \text{NH} \\
| \qquad\quad | \\
\text{HC}\!-\!\!-\!\text{CH} \\
| \qquad\quad | \\
\text{H}_2\text{C} \qquad \text{CH}\!-\!(\text{CH}_2)_4\!-\!\text{COOH} \\
\diagdown \quad \diagup \\
\text{S}
\end{array}
$$

BIOTOPE. See *Biocoenosium*.

BIOTYPE. (1) = PHYSIOLOGIC RACE.
(2) One individual, or a group of individuals with the same genetic make-up.

BIPAROUS. See *Dichasium*.

BIPARTITE. Split into two, nearly to the base.

BIPINNATE. Said of a compound leaf, which has its main segments pinnately divided.

BIPINNATIFID. Said of a pinnatifid leaf, whose parts are themselves pinnatifid.

BIPOLAR. (1) At the opposite ends of a bacterial cell.
(2) Having two sexual sorts or phases.
(3) Of a spore germinating by two germ-tubes, one at each end.

BIREFRINGENCE. Having two or more refractive indices, due to the asymmetry of the molecular structure. The study of the birefringent properties of a material helps to determine its molecular configuration at rest or under stress.

BISACCATE. Having two of the sepals each with a small pouch at its base.

BISECT. A drawing showing the profiles of the shoots and roots of plants growing in their natural positions.

BISERIAL, BISERIATE. (1) In two rows.
(2) Of a vascular ray two cells wide.

BISEXUAL. Bearing both male and female sex organs, *e.g.* a flower having anthers and ovaries.

BISPORAGIATE. Said of a strobilus having megasporophylls, and microsporophylls, with microsporangia and megasporangia.

BISULCATE. Marked by two grooves.

BISYMMETRIC. Symmetric in two planes, at right-angles to each other.

BITERNATE. Divided into three parts, each of which is divided into three.

BITUNICATE. Having two walls.

BIUNCIATE. Having two hooks.

BIURET TEST. A test for protein. Protein in the presence of strong alkaline solution, and copper ions (as copper sulphate) give a pinkish-mauve colour.

BIVALENT. A pair of homologous chromosomes which are united at the first meiotic division, usually by a chiasma. A *ring bivalent* has chiasmata at both ends of the chromosome; a *rod bivalent* has a chiasma at one end only; an *unequal bivalent* has the chromosomes of unequal size.

BIVERTICILLATE. Having branches at the two levels in the same species.

BIXACEAE (EP, H). A family of the Parietales (EP), Bixales (H). Found in the tropics. They are small trees or shrubs with alternate leaves with entire stipules. The flowers are regular and bisexual, borne in panicles. The calyx and corolla have 5 free lobes. There are many stamens with horse-shoe shaped anthers. The superior ovary is of 2 fused carpels, having 1 chamber with parietal placentation, bearing many anatropous ovules. The style is simple. The fruit is a capsule, splitting between the placenta. The seeds have red, fleshy papillae, and a starchy endosperm.

BIXALES (H). An order of the Lignosae. They are trees or shrubs, usually with alternate leaves with simple stipules. The flowers are mostly hypogynous, and rarely perigynous. They are bisexual or unisexual and actinomorphic. The sepals are imbricate or valvate, and petals may, or may not be present. There are many to few stamens which are mostly free. The ovary is superior with parietal (rarely basal) placentation. The seed has a small embryo and copious endosperm.

BIXINEAE (EP). A family of the Parietales (BH). It includes the Bixaceae, Flacourtiaceae, and Cochlospermaceae of EP.

BLACK EARTH. See *Chemozerns*.

BLACKMAN REACTION. A term which includes all the reaction in the photosynthetic process which do not need light (dark reactions). Named after the discoverer.

BLADDER. A modified leaf found on the bladderwort. It is used to catch small aquatic animals.

BLADE. (1) The flattened part of the thallus of the larger sea-weeds.
(2) The flattened part of a leaf, sepal, or petal.

BLASTEMA. The axial part of an embryo gymnosperm or angiosperm, but not the cotyledon.

BLASTENIOSPORE. A lichen spore which is divided by a cross-wall; and the two cells are connected by a tube.

BLASTIC ACTION. The catalytic action initiated by light, and stimulating the division and enlargement of cells.

BLASTO-. A prefix from the Greek 'blastos', which means 'bud'.

BLASTOCHORE. A plant which is dispersed by off-shoots.

BLASTOCLADIACEAE. The single family of the Blastocladiales.

BLASTOCLADIALES. An order of the Phycomycetes. There is a true mycelium, which has the lower hyphae broader than the others. Asexual reproduction is by posteriorly flagellate zoospores, produced in apical sporangia. Sexual reproduction is anisogamous, and the gametes uniflagellate. The biflagellate zoospores germinate without forming resting spores.

BLASTOGENESIS. Transmission of inherited characters by the germ-plasm only.

BLASTOGENIC. Related, in any way, to the germ-plasm. Belonging to the hereditary characteristics, due to the constitution of the germ-plasm.

BLASTOPARENCHYMATOUS. Said of an algal thallus which consists of filaments joined side by side, and not recognizable as separate filaments.

BLASTOSPORE. A fungus spore produced by budding, *e.g.* in yeasts.

BLATTIACEAE (H). See *Sonneratiaceae*.

BLEEDING. The exuding of sap from a wound. BLEEDING PRESSURE = ROOT PRESSURE.

BLEMMATOGEN. A layer of hyphae, usually with thickened walls, which covers the button of an agaric, and sometimes forms the veil.

BLENDING INHERITANCE. The inheritance of characters so that the off-spring, and successive generations are intermediate between the original parents. This is because the character is controlled by several genes.

BLEPHAROPLAST. A granule at the base of a flagellum. It is possibly a centrosome.

BLOOM. Grains, short rods, or crusts of wax on the surface of some leaves and fruits.

BLUE-GREEN ALGAE = MYXOPHYCEAE.

BLUNT. Having a rounded end.

BODY CELL. A cell in the pollen grain of the Gymnosperms which releases the male nuclei.

BOLE. The trunk of a tree.

BOLETACEAE. A family of the Agaricales. The fruiting body bears the hymenium on pores, and is fleshy with a stipe which is usually central. Most of the family grow on the ground. The family is sometimes raised to the status of an order or sub-order.

BOLETALES = BOLETINEAE = BOLETACEAE.

BOLL. The fruit of the cotton plant.

BOMBACACEAE (EP, H). A family of the Malvales (EP), Tiliales (H). BH include it in the Malvaceae. They are trees found in the tropics, especially of America. The large, thick trunks sometimes have water-storage tissue. The leaves are simple or palmate, with deciduous stipules. The flowers are bisexual, often large and usually regular. The 5 calyx lobes are fused and valvate, often with an epicalyx. The 5 free petals are convoluted and asymmetrical. There are 5 to many anthers, which may be free or united into a tube. There are 2-5 fused carpels in the superior ovary which is multilocular. There are 2 to many erect anatropous ovules in each loculus. The fruit is a capsule, often with the seeds embedded in hairs arising from the wall. There is little or no endosperm, and the cotyledons are flat, contorted, or plicate.

BOMBYCINE. Like silk.

BOND ENERGY. The energy holding the atoms of a molecule together. It can be passed from one molecule to another to form new bonds.

BONNETIACEAE (H). A family of the Theales. The leaves are alternate. The flowers are in terminal panicles or racemes and have contorted petals. The fruit has 7 loculi.

BOOTED = PERONATE.

BORAGINACEAE (EP, BH, H). A family of the Tubiflorae (EP), POLEMONIALES (BH), BORAGINALES (H). They are found in tropical and temperate regions, especially around the Mediterranean. They are mostly perennial herbs, but some are trees or shrubs, including a few climbers. The leaves are usually alternate and without stipules. The plants are generally hairy. The inflorescence is usually a scorpiod cyme. The flowers are bisexual and usually regular. The 5 fused calyx lobes are imbricate or open (rarely valvate), with the odd sepal posterior. The 5 corolla lobes are fused into a funnel or tube, and are imbricate or convoluted, bearing the 5 free stamens alternate to them. The anthers are introrse. The superior ovary, of 2 fused carpels, is borne on a hypogynous disk. It usually has 4 loculi (rarely 2-10), usually with a 'false' septum, and with a gynobasic, entire or lobed style. There is one erect, ascending, or horizontal anatropous ovule in each loculus. The fruit is of 4 achenes (nutlets) or a drupe. The seeds usually have little endosperm. The embryo is straight or curved, with the radicle pointing upwards.

BORAGINALES (H). An order of the Herbaceae, containing the single family—the Boraginaceae *(q.v.)*.

BORAGINEAE (BH) = BORAGINACEAE.

BORDEAUX MIXTURE. A fungicide made of copper sulphate and quick lime. A common formula is 4 lb. copper sulphate, 4 lb. quick lime, dissolved in 50 gal. water.

BORDER PARENCHYMA. A sheath of one or more layers of parenchyme surrounding a vascular bundle.

BORDERED PIT. A thin area in the wall between two vessels or tracaids surrounded by overhanging rims of thickened wall.

BOSS. A protuberance.

BOSTRYX. A cymose inflorescence having the flowers all born on the same side of the axis.

BOTHRODENDRACEAE. A family of the Lepidodendrales, found from the Devonian to the early Upper Cretaceous. They were large trees branching dichotomously at the top and bottom. The trunk bore small scale leaves in low spirals. The vascular system was an exarch protostele or siphonostele. Secondary thickening has been recorded. The species were heterosporous.

BOTRYOPTERIDACEAE. The single family of the Botryopterineae.

BOTRYOPTERIDINEAE. A sub-order of the Coenopteridales. The stem is monostelic. The leaves are either not pinnate, or the pinnae are at an angle to the plane of the leaf-blade. The sporangia are borne directly on the rachis or on unflattened pinnules, in stalked clusters.

BOTRYOSE, BOTRYOID, BOTRYTIC. Like a bunch of grapes; racemose.

BOTRYOSPHAERIACEAE. A family of the Pseudosphaeriales. The large stroma of the ascocarp bursts through the bark, forming a spherical perthecium-like body, covered by a hard, thin black layer. The asci are broadly club-shaped.

BOTTOM YEAST. A yeast that collects at the bottom of the vessel during alcoholic fermentation.

BOTULIFORM. Cylindrical, with rounded ends.

BOULDER CLAY. A clay soil, that has been transported by glaciers, and contains large stones.

BOUND WATER. Water held by adsorptive, or other physical properties.

BOUQUET STAGE. When the chromosomes lie in loops with their ends near one part of the nucleus wall, during zygotene and pachytene.

BOURGEON. To bud or sprout.

BRACHIALIS. Latin meaning 'a cubit long'.

53

BRACHIATE, BRACHIFEROUS. Branched; having widely spreading arms.

BRACHY-. A prefix meaning 'short'. From the Greek 'brachus'.

BRACHYBLAST. A short branch of limited growth, bearing leaves, and sometimes flowers and fruits; a spur.

BRACHYCLADOUS. Having very short branches.

BRACHYFORM. Of the life-cycle of some autoecious rusts which lack an acedial stage.

BRACHYMEIOSIS. (1) A simplified form of meiosis, which is completed in one division.
(2) The second reduction division in the production of ascospores.

BRACHYSCLEREIDE = STONE CELL.

BRACT. (1) A small atypical leaf subtending a flower-bud in its axil.
(2) Small leaf-like organ around the sex-organs of the liverworts.
(3) A sterile leaf subtending the sporangiophore of some Equisetinae.
(4) A sterile 'leaf' in the strobilus of the Lycopodinae.

BRACT SCALE. The small outer scale at the base of the large cone scale of conifers.

BRACTEAL LEAF. A general term for bracts and bracteoles.

BRACTEATE. Having bracts.

BRACTEODY. The replacement of flower members by bracts.

BRACTEOLE. A leaf, generally small, on a flower stalk.

BRACTEOMANIA. The abnormal production of large numbers of bracts, often at the expense of normal flowers.

BRACTEOSE. Having conspicuous bracts.

BRADY-. A prefix from the Greek 'bradys' meaning 'slow'.

BRADYSPORE. A plant which liberates its seeds slowly.

BRAK SOIL. An alkaline soil found in South Africa. The dominating method of formation is the extreme temperature changes.

BRANCH ABSCISSION. The shedding of branches by a special separation.

BRANCH GAP. An area of parenchyma developing in a siphonostele immediately above the vascular tissue going into a branch.

BRANCH TENDRIL. A tendril formed from a modified branch.

BRANCH TRACE. Branches of the vascular tissue running to a branch.

BRAND SPORE. A thick-walled resting-spore formed by the Ustilaginales. They are dark coloured and found in sooty masses.

BRAVAISITE. See *Illite*.

BREAKING. The development of striping in tulip flowers A virus infection.

BREAKING OF MERES. The sudden development of large masses of blue-green algae in small masses of fresh water.

BREATHING ROOT. A root produced by large plants growing in mud; it projects above the mud and water allowing the air to reach the root below.

BREVI-. A Latin prefix meaning 'short'.

BREVICIDAL DEHISCENCE. Said of a fern sporangium when the annulus is interrupted by the stalk.

BREVICOLLATE. Short-necked.

BRIDGE. See *Chromatid.*

BRIDGING HOST, BRIDGING SPECIES. An intermediate host, whereby a specialized parasite passes from a susceptible to a resistant host.

BRISTLE. (1) A stiff hair.
(2) A long, hollow outgrowth from the cell-wall of some algae.

BROAD-LEAFED TREE. Any tree that is not a conifer.

BROAD RAY. A vascular ray, many cells thick, whose cells are round in transverse section.

BROKEN-BOND WATER. Water held at the surface of soil particles by mechanical forces.

BROMATIA. Rounded swelling at the tips of the hyphae of fungi which are cultivated by ants for food.

BROMELIACEAE (EP, BH, H). A family of the Farinosae (EP), EPI-GYNAE (BH), BROMELIALES (H). Except for one genus, confined to tropical America. Many members are terrestrial, but most are epiphytes. The stem is usually reduced with a rosette of fleshy, strap-like, spiny leaves. The adventitious roots are mainly for support, as water is stored at the bases of the leaves, and absorbed by glandular hairs on them. The inflorescence is terminal emerging through the rosette of leaves. The bracts are coloured. The flowers are usually bisexual and regular. The perianth is in 2 whorls of 3 lobes each, and the members of each whorl may be free or fused. The outer whorl is sepaloid and persistent, with the inner one petaloid. The 6 free stamens are often epipetalous, and have introrse anthers. The ovary is of 3 fused carpels, and may be inferior to superior; it has 3 loculi with many anatropous ovules on axile placentas in each. There is 1 style with 3 stigmas. The fruit is a berry or capsule, and in the latter case the seeds are light or winged. The embryo is small, and the endosperm is mealy.

BROMELIALES (H). An order of the Calyciferae (H) containing the single family, the Bromeliaceae *(q.v.).*

BRONCHONEMA. Takes place in cell-division when the spireme thread becomes looped.

BROOD BUD. (1) A small multicellular organ serving for vegetative reproduction in some of the Red Algae.
(2) A bulbil of the Bryophyta.
(3) See *soridium.*

BROOD CELL. A naked or walled cell which is produced asexually, and separating from the parent plant to give rise to a new individual.

BROOD GEMMA. A multicellular body formed asexually and separating from the parent plant to form a new plant.

BROWN ALGAE = PHAEOPHYTA.

BROWN EARTH. A soil produced on loams and clays, with a fairly low air-content. Though typically acid, the amount of exchange calcium is fairly high. There are no sharp horizons in the profile.

BROWNIAN MOVEMENT. The continuous movements of the particles in a colloidal solution, caused by the unbalanced bombardment with the molecules of the liquid.

BRUMALIS. Latin for 'winter'.

BRUNELLIACEAE (EP) (H). A family of the Rosales (EP) Dilleniales (H). (BH) include it in the Simarubaceae. There is one genus *Brunellia*, found from

Peru to Mexico. They are trees or shrubs, with the leaves opposite or in whorls. The flowers are monochlamydous and unisexual, with the floral parts in fours, fives, or sevens. There are 5-2 carpels each with 2 pendulous ovules. The fruit is a capsule, and the seeds contain endosperm.

BRUNIACEAE (EP, BH, H). A family of the Rosales (EP, BH), Hamamelidales (H). They are heath-like shrubs with alternate exstipulate leaves. The bisexual flowers are usually regular, and in spikes or heads. The flower parts are (except for the ovary) in fives, and the ovary is semi-inferior to inferior. The stamens are in one whorl. The ovary consists of 3-2 fused carpels each with 3-4 ovules, or with 1 ovule. The fruit is a capsule with 2 seeds, or a nut with 1 seed.

BRUNNEUS. Latin meaning 'brown'.

BRUNONIACEAE (EP, H). A family of the Campanulatae (EP), Goodeniales (H). BH include it in the Goodeniaceae. There is one genus, *Brunonia*, found in Australia and Tasmania. This is a herb with radicle, entire leaves without stipules. The blue, bisexual flowers are in heads. The calyx has 5 fuse lobes, as does the corolla. The stamens are free, but the anthers are connate in a tube around the style. The superior ovary has 1 loculus, and contains 1 ovule. The fruit is an achene, and the seed has no endosperm.

BYRALES = EUBRYALES.

BRYOPHYTA. The Liverworts and Mosses. These are simple green plants, which never have a vascular system. The sex-organs are multicellular, and have an outer layer of sterile cells. There is always an alternation of generations between morphologically distinct gametophyte and sporophyte. The sporophyte is attached to the gametophyte throughout the development of the sporophyte.

BRYOPSIDACEAE. A family of the Siphonales. There are no septa in the thallus which is differentiated into a prostrate rhizome-like part, and an erect part with pinnate branches. The pinnules may break off during asexua, reproduction. No zoospores are known. Sexual reproduction is anisogamousl the gametes being produced in the pinnules, or in branches from them.

BUCKLE = CLAMP CONNECTION.

BUD. A compacted stem with leaves and sometimes flowers. This elongates to produce the mature shoot and its attached organs.

BUDDING. (1) The production of buds in general.
(2) Grafting a bud from a stock to a scion.
(3) The asexual reproduction of unicellular fungi, *e.g.* yeasts, or spores, by the development of a new cell from the outgrowth of the parent.

BUDDLEIACEAE (H). A family of the Loganiales. The corolla lobes are imbricate (rarely contorted). There are 4 or more stamens. There are usually many ovules in each ovule or loculus. The fruit is usually a septically dehiscent capsule, and the seeds are often winged. The indumentum, if present is glandular, stellate, or lepidote.

BULB. An organ of storage and vegetative reproduction. It consists of a flattened stem bearing fleshy leaves, or leaf-bases with buds in their axils, and scale-leaves.

BULBIL. (1) A small bulb.
(2) Outgrowths on the stem of some species of *Lycopodium* functioning for vegetative reproduction. They are probably modified leaves.
(3) An organ of asexual reproduction produced on the rhizoids of the Charales.
(4) A small sclerotium-like body, of a few cells produced by some fungi.

BULBILLATE. Of the stipe of some fungi when it has a small, ill-defined bulb-like structure at the base.

BULBOUS. (1) Bearing bulbs.
(2) Swollen like a bulb.

BULBUS. The enlargement at the base of the stipe of some agarics.

BULLATE. (1) Blistered, puckered, bubble-like.
(2) Of a fungus pileus, having a rounded projection at the centre.

BULLER PHENOMENON. In Basidiomycetes and Ascomycetes. The making of a unisexual mycelium, or fruit-body rudiment, diploid by a bisexual mycelium.

BULLIFORM CELL = MOTOR CELL.

BUNDLE = VASCULAR BUNDLE.

BUNDLE END. The simplified ending of a vascular bundle in the mesophyll of a leaf.

BUNDLE SHEATH. The sheath of parenchyma around the vascular bundle of a leaf. The elongated cells conduct materials between the bundle and the leaf parenchyma.

BUR, BURR. A hooked fruit.

BURDO. A graft-hybrid thought to have been formed by the fusion of vegetative nuclei from the stock and scion.

BURGUNDY MIXTURE. A fungicide similar to Bordeaux mixture, but sodium carbonate replaces the lime.

BURMANNIACEAE (EP, BH, H). A family of the Microspermae (EP, BH), Burmanniales (H). These are mostly herbaceous saprophytes on forest plants of the tropics and sub-tropics. The regular perianth is in 2 whorls of 3, the parts being fused, and with short lobes. There are 3 or 6 free anthers. The inferior ovary has 3 fused carpels, with 1 or 3 loculi. The fruit is a capsule having many seeds which are endospermous.

BURMANNIALES (H). An order of the Corolliferae. They are small herbs, often saprophytes with reduced leaves. The perianth is tubular with the outer lobes valvate. There are 6 or 3 stamens. The ovary is inferior with 1 or 3 loculi. The ovules and seeds are numerous and small. There is little endosperm.

BURSERACEAE (EP, BH, H). A family of the Geraniales (EP, BH), Rutales (H). They are tropical shrubs or trees, usually with alternate compound leaves which are sometimes dotted. Stipules are rarely present. The flowers are small, with a disk, and usually unisexual, with the parts in fours or fives. The petals are imbricate or valvate, and when both whorls of stamens are present, the outer whorl is opposite the petals. The ovary has one style, and consists of 5 to 3 fused carpels. There are many loculi usually with 2 ovules in each. The fruit is a drupe or capsule. The seeds have no endosperm. Many species, *e.g.* frankincense, are useful because of their resins.

BURSICULATE, BURSIFORM. Shaped like a bag.

BUSH. A low woody plant with a number of branches at, or near, ground-level.

BUTOMACEAE (EP, H). A family of the Helobieae (EP), Butomales (H), (BH) include it in the Alismataceae. Water or marsh herbs, with leaves of various types. The inflorescence is usually a cymose umbel of regular, bisexual flowers. The perianth is in 2 whorls of 3 free members; the outer whorl is sepaloid, and the inner is usually petaloid. There are 9 to many stamens, with introrse anthers. The gynecium consists of 6 to many free, superior carpels, each with many anatropous ovules scattered over the inner walls, except on the mid-rib and edges. The fruits are follicles. The seed has no endosperm, and the embryo is straight, or horse-shoe shaped.

BUTOMALES (H). An order of the Calyciferae. They are perennial aquatic herbs in fresh, or salt water. The leaves are radical or on the stem, and are alternate or whorled. The flowers are showy, or small, and are bisexual or unisexual, varying from hypogynous to epigynous. The perianth is in 2 whorls, the outer being sepaloid, and the inner petaloid, or absent. There are α-3 free stamens. The gynecium is of free or fused carpels. If they are fused, the ovary is inferior. There are many ovules scattered on the walls of the pericarp. The seeds have no endosperm.

BUTTERFLY FLOWER. A flower that is pollinated by butterflies.

BUTTON. The young fructification of an agaric, before the cap has broken away from the stipe, and the pileus is exposed.

BUTTRESS ROOT. A root, which is often adventitious and above ground, giving additional support to the stem or trunk.

BUTYRIC ACID. $C_4H_9.COOH$. An acid produced during the anaerobic respiration of carbohydrates. Smells of rancid butter.

BUTYROUS. Like butter.

BUXACEAE (EP, H). A family of the Sapindales (EP), Hamamelidales (H). BH include it in the Euphoebiaceae. They are evergreen shrubs with leathery leaves which lack stipules. There is no latex. The unisexual flowers are in heads or spikes, and have no petals. There are many to 4 free stamens. The ovary is superior consisting of 3 fused carpels. There are 3 styles which are persistent on the fruit. There are 2 or 1 pendulous anatropous ovules in each loculus, and each has a dorsal raphe. The fruit is a capsule or drupe, and the seed has endosperm. The seed may or may not have a caruncle.

BUXBAUMIALES. An order of the Eubrya. The capsule is asymmetrical and the peristome has 3 to 6 layers.

BYBLIDACEAE (H). A family of the Pittosporales. They are herbs or under-shrubs with linear leaves which are glandular with soft, scattered hairs. The sepals are imbricate. The petals are imbricate or contorted and larger than the sepals. There are 5 stamens with anthers opening by pores.

BYSSACEOUS, BYSSOID. Cotton-like; made up of delicate threads.

BYSSISEDE. Said of a fungus fruit-body growing on a cottony mass of hyphae.

C

CAATINGA FOREST. Forests, found in Brazil, where the leaves fall in the dry season.

CABOMBACEAE (H). A family of the Ranales, found in America, India, Australia, and tropical Africa. It is an aquatic family. There is no disk, and the torus is sometimes enlarged to enclose the carpels. The stamens are centrifugal, or few, with extrorse anthers. There are 3 sepals and petals. The ovule are parietal, and spread over the abaxial wall of the carpels. The seeds are not arillate.

CACTACEAE (EP, BH, H). A family of the Opuntiales (EP), Ficoidales (BH), Cactales (H), found mainly in the dry regions of tropical America. The stems are fleshy, with water-storage tissue, and they rarely have green leaves; instead of them, they have thorns. The large, coloured, hermaphrodite flowers are usually solitary, and borne on or near groups of thorns or hairs. They are regular or zygomorphic. The flower parts are spirally arranged often up to the side of the ovary. There are many perianth segments grading from sepals to petals. The many stamens are epipetalous. There are 4 to many fused carpels in the inferior ovary, which has parietal placentation, and contains many anatropous ovules. The style is simple. The fruit is a berry, with the flesh derived from the funicles. The seeds may or may not have endosperm.

CACTALES (H). An order of the Lignosae. See *Cactaceae*.

CACTACEAE (BH) = CACTACEAE (EP).

CADUCOUS. Non-persistent; falling readily. Said of spores and floral parts.

CAECUM. An extension of the embryo-sac into the endosperm.

CAEOMA. An aecium, with no peridium, with or without paraphyses, or it is represented by a sterile ring of hyphae.

CAERULEUS. Latin for 'pale blue'.

CAESALPINIACEAE (H). A family of the Leguminales. The flowers are more or less zygomorphic. The petals are imbricate with the upper petal inside the two lateral petals. The anthers dehisce lengthwise, or by lateral pores.

CAESIOUS. Having a bluish-grey waxy bloom.

CAESIUS. Latin for 'lavender-coloured'.

CAESPITOSE, CESPITOSE. In tufts.

CAFFEINE. A purine in coffee beans, tea leaves, and cacao beans. An oxidation product of the methyl derivative of purine.

CALAMIFEROUS. Having a hollow stem.

CALAMITACEAE. A family of the Equisetales. A fossil family found in the Upper Devonian, Carboniferous, and Triassic. The sporophyte was tree-like, with secondary thickening in the stem and root. The strobili had alternate whorls of peltate sporangiophores and sterile bracts.

CALATHIDE. (1) The involucre of a capitulum.
(2) The capitulum itself.

CALCARATE, CALCARATUS. Having a spur.

CALCAREOUS. Covered with, or containing lime.

CALCAREOUS PAN. A hard layer of limey material, more or less impermeable to water, formed below the surface of the soil, and affecting the water-supply of plants growing above it.

CALCEIFORM, CALCEOLATE. Shaped like a shoe.

CALCICOLE. A plant which goes best on soil containing lime.

CALCICOLOUS. Living on chalky soils.

CALCIFICATION. The accumulation of calcium carbonate, on or in the cell-walls.

CALCIFERAE. Monocotyledons with a distinct calyx and corolla.

CALCIFUGE, CALCIPHOBE, CALCIFOBOUS. A plant species which does not thrive on a lime soil. It prefers an acid soil.

CALCIPHILOUS. Chalk-loving.

CALICHE. A red soil produced by the effect of heat on grassland. The calcium carbonate horizon becomes thickened and may be hardened.

CALINE. A growth-promoting substance. Rhizocalines promote root growth, and caulocalines promote stem growth.

CALLI- A Greek prefix meaning 'beautiful'.

CALLITRICHACEAE (EP, H). A family of the Gereniales (EP), Lythrales (H), and included in the Haloragidaceae by (BH). It contains one genus *Callitriche*. The unisexual flowers have neither sepal or petals, but are subtended by two bracts. The male flower consists of 1 stamen, and the female of 2 fused carpels, each divided by a septum to form 4 loculi in all, each containing 1 pendulous ovule. The fruit is a schizocarp, and the endosperm is fleshy.

CALLOSE. (1) A carbohydrate, which is insoluble in cupraammonia, but soluble in 1% solutions of caustic alkalies. It is deposited seasonally or permanently on sieve plates, causing them to stop functioning. It is also found in calcified walls, and in the cells of some algae.
(2) Hard or thick, and, sometimes, rough.

CALLOSITY. See *Callus* (3).

CALLUS. (1) A more or less corky secondary tissue developed by woody plants over a wound. It is usually derived from a cambium.
(2) The swollen base of the inferior palea of grasses, next to the axis.
(3) A mass of material formed on a cell-wall around the germ-tube of a parasitic fungus.

CALOBRYACEAE. The single family of the Calobryales.

CALOBRYALES. An order of the Hepaticae. The moss-like gametophore is differentiated into stem and leaves. The antheridia are stalked, and ovoid. The neck of the archegonia have only 4 vertical rows of cells. The sporophyte has an elongated capsule with a jacket-layer 1 cell thick, except at the apex.

CALORIE. The amount of heat required to raise the temperature of 1 g. of water through 1°C. The calorie used in heat values of food is 1000 of the calories defined above.

CALVESCENT, CALVOUS. Naked, bald, or becoming so.

CALYC-. A Greek prefix meaning 'a cup'.

CALYCANTHACEAE (EP, BH, H). A family of the Ranales (EP, BH) Rosales (H). They are usually aromatic shrubs, with opposite, simple leaves, without stipules. The perianth has many parts arranged in a spiral, with a transition from sepals to petals. There are many stamens with the inner ones sterile. There are many free carpels in the gynecium, each with 2 anatropous ovules, and in a hollow axis. The fruits are achenes enclosed in the receptacle. There is little endosperm. The embryo is large with the cotyledons wound in a spiral.

CALYCANTHEMY. An abnormal condition in which the sepals become petalloid.

CALYCERACEAE (EP, BH, H). A family of the Campanulatae (EP), Asterales (BH), Valerianales (H). They are herbs, with alternate or radical leaves without stipules. The flowers are in heads with an involucre of bracts. The flowers are bisexual or unisexual, and are regular or zygomorphic. The calyx is leafy, and the flowers valvate (H) or open (EP, BH). The anthers are in 1 whorl, with the filaments united, and are free, or slightly coherent at the base. The gynecium has 1 loculus with 1 pendulous anatropous ovule. The fruits are like achenes, sometimes more or less united, and covered by a persistent calyx. The embryo is straight with a slight endosperm.

CALYCIFERAE (H). The division of the Monocotyledons whose members have a distinct calyx. The root-stock is always a rhizome.

CALYCIFLORAE (BH). The third series of the Polypetalae. The corolla usually has distinct petals, with stamens which are perigynous or epigynous.

CALYCIFORM, CALYCULAR. Cup-like.

CALYCINAE (BH). The fourth series of the Monocotyledons. The perianth is sepaloid, and may be membranous. The ovary is usually free, and superior, and there is abundant endosperm.

CALYCINE. Relating to, or belonging to the calyx.

CALYCOID. Like a calyx.

CALYCULE, CALYCULUS. (1) A group of leaf-like appendages below the calyx.
(2) Sometimes for the epicalyx.
(3) (Of Myxomycetes) a cup- or calyx-like appendage below the sporangium.

CALYPTRA. (1) A cap.
(2) A membranous cap on the sporangium of liverworts or mosses which is derived from the archegonial wall.
(3) The root cap.
(4) A thickened wall on the terminal cell of a filament of the Myxophyceae.

CALYPTRATE. Capped.

CALYPTROGEN. A meristematic tissue from which the root-cap in many plants is formed.

CALYX. The outer whorl of a flower made up of sepals which are usually green, and protect the flower in the bud.

CALYX TUBE. (1) A tube formed by the fusion of at least the lower part of the calyx.
(2) The hollowed receptacle of a perigynous flower, on which the petals and stamens grow.

CAMBIFORM CELL. An elongated, pointed parenchymatous cell found in the phloem.

CAMBIUM. A cylinder, strip, or layer of meristematic cells, which divide to give cells which ultimately form a permanent tissue. The primary cambium in the stem and root gives rise to xylem and phloem, and in woody stems a secondary one produces bark. Between the xylem and phloem of stems and roots of dicotyledons lies the *intrafascicular cambium (fascicular cambium)* derived from the apical meristem, and the *interfascicular cambium* is formed in the parenchymatous tissue between the bundles, by cells which have become secondarily meristematic.

CAMBIUM INITIAL. One of the permanently meristematic cells of a cambium.

CAMBRIAN. A geological period extending from 500 to 420 million years ago.

CAMELINUS. Latin meaning 'tawny'.

CAMNIUM. A succession of plants caused by cultivation.

CAMPANALES (BH, H). An order of the Gamopetalae (BH), Herbaceae (H). The flowers are actinomorphic to zygomorphic. The stamens are free, or inserted low down on the corolla. The anthers are free to connivent. The ovary is inferior, rarely superior, with 2-6 loculi, with usually many ovules in each loculus.

CAMPANULACEAE (EP, BH, H). A family of the Campanulatae (EP), Campanales (BH). The leaves are usually alternate, and without stipules. The bisexual flowers are regular or zygomorphic, in fives, and often showy. The petals are usually united, with the stamens free or united with introrse anthers. The ovary is inferior and of 2-5 fused carpels, each with many ovules. The fruits are capsules or like berries, and the seeds have endosperm.

A family of the Campanales (H), with an actinomorphic corolla, with the anthers free from each other.

CAMPANULATAE (EP). An order of the Sympetalae. They are usually herbs, and rarely woody, with the flower parts in fives, with one whorl of stamens and usually fewer carpels. The anthers have 2 loculi, and are often united. The ovary is inferior or superior with many loculi, and with many to 1 ovules in each, or with 1 loculus with 1 ovule.

CAMPANULATE. Bell-shaped.

CAMPESTRIS. Latin for 'growing in fields'.

CAMPYLOTROPOUS. Of an ovule, curved, so that the chalaza and the micropyle do not lie in a straight line.

CANAL. An elongated intercellular space often containing secretions of various kinds, *e.g.* oils, resins.

CANAL CELL. The initial cell which later divides to fill the central canal of the neck of the archegonia of Bryophyta and Pteridophyta. These cells disintegrate when the archegonium is ripe.

CANALICULATE. Having a groove running lengthwise.

CANCELLATE. Being latticed: in a net-work.

CANDIDUS. The Latin for 'pure white'.

CANDOLLEACEAE = STYLIDIACEAE.

CANELLACEAE (H) = WINTERACEAE. A family of the Magnoliales. These are glabrous aromatic trees, with glandular leaves. The actinomorphic flowers are in cymes, with persistent bracts. There are 4-5 free, imbricate sepals, and up to 20 hypogenous stamens with connate filaments. The ovary is superior and unilocular with 2-5 parietal placentas, containing many subanatropous ovules. The fruit is a berry with 2 or more seeds. The endosperm is oily, and the embryo is straight.

CANESCENT. (1) Becoming hoary or grey.
(2) Being hoary due to the presence of short hairs.

CANKER. A plant disease, caused by bacteria or fungi, giving a limited necrosis of the cortical tissue.

CANNABIACEAE (H). A family of the Urticales. Erect or climbing herbs, often with a fibrous stem. The leaves are alternate or opposite, and sometimes lobed. A calyx is present, the leaves are stipulate. There is 1 ovule, and the anthers are erect in the bud. The seed has a fleshy endosperm with a curved or coiled endosperm.

CANNACEAE (EP, H). A family of the Scitamineae (EP), Zingiberales (H). BH include in the Scitamineae. It contains one genus *Canna*. The androecium consists of 1 stamen, which has 1 loculus, the rest are petalloid. The sepals are free, or at most connivent. There are many ovules in each loculus. The leaves have no ligule, and are not pulvinate.

CANOPY. The branches, leaves etc. formed by woody plants, some distance above the ground.

CANTHARELLACEAE. A family of the Polyporales. The fructification is funnel-shaped to clavate, and is fleshy. The hymenium is on the upper surface, and is smooth, and laddered, or with low, broad, rounded, longitudinal ridges. The basidia are stichobasidial, with white, or light-coloured basidiospores.

CANTHARELLALES. The Cantharellaceae with parts of the Thelophoraceae, Hydnaceae, and Agaricaceae.

CANUS. Latin for 'grey-white'.

CAP. (1) Pileus.
(2) A strand of sclerenchyma, often present outsidê the phloem, seen as a crescent in transverse section.

CAP CELL. The cell at the apex of the fern antheridium. It is shed when the antherozoids are liberated.

CAPILLACEOUS. Like a hair.

CAPILLARY. Hair-like: having a small diameter.

CAPILLARY SOIL WATER. Water held between the soil particles by capillarity. It is available to the plants.

CAPILLIFORM. Like a hair.

CAPILLITIUM. (1) A mass of threads.
(2) A mass of sterile hyphae in the fruit bodies of some fungi.
(3) Strands of protoplasm in the fruit bodies of the Myxomycetes.

CAPITALIST. A plant which has stored food reserves.

CAPITATE. Having a head: in a head: head-like.

CAPITELLUM. A little head.

CAPITULIFORM. Having the characters of a dense head of flowers.

CAPITULUM. (1) A head of flowers.
(2) A racemose inflorescence with the axis flattened to form a disk. The flowers are sessile with the oldest outside, and enclosed in an involucre of bracts.
(3) There are primary and secondary ones in the globules of the Charales, and they bear the antheridial filaments.
(4) A globose apical lichen apothecium.

CAPNODIACEAE. A family of the Erisiphales or Dothideales. The external mycelium is dark with the hyphae joining laterally to form sheets. The conidia are in elongated pycnidia which open by pores at the apices. The perithecia are dark, with, or without apical pores. The perithecial wall is made of rounded cells, or of elongated hyphae, joined by slime. These are the sooty moulds, which are saprophytic on the 'honey-dew' formed by aphids on leaves. The fungus makes the leaves black.

CAPPARIDACEAE (EP, BH, H). A family of the Rhoeadales (EP), Parietales (BH), Capparidales (H). These are xerophytic herbs or shrubs with simple to palmate leaves, which are reduced and often inrolled. Stipules are often modified as thorns or glands. The bisexual flowers are usually in racemes. The calyx is in 2 whorls of 2, and there are 4 petals. There is a great variation in the number of anthers (which are 2 locular (H)). The superior ovary is typically of 2 capels fused together but may be up to 12 (mostly unilocular (H)). The fruit is a siliqua, nut or berry or drupe. There is no endosperm, and the embryo is folded in various ways (more or less curved (H)).

CAPPARIDALES (H). An order of the Lignosae. They are woody to herbaceous with hypogynous to subperigynous flowers which are often somewhat zygomorphic. Petals may or may not be present. The stamens are usually free, and are from many to few. The ovary is often stipulate and often of 2 fused carpels, mostly with parietal placentation. There is little or no endosperm, and the embryo is curved or folded. The leaves are mostly alternate, simple or digitate, rarely with stipules.

CAPREOLATE, CAPREOLATUS. Latin meaning 'having tendrils'.

CAPRIFICATION. (1) The formation of an inflorescence of a number of flowers on a common receptacle which grows upwards to enclose the flowers except for a small pore at the apex.
(2) The fertilization of fig flowers by the fig insects,—a family of Chalcids (*Agagonidae*).

CAPRI FIG. A race of fig that does not produce edible fruit, but provides food for the wasps which pollinate the figs.

CAPRIFOLIACEAE (EP, BH, H). A family of the Rubiales (EP, BH) Araliales (H). BH include the Adoxaceae (EP). These are trees or shrubs having leaves with very small or no stipules. The bisexual flowers are regular or zygomorphic, and usually in cymes. The 5 petals are united into a tube (H), free (BH, EP), with the anthers inserted on them. The inferior ovary consists of 2-5 fused carpels, each being multilocular, with 1 to many pendulous ovules in each on axile placentas. The fruit is a berry, drupe, or capsule. The embryo is small in a fleshy endosperm.

CAPSULE. (1) A dry indehiscent fruit consisting of more than carpel.
(2) The part of a bryophyte sporangium that contains the spores.
(3) The mucilage surrounding a bacterial cell.

CAPUSIACEAE (H). A family of the Celastrales. The carpels are enclosed by an enlarged disk to form a false fruit. The carpels are buried in the disk, with the 5 styles and stigmas joined to its inner side. The stipules are minute and caducous. The flowers are bisexual; the bracts are not tubular, and the 5 stamens are free.

CARBOHYDRATE. A compound of carbon, hydrogen, and oxygen only, with twice as many hydrogen atoms as oxygen atoms in the molecule. They have the general formula $(C_xH_{2y}O_y)$, *e.g.* glucose, starch, and cellulose.

CARBONACEOUS. Like a cinder; dark-coloured, and broken easily.

CARBON ASSIMILATION = PHOTOSYNTHESIS.

CARBON CYCLE. The circulation of carbon in nature. Atmospheric carbon dioxide is built-up into the tissues of green plants. When they die, or are eaten by animals which also die, the dead bodies, and the excretory products and faeces of the animals are decomposed by bacteria, releasing carbon dioxide which is used again by green plants.

CARBONIC ANHYDRASE. The enzyme that catalyses the formation of bicarbonate ions from carbon dioxide and water:—
$$CO_2 = H_2O = H^+ + HCO_3^-$$

CARBONIFEROUS. A geological period extending from about 270 to 220 million years ago.

CARBON-NITROGEN RATIO. Written C/N ratio. The relation between the carbon and nitrogen content of a plant.

CARBONACEOUS. Black and hard, looking like charcoal.

CARBOXYDISMUTASE. An enzyme which catalyses the combination of ribulose-diphosphate and carbon dioxide to form two molecules of phosphoro-glyceric acid.

CARBOXYLASE. An enzyme capable of splitting carbon dioxide from a substrate. *E.g.* $CH_3.CO.COOH = CH_3CHO + CO_2$.

CARCERULUS. A fruit which splits at maturity into several one-seeded portions.

CARDINAL POINTS OF TEMPERATURE. The minimum point is when growth begins; the optimum when growth is best: and the maximum when growth stops.

CARDIOPTERIDACEAE (H). A family of the Celastrales. The sepals are imbricate, and the petals are imbricate, rarely contorted. The cordate leaves are 3-5 nerved from the base, and have no stipules. There is no disk. The stigmas are very unequal. Rarely the fruit is enclosed by an enlarged calyx. These are climbing herbs with a milky juice.

CARICACEAE (EP, H). A family of the Parietales (EP), Cucurbitales (H); BH include this order in the Passiflorales. They are small trees, with a terminal crown of palmate or digitate exstipulate leaves, the flowers are regular in loose axillary inflorescences, and are unisexual with 5 or multiples of 5 parts. The corolla is twisted in the bud. The male flower has a long corolla tube, with 10 stamens inserted on it. The female flower has a short corolla tube with a superior ovary of 5 fused carpels, which has 1 or 5 loculi. The style is short with 5 stigmas. There are many anatropous ovules, usually on parietal placentas. The fruit is a berry, and the endosperm is oily.

CARIES. Decay.

CARINA. A keel, as in the flowers of legumes, consists of 2 fused lower petals which enclose the stamens and stigma. It may play a part in pollination.

CARINAL CANAL. A vertical canal on the inner side of the metaxylem of the Equisetales. It is formed by the disintegration of the protoxylem.

CARINATE. Like a boat; keeled.

CARIOSE, CARIOUS. Looking as if it is decayed.

CARNEUS. Latin for 'flesh coloured'.

CARNIVOROUS PLANTS. Plants which are capable of catching and digesting insects or other small animals.

CARNOSE, CARNOSUS (Latin), CARNOUS. Fleshy in texture.

CARNULOSE. Somewhat fleshy.

CAROTENE (CAROTIN). An orange-coloured hydrocarbon found on chloroplasts and other plastids. They are precursors of vitamin A. $C_{40}H_{56}$

CAROTINOIDS = CAROTENOIDS.

CARPEL. The megasporophyll of an angiosperm. It bears the stigma, and frequently an elongated style, and encloses the ovules. The carpels may be separate or united. The total of the carpels in a flower is the ovary.

CARPELLARY SCALE = BRACT SCALE.

CARPELLOID. Said of some member of a flower which is in part changed into a carpel.

CARPID. A little carpel.

CARPOCEPHALUM = ARCHEGONIOPHORE. The stalked female branches of the Marchantiaceae.

CARPOGENOUS. Living on fruit, parasitically, or saprophytically.

CARPOGONIAL FILAMENT. A specialized branch-filament bearing the carpogonium in the Florideae.

CARPOGONIUM. (1) The female sex-organ of the Floridae.
(2) The female sex-organ of some fungi, *e.g.* in the Erisiphaceae.

CARPOMYCETEAE, CARPOMYCETES. The fungi with fruit bodies; the higher fungi, *i.e.* the Ascomycetes and Basidiomycetes.

CARPOPHORE. (1) A raised part of the receptacle bearing the carpels and stamens.
(2) The stalk of the sporocarp.
(3) Sometimes used to mean the fruit-body of the higher fungi.
(4) The forked stalk from which the mericarps of some Umbelliferae are suspended.

CARPOPHYLL. A carpel.

CARPOSPORANGIUM. A sporangium in which the carpospores are fromed. It is characteristic of the Rhodophyceae.

CARPOSPORE. A rounded, uninucleate non-motile spore formed from the direct or indirect division of the zygote of the Rhodophyceae.

CARPOSPOROPHYTE. A term used for the cystocarp of the Rhodophyceae, when it is considered to be an asexual spore-producing generation, parasitic on the sexual generation.

CARPOSTROTE. A plant which migrates by means of its fruit.

CARPOTROPHIC. The movement of the flower-stalk after fertilization to bring fruit into a favourable position for the ripening and/or dispersal of the seed.

CARR. A community of woody plants on drying fen-land.

CARRIER. A plant infected with a virus, and showing no symptoms, but capable of infecting another plant, especially through an insect vector.

CARTILAGENOUS. Hardened, and tough, but capable of being bent.

CARTONEMATACEAE (H). A family of the Commclinales, containing an Australian genus, *Cartonema*. There are glandular hairs, and the flowers are spicate. The leaf-sheaths are closed. A calyx and corolla are present. The ovary has 3-2 loculi with few to 1 ovules on axile placentas.

CARUNCIE. An outgrowth near the micropyle and hilum of the seed.

CARYALLAGIC. Of reproduction, involving a nuclear change.

CARYO- = KARYO-.

CARYOCARACEAE (EP, H). A family of the Parietales (EP), Theales (H). BH include it in the Ternstoemiaceae. These are trees or shrubs with leaves having deciduous stipules. The leaves are compound, and opposite or alternate (EP), or only alternate (H). The bisexual flowers are in racemes. The calyx has 5-6 fused lobes, and the 5-6 petals are fused, and imbricate. The numerous stamens are united, or in 5 bundles. The superior ovary has 4, or 8-20 free carpels, with as many styles. Each ovary has 1 pendulous ovule. The fruit is usually a drupe with an oily mesocarp, and a woody endocarp which splits into 4 mericarps. There is sometimes a leathery schizocarp. There is little or no endosperm. The embryo has a large radicle and small inflexed cotyledons.

CARYOID. A small mass of protein found in some algal cells.

CARYOPHYLLACEAE (EP, EP, H). A family of the Centrospermae (EP, BH) Caryophyllales (H). There are 3 or more sepals. The stamens, if the same number are opposite the sepals, but are often twice as many. They are hypogynous. The ovary has free-central placentation. The ovary is unilocular or imperfectly more locular, and superior.

CARYOPHYLLALES (H). An order of the Herbaceae. These are herbaceous and becoming fleshy. The regular bisexual flowers are hypogynous to perigynous. The stamens are mostly definite in number. The ovary is syncarpous, with axile to free-central placentation. The seeds have copious endosperm, with mostly a curved embryo. The leaves are mostly opposite, or in whorls. There may or may not be stipules.

CARYOPHYLLATUS. Latin for 'having a long claw'.

CARYOPHYLLINAE (BH). An order of the Polypetalae. The flowers are regular with a calyx of 2-5 (rarely 6) free sepals, and the petals are usually of the same number. There are usually the same number of stamens, or twice as many. The ovary is unilocular, or imperfectly 2-5 locular. The placentation is mostly free central, or rarely parietal. The embryo is usually curved in a powdery endosperm.

CARYOPSIS. An achene in which the pericarp layers are fused; typical of the grasses.

CASPARIAN STRIP, CASPARIAN BAND, CASPARY'S BAND. A band of suberised material around the radial walls of the individual cells of the endodermis in their primary condition. The material is impervious to water.

CASSIDEOUS. Helmet-shaped.

CASTANEOUS. Chestnut-coloured.

CASUAL. (1) A plant occurring in a plant community of which it is not a regular inhabitant.
(2) An occasional weed of cultivation.

CASUARINACEAE (EP, BH, H). A family of the Verticillatea (EP), Unisexales (BH), Casuarinales (H). There is one genus, *Casuarina*. See *Casuarinales (H)*.

CASUARINALES (H). An order of the Lignosae. These are tree shrubs with jointed branches, bearing reduced connate leaves. The flowers are unisexual, the male flowers being in spikes, and the female ones in heads. There is neither calyx nor petals. The male flowers have 1 stamen. The ovary is superior and 1 locular having 2 ovules inserted above the base of the ovary. The fruits are in cone-like heads, and the seeds have no endosperm.

CATA- = KATA.

CATABOLISM, KATABOLISM. The break-down of complex molecules to simpler ones releasing energy.

CATACOROLLA. A second corolla formed externally to the true one.

CATADROMOUS = KATADROMOUS.

CATALASE. An enzyme that breaks down hydrogen peroxide (which is toxic) to water and oxygen. During this reaction it is slowly destroyed itself. Its prosthetic group contains iron, and its maximum activity is between pH 6·5-7.

CATAPHORESIS = ELECTROPHORESIS. The migration of colloidal particles in an electric field, especially towards the cathode.

CATAPHYLL, CATAPHYLLARY LEAVES. A simplified form of a leaf, *e.g.* a scale leaf, bud scale, or cotyledon.

CATAPULT MECHANISM. A method of seed dispersal depending on the jerking of a long stalk swaying in the wind.

CATA-SPECIES. A species of the Uredinales which lacks a pycnidial stage.

CATATHECIUM, CATOTHECIUM. A name sometimes given to the thyriothecium of the Trichothyiaceae.

CATECHOL. A dihydroxyphenol.

CATENA. A gradual succession of soil types over an area, usually due to variation in drainage. Used in slightly different meanings.

CATENARIACEAE. A family of the Blastocladiales. The plant body is at first a tubular coenocyte, but later forms zoosporangia alternately separated by vegetative cells. Rhizoids develop on the vegetative cells, or at all points. Resting spores develop instead of zoosporangia, occasionally.

CATENATE, CATENULATE. In chains, *e.g.* of spores.

CATENATION. The arrangement of chromosomes in chains or rings.

CATENULIFORM. Like a chain.

CATKIN. A pendulous spike, with many simple, usually unisexual flowers. It is usually bracteate, and falls as a unit when the pollen or seeds have been shed.

CATOTHECIUM = CATATHECIUM. An inverted perithecium with the asci hanging from its base.

CAUDA. A tail-like appendage.

CAUDATE. Having a tail.

CAUDEX. A trunk or stock.

CAUDICLE. A stalk of mucilaginous threads holding the pollen mass of orchids to the rostellum.

CAUL-. A Latin prefix meaning 'stem'.

CAULERPACEAE. A family of the Siphonales. There is one genus *Caulpera*. The thallus is one-celled, with a rhizome-like section bearing root-like appendages on the lower surface, and shoot-like appendages on the upper surface. Asexual reproduction is by breaking of the thallus. Sexual reproduction is by isogamous, or anisogamous biflagellate gametes, formed in the upright branches.

CAULESCENS (Latin), CAULESCENT. (1) Having a stem.
(2) Becoming stemmed.

CAULICLE. A small stalk.

CAULICOLE, CAULICOLOUS. Living on herbaceous stems.

CAULIFLORY. Bearing flowers on old stems.

CAULINE. (1) Growing from the stem, in contrast to growing from the base of the plant.
(2) Appertaining to the stem.
(3) Formed from the internal tissues of the stem.

CAULINE BUNDLE, CAULINE VASCULAR BUNDLE. A vascular bundle formed entirely from the tissue of the stem.

-CAULIS. A Latin prefix meaning '-stemmed'.

CAULOBACTERIACEAE. A family of iron bacteria. They are round to kidney-shaped cells, having the axis at right-angles to the stalk of ferric hydroxide which it produces, and by which it is attached.

CAULOCALINE. See *Caline*.

CAULOCARPIC. Said of a plant, which, after flowering survives for the rest of the year, to flower in the next or subsequent years.

CAULOCYSTIDIUM. A cystidium on a stipe.

CAULOME. An organ that is morphologically a stem, or functions as a stem.

CAVERNICOLOUS. Living in caves.

CAYTONIALES. An order of the Gymnospermae, which lived from the Permian to the Jurassic eras. There were no stobili, and the seeds were enclosed in rolled sporophylls.

CECIDIUM. A plant gall caused either by an insect or a fungus.

CELASTRACEAE (EP, BH, H). A family of the Sapindales (EP), Celastrales (BH, H). These are tree, shrubs, or climbers, with simple leaves that are not lepidote. The flowers are bisexual (usually), small and regular. The flowers are not enclosed by a bracteole. The calyx has 4-5 free or united lobes. The corolla has 4-5 free petals. There is usually a well-marked disk which is annular or of separate glands, bearing 4-5 free stamens. The carpels or ovary are not enclosed by the disk. There are no staminodes. The gynecium is of 2-5 fused carpels with 2-5 loculi. There are usually 2 anatropous, or erect ovules in each loculus. The fruit is of various forms, and the seed usually contains endosperm.

CELASTRALES (BH, H). An order of the Thalamiflorae (BH), Lignosae (H). These are trees, shrubs or climbers, with alternate or opposite small leaves without glands. If stipules are present, they are small. The flowers are regular, and mostly bisexual. The calyx is imbricate or valvate, and the petals, which are usually present, are imbricate to partially connate. There are a definite number of petals which alternate with the petals. A disk is often present. The ovary is superior, or partially immersed in the disk. There are 1-2 erect or pendulous ovules in each loculus. The seeds contain endosperm.

CELASTRINEAE (BH) = CELASTRACEAE.

CELL. (1) A nucleus, with the cytoplasm with which it is in intimate contact; in plants it is usually bounded by a definite cell-wall.
(2) The cavity containing pollen in an anther.
(3) A chamber of an ovary.

CELL COMPETITION. See *Renner effect*.

CELL-DIVISION. The division of the nucleus and cytoplasm of a cell to form two new individuals.

CELL-INCLUSION. Any non-living material in the cytoplasm.

CELL-MEMBRANE. See *Plasma-membrane*.

CELLOBIOSE, CELLOSE. β-glucose—β-glucoside. A reducing sugar.

CELL ORGAN. A specialized part of the protoplasm of a cell having a particular function.

CELL PLATE. A fine membrane formed across the equator of the spindle during cell-division. It forms the basis of the new cell-wall, between the daughter cells.

CELL SAP. A solution of organic and inorganic substances in the vacuole of a plant cell.

CELL THEORY. The theory that all living things are made up of cells that divide during growth and reproduction. (Schleiden and Schwann 1838-9).

CELL TISSUE. A group of cells formed from the division of one or a few cells, and functioning as a whole.

CELLULAR SPORE. A multicellular body, released as a single unit like a spore, but each cell is capable of germinating separately.

CELLULASE. An enzyme capable of digesting cellulose into simpler units. Associated with saprophytes and parasites, and particularly important when produced by the microflora of the herbivore's large intestine.

CELLULIN. A carbohydrate characteristic of the Leptomitales.

CELLULOSANS. A group of carbohydrates similar to cellulose, but with shorter chains, mainly of xylose residues, but may contain uronic acid and mannose units.

CELLULOSE. A condensation product of a various number of glucose units, giving a fibrous structure. The main constituent of plant cell-walls (only a few fungi).

CELLULOSE TRABECULAE. A strand of cellulose crossing the lumen of a cell.

CELL WALL. The bounding layer of plant cells. It may be made of cellulose or chitin. It may be changed by secondary deposits of lignin between the cellulose micellae. It gives mechanical support to the cell.

CEMENTATION. The union of fungal hyphae by means of a sticky substance.

CENSER MECHANISM. A method of seed dispersal whereby the seeds are shaken out of pores in the fruit when it sways in the wind.

CENTONATE. Blotched with different colours, and so looking like a patchwork.

CENTRAL BODY. The colourless inner portion of the protoplast of the Myxophyceae. It contains nucleic acid, and may function as a nucleus. See *Coenocentrum*.

CENTRAL CELL. (1) The cell in the centre of developing aechegonium of the Bryophyta and Pteridophyta. It divides to form the primary canal-cell and primary ventral cell.
(2) The cell at the base of the archegonium in the Gymnospermae, containing the egg and ventral canal-cell.

CENTRAL CYLINDER. See *Stele*.

CENTRALES. The Centric diatoms. An order of the Bacillariophyceae which are radially symmetrical about a central point. Generally the protoplasts have many chromatophores, and the valves never have a raphe or pseudoraphe. Auxospores are never formed by the conjugation of 2 cells.

CENTRE OF ORIGIN, THEORY OF. The theory that plant species originated in the area where they show the greatest genetical diversity.

CENTRIC. Circular in section, with tissues distributed evenly around a central point.

CENTRIC OÖSPHERE. An oösphere of a fungus which has one or more layers of oil-droplets surrounding the central protoplasm.

CENTRIFUGAL. Away from the centre.

CENTRIFUGAL INFLORESCENCE. See *Cyme*.

CENTRIFUGAL THICKENING. The deposition of material on the outside of the cell-wall. This only occurs if the cell is free from its neighbours, *e.g.* the markings on pollen grains.

CENTRIFUGAL XYLEM. A xylem in which differentiation occurs away from the centre of a stem of root.

CENTRIOLE. A central granule within the centrosome.

CENTRIPETAL. Towards the centre.

CENTRIPETAL INFLORESCENCE. See *Raceme*.

CENTRIPETAL THICKENING. The deposition of wall material on the inner side of a cell-wall.

CENTRIPETAL XYLEM. A xylem in which differentiation occurs from the centre of the stem or root.

CENTRODESMUS = ATTRACTION SPINDLE.

CENTROLEPIDACEAE (EP, BH). A family of the Farinosae (EP), Glumaceae (BH). Small grass-like herbs with spikes of small, bisexual or unisexual flowers, which are naked or with 1-3 hair-like structures around them. There are 1 or 2 free stamens; the gynecium is superior with 1 to many loculi each with a pendulous orthotropous ovule.

CENTROGENE. A piece of a centromere which is broken by misdivision or X-ray, and is self-propagating.

CENTROLEPIDACEAE (H). A family of the Jucales (H). The leaf-blades are not (or rarely) reduced. The flowers are bisexual or if unisexual the plants are cushion-like, and the flowers are solitary. There is no perianth, and the 1-2 stamens have versatile anthers. There is 1 pendulous ovule in each loculus.

CENTROMERE. See *Spindle attachment*. The self-propagating particle in the chromosome whose activity in protein organization determines certain movements of the chromosome.

CENTROPLASM. A mass of protoplasm in the centre of the cells of one of the Myxophyceae. Its function is obscure.

CENTROSOME. A minute protoplasmic cell-inclusion associated with the nucleus, and dividing with it. In plants, they seem to be confined to the Thallophyta.

CENTROSPERMAE. An order of flowering plants. Sepals and petals are present. The ovary is superior with 1 loculus, and the seeds attached at the base, or to a projection from the base. The embryo is curved or coiled.

CENTRUM. A group of asci and nutritive cells associated with them occurring in the perithecium of some Pyrenomycetes.

CEPACEOUS. Tasting or smelling of onions.

CEPHAL-. A Greek prefix meaning 'head'.

CEPHALOBRANCHIAL. Used to describe a chromosome which has a small rounded projection at one end.

CEPHALODIUM. A gall-like outgrowth on a lichen, containing algal cells and fungal hyphae.

CEPHALOTACEAE (EP, H). A family of the Rosales (EP), Saxifragales (H). (BH) include it in the Saxifragaceae. The leaves are of two kinds, some are modified as pitchers with lids. The calyx is coloured, and there are no petals. There are 12 stamens with a setose glandular disk within them. The carpels are free, or united at the base.

CEPHALOZIACEAE. A family of the Acrogynae. The lateral leaves are entire or with lobes of equal size, or differing in form from the ventral leaves. The perianth is usually triangular with two angles towards the dorsal side of the thallus.

CERACEOUS, CEREOUS. Wax-like in texture and colour.

CERAMIALES. An order of the Florideae. An auxillary cell is formed after fertilization. They are tetrasporophytic, with the auxilliary cell being borne directly on the supporting cell of the carpogonial filament.

CERANOID. Having horn-like branches.

CERASIN. An insoluble constituent of gums, which swell in water.

CERAT-. Greek prefix meaning 'horn'.

CERATIOMYXALES. The single order of the Ceratiomyxomycetidae.

CERATIOMYXOMYCETIDAE. A sub-class of the Myxomycetes which bear spores on the outside of the fructification.

CERATOBASIDIACEAE. A family of the Tulasnellales. The basidium is aseptate. The basidial primodium is subglobose, pear-shaped, or club-shaped, with 4 stout cells arising from the top and narrow to the tip. Each bears a basidiospore. These cells are cornute or flexuous. There are no cross-walls between the primorium and cells on it.

CERATOPHYLLACEAE (EP, BH, H). A family of the Ranales (EP, H), Monochlamydeae. There are no petals or disk. The torus is sometimes enlarged with enclosed carpels. The stamens are many or few. The ovules are pendulous from the apex of carpels. The seeds are not arillate, and contain no endosperm. These plants are aquatic, and submerged.

CERATOPHYLLEAE (BH) = CERATOPHYLLACEAE.

CERATSTOMATACEAE. A family of the Sphaeriales. The perithecia have a fairly long, distinct neck, and leathery walls. There are paraphyses, and the ascus wall does not autodigest. The ascospores have various numbers of cells. These are saprophytes on wood and bark (sometimes on herbaceous stems).

CERCIDIPHYLLACEAE (EP, H). A family of the Ranales (EP), Magnoliales (H). (BH) include it in the Magnoliaceae. These are deciduous trees with opposite or alternate leaves which have stipules adnate to the petiole. The flowers are dioecious. The male are subsessile, single, or in groups in the axils of the leaves. There are 15-20 stamens. The female flowers are stalked with 4 small sepals. There are 4-6 free carpels each with 2 rows of descending anatropous ovules. The fruit is a cluster of follicles. The seed has a woody endocarp, and is square and winged at one end. The embryo has flat cotyledons, and there is copious endosperm.

CEREALS. Members of the Graminae whose grain is used for human consumption, *e.g.* wheat, barley, oats, rye, etc.

CEREBRIFORM. Convoluted like the surface of the brain.

CERIFEROUS. Wax-producing.

CERNUOUS, CERUUS. Latin meaning 'hanging, drooping, nodding'.

CERTATION. Competition between pollen grains, placed at the same time on the same stigma. They have different genetic constitutions giving unequal chances of bringing about fertilization.

CERVIVE. Dark-tawny.

CESPITOSE = CAESPITOSE.

CHAETOMIACEAE. A family of the Sphaeriales. There are long ostiolar hairs. The spores are mostly lemon-shaped, dark-coloured and one-celled. They live on damp straw, paste board, manure etc.

CHAETOPHORACEAE. A family of the Ulotrichales. The pseudoparenchymatous tissue may be formed by the pressing together of the branching filaments. The terminal cell may be elongated into a hair. The cells usually have one flattened central chloroplast. Asexual reproduction is by zoospores, aplanospores or akinetes. Sexual reproduction is usually isogamous, but may be anisogamous, or oögamous.

CHAETOPLANKTON. Small aquatic organisms whose power to float is aided by small spiny outgrowths.

CHAFFY. Covered with brown-to-grey scales.

CHAILLETIACEAE (H). See *Dichapetalaceae*. A family of the Rosales. The leaves are simple, alternate (rarely opposite), and are stipulate. The sepals, petals, and stamens are hypogynous. The petals are often deeply bilobed or bipartite. The 3-5 stamens are free or united.

CHAISTOBASIDIUM. A club-shaped basidium, with the nuclear spindles across it at the same level.

CHALAZA. The base of the nucellus of an ovule.

CHALAZOGAMY. The entry of a pollen tube through the chalaza of an ovule.

CHALICIUM. A gravel-slide formation.

CHALK-GLAND. A secretory organ on some leaves around which calcium carbonate becomes deposited.

CHAMAE-. A Greek prefix meaning 'ground'.

CHAMAEPHYTE. A plant which perennates with its buds at or just above the soil-surface.

CHAMAESIPHONACEAE = CHAMAESIPHONALES.

CHAMAESIPHONALES. An order of the Myxophyceae, having the regular formation of endospores. The cells are solitary, or in non-filamentous colonies, sometimes tending to be filamentous.

CHAMBERED. Said of an ovary divided by incomplete partitions extending inwards from the walls.

CHANNELLED. Hollowed like a gutter.

CHARACEAE. The single family of the Charales.

CHARACIACEAE. A family of the Chlorococcales. The cells are sessile and elongate, and are solitary or formed in radiate colonies. The cells are multinucleate usually, and most genera have a single central, flat chloroplast, with 1 or more pyrenoids. Asexual reproduction is usually by zoospores, but may be by aplanospores. Sexual reproduction is usually by biflagellate gametes which are equal or unequal.

CHARACTER. (1) Any well-defined feature which distinguishes one species from another.
(2) Genetically, any feature which is transmitted from parent to offspring.

CHARALES. The single order of the Charophyceae.

CHAROPHYCEAE. A class of the Chlorophyta. The thallus is erect and branched with a regular succession of nodes and internodes. Each node bears a whorl of branches of limited growth, in the axils of which arise branches of unlimited growth. Sexual reproduction is oögamous. The oögonia are one-celled and solitary, surrounded by a sheath of spirally arranged sterile hairs. The antheridia are one-celled and united into uniseriate filaments. Several of these filaments are surrounded by a common envelope composed of 8 cells.

CHARTACEOUS. Of a papery texture.

CHASMOCLEISTOGAMIC. Producing chasmogamous and cleistogamous flowers.

CHASMOGAMIC. Said of a flower that opens before pollination.

CHASMOPHYTE. A plant growing on rocks, rooted in debris in the crevices.

CHEILOCYSTIDIUM. A cystidium at the edge of the pileus surface.

CHELATE. The bonding of a metal ion on to an organic molecule from which it can be released. *E.g.* iron is only slightly available to plants growing in calcarious soils, as it is in the form of ferric hydroxide which is relatively insoluble. In a case like this it can be added as chelate sequestrene 138Fe.

CHEMOAUTOTROPHIC. Said of bacteria which obtain energy by the oxidation of simple inorganic molecules. *E.g.* The oxidation of hydrogen sulphide to sulphur by *Thiobacillus* spp.

CHEMONASTY. The change in the position or form of a plant organ in relation to a diffuse chemical stimulus.

CHEMOSYNTHETIC. See *Chemoautotrophic*.

CHEMOSYNTHESIS. The formation of organic material by some bacteria by means of energy derived from chemical change, in contrast to the transfer of energy by complex chemical reactions as in respiration.

CHEMOTAXIS. A taxis in response to a chemical stimulus.

CHEMOTROPISM. A tropic response along a chemical gradient.

CHENA. The burning of forest for a few crops.

CHENOPODIACEAE (EP, H). A family of the Centrospermae (EP), Chenopodiales (H). (BH) include the Basellaceae. These are nearly all salt-loving herbs. Stipule are absent. There are 3-5 lobes to the calyx which may be fused. There are no petals. The stamens are as many as, or fewer than, the perianth lobes, and bear anthers which open by longitudinal slits. The ovary is unilocular and usually superior, containing 1 basal campylotropous ovule. The bent or spirally twisted embryo usually surrounds the endosperm. The fruits are free and not united into a fleshy mass.

CHENOPODIALES (H). An order of the Herbaceae. The leaves are alternate or opposite, and its members are found mainly in dry regions. The order is like the Polygonales, but the stipules are absent or small. There are many to 1 free or connate carpels. The embryo is curved around the endosperm.

CHERNOZEM = BLACK EARTH. A grassland soil developed under fairly dry conditions so that there is little leaching. There is an accumulation of calcium carbonate at about 6 feet.

CHERSIUM. A dry waste formation.

CHESTNUT SOIL. A grassland soil developed under arid conditions. There is no leaching, and it carries a shallow-rooting vegetation. The calcium carbonate layer is near the surface.

CHIASMA (pl. CHIASMATA). The joint between two chromatids of two homologous chromosomes resulting in an interchange of chromatic material.

CHIMAERA. A plant having tissues of more than one genetic type. This may be caused by the mutation or abnormal distribution of chromosomes in one cell early in development, thus affecting tissues derived from it. It can be produced by grafting different types of plants—a graft hybrid.

CHIMNEY. An upgrowth of epidermal cells above a stoma, forming a long pore.

CHIMONOPHILOUS. Growing chiefly during the winter.

CHIMOPELAGIC PLANKTON. Plankton occurring only during the winter.

CHIN. Found in some of the orchids. An axial protuberence which grows in such a way that the sepals appear to arise from the labellum.

CHIONIUM. A snow-formation.

CHIROPTEROPHILOUS. Pollinated by bats.

CHI-SQUARED. The ratio of an observed sum of squares to the appropriate or corresponding variance as fixed by a hypothesis.

CHITIN. A major constituent of fungal cell-walls made up of N-acetyl-2-glucose amine units.

CHLAENACEAE (EP, BH). A family of the Malvales (EP), Guttiferales (BH). These are trees with alternate, entire stipulate leaves. The regular, bisexual flowers are solitary or two in an involucre. The calyx is of 3-5 free sepals, and the corolla of 5-6 free petals. There are 10 to many free stamens. The superior ovary has 3 carpels, and consists of 3 loculi each with 2 ovules. The fruit is a capsule of 3-1 loculi. The seed contain endosperm.

CHLAMYDEOUS. Having a perianth.

CHLAMYDOBACTERIACEAE. A family of iron bacteria which have a sheath impregnated with ferric and/or manganese oxides. Reproduction is by swarm spores which have a tuft of flagella at one end.

CHLAMYDOCYST. A two-walled resting zoosporangium in a hypha of the Blastocladiaceae.

CHLAMYDOMONADACEAE. A family of the Volovocales. All are unicellular. There is definite cell-wall, except for those which have a wall of two overlapping halves. Some have many contractile vacuoles. The cells have 2 or 4 flagella. Asexual reproduction is by zoospores, and sexual reproduction is usually isogamous, with a few anisogamous, and 1 oögamous species.

CHLAMYDOMORPHOUS SOIL. A soil in which each grain has a colloidal coating and the grains are bound together in a loose aggregate.

CHLAMYDOSPORE. A thick-walled asexual spore which serves to survive adverse conditions. It is intercalary or terminal and non-deciduous, being made-up of one or more cells.

CHLAMYDOZOA. An obsolete phylum used to include the viruses.

CHLEDIUM. A formation on waste ground.

CHLOANTHACEAE (H). A family of the Verbenales (H). The leaves are opposite or verticillate. The indumentum is often dense consisting of stellate or dendriform (rarely scaly) hairs. The inflorescence is often a head or a corymb, or the flowers axillary. The calyx is actinomorphic and the corolla more-or-less zygomorphic. The ovary has 9-2 loculi each with usually 2 anatropous ovules laterally inserted on the central axis. The seeds contain endosperm.

CHLORANTHACEAE (EP, BH, H). A family of the Piperales (EP, H), Micrembryae (BH). They are herbs, shrubs or trees with opposite stipulate leaves. The unisexual or bisexual flowers are in spikes or cymes, sometimes with a sepaloid perianth. The 3-1 stamens are united to each other and to the ovary. The ovary is unilocular and superior (EP, BH), inferior (H), and contains a few (EP, BH), 1 (H) pendulous orthotropous ovules. The seed contains endosperm which is oily, and the embryo is minute.

CHLORANTHY. Said of an abnormal flower in which all the parts have developed as leafy structures.

CHLORENCHYMA. Unlignified tissue containing chloroplasts.

CHLORIN. A derivative from chlorophyll. The chlorophyll loses the magnesium, phytol and methyl alcohol, and the 5-carbon ring is open.

CHLOROBACTERIACEAE. A family of the sulphur bacteria having a pigment like chlorophyll. They photosynthesize in hydrogen sulphide, depositing sulphur outside the cell, but they do not produce oxygen.

CHLOROCOCCACEAE. A family of the Chlorococcales. The members are unicellular. The cells are more-or-less spherical, usually uninucleate, and apparently haploid. Asexual reproduction is by zoospores or aplanospores, and sexual reproduction by biflagellate gametes.

CHLOROCOCCALES. An order of the Chlorophyceae. The cells are solitary or in non-filamentous colonies of a definite or indefinite number of cells. The cells are uninucleate, or multinucleate, but never divide vegetatively. Asexual reproduction is by zoospores or autospores, and sexual reproduction, when present is by biflagellate gametes.

CHLOROPHYCEAE. A class of the Chlorophyta. The plant body is unicellular or multicellular, but never grows by an apical cell. The sex-organs are unicellular and are freely exposed, being covered rarely by a sheath of sterile cells after fertilization.

CHLOROPHYLL. The fundamental green pigment of photosynthesis. It is localized on the chloroplasts and is made up of chlorophyll a ($C_{55}H_{72}O_5N_4Mg$), and chlorophyll b ($C_{55}H_{70}O_6N_4Mg$).

Chlorophyll a.

CHLOROPHYLLASE. An enzyme occurring in association with chlorophyll and able to hydrolyse the phytol group in the chlorophyll, and hence decompose it.

CHLOROPHYLLINS. The two acid precursors of chlorophyll a and chlorophyll b respectively.

CHLOROPHYLLOGEN. A hypothetical precursor of chlorophyll formed in darkness but changed to chlorophyll in light.

CHLOROPHYLLOSE, CHLOROPHYLLOUS. Containing chlorophyll.

CHLOROPHYTA. A division of the Algae. Chlorophyll is the photosynthetic pigment, and most of them accumulate starch. The motile reproductive cells have 2–4 anterior flagella which are mostly of equal length. The sex organs are unisexual, and the pairing gametes are of equal or unequal size.

CHLOROPLAST. A variously-shaped plastid made up of lamellated grana with chlorophyll distributed on them. The grana are interspersed in the non-lamellate stroma. They are embedded in the cytoplasm.

CHLOROSIS. An abnormal yellowing of leaves due to the reduction of the chlorophyll content below the normal. It is due to a deficiency of certain elements, *e.g.* iron or magnesium, attack by parasites, *e.g.* virus yellows, lack of light, drought etc.

CHLOROSTATOLITH. A chloroplast containing starch, and acting as a statolith.

CHOANEPHORACEAE. A family of the Mucorales. They have large sporangia with a columella and sporangioles crowded on to the swollen apex of a sporangium.

CHOLINE. Ethanol trimethyl-ammonium hydroxide. It is found in many seeds, and with fatty acids, glycerol, and phosphoric acid yields lecithins, from which it is liberated by hydrolysis.

$$\left[\begin{array}{c} CH_2OH \\ \\ CH_2-N-CH_3 \\ \end{array} \begin{array}{c} \\ CH_3 \\ CH_3 \\ CH_3 \end{array} \right] OH$$

Choline

CHOMOPHYTE. A plant growing on rocky ledges littered with detritus.

CHONDRIOCONT, CHONDRIOKONT. A mitochondrium which has the form of a rod or thread.

CHONDRIOLYSIS. The dissolution of the mitochondria.

CHONDRIOMES = MITOCHONDRIA.

CHONDRIOMITE. A chondriosome having the form of a chain of granules.

CHONDRIOSOMAL MANTLE. An accumulation of chondriosomes surrounding a dividing nucleus.

CHONDRIOSOME = MITOCHONDRIA.

CHONDROID. Said of the medulla of a lichen when it is hard and tough, consisting of thick-walled hyphae in very firm association.

CHORDARIALES. An order of the Haplostichinae. The thallus is filamentous and branched, but not compacted into a pseudoparenchyma. The gametophytes are isogamous.

C HORIZON. The material from which a soil is derived.

CHORIOPETALOUS. Polypetalous.

CHORISIS. The splitting into two or more lobes; often applied to abnormalities.

CHRESARD. The total amount of water in the soil which is available to plants.

CHROMASIC. An increase in the chromatin.

CHROMATIC ADAPTATION. A variation in the coloration in relation to the amount of light reaching a plant.

CHROMATID. A half chromosome during duplication in early prophase and metaphase of mitosis and between diplotene and the second metaphase of meiosis. After these stages it is called a daughter chromosome.

CHROMATIN. A nucleoprotein, staining with basic dyes, and forming part of chromosomes. It is made up of two proteins, one of which is histone, and DNA and RNA.

CHROMATOLYSIS. The dissolution of chromatin in injured cells.

CHROMATOPHORE = CHROMOPLAST.

CHROMATOPLASM. The outer region of the cytoplasm of the Myxophyceae. It contains the cell-pigment.

CHROMATOSPHERITE = NUCLEOLUS.

CHROMATIDIA. Pieces of chromatin not aggregated into a nucleus, but lying free in the cytoplasm.

CHROMIDIUM. An algal cell in the thallus of a lichen.

CHROMIOLE. One of the deeply staining granules of which chromatin is composed.

CHROMOCENTRE. Any accumulation of chromatin in the nuclear reticulum, staining more darkly than the rest of the chromatin network in the resting cell.

CHROMOCHONDRIA. Mitochondria concerned with pigment formation.

CHROMOGEN. A pigment-producing organism.

CHROMOGENIC, CHROMOGENEOUS. Colour-producing.

CHROMOMERE. Dense granules of chromatic material which appear on the chromosomes during the early stages of meiosis. They are characteristic in number, position and size.

CHROMONEMA. The longitudinal strands making up a chromatid. They are about 200Å across, with an inner core of DNA with an outer protein sheath.

CHROMOPAROUS. Said of colourless bacteria which secrete a colourless material.

CHROMOPHIL. Staining well and easily with stains used in microscopy.

CHROMOPHOBE. Not staining, or staining with difficulty with stains used in microscopy.

CHROMOPHOROUS. Of bacteria having colouring materials as part of their make-up.

CHROMOPLASM. The pigmented outer layer of the protoplasm of a myxophycean cell.

CHROMOPLAST. A plastic containing one or more pigments including chloroplasts.

CHROMOSOMAL CHIMAERA. A chimaera in which the nuclei do not all contain the same number of chromosomes.

CHROMOSOME. One of the deeply staining rod-like structures in the cell-nucleus carrying the genes, and containing DNA. There is a constant number for any particular species. They undergo a co-ordinated cycle of reproduction during mitosis and meiosis.

CHROMOSOME ARM. One of the two parts of a chromosome to which the spindle fibre is attached along the side.

CHROMOSOME COMPLEMENT. The set of chromosomes characteristic of a cell of a species.

CHROMOSOME CYCLE. The complete changes in the chromosomes during the life-cycle.

CHROMOSOME MAP. A diagram of a chromosome showing the position of the genes. It is obtained by a statistical analysis of the frequency of crossing-over.

CHROMOSOME MATRIX. A sheath of weakly staining material around the more stainable substance of a chromosome.

CHROMOSOME SET. The complete haploid set of chromosomes characteristic of a species.

CHROMULE. A general term for plant pigments.

CHROMULINEAE. A sub-order of the Chrysomonadales, distinguished by the cells having a single long flagellum.

CHRONISPORE. A resting spore.

CHROOCOCCACEAE = CHROOCOCCALES.

CHROOCOCCALES. An order of the Myxophyceae. The cells are single or united in non-filamentous colonies. The only regular method of reproduction is by cell-division or breaking of the colony.

CHRYSOCAPSALES. An order of the Chrysophyceae. The immobile vegetative cells are united in palmelloid colonies by a common gelatinous matrix. Any of the cells may divide, but division may be confined to one end. The cells of most, if not all, genera can become flagellate without any further change. Some genera produce statospores.

CHRYSOCHROME. The golden-brown pigment of the Chrysophyceae.

CHRYSOGONIDIUM. A yellow algal cell in a lichen.

CHRYSOMONADALES. An order of the Chrysophyceae. The cells are motile during the vegetative phase or may have temporary amoeboid or rhizopodal stages. The motile cells may be solitary or united into colonies of a definite shape. The cells may be naked or with an envelope completely or incompletely surrounding the protoplast.

CHRYSOPHYCEAE. A class of the Chrysophyta. The cells have a small number of yellow-gold-brown chromatophores, and are usually without pyrenoids. The food stored as fat, leucosin, never as starch. The cells are flagellate or immobile, with or without a wall. They may be single or in colonies. The motile vegetative cells and zoospores have 1-2 flagella at the anterior end. Some produce endoplasmic statospores enclosed by a wall of two parts which are impregnated with silica and have a terminal pore. Sexual reproduction has not been established.

CHRYSOPHYTA. A division of the Thallophyta. The pigment is in chromatophores most of which are carotinoids. The food is stored as oils and leucosin, never as starch. The cell-wall is usually in two overlapping halves. The cells are flagellate, or non-flagellate, solitary or colonial. Asexual reproduction is by zoospores or aplanospores, while some genera produce statospores. Sexual reproduction, if present is isogamous, by zoogametes or aplanogametes.

CHRYSOSPHAERALES. An order of the Chrysophyceae. The members are unicellular or form non-filamentous colonies. The protoplast is not metamorphosed directly into a motile state.

CHRYSOTRICHALES. An order of the Chrysophyceae. The cells form a branching filamentous thallus, and statospores are produced.

CHYLOCAULY. Having a succulent stem.

CHYLOPHYLLY. Having succulent leaves.

CHYTRIDIALES. An order of the Phycomyceteae. The plant body is unicellular, never forming a true mycelium. Most, or all of the plant body forms a sporangium or gametangium. There is no alternation of generations. There may be several reproductive organs. Asexual reproduction is by anteriorly uniflagellate zoospores which mostly lack a nuclear cap. Sexual reproduction is by flagellate isogametes.

CICATRIX. A scar.

CILIATE. Having fine hairs or projections.

CILIATULATE. Having widely dispersed cilia.

CILIOLATE. Fringed with very short fine hairs.

CILIUM. A hair-like outgrowth; sometimes used for a flagellum, which is longer.

CINCINNAL, CINCINNATE. Curled, or rolled around.

CINCINNUS. A cymose inflorescence having the lateral axes arising on alternate sides of the relatively main axis.

CINEREOUS, CINERUS. Latin meaning 'ash-grey'.

CINGULATE. Edged all round.

CINGULUM. An open hoop-like band attached to the edge of the silicified part of a valve on the cell-wall of diatoms.

CINNABARINE. Bright orange-red.

CINNAMONEOUS. Light yellowish-brown.

CIRCAESTERACEAE (H). A family of the Berberidales (H). These are annual herbs with a rosette of slightly spiny dentate leaves. The bisexual axillary flowers are borne singly. There are 2 valvate sepals, which are valvate. The 2 stamens have anthers which open by longitudinal slits. The single carpel contains pendulous ovules.

CIRCINATE. Coiled like a watch-spring.

CIRCULATION. The movement of protoplasm around the inside of a cell.

CIRCUM-. A prefix meaning 'all around'.

CIRCUMCINCT. Having a band around the middle.

CIRCUMNUTATION. See *Nutation*. The rotation of the tip of a growing stem so that it traces a helical curve in space.

CIRCUMSCISSILE. Opening all round by a transverse split.

CIRRATE, CIROSE. Rolled around, or becoming so; like a waved hair.

CIRRHIFEROUS, CIRRHOSE. Having tendrils.

CIRRHUS, CIRRUS. (1) A tendril.
(2) A curl-like tuft; a tendril-like mass or 'spore horn' of forced-out spores.

CISTACEAE (EP, H). A family of the Parietales (EP), Bixales (H). They are shrubs or herbs with opposite (rarely alternate) leaves. The bisexual regular flowers are solitary or in cymes. There are 5 free sepals, which are contorted. There are 5, 3, or no free petals. There are usually many stamens, on a disk developing in descending order. The anthers open by longitudinal slits. The superior ovary has 10, 5, 3, or 1 loculi bearing many or 2 orthoptropous or anatropous ovules on parietal placentas. The fruit is a capsule. The seeds have endosperm and curved embryo.

CISTERNAE. Enclosed spaces in the endoplasm of a cell, which are separated from the ground substance by the cytoplasmic membranes.

CISTINEAE (BH) = CISTACEAE.

CITRIC ACID.

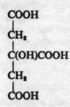

$$
\begin{array}{l}
\text{COOH} \\
| \\
\text{CH}_2 \\
| \\
\text{C(OH)COOH} \\
| \\
\text{CH}_2 \\
| \\
\text{COOH}
\end{array}
$$

CITRIC ACID CYCLE. The method by which acetyl-S-coenzyme A is broken down to carbon dioxide and water during cellular respiration. The process takes place on the mitochondria. Basically the process involves the formation of citric acid with six carbon atoms from oxaloacetic acid with four carbon atoms and acetyl-S-coenzyme A with two. This process releases two molecules of carbon dioxide, resulting again in oxaloacetic acid.

CITRIFORM. Lemon-shaped.

CITRINE, CITRON-COLOUR. Lemon-coloured.

CITRININ. An antibiotic produced by some *Penicillium* spp.

CITRULLINE. A precursor of arginine. It is synthesized from ornithine.

CLADAUTOICOUS. Said of a moss that bears the antheridia on a separate branch.

CLADOCARPOUS. (1) Having the fruit at the end of a lateral branch.
(2) Said of mosses bearing the sporangia on very short branches.

CLADOCHYTRIACEAE. A family of the Chytridiales. The plant-body branches through several cells, swelling at various points. There are many terminal or intercalary thin-walled sporangia in which the uniflagellate zoosporangia are formed. Sexual reproduction is rare and by uniflagellate zoogametes. The resting spores are thick-walled, and apparently not produced sexually.

CLADODE. A flattened photosynthetic stem.

CLADOGENOUS. Producing the flowers at the end of a branch.

CLADOPHORACEAE. A family of the Cladophorales. The cells are multinucleate and rarely more than eight times as long as wide. The chloroplasts are not in transverse bands, and the filaments are simple or branched. Asexual reproduction is usually by zoospores, and sexual reproduction is by biflagellate isogametes.

CLADOPHORALES. An order of the Chlorophyceae. The multinucleate cells are in simple or branched filaments, each with many disk-like chloroplasts. Asexual reproduction is by zoospores, aplanospores or akinetes, while sexual reproduction is isogamous or oögamous.

CLADOPHYLL = CLADODE.

CLADOPTOSIS. The shedding of branches.

CLADOSIPHONIC. A siphonostele having branch gaps, but no leaf gaps.

CLADOXYLACEAE. The single family of the Cladoxylales.

CLADOXYLALES. An order of the Primofilices which are found from the Lower Carboniferous to Mid-Devonian. The stems were dichotomously branched with many small leaves which were similarly branched on one plane. There were sterile and fertile leaves, the final branches of the fertile ones ending in a single sporangium. The stems were polystelic, each stele being an actinostele.

CLAMP-CONNECTION (CONNEXION). An outgrowth from the ultimate cell of a hypha forming a connection by fusion with the penultimate cell which is formed during cell-division. They are found in all the Basidiomycetes, except the Uredinales.

CLAN. A small group of individuals covering a small area, and derived from the same parent either by vegetative propagation, or from seed.

CLASS. A group immediately below a phylum, having the characteristics as defined by the classifier.

CLASTIDIACEAE. A family of the Cyanophyceae. They are epiphytic with long cells in a thin gelatinous sheath. The entire protoplast divides to form chain(s) of round endospores.

CLATHRACAE. A family of the Phallales. The fertile part of the gleba is on, or between, a mass of radiating or anastomosing hyphae. The young gleba is in a system of branches and plates which disintegrate to form an evil-smelling mass. There may or may not be a stipe.

CLATHRATE, CLATHROID. Like a net.

CLAUSTULACEAE. An order of the Phallales, in which the ovoid receptacle remains enclosed in the volva until the gleba is mature.

CLAVA. The club-like fruit-body of some fungi.

CLAVACIN = CLAVIFORMIN. An antibiotic produced by some *Penicilium* spp.

CLAVARIACEAE. A family of the Polyporaceae. They are saprophytic with the hymenium on all sides of an erect thin, or club-shaped, or much-branched fruit-bodies which are fleshy or leathery. The basidia are oval or club-shaped and stichobasidial or chiastobasidial. The basidiospores germinate directly.

CLAVATE. Club-shaped.

CLAVICIPITACEAE. A family of the Sphaeriales. The perithecia are without a well-developed wall and are embedded in the stroma. There are no paraphyses, and the ascospores are slender.

CLAVULATE. Slightly club-like.

CLAVUS. An obsolete term for the sclerotium of ergot.

CLAW. The narrowed base of a petal.

CLEARING AGENT. A reagent used in microscopy for treating sections to make them capable of examination by transmitted light.

CLEFT. Divided half-way down, but not to the midrib.

CLEISTOCARP. A fungus fruit-body with no special opening.

CLEISTOCARPIC, CLEISTOCARPOUS. Said of a moss capsule which does not open by a lid.

CLEISTOGAMY. Pollination and fertilization before the flower has opened.

CLEISTOTHECIUM = CLEISTOCARP.

CLETHRACEAE (EP, H). A family of the Ericales (EP, H). (BH) place this family in the Ericaceae. They are shrubs and trees with alternate leaves, with the flowers in racemes or panicles. The perianth consists of 2 whorls (calyx and corolla) each of 5 free segments. The anthers are bent outwards in the bud, and open by terminal pores. The pollen is in single grains. The superior ovary has 3 loculi, and the style has 3 stigmas. The fruit is a capsule, and the seeds are endospermous.

CLIMACTERIC PHASE. The phase in the life of a fully-grown fruit when the rate of respiration increases. It is immediately previous to the beginning of senescence.

CLIMATIC CLIMAX, CLIMATIC COMMUNITY. A plant community determined, and maintained in a given area by the prevailing climatic conditions.

CLIMATIC FACTOR. A condition such as rainfall, temperature etc. which plays a controlling part in determining the vegetation of an area.

CLIMAX. The stable vegetation of an area under the prevailing environmental conditions.

CLIMAX ASSOCIATION, CLIMAX COMMUNITY. The plant association which is permanently established in any given habitat.

CLIMAX DOMINANT. A species which dominates a climax community.

CLIMAX SPECIES. Any species of plant which is a characteristic member of a climax community.

CLINANDRIUM. The anther-bed of orchids.

CLINE. A morphological or genetical gradation of a species in a geographical area.

CLINOSTAT. A rotating disk to which a plant can be attached so that it can receive an equal amount of a stimulus on all sides. Rotating in a horizontal plane a clinostat is used to study phototropic responses, and vertically to study geotropism.

CLONE. The entire vegetatively produced descendants from a single original seedling.

CLOSED (VASCULAR) BUNDLE. A vascular bundle that does not have a cambium, and therefore cannot increase in diameter by normal secondary thickening.

CLOSED COMMUNITY. A plant community which completely covers the ground it occupies, and consequently preventing the introduction of new species.

CLOSING LAYER. A layer of closely packed cells over a lenticel preventing diffusion of water vapour and gases.

CLOSING MEMBRANE. The thin wall between adjacent pits in vessels or tracheids.

CLOSTEROSPORE. A multinucleate phragmospore.

CLUB MOSS. See *Lycopodales*.

CLUSTER. A general term for a closely crowded inflorescence of small flowers.

CLUSIACEAE (H). A family of the Guttiferales. These are trees, shrubs or herbs with a resinous juice. There are no stipules. The flowers are mostly unisexual, and the stamens are often in bundles. The seeds have no endosperm and are often arillate.

CLUSTER-CUP = AECIUM.

CLYPEATE. (1) Shield-shaped.
(2) Having a clypeus.

CLYPEUS. A shield-like stromatic growth, with or without host tissue, over one or more perithecia.

CNEORACEAE (EP, H). A family of the Geraniales (EP), Celastrales (H)· (BH) include this family in the Simarubaceae. These are trees, or at least woody, but never climbers. There are no stipules and no disk. The petals are imbricate (rarely contorted), and are free. The sepals are imbricate, and the peduncle is adnate to the petiole. The ovules are pendulous from the apex of the loculi, later producing a hooked embryo. The fruit is rarely enclosed by the calyx.

CO I. See *Diphosphopyridine nucleotide*.

CO II. See *Triphosphopyridine nucleotide*.

COACERVATE. Massed: heaped together.

COBAEACEAE (H). A family of the Bignoniales. The alternate leaves are pinnately compound, with the terminal pair of leaflets forming tendrils. Stipules are present. The corolla is regular with contorted lobes. The 5 stamens are all fertile with the anther-loculi parallel and not separated. The ovary has 3 loculi.

COADAPTATION. A correlated change in two organisms or organs that are mutually dependent.

COADNATE. Joined.

COADUNATE = CONNATE.

COAGULATION. The solidifying, or setting of protein by heat or poisons. This change is irreversible.

COAL BALL. A calcareous nodule found in some coal seams, and containing petrified plant remains.

COALESCENT. Joining together.

COARCTATE. Crowded together.

COAT. (1) An integument of an ovule.
(2) = TESTA.

COCAINE. An alkaloid derived from the leaves of *Erthroxylum coca*.

COCCINEUS. Latin for 'Scarlet'.

COCCOID. Being unicellular and motionless when vegetative, but releasing motile spores.

COCCULE. A piece of a dividing coccus.

COCCUS. (1) A spherical bacterium.
(2) A mericarp.

COCHLEA. A tightly coiled legume.

COCHLEAR, COCHLEARIFORM. Spoon-shaped.

COCHLEATE. Coiled like a snail's shell.

COCHLOSPERMACEAE (EP, H). A family of the Parietales (EP), Bixales (H). (BH) include it in the Bixineae. These are trees or shrubs with alternate leaves which are usually lobed. The bisexual flowers are more or less regular and in racemes. The 4-5 sepals are free and imbricate or open in the bud. The 4-5 free petals are contorted. The stamens are numerous and have anthers which open by short terminal slits. The superior ovary has 3-5 fused carpels each with many ovules on axile or parietal placentas. The fruit is a capsule, and the endosperm is oily.

CODIACEAE. A family of the Siphonales. The plants have a definite macroscopic form made up of the interwoven branches of the thallus. Reproduction is sexual and anisogamous, the gametes being produced in distinctive gametangia.

CO-DOMINANT. One of two or more species which together dominate a plant community.

CODONIACEAE. A family of the Anacrogynae. The gametophyte of most members is a simple thallus, a few have a foliose one. The capsule is spherical. Many have a basal elaterophore or a radiate arrangement of elaters at the base of the capsule.

COELOMOMYCETACEAE. A family of the Blastocladiales which are parasitic on insect larvae (especially mosquito). The thallus has no cell-walls, and no rhizoids. The ends branch and the tips enlarge breaking off to form a thick-walled resting spore, which breaks open by a longitudinal slit. The zoospores have 1 flagellum posteriorly with at most, an imperfectly formed nuclear cap. There are no thin-walled sporangia or sexual stage.

COELOMYCETES. A group of Fungi Imperfecti which form their spores in a cavity of the substrate.

COELOSPERMOUS. Bearing boat-shaped seeds.

COENO-. A prefix meaning 'living together', *e.g.* multinucleate.

COENOBIUM. A group of algal cells having a definite number and organization. It behaves as an individual, often shows polarity, and gives rise to identical daughter coenobia.

COENOCENTRUM. A small deeply staining body at the centre of the multinucleate oösphere of the Peronosporales, to which the egg-nucleus goes.

COENOCYTE. A multinucleate, vegetative protoplast.

COENOCYST. A multinucleate aplanospore.

COENOGAMETANGIUM. A gametangium in which a coenogamete is formed.

COENOGAMETE. A multinucleate gamete produced by the Mucorales, and on fusion give rise to a coenozygote.

COENOPTERIDALES. An order of the Primofilices. The stems are mono-stelic, and the leaves either have no pinnae, or if they have the pinnae are at an angle to the plain of the leaf-blade. The sporangia are in stalked clusters borne directly on the rachis or on unflattened pinnules.

COENOPTERIDEAE = PRIMOFILICES.

COENOSPECIES. A species that is incapable of genetic recombination with other similar species.

COENOZYGOTE. See *Coenogamete.*

CO-ENZYME. An organic substance essential to the activity of a particular enzyme. The co-enzyme can sometimes be separated from the active group of the enzyme.

COENZYME I = diphosphopyridine nucleotide, **COENZYME II** = triphos-phopyridine nucleotide. These are dinucleotides containing adenine and nicotamide, acting as hydrogen carriers when substances are oxidized by dehydrogenases.

COENZYME A. A coenzyme necessary for the oxidative decarboxylation of pyruvate.

$$CH_3COCOO^- + CoA + 2H_2O - 2H \rightarrow CH_3COO-CoA + HCO_3^- + H_2$$
pyruvate   acetyl CoA

also:

Acetyl phosphate + CoA $\rightarrow$ acetyl CoA + inorganic phosphate then:
Acetyl CoA + oxalacetate $\rightarrow$ citrate + CoA

COHERENT. United, but very slightly, so that the joined organs appear free when viewed superficially.

COHESION. The fusion of organs during growth, usually of members of the same whorl of a flower.

COHESION MECHANISM. Any mechanism that depends on the cohesive properties of water, *e.g.* the dehiscence of a sporangium which bursts open when the stains set up by the drying out of one part overcome the water's cohesive properties.

COHORT. A group of allied families, *i.e.* an order.

COINCIDENCE, COEFFICIENT OF. The proportion of the number of observed double cross-overs to the expected number from the random combination of single cross-overs between 3 or 4 linked genes.

COLACIALES. An order of the Euglenophyceae. The immobile cells are permanently in a capsule and united to form a branched palmelloid colony. Temporary uniflagellate, naked stages may be formed.

COLCHICINE. $C_{22}H_{25}O_6N$. An alkaloid from the corm of the autumn crocus *Colchicum autumnale*. It causes abnormal division of nuclei, resulting in an increase of the chromosome number form a polyploid.

COLD TREATMENT. See *Vernalization*.

-COLE, -COLOUS. Suffix meaning 'living on': Inhabiting.

COLEOGEN. The layer of meristematic cells from which the endodermis is derived.

COLEOPTEROID. Looking like a beetle.

COLEOPTILE. The first leaf of a grass seedling. It appears above the ground first as a sheath around the plumule, and contains little chlorophyll.

COLEORHIZA, COLEORRHIZA. A protective layer of cells around the radicle of grass seedlings.

COLEOCHAETACEAE. A family of the Ulotrichales. The cells have 1 nucleus with a single plate-like parietal chloroplast, and may be single or united to form branching filaments. Some or all of the cells bear seta(e). Most genera produce zoospores, and sexual reproduction is isogamous or oögamous.

COLLABENT. Collapsing inwards.

COLLAPSING. Falling together into a brush-like form.

COLLAR. The junction of root and shoot.

COLLARIATE. Having a collar.

COLLATERAL. (1) Running side-by-side.
(2) Individuals in a family, but not related by direct descent.

COLLATERAL BUD. An accessory bud lying beside the axillary bud.

COLLATERAL BUNDLE. A vascular bundle with a strand of xylem, with a strand of phloem external to it on the same radius.

COLLECTING CELL. A thin-walled cell in the mesophyll of a leaf below the palisade layer, and in contact with it. The collecting cell transports elaborated food-materials from the palisade.

COLLECTIVE FRUIT. A single fruit formed from several flowers.

COLLENCHYMA. A mechanical tissue of long cells with cellulose cell-walls thickened at the corners with cellulose. The cells are living and can elongate.

COLLET = COLLAR.

COLLETERS. Glandular hairs.

COLLICULOSE, COLLICULOUS. Having rounded swellings.

COLLINUS. Latin for 'living on low hills'.

COLLOID. Solid particles suspended in a liquid. The particles are electrically charged and between 10^{-7} to 10^{-5} cm. in diameter. Because they are electrified at this size, they can be separated by electrophoresis except at the isoelectric point.

COLLUVIUM. The soil accumulated at the foot of slopes.

COLONIZATION. The occupation of bare soil by seedlings or sporelings.

COLONIST. A weed of cultivated land.

COLONY. (1) (Of bacteria, yeasts and fungi). A group of individuals derived from one, or a few related, spores.
(2) A group of individuals of the same species invading a new environment.

COLOPHONY. A form of resin.

COLUMELLA. (1) The central axis of a fruit.
(2) The sterile central axis in the fruit-body of the Gasteromycetes.
(3) The domed pillar in the sporangium of the Mucorales.
(4) The central column of sterile tissue in the sporangium of liverworts and mosses.
(5) A similar tissue in the sporangium of *Hornea*.
(6) The central part of a root-cap which contains statoliths.

COLUMELLIACEAE (EP, BH, H). A family of the Tubiflorae (EP), Personales (BH), Personales (H). These are shrubs with opposite leaves that lack stipules. The bisexual flowers are in cymes. There are 5 sepals and petals, the former being free and the latter fused. There are 2 free stamens, each with a single twisted anther. The ovary is superior (EP), inferior (H), of two fused loculi, which lie to right and left of the floral axis. There are many anatropous ovules. An endosperm is present. The fruit is a capsule enclosed in the calyx.

COLUMN. The combined style and stigma, or probably an outgrowth of the axis in the flowers of orchids.

COLUMNIFERAE = MALVALES.

COMA. (1) A tuft of hairs, attached to the testa of a seed.
(2) A tuft of leaves at the tip of a moss stem.

COMAL TUFT. A bunch of leaves at the end of a twig.

COMBRETACEAE (H). A family of the Myrtales. The leaves rarely have parallel nerves running lengthwise, and are not dotted with glands. The anthers open by lengthwise slits. The ovaries have a single loculus with the ovules suspended from the apex.

COMATE = COMOSE.

COMATE DISSEMINULE. A seed or fruit with fine long hairs which aid in wind-dispersal.

COMBRETACEAE (EP, BH). A family of the Myrtiflorae (EP), Myrtales (BH). The family contains many trees and shrubs which are climbers, with alternate or opposite leaves which lack stipules. The flowers are usually sessile and in racemes; they are usually bisexual and regular. Typically the flower has 5 free (usually valvate) sepals, and 5 free, or no petals. There are 2 whorls of 5 stamens (rarely many). The inferior ovary has 1 loculus with 2-5 anatropous pendulous ovules, and has a disk at its summit. The dry fruit contains 1 seed and is winged. The seed has no endosperm and the cotyledons are usually spirally twisted.

COMMELINACEAE (EP, BH, H). A family of the Farinosae (EP), Coronarieae (BH), Commelinales (H). These are herbs with jointed stems and alternate, sheathing leaves. The bisexual, usually regular flowers are in a cincinnus. There are 5 free sepals and petals, the latter are rarely fused. The 6 free anthers are in 2 whorls, but may be absent or as staminodes. The ovary is 2-3 locular with a few or 1 orthoptrous ovules on axile placentas in each, and is superior. The fruit is a capsule. The seeds have copious endosperm, and often have an aril.

COMMELINALES (H). An order of the Calyciferae. These are terrestrial (rarely aquatic) herbs whose leaves have a closed sheath (rarely not sheathing). The flowers are regular or zygomorphic, mostly bisexual, showy and usually in cymes or panicles. The perianth is in 2 whorls of typical sepals and petals. There are 3, frequently free petals. There are 6 or 3 stamens with anthers that open lengthwise or by pores. The superior ovary is of united carpels with 1 style. There are many to 1 ovules on axile or parietal placentas. The fruit is a capsule or baccate. The seeds contain endosperm and have an embryostega.

COMMENSALS. Two organisms living in association with each other, not parasites or symbionts, but mutually benefiting from the association.

COMMISSURAL STRAND. Found in the Marattiales. A strand of vascula tissue running diagonally across the pith at the nodes.

COMMISSURE. (1) A bridge between 2 organs, or parts of organs.
(2) A junction between 2 cells.
(3) The joint between the lid and the mouth of a moss sporangium.
(4) The joint between carpels.
(5) The line where the antical and postical lobes of the leaf of a liverwort join.
(6) A general term for any seam or joint.

COMMIXT. Mixed with.

COMMON BUNDLE. A vascular bundle running through the stem and leaf.

COMMON RECEPTACLE. The receptacle of all flowers on a head.

COMMUNIS. Latin for 'social, general'.

COMMUNITY. A distinct unit of vegetation consisting of 2 or more species.

COMOSE. Having tufted hairs.

COMPAGINATE. Joined tightly together.

COMPANION CELL. A nucleated cell with dense cytoplasm, associated with a sieve-tube, with which it is derived from a common parent cell. It apparently plays some part in the functioning of the sieve-tube. Companion cells are found only in Angiosperms.

COMPATIBLE. Capable of self-fertilization.

COMPENSATION PERIOD. The time needed for a green plant to be in light to make-up the carbohydrate lost during respiration in the dark.

COMPENSATING POINT. The light-intensity at which the production of carbohydrate by photosynthesis exactly balances that lost by respiration, so that neither carbon-dioxide nor oxygen are released, or absorbed. This assumes that the other factors are constant.

COMPENSATING TONGUE, COMPENSATION STRAND. A strand of vascular tissue passing from one vascular ring to another. Found in the siphonosteles of some ferns.

COMPETITION. The struggle between organisms for the necessities of life.

COMPITAL. Where veins intersect at an angle.

COMPLANATE. Flattened.

COMPLEMENT. (1) A heat-stable substance which extends the activity of α-amylase and β-amylase so that they will hydrolyse dextrin.
(2) A group of chromosomes derived from one nucleus, and consisting of one or more sets.

COMPLEMENTARY CHROMATIC ADAPTATION. The theory, applied to the Myxophyceae and Rhodophyceae, that the colour of the light-absorbing pigment is complementary to the colour of the light in which the plant is living. *E.g.* plants growing in blue light are red, and those in red light are green.

COMPLEMENTARY FACTOR. A factor in inheritance, which, with other similar factors leads to the appearance of some character in the offspring.

COMPLEMENTARY GENES. Genes which together produce an effect which is qualitatively distinct from the effects of any of them separately.

COMPLEMENTARY SOCIETY. A community of 2 or more species, occupying the same area, but not competing with each other, *e.g.* they may vegetate at different times, or have their roots at different levels.

COMPLEMENTARY TISSUE. Loose, thin-walled cells in the cavity of a lenticel, thus allowing the diffusion of vapours and gases.

COMPLETE FLOWER. A flower having calyx, corolla, stamens and carpels.

COMPLEX CHARACTER. A difference in phenotype determined by more than 1 gene, and consequently not transmitted to offspring as a unit.

COMPLEX TISSUE. A tissue made up of cells of more than one type.

COMPLICATE. Folded itself.

COMPOSITAE (EP, BH, H). A family of the Campanulatae (EP), Asterales (BH). (H) places them as the single order of the Asterales (*q.v.*).

COMPOUND. Made up of several similar parts.

COMPOUND FRUIT = AGGREGATE FRUIT.

COMPOUND HEAD. A head of flowers, made up of many single-flowered heads, each with its own involucre.

COMPOUND-INTEREST LAW. The rate of growth of a plant is proportional to the amount of plant material that is already present.

COMPOUND LEAF. A leaf whose petiole bears several leaflets.

COMPOUND OÖSPHERE. A multinucleate body in an oögonium probably composed of a number of female gametes which have not become individuals.

COMPOUND PYRENOID. A pyrenoid of two closely associated portions.

COMPOUND UMBEL. A raceme consisting of a large number of heads rising very close together at the end of a main branch. Each bears a number of flower stalks similarly arranged, each stalk bearing one flower. The whole generally forms a flat head.

COMPRESSED. Flattened.

COMPRESSION FLANGE. A group of turgid parenchymatous cells on the convex side of a coiled tendril.

COMPRESSION WOOD. Wood or dense structure formed at the bases of some trees and on the underside of branches.

CONCATENATE. In chains.

CONCAVE. Hollowed out; like a saucer.

CONCENTRATE. Arranged around a common centre, or having a common centre.

CONCENTRIC BUNDLE. A vascular bundle in which one tissue surrounds another. See *Amphicribal* and *Amphivasal*.

CONCEPTACLE. A flask-shaped cavity in the thallus of the Fucales in which the sporangia develop.

CONCHATE, CONCHIFORM. Like a scallop shell.

CONCINNUS. Latin for 'neat'.

CONCOLOR. Latin meaning 'of uniform colour'.

CONCRESCENCE. Two organs that when immature are separate, become fused as they mature. They may form a single structure, but may be functionally separate, *e.g.* fused stamens and petals.

CONCRESCENT. Becoming joined.

CONCRETE. (1) Growing together to form a single structure.
(2) Adhering closely to anything.

CONDENSATION. Crowded vertically, due to the absence or suppression of internodes.

CONDENSED. Said of an inflorescence in which the flowers are crowded together, and are nearly or quite sessile.

CONDUCTING STRAND = VASCULAR BUNDLE.

CONDUPLICATE. (1) Folded lengthwise.
(2) Said of a cotyledon which is folded longitudinally about the radicle.

CONFERTED. Crowded: close together.

CONFIGURATION. An association of two or more chromosomes at meiosis which separate independently of other associations at anaphase.

CONFLUENT. Blending: running into one another

CONGRESSION. The movement of chromosomes on to the metaphase plate, especially at the first meiotic division

CONE = STROBILUS.

CONFERTUS. Latin meaning 'crowded'.

CONFERVOID. Composed of threads.

CONGESTED. Packed close together.

CONGENITAL. Grown to.

CONGLOBATE. Of the bases of stipes, when they are massed into a fleshy ball.

CONGLOMERATE. Clustered.

CONGLUTINATE. Stuck together in a sticky mass.

CONGREGATE. Collected into a dense group.

CONIDIAL. Referring to, or pertaining to a conidium.

CONIDIOLE. A small conidium, usually budded from another conidium.

CONIDIOPHORE. A simple or branched hypha, bearing conidia.

CONIDIOSPORANGIUM. A sporangium which will, under certain conditions germinate directly.

CONIDIUM. An asexual spore cut off from the end of a hypha (conidiophore).

CONIDIUM VERUM. An asexual spore which comes away from the conidiophore when mature.

CONIFERAE. A class of the Gymnospermae. They are mostly evergreen trees, with monopodial branching and pointed leaves which are frequently borne on short shoots. Resin canals occur throughout the plant (except *Taxus*). The flowers are in perfect or imperfect cones, which are always unisexual but may be monoecious or dioecious, and are never terminal on the main branch.

CONIFEROPSIDA. A group containing the Ginkgoales, Cordaitales and the Coniferae.

CONIFEROUS. Cone-bearing.

CONIFERYL ALCOHOL. $C_6H_3(OH)(OCH_3)(CH:CH.CH_2OH)$. Derivatives of coniferyl alcohol possibly one of the primary units from which the lignin molecule is constructed.

CONIOMYCETES. A mixed group of fungi including Coelomycetes, smuts, and rusts.

CONJUGATE. Joined, occurring in pairs.

CONJUGATE DIVISION. Division of a pair of associated nuclei at the same time as in the Basidiomycetes.

CONJUGATE NUCLEI. Two nuclei in one cell, which undergo division.

CONJUGATION. (1) The pairing of gametes (usually isogametes) or zygotes; or the fusion of pairs of nuclei.
(2) The lateral association of chromosomes in the early prophase of meiosis.

CONJUGATION TUBE. A tubular outgrowth formed by the fusion of two lateral outgrowths, one from each of a pair of copulating cells. It is through this tube that the male and female (or $+$ and $-$ strains) gametes fuse.

CONJUCTIVE PARENCHYMA. Parenchyma formed between specialized tissue, *e.g.* between vascular strands.

CONK. The fruit-body of a wood-attacking fungus, especially a polypore.

CONNARACEAE (EP, BH, H). A family of the Rosales (EP, BH), Dilleniales (H). These are mostly twining shrubs with alternate, exstipulate leaves (or winged and adnate to the petiole (H)). The regular flowers are in panicles. The calyx is of 5 free or fused sepals. There are 5 free petals, and 10 or 5 free stamens (fused to the gynecium (H)). The ovary consists of 5, 4, or 1 free carpels, each with 2 erect orthopterous ovules. The fruit usually is 1 follicle with 1 seed. The seeds are often arillate.

CONNATE. Having similar parts joined by growth.

CONNECTING BAND = CINGULUM.

CONNECTING THREAD. A strand of protoplasm passing through fine pores in cell-walls, thus joining the protoplasm of two adjacent cells.

CONNECTIVE. (1) The prolongation of the filament into the anther.
(2) = DISJUNCTOR (Fungi).

CONNIVENT. Touching, but not fused.

CONSOCIATION. An association in which one species is distinctly dominant.

CONSOCIES. A stage in the development of a consociation, before the conditions have become stabilized.

CONSORTISM, CONSORTIUM. The mutual relationship, as between the fungus and alga in a lichen thallus.

CONSTRICTION. Part of a metaphase chromosome which is in a fixed position, and not twisted into a spiral.

CONSTIPATE. Crowded together.

CONSTRICTED. (1) Narrowed suddenly.
(2) Narrowed suddenly at one or more points along the length.

CONTABESCENT. Abortion of pollen and stamens.

CONTACT POINTS. Places where chromosomes first come together in pairing at zygotene.

CONTAMINATED. (1) Bearing, or mixed with a pathogen, *e.g.* fungus spores on seeds.
(2) Bacteral or fungus cultures, containing contaminants.

CONTEXT. The tramal or fleshy tissue of a hymenomycete fruit-body, but not the hymenium or cortex.

CONTERMINOUS. The marginal ray cells when they form an uninterrupting row.

CONTIGUOUS. Touching, but not united.

CONTINGENT. Touching.

CONTINGENCY TABLE. A table for checking the independence of two classifications by showing their frequencies simultaneously.

CONTINUOUS. (1) Having an uninterrupted outline or contour.
(2) Having a smooth surface.
(3) Without septa.
(4) Of hyphae; spores having no septa; of a stipe, one with the tissue of the pileus or periderm, joining with it.

CONTINUOUS VARIATION. Slight variation between a number of individuals of the same lineage, so that the differences grade into each other.

CONTORTED. Said of sepals and petals in the bud, twisted so that they overlap on one side only.

CONSTRICTION. (1) A narrowing.
(2) A narrow region on a chromosome which is usually difficult to stain.

CONTORTAE (EP) = GENTIANALES.

CONTORTED. Twisted together.

CONTORTED AESTIVATION. The arrangement of the perianth in the bud, when all the perianth segments overlap by either their right-hand or left-hand edges.

CONTORTODUPLICATE. Twistedand folded.

CONTRACTILE ROOT. A rather fleshy root, which becomes transversely corrugated as it ages, thus shortening, and d¬agging the plant deeper into the ground.

CONVERGENT IMPROVEMENT. The simultaneous improvement of two inbred strains by back-crossing their hybrids to each strain in separate lines, and selecting for the desired features of each strain in that line of which it is not the recurrent parent.

CONVERGING. Having the tips coming together gradually.

CONVEX. Rounded.

CONVOLUTE. Rolled around, coiled or folded, so that one half is covered by the other.

CONVOLUTE AESTIVATION = CONTORTED AESTIVATION.

CONVOLVULACEAE (EP, BH, H). (BH) include the Nolanaceae. A family of the Tubiflorae (EP), Polemoniales (BH), Solanales (H). These are herbs, shrubs, rarely trees, and many are climbers. The roots or stems are often storage organs. The alternate leaves are rarely stipulate, but often have accessory axillary buds. The regular flowers are bisexual, with the perianth segments in 2 whorls of 5. The sepals are rarely fused, while the petals always are. The 5 free stamens are alternate with the petals, and borne fused to the base of the petals. The superior ovary has 2 fused carpels (rarely 3-5) and is on a honey-secreting disk. Sometimes only the styles are fused. There are 2 loculi with 1-4 (usually 2) ovules in each. The placentation is axile (ovules arise erect from the base of the loculi (H)). The fruit is a berry, nut or capsule. The embryo is folded or much incurved, and there is little or no endosperm.

CO-ORIENTATION. See *Orientation*.

COPAL. A hard resin.

COPROPHILOUS. Living on dung.

COPSE. A wood of small trees that are periodically cut.

COPULATION. The fusion of gametes.

CORALLINACEAE. A family of the Cryptonemiales which have the plant-body strongly impregnated with calcium carbonate. They are important reef-builders.

COPULATION TUBE = CONJUGATION TUBE.

CORALLOID. Much-branched: looking like coral.

CORBICULAE. Paraphysis-like structures around the telia of certain rusts.

CORDATE. Heart-shaped.

CORDAITALES. An order of the Coniferopsida. They were tall woody trees of the Carboniferous era. The leaves were slender and simple, and the inflorescence was a loose cone.

CORDIFOLIUS. Latin meaning 'having heart-shaped leaves'.

CORDIFORM. Said of an ovate leaf with a pointed tip and a heart-shaped base.

CORE. (1) = CENTRUM.
(2) The plant material that forms the inner part of a periclinal chimaera.

COREMIFORM. Forming a tight bundle of elongated elements.

COREMIUM. (1) A rope-like strand of hyphae.
(2) A tightly packed group of conidiophores, bearing conidia which form a head.

CORIACELLATE. Slightly leathery in structure.

CORIACEOUS. Leather in structure.

CORIARIACEAE (EP, BH, H). A family of the Sapindales (EP), Coriariales (H). (BH) place it as an anomalous order at the end of the Disciflorae. These are shrubs with opposite or whorled, simple leaves without stipules. The buds are scaly, and the regular, bisexual or unisexual (H), inconspicuous flowers are borne in racemes. All the flower-parts are free. There are 5 sepals and petals, 2 whorls of 5 stamens, and 5 superior carpels, each with a single pendulous ovule. The endosperm is thin, and the embryo straight. The petals are persistent, and become fleshy enclosing the carpels to form a pseudo-drupe.

CORIAREAE (BH) = CORIARIACEAE.

CORIARIALES (H). An order of the Lignosae, with one family, the Coriariaceae (*q.v.*).

CORK. A layer of suberized, dead cells, cut off by a phellogen on the outside of the stem or root of a woody plant. The layer is protective and impervious to gases and water vapour. It replaces the epidermis.

CORK CAMBIUM = PHELLOGEN.

CORK CRUST. A thick layer of corky cells, made up of large soft-walled cells with narrow strips of flattened cells interspersed between them.

CORK FILM. A layer of flattened corky cells 2-3 cells thick.

CORK WART. A small corky growth on the surface of a leaf.

CORK WOOD. A wood of low specific gravity, due to the presence of many large thin-walled parenchymatous cells.

CORK. (1) A bulbous swollen stem base, bearing scale leaves and adventitious roots. It is a storage organ, and acts as a means of vegetative propagation.
(2) The bulbous stem of *Isöetes*.

CORMOPHYTE. A plant with a stem, root, and leaf.

CORMUS. A plant body developing a definite shoot system.

CORN. Wheat, possibly barley or oats (in United Kingdom), maize (in America).

CORNACEAE (EP, BH, H). A family of the Umbelliflorae (EP), Umbellales (BH), Araliales (H). There are no stipules. The flowers are in cymes or head, sometimes with petalloid bracts. The petals are free, valvate, or absent, and rarely joined at the base. The stamens are free from the petals (if the latter are present). The ovules are enclosed in one integument.

CORNEOUS. Horny in texture.

CORNICULATE, CORNIFORM, CORNUTE. Horned; shaped like a horn.

COROLLA. A collective term for the petals of a flower.

COROLLIFERAE (H). A division of the Monocotyledons, whose members do not have a distinct calyx, but the whorls of the perianth are similar.

COROLLIFLORAE = SYMPETALAE.

COROLLINE. Appertaining to the corolla.

CORONA. (1) A trumpet-like outgrowth of the perianth, *e.g.* of a Daffodil.
(2) A ring of leafy outgrowths from the petals.
(3) A five-celled cap of small cells on the oögonium of the Charales.
(4) The CORONA INFERIOR and CORONA SUPERIOR are two rings around the top of the stalk of *Acetabularia*, between which the plate-like gametangia emerge, and within which is a ring of sterile hairs.

CORONARIEAE (BH). A series of the Monocotyledons. The inner whorl of the perianth is petalloid. The ovary is usually free and superior, and the seeds contain abundant endosperm.

CORONATE. Crowned; having a corona.

CORPUSCLE. Any small cell-inclusion.

CORRELATED RESPONSE. The change in one character due to selection for another. This may be due to a single gene affecting more than one character, or to linkage between two genes.

CORRELATION. (1) The balance, and interdependence between the organs of a plant.
(2) The interdependence between any two variables, *e.g.* between the size of the plant and the amount of nutrients available.
(3) The mutual relationship between two or more organisms.
(4) The statistical analysis of any of the above.

CORRELATION COEFFICIENT. The ratio of the covarience of two variates to the geometrical mean of their variances. It ranges from $+1$ for complete interdependence, to -1 for complete independence.

CORRUGATE, CORRUGATED. Having a ridged or wrinkled surface.

CORSIACEAE (H). A family of the Burmanniales (H). (EP) include it in the Burmanniaceae. The perianth segments are zygomorphic, one of the outer perianth segments is large and ovate-cordate, the remainder are linear. There are 6 stamens. The ovary has 1 loculus, with parietal placentas well-extended into the cavity.

CORSINIACEAE. A family of the Marchantiales. The branches of the female gametophyte have a linear series of circumscribed receptacles. There are no elaters in the capsule.

CORTEX. (1) The tissue in a stem or root between the vascular bundles and the epidermis. Typically it is parenchyma.
(2) The outer layer of slightly thickened, closely packed cells of the thallus of some brown and red algae, fungi, and lichens.

CORTICAL. (1) Relating to the cortex.
(2) Relating to the bark.
(3) Living on bark.

CORTICAL BUNDLE. A vascular bundle in the cortex.

CORTICATE. (1) Having a cortex.
(2) Covered with an unbroken sheath of interwoven hyphae.

CORTICATION. In some Algae, the production of a cylinder of cells around a central axis by a meristematic cell on the outside of the cylinder.

CORTICOLOUS. Living on bark.

CORTINA. A partial veil, or part of one, covering the mature gills of some agarics.

CORYLACEAE (H). A family of the Fagales. The male flowers have no calyx, and 3 or more stamens. The calyx of the female flower is joined to the ovary, which is therefore inferior. The ovary has 2 loculi, each with 1 ovule.

CORYMB. A raceme with the lower flower-stalks longer than those above, so that all the flowers are at the same level.

CORYNELIACEAE. A family of the Dothideales.

CORYNOCARPACEAE (EP, H). A family of the Sapindales (EP), Celastrales (H). (BH) include it in the Anacardiaceae. These are trees or shrubs with alternate, leathery leaves. The bisexual flowers are in panicles. There are 2 whorls of stamens, the inner being staminodes. The ovary has 2 fused carpels, one of which is fertile, and contains 1 pendulous ovule. The fruit is a compressed drupe. The seeds have no endosperm.

COSCINOCYSTIDIUM. A cystidium which clearly projects.

COSCINOID. A pitted conducting element in *Linderomyces*.

COSTA. (1) A general term for a midrib or vein.
(2) The midrib of the thallus of a liverwort.
(3) A rib or valve of a diatom.

COSTATE. Ribbed.

COTTONY. With soft long hairs.

COTYLEDON. (1) The leaf formed directly from the embryo of an angiosperm or gymnosperm. There may be one (in monocotyledons) two (in dicotyledons), or several (in gymnosperms). They act as storage organs in non-endospermous seeds, and as the first photosynthetic organs in endospermous seeds. It may be an absorbtive organ in some seeds, *e.g.* the Graminae.
(2) The first leaf develops from the embryo sporophyte of the Pteridophyta.

COTYLEDONOIDS = PROTONEM.

COTYLIFORM. Disk-shaped.

COUMARIN. A lactone of *o*-coumaric acid. It is responsible for the odour of sweet vernal-grass, and tonka beans used in perfumery.

COUPLING. The presence of two given genes in the same chromosomes in a double heterozygote.

C.O.V. = CROSS OVER VALUE.

COVARIANCE. The mean of the products of corresponding deviation of two variates from their individual means. Estimated as the ratio of the sum of cross-products, to the corresponding number of degrees of freedom.

COVER CELL. One of four cells closing the neck of the archegonium of the Bryophyta and Pteridophyta.

CRADINA. An enzyme able to break down proteins, present in the juice of stems, leaves and fruits of figs.

CRASSULA. A horizontal rod or band of thickening, consisting of pectic materials, or cellulose, occurring between pits in the walls of tracheids and vessels.

CRASSULACEAE (EP, BH, H). A family of the Rosales (EP, BH), Saxifragales (H). These are mostly xerophytic perennials, with fleshy leaves which are not modified into pitchers. The bisexual (rarely unisexual) flowers are in cymes, and are regular. Each whorl of flower parts may contain 3-30 members. All the flower parts are free, and there are 2 whorls of stamens. The ovary is superior, with the carpels slightly fused at the base, containing many ovules. The fruits are usually follicles. The seeds contain little or no endosperm.

CRASSUS. Latin meaning 'thick'.

CRATERIFORM. Cup-shaped.

CRATICULAR STAGE. A stage produced by some pennate diatoms under adverse conditions. The protoplast produces several sets of new half-walls without escaping from the original wall surrounding the cell.

CREEPING. Of a stem, growing along the surface of the ground and rooting at the nodes.

CREMNIUM. A cliff formation.

CREMOCARP. A fruit splitting into two or more one-seeded portions.

CRENATE. Edged with rounded teeth that point forward.

CRENID ACID. One of the acids in humus.

CRENIUM. A spring formation.

CRENULATE. Finely crenate.

CREST. A ridge of outgrowth.

CRETACEOUS. (1) A geological period from 140 to 70 million years ago. (2) Chalky.

CRIBOSE = CRIBRIFORM.

CRIBRIFORM. Sieve-like.

CRINITUS. Latin meaning 'having soft hairs'.

CRISP, CRISPATE, CRISPED. Curled, looking like crumpled paper.

CRISS-CROSS INHERITANCE. This has occurred when all the male offspring of a cross have a character shown only by the female parent, and all the female offspring have a character shown only by the male parent.

CRISTA. (1) A ridge-like membrane, running the length of some bacterial cells. (2) Paired membranes, found in mitochondria, and separated by cistera-like spaces. They may be branched. Some workers consider them to be a single compound membrane, rather than being separated by a space.

CRISTATE, CRISTATUS. Latin meaning 'crested'.

CROSS. (1) The act of fertilization between two individuals of different breeds or races. (2) An individual produced by such a fertilization.

CROSS-BREEDING. See *Cross* (1). Applies particularly if carried-out deliberately by man.

CROSS-FERTILIZATION. The fertilization of the female gamete of one individual by the male gamete of another individual.

CROSSING-OVER. The exchange of corresponding segments of two chromatids of homologous chromosomes during meiosis. This is caused by the breaking and reunion of the chromatids, and results in the independent segregation of the genes.

CROSSOSMATACEAE (EP, H). A family of the Rosales (EP), Dilleniales (H). (BH) include it in the Dilleniaceae. These are shrubs with stiff alternate leaves (rarely opposite), which are often strong and parallel. Stipules are either absent, or winged and adnate to the petiole. The stamens are perigynous and joined to the calyx-tube. The seeds are kidney-shaped, with a rich endosperm, and often have an aril.

CROSS-OVER UNIT. A 1% frequency of interchange between a pair of linked genes.

CROSS-OVER VALUE. The percentage of gametes that show crossing-over of a particular gene.

CROSS-POLLINATION. The transfer of pollen from the anther of one individual to the stigma of another.

CROSS-TIES. Small veins in a leaf running straight between the larger veins, giving a ladder-like appearance.

CROTOVINAS. Channels in the soil, caused by burrowing animals of all types, and which have become filled with surface soil.

CROWN. A very short root-stock.

CROWNED. (1) Having a terminal outgrowth, *e.g.* a group of hairs.
(2) Having an appendage on the upper surface of the leaf or petal.

CROZIER. The hook formed by an ascogenous hypha previous to ascus development.

CRUCIATE, CRUCIFORM. Cross-shaped.

CRUCIALES (H). An order of the Herbaceae. These are mostly herbs with alternate leaves without stipules. The flowers are usually regular, bisexual and usually borne in racemes or corymbs. There are 2 whorls of free sepals, each whorl as 2 segments. There are 4 free petals, and 6 anthers, in 2 whorls of 2 and 4, the outer stamens being shorter than the inner. The ovary is superior, with 2 fused carpels, and parietal placentas. Although the 2 carpels are transverse, the ovary is separated by a vertical replum (false septum) derived from the placenta. The style is very short, and bears 2 stigmas. The fruit is a siliqua or silicula, and the seeds contain no endosperm.

CRUCIFERAE (EP, BH, H). A family of the Rhoeadales (EP), Parietales (H). The single family of the Cruciales (H) (*q.v.*).

CRUDE FIBRE. The residues in the soil formed from the woody parts of plants.

CRUCIFORM DIVISION = PROMITOSIS.

CRUENTUS. Latin meaning 'blood-coloured'.

CRUMPLED = CORRUGATE.

CRUSTACEOUS, CRUSTOSE. (1) Hard and brittle.
(2) Said of a lichen thallus lying close and tightly fixed to the substratum.

CRYMIUM. A polar, barren formation.

CRYOPLANKTON. Algae which live on the surface of snow and ice in polar regions and on high mountains.

CRYPTERONIACEAE (H). A family of the Cunoniales (H). There are no stipules, petals, or disk. The style is single, the anther loculi oblique, and the seeds have no endosperm.

CRYPTO-. A prefix meaning 'hidden'.

CRYPTOBIOTIC. Living hidden.

CRYPTOGAMAE (-IA). The plants that do not produce seeds.

CRYPTOGAMAE VASCULARES = PTERIDOPHYTA.

CRYPTOMERE. A genetic factor which is not seen.

CRYPTOMERISM. A failure of characters to show in the offspring, which nevertheless contain the corresponding hereditary factor.

CRYPTOMONADALES. An order of the Cryptophyceae, containing all the motile genera.

CRYPTONEMIALES. An order of the Florideae. They are the only tetrasporophytic Florideae with axilliary cells in a special filament of the gametophyte.

CRYPTOPHYCEAE. A class of the Pyrrophyta. There are usually two or more brownish chromatophores, with or without pyrenoids. The reserve food is usually starch, or starch-like. The motile cells are compressed They have 2 flagella, which are of slightly different length, and inserted terminally or laterally.

CRYPTOPHYTE. A plant which forms dormant buds below the soil-surface.

CRYPTOPLASM. That part of the cytoplasm that appears to lack granules.

CRYPTOPODSOL. A soil which does not appear to be a podsol but can be shown to be so by chemical analysis.

CRYPTOSTOMA. A flask-shaped cavity in the thallus of some of the larger brown algae. It contains mucilage-secreting hairs.

CRYSTAL SAC. A cell containing large numbers of calcium oxalate crystals.

CRYSTALLOID. A crystal of protein, found extensively in the cells of seeds and other storage organs.

CTENOID. Comb-like.

CTENOLOPHONACEAE (H). A family of the Malphigiales (H). The leaves are opposite and simple, and without glands. The sepals are imbricate (rarely valvate) and lack glands. The fruit is usually a drupe. The seeds lack endosperm.

CUCULLATE. Hooded.

CUCURBITACEAE (EP, BH, H). A family of the Cucurbitales (EP, H), Passiflorales (BH). There are usually 3 stamens, rarely 1-5, and 1 anther always has 1 loculus, the others being bilocular. The anthers are often flexuose, or conduplicate.

CUCURBITACEOUS. Gourd-like.

CUCURBITALES (EP, H). An order of the Sympetalae (EP), Lignosae (H). These are mostly herbaceous, climbing by tendrils. The flower parts are typically in fives. They are usually bisexual and regular, with a cup-shaped axis. The calyx lobes are imbricate. The petals are free, united, or rarely absent. There are 5 stamens (many-to-few (H)), at the edge of the axis, and they may be free or united. The inferior ovary is usually of 3 fused carpels, and 3 loculi. There are many ovules; the fruit is a berry, and the seeds have little or no endosperm.

CULM. The stem of a grass or sedge.

CULMICOLE. Growing on grass stems.

CULTRIFORM. Shaped like a knife.

CULTURE. The growth of organisms, especially microorganisms, under clearly defined conditions which are usually artificial. Such methods may be used experimentally, *e.g.* with fungi and bacteria, or horticulturally, *e.g.* with mushrooms and tomatoes.

CUMULATE. Heaped in a mass.

CUMULATIVE FACTORS. See *Polymeric genes*.

CUNEAL, CUNEATE, CUNEIFORM. (1) Wedge-shaped.
(2) Wedge-shaped, and attached by the point.

CUNONIACEAE (EP, H). A family of the Rosales (EP), Cunoniales (H). (BH) include it in the Saxifragaceae. These are shrubs or trees with opposite or whorled leathery leaves, having stipules which are never joined to the petiole. The flowers are small and usually bisexual. There are 4-5 free sepals, and 4-5 free petals (these may be absent). The stamens are free and vary in number. They are all fertile, and the filaments are not toothed at the apex. There is a disk. The ovary is superior, with 1 fused (rarely free) carpels. It is bilocular with many to 2 ovules in 2 rows in each loculus. The fruit is a capsule, rarely a drupe or nut. The seeds contain endosperm.

CUP. (1) A hollow floral receptacle.
(2) An apothecium.

CUPPING CELL. A swollen hyphal attachment formed by some fungi which parasitise other fungi. Food materials from the host are accumulated in it.

CUPREOUS. Copper-coloured.

CUPULAR, CUPULATE. Cup-shaped.

CUPULE. (1) A depression on the surface of the gametophyte of some Marchantiales, the cells of which produce the gemmae.
(2) A cup surrounding the fruit of some trees.

CUPULIFERAE (BH) = BETULACEAE (EP) + FAGACEAE (EP).

CURTAIN = CORTINA.

CURVATURE. A change in the direction of growth of part of a plant, due to one side growing faster than the other.

CURVEMBRYAE (BH). A series of the Monochlamydeae (BH). These are terrestrial plants with bisexual flowers (usually), generally having the same number of perianth segments as stamens. There is usually 1 ovule, and the embryo is curved in a floury endosperm.

CUSCUTACEAE (H). A family of the Polemoniales. These are leafless, parasitic herbs, with thin, climbing. The corolla lobes are imbricate, rarely contorted. The stamens have scales between the filaments. There are often 2 styles which are mostly separate.

CUSHION. The central portion of the prothallus of a fern. It is several cells thick and bears rhizoids and sex-organs.

CUSP. A sharp point, as on teeth.

CUSPIDATE. Having a rigid point.

CUTICLE. A non-cellular waxy layer secreted by the epidermis. It protects the surface and reduces water-loss.

CUTICULAR DIFFUSION. The passage of gases through the cuticle of a plant.

CUTICULARIZATION, CUTINIZATION. The formation of a cuticle.

CUTICULAR TRANSPIRATION. The loss of water-vapour from a plant through the cuticle.

CUTIN. The substance of plant cuticle deposited on, or in, the outer layer of cell walls. It is formed by the oxidation and condensation of fatty acids.

CUTINIZATION. The formation of cutin.

CUTLERIALES. An order of the Isogeneratae. The thallus is a flattened blade or disk. Growth is entirely, or partially trichothallic. The sporophytes produce unilocular sporangia only. The gametophytes are heterothallic and markedly anisogamous.

CUTTING. A method of vegetative propagation, whereby pieces of stems or roots are planted, and develop into new plants.

CYANASTRACEAE (EP). A family of the Farinosae (EP). (BH) include it in the Pontederiaceae. These are herbs with tubers or tuberous rhizomes. The bisexual, regular flowers are in racemes or panicles. The perianth consists of 2 whorls of 3 segments, all fused. There are 6 fused stamens. The inferior ovary consists of 3 fused carpels. It has 3 loculi each with 2 ovules. The fruit is one-seeded.

CYANEUS. Latin meaning 'bright blue'.

CYANOPHYCEAE = MYXOPHYCEAE.

CYANOPHYTA. A division of the Algae. They contain blue phycocyanin and chlorophyll distributed throughout the peripheral protoplasm. The nucleus lacks a membrane and there are no nucleoli. There is no sexual reproduction.

CYANOPLAST. A small pigmented granule in the cytoplasm of the cells of the Myxophyceae.

CYATHEACEAE. A family of the Leptosporangiatae. The sori are gradate, and may be marginal or superficial, with or without an indusium. The sporangia dehisce transversely by an obliquely vertical annulus. Some members are tree-like.

CYATHIFORM. Like a conical, narrow cup. It is sometimes stalked.

CYATHIUM. An inflorescence which is reduced to look like a single flower.

CYBELE. A flora.

CYCADACEAE. A family of the Gymnospermae. They flourished in the Triassic and Jurassic periods. Nine genera now survive. The stem is usually short and stout, but may be tree-like. It is usually unbranched. The leaves are usually in a crown, the lower part of the stem being covered with scales, both being spirally arranged. The flowers are dioecious and in cones, which are terminal. The male gametes are motile. The stem tissue is in concentric rings around a pith, and may show secondary thickening.

CYCADALES = CACADACEAE. An order of the Gymnospermae.

CYCADOPSIDA. The cycads and the Cycadeoids.

CYCLANTHACEAE (EP, BH, H). A family of the Synanthae (EP), Nudiflorae (BH). The single family of the Cyclanthales (H) (*q.v.*).

CYCLANTHALES (H). An order of the Corolliferae (H). These are herbs or small shrubs. They are climbers, epiphytes, or palm-like. The leaves are often deeply lobed. The small flowers are crowded into a spadix with male and female flowers alternating. The spadix is surrounded by a bract. The multiple fruit is fleshy. The seeds have a small embryo and an endosperm.

CYCLIC. Having floral parts in whorls.

CYCLO-. A Greek prefix meaning 'a circle'.

CYCLOSIS. The circulation of protoplasm within a cell.

CYCLOSPOROUS. Having the embryo coiled around the endosperm.

CYCLOSPOREAE. A class of the Phaeophyta. The thallus is parenchymatous, growing by an apical cell. There is no alternation of generation. The plants are sporophytes, producing spores, which act as gametes, in unilocular sporangia. The sporangia are in conceptacles which are usually at the tips of the branches, *i.e.* in receptacles. The plants are oögamous.

CYLINDROCAPSACEAE. A family of the Ulotrichales. The filaments are unbranched, and the cells have concentrically stratified walls. Sexual reproduction is oögamous.

CYMBIFORM. Boat-shaped.

CYME. An inflorescence in which the terminal bud is a flower-bud, *i.e.* it is a sympodium, and any subsequent flowers are formed in a similar way at the ends of lateral branches.

CYNOCRAMBACEAE (EP, H). A family of the Centrospermae (EP), Chenopdiales (H). (BH) include it in the Urticaceae. It contains one genus *Theligonum (Cynocrambe)* and the position of the family is very anomalous. These are herbs with fleshy stipulate leaves. The flowers are unisexual. The male flowers are opposite the leaves with 2-5 free perianth segments. There are 10-30 free stamens with anthers opening by longitudinal slits. The female flowers are in 3-flowered cymes, with perianths of 3-4 free segments. The superior ovary is of 1 carpel, 1 loculus, with a basal style. The fruit is a drupe, and the seed contains endosperm.

CYNOMORIACEAE (EP). A family of the Myrtiflorae (EP). (BH) include it in the Balanophoraceae. They have brownish rhizomes, and are totally parasitic herbs. The bisexual flowers have epigynous stamens. The 1 ovule is pendulous.

CYPERACEAE (EP, BH, H). A family of the Glumiflorae (EP), Glumaceae (BH). The single family of the Cyperales (H) (*q.v.*).

CYPERALES (H). An order of the Glumiflorae. These are perennials (rarely annuals) with rhizomes. The stems usually have pith (rarely hollow). The leaves are grass-like, sheathing at the base, but rarely ligulate. The flowers are in heads or small spikes, each within a glume-like bract, and are bisexual or unisexual. The perianth is reduced to scales, bristles, or is absent. There are usually 3 stamens with bilocular basifixed anthers. The ovary is superior, with 1 loculus, and 1 erect ovule. The fruit is nut-like, indehiscent, with endospermous seeds.

CYPHELLA. Small circular depressions in the under surface of some foliose lichens. They are aerating organs.

CYPRESS KNEE. A vertical upgrowth from the roots of the swamp cypress. It is an aerating organ.

CYPSELA. A 1-seeded fruit, formed from a syncarpous, inferior ovary.

CYRILLACEAE (EP, BH, H). A family of the Sapindales (EP), Olacales (BH), Celastrales (H). These are evergreen shrubs. The alternate leaves lack stipules. The bisexual flowers are regular and borne in racemes. There is no disk. The sepals are 5 in number, free, and persistent. They are valvate. The petals are 5 in number, and imbricate. The stamens are free, in 1 or 2 whorls of 5 each, with introrse anthers. The superior ovary is of 5-2 carpels, and multilocular with equal stigmas. Each loculus has 1 (rarely (2-4) pendulous anatropous ovule. The raphe is dorsal. The embryo is straight embedded in endosperm.

CYRTANDRACEAE = GESNERIACEAE.

CYST. A thick-walled resting spore, or sporangium.

CYSTEINE. A sulphur-containing amino acid.

$$CH_2—SH$$
$$|$$
$$CH—NH_2$$
$$|$$
$$COOH$$

CYSTIDIOLE. A sterile structure in the hymenium of some Basidiomycetes. It is found at the same level as the basidia. It is thin-walled, little differentiated, and wider than a paraphysis.

CYSTIDIUM. A sterile, generally light-coloured, cells, projecting from the hymenium of Basidiomycetes. They are frequently swollen.

CYSTINE. An amino acid formed by the polymerization of two cysteine molecules.

$$HOOC—CH—CH_2—S—S—CH_2—CH—COOH$$
$$|\qquad\qquad\qquad\qquad\qquad |$$
$$NH_2\qquad\qquad\qquad\qquad\qquad NH_2$$

CYSTOCARP. The fruiting body of the red algae.

CYSTOGENOUS. Producing a cyst.

CYSTOLITH. A crystal, or concretion of calcium carbonate crystals in a cell.

CYSTOSORUS. A group of united cysts found in the Chytridiales.

CYSTOSPORE. An encysted zoöspore formed in the Chytridiales.

CYTASE. A general term for any enzyme which breaks down cellulose.

CYTINACEAE (H). A family of the Aristolochiales. (BH) include Rafflesiaceae and Hydnoraceae of (EP). The flowers are unisexual. The anthers are sessile in 1-3 series around a fleshy column. The ovary is inferior, or semi-inferior with parietal or apical placentas. The fruit is fleshy, indehiscent, or opening irregularly. These are parasites with scale-like leaves.

CYTOCHEMISTRY. The chemistry of cells.

CYTOCHROMES. Cytochrome a, b, and c, are haemochromogens widely occurring in cells and acting as oxygen carriers during cellular respiration.

CYTODE. A mass of cytoplasm, not containing a nucleus.

CYTODIAERESIS = MITOSIS.

CYTOGAMY. The conjugation of cells.

CYTOGENESIS. The formation and development of cells.

CYTOGENOUS. Cell forming, or producing cells.

CYTOKINESIS. The division of the cytoplasm during cell-division.

CYTOLOGY. The study of cells in all its aspects, particularly as individuals.

CYTOLYMPH. Cell-sap.

CYTOLYSIN. Any substance causing the breakdown of cells.

CYTOLYSIS. The break-down of cells, especially the cell-wall.

CYTOME. The whole of the chondriosomes present in a cell.

CYTOMICROSOMES = MITOCHONDRIA.

CYTOMORPHOSIS. The changes that take place in a cell from its formation to death; or the changes that take place in a successive generation of cells derived from one individual.

CYTOPHARNYX. The narrow neck at the opening of the gullet of the Euglenophyta.

CYTOPLASM. The protoplasm of a cell, excluding the nucleus.

CYTOPLASMIC INHERITANCE. The transmission of hereditary characters through the cytoplasm, rather than the nucleus.

CYTOPLASMIC STAIN. A stain that stains cytoplasm, and its inclusions, in contrast to nuclear material.

CYTOSINE. One of the pyrimidine bases found in DNA.

$$
\begin{array}{c}
NH_2 \\
| \\
C \\
/ \ \ \backslash \\
N \quad\ \ CH \\
| \qquad\ \ \| \\
C \quad\ \ CH \\
\ \backslash \ \ / \\
HO \ \ N
\end{array}
$$

CYTOSOME. (1) The cytoplasm of a cell.
(2) The area on the surface of the cell of the Euglenophyta through which the food-particles are ingested. It is usually in the form of a gullet.

CYTOTAXIS. The reorientation of cells resulting from an internal or external stimulus.

CYTOTAXONOMY. The classification of plants based on a study of cell-structure.

CYTOTROPISM. The reaction of response to the stimulus of mutual attraction or repulsion between two cells.

CYTTARIACEAE. A family of the Pezizales. The relationships are vague. The many apothecia are embedded in a fleshy stroma, found outside the twig of the host (*Nothofagus*, the Southern beech of America). The basal part of the stroma of some species produce organs like spermogonia with sperm cells.

D

DACRYMYCETACEAE. The single family of the Dacrymycetales.

DACRYMYCETALES. An order of the Basidiomycetales. The distal ends of the hypobasidium has two divergent epibasidia, which are not covered by sterile tissue. The fruiting-bodies are small and gelatinous to waxy.

DACRYOID. Tear-shaped; pear-shaped.

DACTYLOID. Finger-like.

DAEDALEOUS. Having an irregularly wrinkled, or plaited surface.

DAMMAR. A hard resin.

DAMPING OFF. A disease which attacks young seedlings at ground level, causing them to rot and fall over. It is usually caused by species of the fungus *Pythium*, and is favoured by overcrowding in damp conditions.

DARK SEED. A seed which will germinate, only if kept in the dark, when other conditions would normally favour germination.

DAPHNALES (BH). The fifth series of the Monochlamydeae. The perianth is of 1 to 2 whorls and is sepaloid. The stamens are perigynous. The ovary is of 1 carpel containing 1 or few ovules.

DAPHNIPHYLLACEAE (EP, H). A family of the Geraniales (EP), Hamamalidales (H). These are small trees or shrubs, with astipulate leaves, which are simple. The flowers are unisexual, and lack sepals and petals. The ovary is imperfectly bilocular, and the fruit is a 1-seeded drupe.

DARK REACTION. Takes place during the photosynthetic process, when oxygen is evolved from a complex peroxide (formed from water). The carbon dioxide is fixed to form sugars, by combination with the hydrogen. This is a complex reaction involving the reduction and subsequent oxidation of a hydrogen-carrier, and takes place in the stroma of the chloroplast.

DARWINIAN THEORY, DARWINISM. The theory of evolution propounded by Charles Darwin. See *Natural Selection*.

DASYCLADIACEAE. A family of the Siphonocladiales. The central axis of the thallus has transverse whorls of branches from top to bottom, or only at the upper end. All, or only some of the whorls are fertile. The protoplast of a fertile branch may divide directly into biflagellate isogametes, or form 1 or more aplanogametes, which produce biflagellate isogametes on germination.

DASYPHYLLOUS. (1) Having leaves with a thick coat of cottony hairs.
(2) Having thick leaves.
(3) Having crowded leaves.

DATISCACEAE (EP, BH, H). A family of the Parietales (EP), Passiflorales (BH), Cucurbitales (H). These are trees of herbs with simple or pinnate leaves, which lack stipules. The flowers are in racemes or spikes, are regular, and usually bisexual. The male flower has 9-3 calyx lobes which may be free or untied, there are 9-4 free petals, or they are absent. The many or 9-4 stamens have bilocular anthers, and the loculi are straight. The female flower has 8-3 calyx lobes which are united to themselves and to the ovary, no petals, and an inferior ovary consisting of 8-3 fused carpels with free styles. There is 1 loculus with parietal placentation, containing many anastropous ovules. The fruit is a capsule, and the seeds contain little endosperm.

DAUER HUMUS. Humus which is resistant to the attack of micro-organisms.

DAUERMODIFICATION. A lasting inheritable change, possibly cytoplasmic, which is produced by some treatment.

DAWSONIALES. An order of the Eubrya. The gametophyte may be up to 60 cm. high. The capsule is dorsiventral, and the peristome is subtended by a mass of filamentous hairs.

DAY POSITION. The situation adopted by leaves of plants during the day, if they alter this position when the light intensity decreases.

DAY SLEEP. The folding of leaflets of a compound leaf in bright light. This brings the stomatal surfaces together, thus probably reducing the water-loss.

DEALBATE. Whitened, usually by a covering of hairs.

DEASSIMILATION. The utilization of food by a plant.

DEATH POINT. The maximum, or minimum limit of any factor in the internal or external environment, beyond which the cell or organism is killed.

DECA-. A Greek prefix meaning 'ten'.

DECANDROUS. Having ten stamens.

DECAPLOID. Having ten times the haploid number of chromosomes.

DECARBOXYLATION. The loss of carbon dioxide during the metabolism of an acid, *e.g.* pyruvic acid forms acetaldehyde and carbon dioxide:—
$$CH_3CO.COOH = CH_3CHO + CO_2.$$

DECEM-. A Latin prefix meaning 'ten'.

DECIDUOUS. (1) The seasonal shedding of leaves.
(2) Falling off.

DECIPIENS. The Latin meaning 'deceiving'.

DECLINATE. Growing downwards in a curve.

DECLINING. Growing straight downwards.

DECLIVATE, DECLIVOUS. Sloping.

DECOLOURATE, DECOLOURED. Without colour.

DECOMPOSED. Said of a lichen cortex, when it is made up of gelatinous hyphae.

DECOMPOUND. Said of a compound leaf, when the leaflets are made up of distinct parts.

DECONJUGATION. The separation of chromosomes before the end of prophase of meiosis.

DECORTICATED. (1) Devoid of bark.
(2) Devoid of cortex.

DECUMBENT. Lying flat, or being prostrate, but having the tip growing upwards.

DECURRENT. (1) Of a leaf, having the base extending down the stem as two wings.
(2) Of the gills of an agaric when they run some distance down the stipe.

DECURVED. Bent downwards.

DECUSSATE. Said of leaves, when they are opposite, but each pair is at right angle to the ones above and below it.

DEDIFFERENTIATION. The reduction of the cells of a differentiated tissue to a common undifferentiated form.

DEDIPLOIDIZATION. The production of haploid cells (or hyphae) from a dikaryotic diploid cell (or mycelium).

DÉDOUBLEMENT. Branching.

DEFICIENCY. The loss of a terminal acentric segment of a chromosome. Sometimes used to include 'deletion'.

DEFICIENCY DISEASE. An abnormality caused by the absence of an essential chemical element.

DEFINITE. (1) Always of the same number in a given species.
(2) Of a stem, ending in a flower, and so stopping growth.
(3) Of a stem, when the bud grows rapidly to its full length, and then stops.
(4) Said of an inflorescence when all its branches end in flowers, *i.e.* it is cymose.

DEFINITE VARIATION. A change taking place in a definite direction in the history of a race.

DEFLEXED. Bent sharply outwards and downwards.

DEFOLIATION. Shedding leaves.

DEFLOCCULATION. The aggregation of clay particles. This forms a sticky mass, which is difficult to redisperse, thus destroying the soil-structure.

DEGENERATION. (1) The change from a complex to a simpler form during evolution.
(2) The loss of morphological or physiological features by a fungus or bacterium kept in culture.

DEGREE OF FREEDOM. A comparison of items within a body of data, independent of any other comparison used in the analysis. The number of degrees of freedom is the number of independent comparisons that can be made within the body of data.

DEGRESSIVE. A change towards a more simple form. It may be degeneration.

DEHISCENCE PAPILLA. A small rounded projection on the surface of a zoosporangium of the Blastocladiaceae, later becoming a dehiscence pore.

DEHISCENT. Opening spontaneously to release spores or seeds.

DEHYDRASE. An enzyme that catalyses the removal of water, *i.e.* condensation.

$$E.g. \ 2C_6H_{12}O_6 = C_{12}H_{22}O_{11} + H_2O.$$

DEHYDROGENASE. An enzyme that catalyses the removal of hydrogen, and by so doing oxidizes the substrate.

DEHYDROGENATION. A biological oxidation, involving dehydrogenases, when hydrogen is removed from a donor, which is oxidized, to a receptor which is reduced.

$$E.g. \ AH_2 + B = A + BH_2.$$

DELAYED INHERITANCE. When each successive generation has the genotype of the female parent for a particular character.

DELETION. The loss of an intercalary acentric segment of a chromosome.

DELIGNIFICATION. The destruction of lignin in wood by a fungus.

DELIQUESCENCE. (1) The softening and liquefying of a sporangium membrane etc. at maturity.
(2) Said of a stem that breaks into separate branches.

DELPHINIDIN. An anthocyanin. The blue pigment in delphiniums.

DELTOID. Triangular.

DEMATIACEAE. A family of the Moniliales, characterized by the dark-coloured mycelium and spores. The conidiophores are not joined in bundles.

DEMATOID. (1) Black and web-like; having a covering of dark interwoven hyphae.
(2) Said of sooty moulds that are dark and moniloid.

DEMERSUS. Latin meaning 'sub-aquatic'.

DEMICYCLIC. Of Uredinales which have a life-cycle in which no uredospores are produced.

DENATURATION. Reducing the solubility of a protein, by altering its structure, chemically or by heating.

DENDRITIC. (1) Much branched.
(2) Having markings which are tree-like or moss-like.

DENDROGRAM. A two-dimensional 'tree' showing the relationships between organisms—a family tree.

DENDROGRAPH. An instrument which is used to measure the periodic shrinking and swelling of tree trunks.

DENDROID. (1) Tree-like.
(2) Tall, with an erect main stem.
(3) Freely branching.

DENDRON. Greek, meaning 'tree'.

DENDROPHYSIS. A paraphysis-like structure, bearing simple or branched spines.

DENIGRATE. Blackened.

DENITRIFICATION. The decomposition of nitrates in the soil by *Bacterium denitrificans*, resulting in the production of nitrogen.

DENITRIFYING BACTERIA. A group of soil bacteria which break down nitrites and nitrates anaerobically to produce nitrogen.

DENIZEN. A species of plant which can maintain itself in the wild, but was probably introduced by man.

DENS. The Latin for 'tooth'.

DENTATE. Having a toothed margin.

DENTATION. (1) Of a leaf margin, having a small blunt or pointed outgrowth. (2) The collective term for the thickening of the wall projecting into the lumina of ray tracheids in the wood of pines.

DENTICULATE. Of a leaf margin having small teeth.

DENUDATE, DENUDED. (1) Having a worn, or naked appearance. (2) Naked, or hairless, by loss of scales *etc*.

DENUDED QUADRATE. A square piece of ground, from which all the vegetation has been removed to study its recolonization.

DEOPERCULATE. (1) Lacking an operculum. (2) Having an operculum which does not break away spontaneously.

DEOXYNUCLEOSIDE. A 2-desoxyribose sugar with adenine, guanine* cytosine, or thymine attached to it. A base unit of nucleic acid.

DEOXYRIBONUCLEOTIDES. The products of hydrolysis of DNA.

DEOXYRIBOSE. A pentose sugar.

DEOXYRIBOSE NUCLEIC ACID (DNA). The hereditary material of the cell. The molecule contains deoxyribose (a sugar), phosphoric acid, two pyrimidines (thyamine and cytosine) and two purines (adenine and guanine). The pyrimidines and purines are bases. The molecule is a double helix of sugar-phosphate linkages (forming the sides of a twisted ladder) with the bases joined across to form the rungs. Adenine and thyamine are always paired, as are guanine and cytosine. *E.g.*

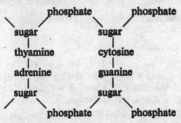

DEPAUPERATE. Diminutive; looking starved and underdeveloped.

DEPENDENT. Hanging down.

DEPILATION. The natural loss of hairs as the plant matures.

DEPLANATE. Flattened.

DEPLASMOLYSIS. The recovery from plasmolysis by the intake of water by osmosis.

DEPRESSED. (1) Flattened. (2) Sunken to become concave.

DEPSIDE. A product from hydroxy-aromatic acids by the condensation of the carboxyl group of one molecule with the phenol group of the second molecule.

They are probably concerned with the oxidation of fats and proteins inside the cell.

DERBESIACEAE. A family of the Siphonales. The thallus is a branched tubular coenocyte, with branches separate from one another. There are many zoospores produced from each sporangium, and they have a whorl of flagella at the anterior end.

DERIVATIVE HYBRID. A hybrid derived by crossing one hybrid with another, or with one of its parents.

DERMAL. Relating to the epidermis or other superficial layer.

DERMATEACAE. A family of the Pezizales. The apothecia are small to medium in size, and mostly embedded in the host tissue through which they burst at maturity. The excipulum is usually pseudoparenchymatous, and dark. They grow mostly, as parasites or saprophytes on woody or herbaceous plants. They are usually fleshy, but may be cartilaginous or leathery.

DERMATOCYSTIDIUM, DERMATOCYST. A cystidium on the cuticle or pellicle.

DERMATOGEN. The external layer of a stem or root apex, one cell thick, and giving rise to the epidermis.

DERMATOPHYTE. A parasitic fungus which causes a disease of skin, hair, or nails.

DERMATOSOME. One of the small pieces into which a cell wall can be resolved by prolonged treatment with dilute hydrochloric acid, followed by heating at 50-60°C.

DERMATOCALYPTROGEN. A meristematic layer in the apex of the root of many dicotyledons, giving rise to the root-cap and dermatogen.

DESCENDING. Growing, or hanging downwards in a curve.

DESCENDING AESTIVATION. Aestivation in which each segment overlaps the one anterior to it.

DESICCATION. Drying up.

DESMARESTIALES. An order of the Phaeophyta. The thallus has a single filament at each growing apex. The thallus shows pseudoparenchymatous cortication behind the filament, giving it a definite macroscopic form. The gametophyte is microscopic and oögamous.

DESMIDIACEAE. A family of the Zynematales. The 'placoderm' desmids. The cells are solitary, in filaments, or amorphous colonies. In all but two genera the cells divide by a median constriction (sinus), into two halves (semicells), joined by an isthmus. The walls are transversely segmented with vertical pores. The contents of conjugating cells usually leave the wall as the zygote is formed.

DESMIDS. See *Desmidiaceae* and *Mesotaeniaceae*.

DESMOKONTAE. A Class of the Pyrophyta. The cell wall is vertically divided into two halves (valves) without being divided into definitely arranged plates. The motile cells have two apically inserted flagella, which are flattened. They differ from each other in their form of movement.

DESMOLASES. Enzymes breaking, or forming a C-C link, without hydrolysis. *E.g.* Carboxylase catalyses the change of pyruvic acid to acetaldehyde and carbon dioxide.

DESOXYRIBOSE. A pentose sugar.

DESOXYRIBOSE NUCLEIC ACID = DEOXYRIBOSE NUCLEIC ACID.

DESTARCHED. Said of a green plant which no longer contains starch. This is brought about by keeping the plant in the dark. This stops photosynthesis, and any starch present will either have been used, or translocated.

DESTHIOBIOTIN. A substance which may replace biotin in the growth yeasts, but in some cases may inhibit the effect of biotin.

DESYNAPSIS. The abnormally early breaking of synapsis in meiosis.

DETERMINANT. (1) With a well-marked edge.
(2) Of an inflorescence that ends in a flower.
(3) An element in a cell supposed to control heredity and development; the initial idea of a gene.

DETERMINATE. (1) Clearly marked; definite.
(2) Definite; ending in a bud.

DETERSILE. Said of fine hairs which can be easily removed, leaving the surface bare.

DEUTEROCONIDIUM. A spore-like cell of dermatophytes, produced by the division of a hemisphere (protoconidium).

DEUTROGAMY. Any process which replaces normal fertilization.

DEUTEROMYCETES = FUNGI IMPERFECTI.

DEUTEROPLASM = METAPLASM.

DEUTOBROQUE. A stage proceeding leptotene in oögenisis. The chromosomes radiate from the nucleolus and wind about just under the nuclear membrane.

DEUTOPLASM = METAPLASM.

DEVELOPMENT. The succession of stages in the life of the plant, as distinct from simple growth.

DEVIATION. The departure of a quantity from its expected value.

DEVONIAN. A geological period 310-270 million years ago.

DEXTRINASE. A plant enzyme which hydrolyses dextrin.

DEXTRINS. A series of polysaccharides intermediate in structure between glucose and starch, with 5-10 glucose units in straight chains. Formed during starch hydrolysis.

DEXTRORSE. Turning or twisting to the right.

DEXTROSE. A hexose sugar (D-glucose), a solution of which bends a beam of polarized light to the right.

DI-. A Greek prefix meaning 'two'.

DIA-. A Greek prefix meaning 'transverse'.

DIADELPHOUS. Of stamens which have the filaments, either in two bundles or in a group with one free stamen.

DIADROMOUS. Said of venation resembling the ribs of a fan.

DIAGEOTROPISM. The growth of a plant member so that it is at right angles to the gravitational field.

DIAGNOSIS. (1) The formal description of a plant, with special reference to the characters that distinguish it from related species.
(2) The identification of a pest or disease.

DIAGONAL. Said of any part of a flower situated in a position other than median or lateral.

DIAHELIOTROPIC. An obsolete term for diaphototropic (*q.v.*).

DIAKINESIS. The last stage in the prophase of meiosis. The nuclear membrane and nucleolus disappear. The bivalents become more contracted, the chiasmata tend to move to the ends of the chromosomes, and the tetrads pass on to the spindle.

DIALYCARPIC = APOCARPOUS.

DIALYPETALOUS = POLYPETALOUS.

DIALYPETALANTHACEAE (H). A family of the Rubiales. There are 4 free petals in 2 whorls. There are 16-25 stamens in 2 whorls, free from the corolla and epigynous. The filaments are fused into a ring at the base. There are many seeds which are pointed at each end.

DIALYSIS. A method of separating small particles (ions or small molecules) from big ones (colloids or large molecules) in solution. This is done by placing the mixed solution in a membrane which has pores large enough for the small molecules to pass through, but which obstruct the passage of the larger ones (*e.g.* cellophane, collodian, parchment). This is then placed in water, and the smaller particles diffuse out into it.

DIALYSTELIC. Having several separate steles.

DIAMETER LAW. The rate of diffusion of a gas through a membrane is proportional to the mean diameter of the pores.

DIAMINES. Compounds containing two amine groups.

DIANDROUS. Having two antheridia, or two stamens.

DIAPENSIACEAE (EP, BH, H). A family of the Ericales (EP, BH, H). Chiefly alpine or arctic undershrubs, with the leaves in rosettes. The actinomorphic, bisexual flowers are solitary or in racemes, without a disk. The 5 calyx lobes are free or fused. The 5 petals are fused. The 5 free stamens are free, epipetalous, and opposite the sepals. There may be 5 staminodes as well. The anthers are transverse, with each lobe opening by a longitudinal slit. The superior ovary is of 3 fused carpels. The placenta is axile, with many anatropous or amphitropous ovules. The style is simple, with a 3-lobed capitate stigma. The fruit is a loculus or capsule. The embryo is cylindrical with a fleshy endosperm.

DIAPHANOUS. Transparent, or nearly so.

DIAPHOTOTROPHIC. Said of a plant member which grows into a fixed position across the direction of incident light.

DIAPHRAGM. (1) A transverse plate of cells across a stem, usually at the nodes.
(2) A cross-partition in the air chambers of hydrophytes, usually perforated.

DIAPORTHACEAE. A family of the Sphaeriales. The stoma is not entirely of fungus structure. The asci have short, or long evanescent stalks, soluble in water, so that the mature asci form a loose central mass. Paraphyses are present. The ascospores are ellipsoid, fusoid, or sometimes allantoid, or long cylindrical, hyaline or coloured. The conidia may be short cylindrical to filiform, or ellipsoid to long cylindrical.

DIARCH. Having two strands of xylem.

DIASCHISTIC. Said of tetrads which divide once transversely and once longitudinally, in meiosis.

DIASPORE. Any unit of dissemination, *e.g.* spore, piece of mycelium, sclerotium etc.

DIASTASE. An enzyme complex (about 5) which breaks down starch to glucose. See *Amylase*.

DIASTER. In cell division, when the daughter chromosomes are situated in two groups near the poles of the spindle, ready to form the daughter nuclei.

DIASTOLE. The growth and expansion of a nucleus from the end of one mitosis to the beginning of the next.

DIATOMIN. The gold-brown pigment said to be present in the chromatophores of the Chrysophyta.

DIATOMS = BACILLARIOPHYCEAE.

DIATROPISM. The growth of an organ or organism so that its axis is a^t right-angles to the line of action of the stimulus.

DIATRYPACEAE. A family of the Sphaeriales, included in the Diaporthaceaea.

DICARPELLARY. With two carpels.

DICARYON = DIKARYON.

DICENTRIC. Of a chromosome or chromatid having two centromeres.

DICHAPETALACEAE (EP). See *Chailletiaceae* (BH, H). These are woody plants with entire stipulate leaves. The flowers are in various forms of heads, regular, bisexual or unisexual. The calyx or petals are in fives, free or united. The 5 free stamens are sometimes epipetalous. The superior ovary is of 2-3 fuse carpels, each with 2 ovules. The flowers are sometimes borne on the leaves. The fruit is a drupe with a 1-2 locular stone. The seed has no endosperm.

DICHASIUM. A cyme, with each branch giving rise to two other branches.

DICHLAMYDEOUS. Said of a flower having a calyx and corolla.

DICHLAMYDEOUS CHIMAERA = DIPLOCHLAMYDEOUS CHIMAERA.

DICHOGAMY. The maturing of the anthers and ovules in the same flower at different times.

DICHOPHYSIS. In the Thelephoraceae, a sterile structure in the hymenium or trama having successively dichotomous branching, thick walls, narrow lumen, and subulate apices.

DICHOPODIUM. A sympodial branch system made up of successive parts of a dichotomizing branch system, of which only one part assists in forming the axis.

DICHOTOMOUS. Forked branching, produced by the subdivision of an apical meristem, to form two branches of the same size. This is a primitive form of branching.

DICLINOUS. Unisexual. Male and female organs born on different plants.

DICOTYLEDONEAE (DICOTYLEDONS). One of the two divisions of the Angiosperms. The embryo has two cotyledons; the leaves are usually net-veined; the stems have open bundles; and the flower parts are usually in fours or fives.

DICRANALES. An order of the Eubrya. The gametophores are erect. The leaves are generally lanceolate, with blades more than two cells thick; with or without a midrib. They are many-ranked. The sporophytes are usually acrocarpous. The capsule has a simple peristome of 16 teeth.

DICTYDINE GRANULES = PLASMODIC GRANULES.

DICTYOSIPHONACEAE. The single family of the Dictyosiphonales.

DICTYOSIPHONALES. An order of the Polystichineae. The thallus has a single apical cell, and is profusely branched and cylindrical. The mature portions are differentiated into two or three regions. The sporophytes usually have only unilocular sporagia. The gametophytes are microscopic and isogamous.

DICTYOSPORANGIUM. (1) A septate sporangium.
(2) A sporangium found in some oömycetes, in which the spores encyst in the sporangium, then emit their contents separately, leaving a network of empty spore-walls.

DICTYOSPORE. A multicellular spore divided by transverse and longitudinal walls.

DICTYOSTELE. A siphonostele with overlapping leaf-gaps. This gives distinct vascular strands called meristeles. The whole is enclosed by an endodermis.

DICTYOTALES. An order of the Isogeneratae. The erect, flattened thallus grows by an apical cell or a marginal row of apical cells. The gametophyte and sporophyte are identical. The sporangia are unilocular and produce 4-8 aplanospores. The gametophytes are oögamous.

DICYOXYLIC. See *Dictyostele*.

DICYCLIC. In two whorls.

DIDIEREACEAE (H). A family of the Sapindales. These are spiny trees. The sepals are petaloid, persistent, and decussate. The ovule is trilocular with one fertile loculus.

DIDYMOCARPACEAE = GESNERIACEAE.

DIDYMOSPORE. A spore consisting of two cells.

DIDYMOUS. (1) In pairs.
(2) Of a fruit composed of two similar parts, slightly attached along the edge.

DIDYNAMOUS. Having two stamens longer than the rest.

DIFFERENTIAL AFFINITY. The failure of two chromosomes to pair at meiosis in the presence of a third, although they pair in its absence.

DIFFERENTIAL HOST. A particular species, or variety of plant whose reaction is used for determining physiological races of parasites.

DIFFERENTIAL SEGMENT. A block of genes in respect of which two pairing chromosomes differ, in a permanent hybrid, in contrast to a *pairing segment* where they pair and cross-over, and are therefore homologous.

DIFFERENTIATION. The process whereby a cell divides to give rise to dissimilar tissues within an individual. The process is repeated by heredity, and is therefore genetically determined.

DIFFLUENT. (1) Disintegrating in water.
(2) Readily becoming fluid.

DIFFORMED, DIFFORM. Of unusual or irregular shape.

DIFFRACT. Said of a surface divided into areolae.

DIFFUSE. (1) Said of a prostrate stem that branches freely and loosely, spreading over a wide area.
(2) Said of parenchymatous cells scattered throughout the xylem.

DIFFUSE GROWTH. The growth of an algal thallus by the division of any of its cells.

DIFFUSE NUCLEUS. The chromatidia sometimes present in non-nucleated cells.

DIFFUSE POROUS. When the vessels are scattered evenly throughout the xylem, or when there is little difference in the size of the vessels formed in different seasons.

DIFFUSE STIMULUS. A stimulus not coming from any fixed position.

DIFFUSE TISSUE. A tissue of single or small groups of cells distributed throughout a distinct tissue.

DIFFUSION. The movement of a substance from a high concentration to a low concentration by the activity of its own molecules or ions. In a solution, both the solute and solvent move, so that it is the *net diffusion* that is measured. *Facultative diffusion* takes place if the membrane through which it is occurring contains substances with which the penetrating particles can combine reversibly to form a more soluble complex.

DIFFUSION CARRIER. A substance to which a dissolved particle becomes attached on one side of a membrane, moving with the dissolved particle through the membrane, and then depositing it on the other side.

DIFFUSION COEFFICIENT. See *Fick's Law.*

DIFFUSION PRESSURE DEFICIT. The osmotic pressure of a solution particularly when comparing it with that of another solution, *e.g.* a comparison of the osmotic pressure of the cell-sap, compared with that of the surrounding medium.

DIFFRACT. Cracked into small areas.

DIGENIC. Of hereditary difference determined by two genes.

DIGESTIVE CELL. One of the cells in the cortex of a root in which the hyphae of an endotrophic fungus are killed and digested.

DIGESTIVE GLAND. A glandular hair characteristic of carnivorous plants, producing enzymes which digest their prey.

DIGESTIVE POUCH. A layer of cells at the apex of a lateral root. They secrete enzymes which help to break down the cortical cells of the parent root, as the lateral root grows through it.

DIGITALIFORM, DIGITATE. Diverging from a central point, like the fingers on a hand.

DIGYNOUS. Having two carpels.

DIHYBRID, DIHETEROZYGOTE. Heterozygous in respect of two genes.

DIHYDRIC ALCOHOL. An alcohol containing two hydroxyl groups, *e.g.*

$$HO—CH_2$$
$$|$$
$$HO—CH_2$$

ethylene glycol

DIHYDROXYACETONE. A triose sugar, with the properties of a ketone.

$$CH_2OH$$
$$|$$
$$CO$$
$$|$$
$$CH_2OH$$

DIKARYON. A fungus hyphae whose cells contain two haploid nuclei that divide simultaneously.

DIKARYOPHASE. A diploid phase having a dikaryon.

DIKONTAN. Having two flagella.

DILACERATE. As if torn into strips.

DILATED. Expanded and flattened.

DILLENIACEAE (EP, BH, H). A family of the Parietales (EP), Ranales (BH), Dilleniales (H). (BH) include the Crossosomaceae. These are woody plants (many lianas), with alternate, usually leathery leaves. When stipules are present they are winged and adnate to the petiole. The flowers are in cymes or solitary. The bisexual flowers are usually regular, with many-5-4-3 free sepals, spirally arranged and imbricate. There are usually 5 free imbricate petals. There are many stamens (rarely 10 or less) which are hypogynous, free, or united at the base. The anthers are usually adnate and versatile. The superior ovary consists of many to 1 free or slightly united carpels. The styles are usually free. There are many to 1 ascending, anatropous ovules with ventral raphes. The seeds are often arrilate, with copious endosperm and a straight embryo.

DILLENIALES (H). See *Dilleniaceae*. An order of the Lignosae.

DILUTED. Pale, and faint coloured.

DIMER. A polymer of two components.

DIMEROUS. Having two members in a whorl.

DIMIDIATE. (1) Lop-sided.
(2) Hood-shaped.
(3) Appearing to lack one half.
(4) Of an anther which is lop-sided due to the absence or abortion of a lobe.
(5) Of a pileus, with one side larger than the other.
(6) Of a lichen perithecium, when the upper side only is enclosed in a wall.

DIMITITIC. Having two different forms of hyphae.

DIMONOECIOUS. Having bisexual flowers as well as, male, female, and neuter ones.

DIMORPHIC, DIMORPHOUS. Having two distinct forms.

DIMORPHIC HETEROSTYLY. When the flowers of the same species have styles of two different lengths.

DINOCAPSALES. An order of the Dinophyceae. Its members are palmelloid, with a temporary motile gymnodinoid stage.

DINOCOCCALES. An order of the Dinophyceae. There is no vegetative cell-division, the new cells being produced by zoospores or aplanospores.

DINOFLAGELLATES. The mobile Dinophyceae.

DINOPHYCEAE. A class of the Algae. The cells have many golden-brown or chocolate-brown discoid chromatophores, with or without pyrenoids. The chromatophores may be absent. The food-reserve is oil or starch. The cell-wall is of cellulose, and in the motile genera, is of articulated plates. Motile cells are formed. The zoospores have a transverse groove (girdle) with two flagella in, or near it. The motile genera reproduce by vegetative division or aplanospores (rarely sexually), and the immobile genera by zoospores or aplanospores.

DINOPHYSIDALES. An order of the Dinophyceae. They are motile, with the wall of a definite number of plates arranged in a specific way, and vertically differentiated into two opposed halves (valves).

DINOSTRICHALES. An order of the Dinophyceae. The immobile cells are more-or-less cylindrical, and joined into branching filaments.

DINUCLEOTIDE. A unit derived from deoxyribose, linked through a phosphate radical. Some are important as prosthetic groups of enzymic globular proteins.

DIOECIOUS. Having male and female sex organs on separate individuals.

DIOSCOREACEAE (EP, BH, H). A family of the Liliiflorae (EP), Epigynae (BH), Dioscoreales (H). These are climbing herbs or shrubs with tuberous roots, or rhizomes. The alternate leaves are net-veined, and often arrow-shaped. The unisexual flowers are regular and in racemes. The perianth has 6 fused lobes. There are 6 free stamens, or 3 stamens and 3 staminodes. The inferior ovary is of 3 fused carpels, with 3 loculi and axile placentation, or unilocular with parietal placentation. There are usually 2 anatropous ovules in each loculus. The fruit is a berry or capsule. The fruit and seeds are often winged.

DIOSCOREALES (H). An order of the Corolliferae. These are herbs or climbers with rhizomes or root tubers. The leaves are alternate, rarely opposite, mostly cordate or ovate, but sometimes digitate. The veins are prominent, and net-like. The flowers are small, bisexual or unisexual, with the perianth segments mostly united, and white or pale-coloured. There are 6-3 stamens. The ovary is usually inferior (rarely superior) and of 3 loculi (rarely 1). The fruit is a capsule (rarely indehiscent or baccate). The seeds are often winged, and contain endosperm.

DIOSE. A carbohydrate of two CH_2O units, *e.g.* $C_2H_4O_2$. *E.g.* $CH_2OH.CHO$, (glycollic aldehyde).

DIPENTODONTACEAE (H). A family of the Olacales. The stipules are small, sometimes very caducous. The sepals and petals are like each other, and there are disk-glands opposite the valvate petals. The flowers are in globose umbels. The ovary has free-basal placentation. There are no staminodes.

DIPEPTIDE. A compound of two amino acid residues, joined by a peptide link.

$$R—CH—CO—NH—CH—R_1$$
$$\quad\;\; NH_2 \qquad\qquad\quad COOH$$

DIPHOSPHOPYRIDINE NUCLEOTIDE (DPN) (COENZYME 1). A coenzyme which is oxidized and reduced during the early stages of respiration. The pyridine ring is the active group. It oxydizes glyceraldehyde-3-phosphate to 3-phosphoroglyceric acid by the removal of hydrogen.

Nicotamide mononucleotide (NMN)

Adeninemononucleotide (adenylic acid).

DIPHOTIC. Having two surfaces unequally lighted.

DIPHYLETIC. Descending from two distinct ancestral groups.

DIPICOLINIC ACID. This is synthesized during the sporulation of bacteria. It is released during germination and apparently is associated with calcium metabolism. Its loss is associated also with the loss of heat resistance.

DIPLANETISM. Producing zoospores of two morphological types.

DIPLECOLOBOUS. Said of an embryo with an incumbent radicle, and with cotyledons folded twice or more.

DIPLOBIONT. A plant which has two different kinds of individual in its life-cycle. If the species is dioecious, there will be three kinds.

DIPLOBIONTAE. A sub-class of the Ascomycetes, in which the diploid generation is as important as the haploid.

DIPLOCAULESCENT. Having a main axis with branches on it.

DIPLOCHLAMYDEOUS = DICHLAMYDEOUS.

DIPLOCHLAMYDEOUS CHIMAERA. A periclinal chimaera consisting of an outer skin of one constituent, and two cells thick, surrounding a core of the other constituent.

DIPLOCHROMOSOME. A chromosome which has divided twice instead of once since the preceding mitosis, the centromere being undivided.

DIPLOCOCCUS. A coccus in which the individuals tend to form pairs.

DIPLODESMIC. Having two parallel vascular systems.

DIPLODIZATION, DIPLOIDIZATION. The conversion of a mycelium containing uninucleate cells into one containing binucleate ones.

DIPLOHAPLONT. Having a life-cycle in which a many-celled haploid generation alternates with a similar diploid one.

DIPLOID. (1) The number of chromosomes found in the zygote.
(2) An organism having two sets of chromosomes.

DIPLOID APOGAMY = EUAPOGAMY. The development of a sporophyte containing diploid nuclei from one or more cells of the gametophyte, without any preliminary fusion of gametes.

DIPLOIDIZATION. The process of fusion of hyphae, followed by the division of nuclei in pairing fungi, whereby haploid cells or mycelia become diploid.

DIPLOKARYOTIC. Having twice the normal diploid number of chromosomes.

DIPLONEMA. (1) The stage in the meiotic division at which the chromosomes are clearly double.
(2) A small diploid plantlet in the life-cycle of some of the Phaeophyta.

DIPLONT. (1) An organism at the diploid stage of the life-cycle.
(2) Having the alternation of a one-celled haploid generation with a many-celled (or coenocytic) diploid one.

DIPLOPHASE. (1) The diploid stage in the life-cycle of plants.
(2) The stage in the life-cycle of the Basidiomycetes when the cells have two nuclei.

DIPLOSIS. The doubling of the chromosome number.

DIPLOSOME. A paired heterochromosome; a double centrosome lying in the cytoplasm.

DIPLOSTEMONOUS. Having two whorls of stamens, each containing the same number as there are petals, and the outer whorl alternating with the petals.

DIPLOTENE. The stage in meiosis when the chromatids become obvious, and the homologous chromosomes begin to separate.

DIPOLAR ION. An ion having both positive and negative charges, *e.g.* on an amino acid in solution.

DIPSACACEAE (EP, BH, H). A family of the Aggregatae (EP), Asterales (BH), Valerianales (H). They are mostly herbs, with opposite exstipulate leaves (some verticillate). The flowers are in cymes or heads. There is often an involucre of 2 united bracteoles beneath the calyx. There are 5 or 4 sepals, and a similar number of petals, which are imbricate. The 4 free stamens are epipetalous. The inferior ovary is of 2 fused carpels, making 1 loculus, and containing 1 pendulous anatropous ovule. The fruit is an achene, usually enclosed in the epicalyx. The seeds are endospermous.

DIPTERIDACEAE. A family of the Leptosporangiatae. The sporangia are borne in sori, and develop simultaneously. There is no indusium, but a mass of capitate hairs. The dehiscence of the sporangium is more or less transverse and is brought about by means of an obliquely vertical annulus.

DIPTEROCARPACEAE (EP, BH, H). A family of the Parietales (EP), Guttiferales (BH), Ochnales (H). (BH) include the Ancistrocladales. These are tall, little-branched trees, which are evergreen with entire stipulate leaves. There are resin canals. The bisexual flowers are regular and borne in racemes. There are 5 free sepals. There are the same number of free petals, which are convoluted, and often connate at the base. There are many, 15-10-5 free stamens, with the anther connectives produced at the apex. The superior ovary is of 3 fused carpels, and 3 loculi, each with many to 2 ovules (H states 2 only). The fruit is usually a one-seeded nut, enclosed in the calyx which is usually enlarged to form wings. The seeds have no endosperm.

DIRACHMACEAE (H). A family of the Tiliales. There are 8 free stamens opposite the petals, and the anthers are bilocular. The ovary has 8 loculi, and is very deeply laterally lobed.

DIRECT ADAPTATION. Any adaptation which does not appear to stand in relation to natural selection.

DIRECT GERMINATION. The germination of a spore to produce a hypha or filament, rather than a sporangium.

DIRECT NUCLEAR DIVISION = AMITOSIS.

DIRECTIVE MOVEMENT. A movement of orientation.

DISACCHARIDE. A carbohydrate of the formula $C_{12}H_{22}O_{11}$, *i.e.* the condensation product of two monosaccharide molecules

$$2C_6H_{12}O_6 = C_{12}H_{22}O_{11} + H_2O.$$

DISARTICULATE. To separate at a joint.

DISC, DISK. (1) An outgrowth from the receptacle of a flower, beneath the carpels or stamens.
(2) The central part of a capitulum.
(3) The portion of an apothecium that bears the asci and paraphyses.

DISCELLACEAE = EXCIPULACEAE.

DISCIFLORAE (BH). A group of dicotyledons, having free petals, a superior ovary, and a disk.

DISC(K) FLORET. A regular, tubular flower, found in the centre of a capitulum when it contains two types of flowers.

DISCIFORM. Round and flat.

DISCOCARP = APOTHECIUM.

DISCOID. (1) = DISCIFORM.

(2) Said of a capitulum with no ray florets.
(3) Said of an algal thallus which is one-cell thick and closely applied to the substratum.

DISCOLICHENS. The group of lichens in which the fungus is a Discomycete.

DISCOLOR, DISCOLOROUS. Not the same colour throughout.

DISCOMYCETES. A subclass, or order, of the Ascomycetes in which the fruiting body is an apothecium.

DISCONTINUOUS DISTRIBUTION. The occurrence of a species in one area, then in another area far from it, but not in any areas between.

DISCONTINUOUS VARIATION. A rare variation; sport; saltation.

DISCOPLANKTON. Plankton in which the cells form thin disks.

DISCRETE. Remaining separate.

DISCRIMENANT FUNCTION. A linear compound of a series of variates, obtained by maximizing the differences between the classes relative to the differences between individuals in the classes, for the particular factor being measured.

DISCUS. The hymenium of an apothecium.

DISJUNCTION. The separation of chromosomes at anaphase.

DISJUNCTOR. A small cell between two neighbouring conidia in a chain. It disintegrates, thus aiding in dispersion.

DISK = DISC.

DISLOCATED SEGMENTS. Homologous pairs of segments, differing in their linear sequence with other segments.

DISOME = BIVALENT.

DISOMIC. Relating to two homologous chromosomes or genes.

DISOMIC INHERITANCE. Arising from the determinate association of chromosomes in bivalents at meiosis.

DISPERSAL. The establishment of individuals in a new area.

DISPIREME. The stage of telophase in which the spireme thread of each daughter nucleus has been formed.

DISSECTED. Cut deeply into narrow lobes.

DISSEMINATE. Scattered.

DISSEMINATION. The spread of a species, usually by spores or seeds.

DISSEMINULE. Any part of a plant that serves for dissemination.

DISSEPIMENT. (1) The wall dividing the loculi of a syncarpous ovary.
(2) See *Trama*.

DISSIMILATION = RESPIRATION.

DISSOCIATION. A mutation or saltation.

DISSOCIATION CONSTANT (K). The extent to which a substance ionizes in solution. When a substance HA in solution ionizes, the free ions and the undissolved molecules are in dynamic equilibrium, thus, $HA \rightleftharpoons H^+ + A^-$.

$$K = \frac{(\text{Concentration } H^+) \times (\text{Concentration } A^-)}{(\text{Concentration } HA)}$$

DISTAL. (1) Farthest from the point of attachment.
(2) Of the part of a chromosome arm, which is further from the centromere than another part.

DISTANT. Widely spaced.

DISTICHOUS. Arranged in two vertical rows.

DISTRIBUTION. (1) = DISSEMINATION.
(2) The occurrence of a species, considered from a geographical point of view.

DISULPHUR (DITHIOL) LINKAGE or BRIDGE. The linkage between two sulphur atoms in a protein molecule.

$$H\overset{/}{C}—CH_2—S—S—CH_2—\overset{/}{C}H$$

DITHALLIC. Said of a mycelium formed from the union of material from two distinct strains.

DITHECAL. Said of an anther with two loculi.

DITHEOUS. Having two chambers.

DIURNAL. Happening during the day, or every day.

DIVARIACTE. Widely spreading; forked.

DIVERGENS. Latin meaning 'separating'.

DIVERGENT, DIVERGING. Said of two or more organs, which spread, so that they are farther apart at the top than the base.

DIVERTICULUM. A pocket.

DIVISION. A major group of plants.

DIVISURAL LINE. The line along which the peristome teeth of a moss split.

DNA = DESOXYRIBOSE NUCLEIC ACID.

DOLABRIFORM, DORABRATE. Shaped like a pick-axe.

DOLIFORM. Barrel-shaped.

DOMATIUM. A cavity formed by a plant in which live insects or mites, apparently in symbiosis with the plant.

DOME. The growing point of the receptacle of a flower.

DOME CELL. The apical cell of the developing antheridium of some ferns. It divides to form the cap cell and the secondary cell.

DOMINANCE. (1) The relationship of two allelomorphs where the single gene heterozygote resembles one of the two homozygous parents, in phenotype, rather than the other.
(2) The prevalence of one or a few species in a community, influencing its general appearance, and the other plants in the community.

DOMINANCE MODIFIER. A gene which modified the extent of dominance of another gene.

DONATIACEAE (H). A family of the Saxifragales (H). The petals are free, or absent. The anthers are bilocular, and there are no staminodes. The carpels are fused to form a superior or inferior ovary. The styles are free, and the fruit is indehiscent.

DORMANCY. A state when the metabolic processes are slowed. This applies especially to respiration.

DORSAL. Upper.

DORSIFIXED. (1) = ADNATE.
(2) Said of an anther, when the filament is attached firmly to one point on the dorsal surface.

DORSIVENTRAL. Said of a flattened plant member, when the upper and lower sides are structurally different.

DOTHIDIACEAE. The single family of the Dothidiales.

DOTHIDIACEOUS. Having the asci in cavities in the stroma.

DOTHIDIALES. An order of the Euascomycetae. They are usually parasitic on leaves. The fruit-body is a perithecium, embedded in the mycelium, from which the periderm is not distinct.

DOTHIORACEAE. A transitional family of the Pseudosphaeriales, having one-celled hyaline spores.

DOUBLE FERTILIZATION. This takes place in the embryo-sac of angiosperms. One male nucleus fuses with the egg-nucleus, and the other fuses with the polar nuclei, before, or after they themselves have fused.

DOUBLE FLOWER. A flower having more than the usual number of petals.

DOUBLE FLOWERING. The abnormal production of flowers in the spring, and again in the autumn.

DOUBLE HETEROZYGOTE. A heterozygote in respect of two genes.

DOUBLE RECESSIVE. A diploid homozygote for a recessive gene.

DOUBLE REDUCTION = NON-DISJUNCTION.

DOWN. A fine coating of soft hairs.

DPN = DIPHOSPHOPYRIDINE NUCLEOTIDE.

DREPANIUM. A monochasial cyme in which all the branches arise from the same side of the relatively main axis.

DRIFT. Changes in the aggregate of genotypes in a small population resulting from the random extinction of allelomorphs of genes in regard to which the population was heterozygous.

DRIODIUM. A dry thicket formation.

DRIP TIP. An elongation of the tip of a leaf, which possibly helps in the shedding of water from the surface.

DROPPER. A young immature bulb.

DROSERACEAE (EP, BH, H). A family of the Sarraceniales (EP, H), Rosales (BH). These are insectivorous, perennial herbs, mostly with sticky glandular hairs on the leaves. The regular bisexual flowers are usually in cincinni, rarely in racemes, or solitary. The 5 calyx lobes are fused; the 5 petals are free and imbricate. There are usually 5 free stamens, ((H) states 20-4). The anthers are extrorse, and the pollen in tetrads. The superior ovary is unilocular, consisting of 5, 3, or 2 fused carpels. The placentation is usually parietal, rarely axial, or free-central. The styles are usually free. There are many to 3 anatropous ovules. The seeds have endosperm and small basal embryos. The fruit is a capsule.

DRUPACEOUS. Like a drupe.

DRUPE. A fleshy fruit, with a thin epicarp, a fleshy mesocarp, and a hard endocarp, containing a single seed. The seed and endocarp form the stone, *e.g.* a plum.

DRUSE. Globose, spiked crystals of calcium oxalate, formed around a organic material, and found in the cortex, pith, and phloem.

DRY SPORE. A spore that separates from the cell producing it, without any slime.

DUAL PHENOMENON = HETEROKARYOSIS. The association of genetically unlike nuclei within a single mycelium.

DUFF. Acid humus, typically formed in free-draining acid soils of low calcium concentration.

DULCIS. Latin meaning 'sweet'.

DUMOSE. Shrubby.

DUPLEX. (1) In two layers, one harder than the other.
(2) The condition of a polyploid when a particular dominant allelomorph is represented twice.

DUPLEX GROUP. The diploid complement of factors and chromosome.

DUPLICATE GENES. Genes which have an identical, but non-accumulative effect.

DUPLICATION. The union of a fragment of a chromosome with a whole chromosome of the same sort.

DURAMEN. An archaic term for heartwood.

DUVET. A soft, thick layer of hyphae, like brushed up cloth.

DY. A type of lake-bottom deposit largely composed of plant detritus, and having a marked effect on the fauna.

DYAD. A pair of cells formed at meiosis, instead of a tetrad.

DYSGENIC. Referring to any factor which tends to decrease the fitness of the race.

DYSTROPHIC. Said of a lake habitat in which the iron and humic acids in the water reduce the dissolved oxygen content.

DYSTROPHY. Insects removing the nectar from a flower by some abnormal method, and consequently not operating the pollinating mechanism.

E

E-. A Latin prefix meaning 'without' or 'out of'.

EBENACEAE (EP, BH, H). A family of the Ebenales (EP, BH, H). These are trees or shrubs with alternate, opposite, or whorled leaves, which are usually entire. The flowers are axillary, solitary, or in cymes, and are regular. They are usually bisexual, with the parts in threes-to-sevens. The calyx lobes are fused and persistent. The corolla lobes are fused and convolute. The stamens are epipetalous, and staminodes are usually present in the female flowers. The ovary is superior and of fused carpels. There are 16-2 loculi, each with 2-1 anatropous, pendulous ovules. The fruits are usually berries, but sometimes dehiscent. The embryo is straight or slightly curved, in abundant cartilagenous endosperm.

EBENALES (EP, BH, H). An order of the Sympetalae (EP, BH), Lignosae (H). These are trees or shrubs with mostly simple, alternate leaves without stipules. The flowers are regular, bisexual or unisexual. The petals are imbricate and fused. The stamens are epipetalous or rarely hypogynous, usually twice as many as the corolla lobes. Petalloid staminodes are often present. The anthers open lengthwise. The ovary is superior, with axile placentation. The seeds contain endosperm.

EBENEOUS. Black as ebony.

EBRACTEATE. Without bracts.

EBRACTEOLATE. Without bracteoles.

ECAD. A plant form which is assumed to be adapted to the habitat.

ECALCARATE. Not spurred.

ECCENTRIC, EXCENTRIC. (1) Situated to one side.
(2) Having fatty droplets lying to one side of a globular structure.

ECCRINACEAE. A family of the Eccrinales. There are no amoeboid spores, and the mature resting-spore has four nuclei.

ECCRINALES. An order of the Phycomyceteae. They are internal or external parasites on Arthropods. The hyphae are unbranched, or branched only slightly, and attached by a holdfast of callose. Macrospores, microspores, or resting spores are produced by the transverse segmentation of the thallus. Thick-walled zygotes are produced by nuclear fusion within the filament.

ECESIS. The invasion of an area by a species which is unable to establish itself there, and consequently dies out in a few generations.

ECHARD. The water present in the soil, which is unavailable to plants.

ECHIN-. A Greek prefix meaning 'spiny'.

ECHINATE. Having an even covering of rather long, pointed bristles or outgrowths.

ECHINULATE. Slightly echinate.

ECOLOGICAL. Referring to the environment of an organ or organism.

ECOLOGICAL FACTOR. Anything in the environment which affects the growth, development and distribution of plants, and hence determines the characteristics of the plant community.

ECOLOGY. The study of organisms in relation to their environment.

ECOPHENE. The range of phenotypes produced by one genotype within a particular habitat. One genotype may be taken to mean a class of closely related genotypes.

ECORTICATE. Lacking a cortex, especially of seaweeds.

ECOSPECIES. A category of variant individuals capable of genetic recombination with other similar groups, but liable to reduce fitness or fertility from it.

ECOSYSTEM. The complete ecological system of an area, including the plant, animals, and the environmental factors.

ECOTONE. A boundary between two plant communities of major rank.

ECOTYPE = ECOSPECIES.

ECRUSTACEOUS. Said of a lichen with no well-defined thallus.

ECT-, ECTO-. A Greek prefix meaning 'outside'.

ECTAL LAYER. A thin membrane at the extreme edge of an excipulum.

ECTOASCUS. An outer ascus.

ECTOCARPACEAE. The single family of the Ectocarpales.

ECTOCARPALES. An order of the Isogeneratae. The branched filamentous thallus has trichothallic growth. The reproductive organs are single in uniseriate rows. The sporophyte forms zoospores or neutral spores, and the gametophyte produces isogametes.

ECTOGENESIS. Variation due to conditions outside the plant.

ECTOGENIC. (1) Of bacteria, living on the outside of a body.
(2) Describes the effect of pollen on the tissues of the female organs of a flowering plant.

ECTOPARASITE. When referring to a plant, a parasite which feeds from inside the host, but the bulk of the plant-body, and reproductive organs are outside the host.

ECTOPHLOCODAL. Living on the outside of bark.

ECTOPHLOIC. Said of a vascular bundle that has phloem only on the outside of the xylem.

ECTOPHYTE. A parasite growing on the surface of its host.

ECTOPLACODIAL. Said of the ostiolar disk, when it is formed from the ectostroma.

ECTOPLASM, ECTOPLAST. The outer, clear cytoplasm lying against the cell-wall.

ECTOSARC = ECTOPLASM.

ECTOSPORE. (1) A spore developed on the outside of a hyphae.
(2) An obsolete term for a basidiospore.

ECTOSTROMA. The part of the stroma formed on the surface of the bark, beneath, or within the periderm. It is typically of fungus tissue only.

ECTOTHECAL. Of ascomycetes, not having the hymenium covered.

ECTOTROPHIC INFECTION. (1) Of root-infecting fungi, growing on the surface of the root, but sending infection hyphae into the host tissue below.
(2) Hyphae running between the epidermal cells.

ECTOTROPHIC MYCORRHIZA. A mycorrhizal association where the bulk of the fungus covers the root, with hyphae growing into the intercellular spaces.

ECTROTROPIC. Curving outwards.

ECTROGELLACEAE. A family of the Lagenidiales, of doubtful validity.

EDAPHIC CLIMAX. A climax community which is determined by some property of the soil.

EDAPHIC FACTOR. Any soil factor which affects the organisms growing on or in it.

EDAPHON. The plants and animals living in a soil, and affecting its nitrogen content.

EDEMA = OEDEMA.

EDESTIN. A protein found in hemp seed, having a molecular weight of about 309,000.

EDGED. Having a margin of a different colour from the rest.

EDULIS. Latin meaning 'edible'.

EFFETE. Functionless due to ageing.

EFFICIENCY INDEX. The rate at which dry matter accumulates in plants.

EFFIGURATE. Having a definite shape.

EFFIGURATIONS. Outgrowths from the receptacle.

EFFLORESCENCE. (1) The production of flowers.
(2) The period of flowering.

EFFUSE. Spread out on a substrate, and often having an ill-defined edge.

EGG APPARATUS. The functional female nucleus, and the two synergidae in the embryo-sac of an angiosperm.

EHRETIACEAE (H). A family of the Verbenales. The leaves are alternate. The corolla is regular, and there are as many stamens as corolla lobes.

EISODAL APERTURE. The enlargement of the stomatal pore nearest to the surface of the leaf.

EJACULATION, EJECTION. The forcible expulsion of spores from a sporangium, or seeds from a fruit.

EKTEXINE. The outer layer of the exine of a pollen grain. It usually has projections on it.

EKTODYNAMORPHOUS SOIL. A soil whose characteristics are primarily affected by external factors.

ELAEAGNAEAE (EP, BH, H). A family of the Myrtiflorae (EP), Daphnales (BH), Rhamnales (H). These are much branched thorny shrubs. The leaves are entire, opposite or alternate, and leathery. They are covered with scaly hairs. The flowers are unisexual or bisexual, in racemes, and the flower parts are in twos or fours. In the male flower, the receptacle is often flat, but in the female or bisexual flower it is tubular. The calyx lobes are valvate or open in the bud. There are no petals. There are as many, or twice as many stamens as sepals. The superior ovary is of 1 carpel, with 1 erect anatropous ovule. The fruit is a drupe, and the seed has little or no endosperm.

ELAEOCARPACEAE (EP). A family of the Malvales (EP). (BH) include it in the Tiliaceae. These are trees or shrubs with alternate or opposite stipulate leaves. The flowers are in racemes, panicles, or dichasia. A disk is usually present. The calyx is valvate and of 5 or 4 free or united sepals. The 5 or 4 petals are valvate or imbricate, usually free, and often absent. There are many free stamens on the disk. The anthers are bilocular, usually opening by 2 pores. The superior ovary consists of many to 2 (rarely 1) free carpels, each with many or 2 anatropous pendulous ovules, with a ventral raphe. The fruit is a capsule or drupe, and the seeds contain a straight embryo in abundant endosperm.

ELAIOPLAST. A plastid which contains stored oil.

ELAISOME. An outgrowth from the surface of a seed, containing fats or oils. They are sometimes attractive to ants, hence aiding in seed-dispersal.

ELAPHOMYCETACEAE. A family of the Aspergillales. These fungi are subterranean, and are probably saprophytic, but may form mycorrhiza. The ascocarp is large, being 2-3 cm in diameter. The periderm is thick and roughened. It contains ascogenous hyphae, asci, and thin-walled cells. All these disintegrate to release the ascospores.

ELATER. (1) In bryophytes, an elongated, spirally thickened cell derived from the sporogenous tissue, and sometimes aiding in spore-dispersal.

(2) One of the four strips derived from the outer layer of the spore wall of *Equisetum*. It is hygroscopic, but its function is obscure.

(3) A spirally thickened structure in the fruit-body of some Lycoperdales.

(4) A free capillitium-thread, *e.g.* in the Myxomycetes.

ELATEROPHORE. (1) A tissue that bears the elaters.

(2) A central column-like mass of sterile tissue in the sporangium of the Jungermanniales.

ELATINACEAE (EP, BH, H). A family of the Parietales (EP), Guttiferales (BH), Caryophyllales (H). These are undershrubs, herbs, or annual water-plants. The leaves are opposite, or whorled, simple with interpetiolar stipules. The bisexual flowers are regular, solitary or dichasia. The flower parts are in whorls of 6-2. The sepals are free or united, and the petals are imbricate. The stamens are in 1 or 2 whorls, being the same number, or twice as many as the sepals. When in one whorl they are opposite the sepals. The ovary is syncarpous and superior, with free styles, and axile placentation. There are many anatropous ovules. The fruit is a capsule, which breaks into seven. The seed is straight or curved, with little or no endosperm.

ELATINEAE (BH) = ELATINACEAE.

ELECTRO-CULTURE. The stimulation of growth of plants by electrical means.

ELECTROLYTE. A substance that ionizes in solution, *i.e.* conducts electricity, when in solution, by the movement of ions.

ELECTRON TRANSFER. The carrying of an electron from one substrate to another, involving the oxidation of one and the reduction of the other, *e.g.* HA + B = HB + A. The enzyme bringing about such a transfer is termed an *electron carrier*.

ELECTROPHORESIS. The migration of charged particles in an electric field· This method is used widely in the separation of the components of a large variety of mixtures.

ELEUTHEROPETALOUS. Polypetalous.

ELLIPTICAL. Narrow, and tapering at each end.

ELONGATE. Drawn out.

ELONGATION STAGE. The period of growth when an organ increases in length, but before secondary thickening begins.

ELUVIATION. The translocation of soil material, either mechanically, or in solution.

EMARGINATE. (1) Lacking a distinct margin.
(2) Having a slight notch at the tip.
(3) As if scooped out at the point of attachment.

EMASCULATE. To remove the anthers from an unopened flower.

EMBDEN-MEYERHOF-PARNAS SEQUENCE. The second phase in glycolysis during aerobic respiration. A molecule of hexose diphosphate is formed into two molecules of triose phosphate. Two energy-rich phosphates are released. ADP is converted to ATP and DPN is reduced, with the formation of pyruvic acid:—

$$C_6H_{12}O_6 - (2P) + 2DPN + 4ADP = 2CH_3CO.COOH + 2DPNH_2 + 4ATP.$$

EMBOSSED. Umbonate.

EMBRYO. A young plant developing from an egg-cell, after fertilization, or without fertilization.

EMBRYO SAC. The female gametophyte (megaspore) of angiosperms. It contains several nuclei derived from the megaspore nucleus, one of which is the egg-nucleus. Two other nuclei fuse to form the primary endosperm nucleus.

EMBRYOGENY. The processes leading to the formation of an embryo.

EMBRYOLOGY. The study of the formation and development of embryos.

EMBRYONIC TISSUE. A tissue made up of actively dividing cells.

EMBRYOPHYTE. A plant which forms an embryo.

EMBRYOSTEGA. A disk-like callus on a seed.

EMBRYOTEGA = EMBRYOSTEGA.

EMERGENCE. An outgrowth of the epidermis and cortex, having no vascular tissue, and not developing into stem, root or leaf.

EMERSED. (1) Amphibious.
(2) Protruding upwards.

EMPETRACEAE (EP, BH, H). A family of the Sapindales (EP), Monochlamydeae (anomalous) (BH), Celastrales (H). These are small shrubs with tough, rolled leaves. There are no stipules, and no disk. The flowers are in racemes, and are usually dioecious, usually being borne on 'short shoots'. There are 3 free calyx lobes, and 3 free petals (H states no petals). The 3 stamens are free, and the ovary consists of 9-2 fused carpels, and is superior.

Each carpel forms a loculus containing one erect, anatropous (or nearly campylotropous) ovule each. The placentation is axile, and there is a ventral raphe. The fruit is a drupe with 9-2 stones. The seed contains endosperm.

EMPTY GLUME = STERILE GLUME.

EMULSIN. An enzyme that breaks down glucosides to give a sugar and an aromatic group.

EMULSION. A colloidal suspension of one liquid in another.

ENATIOSTYLOUS. Said of a flower which has the style(s) projecting from one side.

ENATION. (1) An outgrowth.
(2) Outgrowth from the underside of the veins of a leaf, caused by a virus infection.

ENCHYLEMA. The more fluid constituents of the cytoplasm.

ENCRUSTATION = ADCRUSTATION.

ENCYSTMENT. The formation of a walled, non-motile body from a swimming spore.

ENDARCH. The direction of differentiation of the xylem, when it is cut off from the cambium towards the centre of the axis. The protoxylem is then on the edge nearest the centre of the axis.

ENDECANDROUS. Having eleven stamens.

ENDEMIC. Said of a pest, or disease organism which occurs permanently in an area.

ENDERGONIC REACTION. A reaction that needs energy put into the system for it to take place.

ENDO-. A Greek prefix meaning 'within'.

ENDOASCUS. An inner ascus.

ENDOBASIDIAL. Said of a lichen sporophore, having a secondary sporing branch.

ENDOBASIDIUM. A basidium developing inside the fruit-body.

ENDOBIOTIC. (1) Growing inside another organism.
(2) Formed inside the host cell.

ENDOCARP. The inner layer of a fruit wall (pericarp). It is usually woody.

ENDOCARPOUS. Having the mature hymenium covered over.

ENDOCELLULAR ENZYME. An enzyme which functions within the cell in which it was formed.

ENDOCONIDIUM. An asexual spore of fungi, formed inside a cell or coenocyte, from which it is later extruded.

ENDODERMAL PRESSURE = ROOT PRESSURE.

ENDODERMIS. (1) The cylinder of cells surrounding the vascular bundle of a root. The radial walls are thickened by the Casparian strip, so that substances entering or leaving the vascular bundle have to pass through the protoplasm of the endodermal cells.
(2) A layer of cells delimiting the cortex from the stele.

ENDODYNAMORPHIC SOIL. A soil whose characteristics are mainly determined by the parent material.

ENDOECTOTHRIX. Growing in or on hair.

ENDOENZYME = ENDOCELLULAR ENZYME.

ENDO-FORM. Of the Uredinales where the teliospores are like the aeciospores in form.

ENDOGAMY. (1) In-breeding.
(2) Pollination between two flowers on the same plant.
(3) The union of two sister gametes, both female.

ENDOGELATIN. A gelatinous layer on the inner wall of the macrosporangia of the Fucales. When wet, it swells, forcing out the macrospores.

ENDOGENOUS. Formed inside another organ of the plant.

ENDOGENOUS SPORE. A spore formed within a sporangium.

ENDOGENOUS THALLUS. Said of a lichen thallus, in which the alga predominates.

ENDOGONACEAE. A family of the Mucorales, based on the structure of the sporocarp. They are found as loose to firm sclerotium-like bodies in humus-rich soil, and among mosses. These may contain 'sporangia', chlamydospores, or zygospores, but never all three in the same sporocarp. There are no true sporangia.

ENDOGONIDIUM. A gonidium developing inside a receptacle or gonid angium.

ENDOLITHIC. Growing in the substance of rocks or stones.

ENDOMITOSIS. The doubling of the chromosomes without the division of the nucleus; resulting in polyploidy.

ENDOMYCETACEAE = ENDOMYCETALES. A family of the Endo-mycetales.

ENDOMYCETALES = SACCHAROMYCETALES. An order of the Ascomycetae, in which the zygote develops directly into an ascus. These are always single, and never in an ascocarp. Asexual reproduction is by budding branches, or breaking of the hyphae to form 'oidia'.

ENDOPARASITE. A parasite living inside its host.

ENDOPEPTIDASE = PROTEINASE.

ENDOPERIDIUM. The inner, palisade-like layer of the wall of the fruiting body of the Lycoperdales.

ENDOPHELLODERM. Cells cut off from the inner face of the phelloderm.

ENDOPHYLLOUS. (1) Of a plant member developing in the shelter of a sheathing leaf.
(2) Said of a parasite living inside a leaf.

ENDOPHYTE. A plant living inside another plant, but not a parasite.

ENDOPHYTIC MYCORRHIZA = ENDOTROPHIC MYCORRHIZA.

ENDOPLACODIAL. Having the ostiolar disk developed from the endostroma.

ENDOPLASM, ENDOPLAST. The cytoplasm inside the plasma membrane.

ENDOPLASMIC RETICULUM. The submicroscopic, double-layered membranes, about 100Å thick, consisting of canaliculi and cisternae. They are much-folded in the cytoplasm, and have microsomal particles attached to them; these are usually ribosomes. This association is responsible for the major biosynthesis of the cell.

ENDOPLEURA. The inner seed-coat of the Cycadaceae.

ENDORHIZAL. Monocotyledonous.

ENDORHIZOID. A rhizoid formed at the base of the seta of a bryophyte, and penetrating the gametophyte.

ENDOSAPROPHYTISM. The destruction of the lichen alga by the fungus.

ENDOSCLEROTIUM. A sclerotium of an endogenous spore.

ENDOSCOPIC. Of a plant embryo, which divides transversely at the first division, the inner daughter cell being the embryonic one, and the outer one the suspensor, *i.e.* the embryo points to the base of the archegonium.

ENDOSOME = KARYOSOME.

ENDOSPERM. The nutritive tissue developed in the embryo-sac of flowering plants, from the fusion of one female nucleus with one or more others, or with a male nucleus, or with both. It forms the food-storage tissue of seed. It is absorbed after germination in endospermous seeds, and before the complete development of the embryo in non-endospermous ones.

ENDOSPHAERACEAE. A family of the Chlorococcales. They are unicellular, with large, irregularly-shaped cells. Most are endophytic in marine algae, mosses or angiosperms. The chloroplasts are parietal or central, with many-to-1 pyrenoids. Reproduction is by biflagellate zooids.

ENDOSPORE. (1) A small spore formed in the cells of some Myxophyceae and bacteria. The spore-wall is not fused to the wall of the parent cell.

(2) The innermost layer of the wall of a spore.

ENDOSPOREAE. A sub-class of the Myxomycetae, containing members which form their spores within a sporangium.

ENDOTHECIUM. (1) The inner mass of cells in the early development of the sporangium of the Bryophyta.

(2) The fibrous layer in the wall of an anther.

ENDOTHRIX. Living inside a hair.

ENDOTROPHIC. Having hyphae entering the cells.

ENDOTROPHIC MYCORRHIZA. A mycorrhiza in which the fungus is growing almost completely internally, within the cortex of the root.

ENDOZOIC. (1) Living inside an animal.

(2) The method of seed dispersal, whereby the seeds are swallowed by an animal, and voided unharmed in the faeces.

ENERGID. A volume of protoplasm which is controlled by the enclosed nucleus.

ENGLERULACEAE. A family of the Erisiphales. These are leaf-parasites, whose perithecial cells dissolve into a slime at maturity to expose the asci.

ENLARGEMENT. Primary growth in thickness, before secondary thickening begins.

ENOL FORM. A cyclic molecule with an $-OH$ group on the second carbon atom, *e.g.*

Cytosine

131

ENOLASE. The enzyme catalysing the formation of phosphoenolpyruvic acid from 2-phosphoglyceric acid.

ENDOPHYTOTIC. A plant disease, causing a constant amount of damage from year to year.

ENSATE, ENSIFORM. Sword-shaped.

ENTIRE. Said of the margin of a flattened organ, when it is continuous, not being broken into teeth or lobes.

ENTO-. A Greek prefix meaning 'within'.

ENTOMOGENOUS. Living on, or in insects.

ENTOMOPHILOUS. Pollinated, or having spores distributed by insects.

ENTOMOPHTHORACEAE. The single family of the Entomophthorales.

ENTOMOPHTHORALES. An order of the Phycomycetes, all of which are saprophytes or parasites on insects. The thallus ranges from a coenocyte, in the vegetative condition, to being septate and much-branched, each cell containing a few nuclei. Asexual reproduction is by conidiosporangia, which are usually explosively discharged, and germinate directly. Sexual reproduction is by the apposition of gametangia, and the fusion of equal, or unequal aplanogametes.

ENTOPARASITE. A parasite inside the host.

ENTOPHYLATACEAE. A family of the Chytridiales. Some are saprophytes, and some parasites, mainly on algae. The zoospore cyst becomes emptied, and usually disappears, or remains as an empty cap. This fungus-body enlarges in the host, and sends out rhizoids which may become extensive.

ENTOSTROMA. The part of the stroma within the host plant. It is made up of fungus and host tissue.

ENTOZOIC. Living inside an animal.

ENUCLEATION. The removal of the nucleus of a cell, by manipulation.

ENVIRONMENT. The conditions external, or antecedent to an organism which are related to its development. Its reaction with the genotype determines the phenotype.

ENZYME. An organic catalyst which accelerates a reaction within a cell. All are wholly or partially protein, with, or without a prosthetic group. Most of them are highly specific in their effects, *i.e.* one enzyme will effect only one (or a few) reactions. They will function only in a very limited range of pH and temperature, outside which they are destroyed.

EOCENE. A geological period 60-45 million years ago.

EPACRIDACEAE (EP, H). = EPACRIDEAE (BH). A family of the Ericales (EP, BH, H). These are small trees or shrubs, with small sparingly branched stems. The leaves are narrow, entire, and ridged. They are alternate, rarely opposite or whorled. The flowers are in terminal racemes or spikes, sometimes solitary. They are bisexual and regular, with all the parts in fives. The sepals are free, the petals are fused, and the stamens are epipetalous or hypogynous (hypogynous, H). The anthers open by a central longitudinal slit. The ovary is superior, of 5 fused carpels, opposite the petals, and 5 loculi, with axile placentation, and containing α-1 anatropous, usually pendulous ovules. The fruit is a capsule or drupe. The embryo is straight, and there is copious endosperm.

EPHARMONIC CONVERGENCE. A likeness in external appearance and structure between plants that are not closely related systematically.

EPHEMERAL. A plant which completes its life-cycle in a very short time.

EPHEMERAL MOVEMENT. The movement of a part of a plant which cannot be repeated, *e.g.* opening of a bud.

EPI-. A Greek prefix meaning 'upon'.

EPIASCIDIUM. An abnormal funnel-shaped leaf, with the upper surface of the leaf lining the inside of the funnel.

EPIBASAL CELLS. The upper cells of the embryo of the Bryophyta and Pteridophyta, giving rise to the capsule and part of the seta, or to the stem and cotyledon.

EPIBASIDIUM. Any structure that develops between the hypobasidium and the sterigmata.

EPIBENTHOS. Animals and plants living below the low-tide mark, and above the 100 fathom line.

EPIBIOSES. Living on the surface of water.

EPIBIONTIC. Living on the surface of another organism.

EPIBLAST. A small outgrowth from a grass embryo.

EPICALYX. A series of small bracts growing out from beneath the sepals, and alternating with the sepals.

EPICARP. The outer layer of the pericarp, especially if it can be peeled off as a skin.

EPICORMIC. Said of a branch developing from a dormant bud on the trunk of a tree, which becomes active due to damage to the tree, or some abnormality in the environment.

EPICOTYL. The part of a seedling stem above the cotyledons, but below the first foliage leaves.

EPICTESIS. The ability of living cells to accumulate salts in a higher concentration than that in which they occur in the surrounding solution.

EPIDERMIS. The outer single layer of cells on an organ. The outer wall may be thickened by the production of a cuticle, and the cells may be extended into hairs.

EPIGEAL. (1) Living on the surface of the ground.
(2) Of germinating seeds when the cotyledons are brought above the surface of the ground.

EPIGEAN. Occurring on the ground.

EPIGEIC. Said of a plant having stolons on the surface of the ground.

EPIGENOUS. Said of a leaf fungus which grows on the surface.

EPIGYNAE (BH). The second series of the Monocotyledons. The perianth is partly petaloid; the ovary usually inferior, and the seeds contain abundant endosperm.

EPIGYNOUS. (1) Said of a flower, when the receptacle encloses the carpel(s), so that the other flower parts arise above the carpel(s).
(2) Having the antheridium arising above the oögonium.

EPILITHIC. Growing on the surface of rocks.

EPILITTORAL ZONE. A zone on the coast, next to the zone occupied by plants which cannot withstand exposure to salt.

EPILOBIACEAE = ONAGRACEAE.

EPIMATIUM. The ovuliferous scale of conifers.

EPINASTY. The greater growth of the upper surface of an organ, compared with the lower, causing the organ to bend downwards.

EPIPETALOUS. Growing on the petals.

EPIPHLOEDAL, EPIPHOEDIC. Growing on the surface of bark.

EPIPHRAGM. (1) The membrane stretching across the top of a moss capsule, and is bordered by the peristome teeth.
(2) A membrane over the young fruit-body of the Nidulariaceae.

EPIPHYLLOUS. Growing on a leaf.

EPIPHYSIS. A growth around the hilum of a seed.

EPIPHYTE. A plant growing on another plant, but not deriving any food from it.

EPIPHYTOTIC. An epidemic among plants.

EPIPLASM. The residual cytoplasm left in an ascus after ascospore formation· Initially it is rich in glycogen, but later this is lost.

EPIPTEROUS. Having a wing at the apex.

EPISEPALOUS. Borne on the sepals.

EPISPERM. The outer part of a seed coat.

EPISPORE. The thickened outer coat of a fungus zygote.

EPISTATIC. Said of a character which is dominant to another, to which it is not an allelomorph.

EPISTROMA = ECTOSTROMA.

EPISTROPHA. The positioning of the chloroplasts on the periclinal walls of the palisade cells. This takes place in diffuse light.

EPITHALLINE. (1) Growing on a thallus.
(2) A falsely thalline apothecial edge in lichens.

EPITHALLUS. The upper layer of fungal hyphae in a lichen thallus.

EPITHECA. The outer, and older layer of the half-wall of diatoms and dinoflagellates.

EPITHECIUM. (1) The outer dark-coloured layer of a peridium.
(2) A thin, coloured layer over the asci in an apothecium, particularly in lichens, formed from the tips of the paraphyses.

EPITHELIAL LAYER. A layer of elongated cells, set end-on to the endosperm in a grass fruit, forming the boundary of the scutellum.

EPITHEM. A group of cells in the mesophyll of a leaf, which exude water.

EPITHEM HYDATHODE. A hydathode which is directly connected to the vascular bundle in a leaf.

EPITHET. The second (specific) part of a Latin binomial of a plant. It may also be the third or fourth term.

EPITROPHIC. (1) Having buds on the upper side.
(2) Growing more on the upper than lower side.

EPIXYLOUS. Growing on wood.

EPIZOIC. Growing on a living animal.

EQUAL. Not lop-sided.

EQUALLING. The condition when the tips of several organs are at the same level, but the individual organs are of different size.

EQUINOCTIAL. Said of plant bearing flowers that open and close at definite times.

EQUISETACEAE. A family of the Equisetales. There is one living genus, *Equisetum*. There is no secondary thickening, and no whorls of distinct bracts in the strobilus.

EQUISETALES. An order of the Equisetinae. The stem has longitudinal ridges down the internodes, with whorls (generally) of branches at the nodes, where there are also whorls of scale-like leaves. There is an endarch siphonostele, which is entire at the nodes, but perforated at the internodes. Some genera have secondary thickening. The strobili are whorls of peltate sporangiophores, with homosporous or heterosporous sporangia.

EQUISETINAE. A class of the Pteridophyta. The sporophyte has stem, root, and leaf. The stem is longitudinally ribbed, with leaves and branches alternating at the nodes, and in whorls. There are no leaf-gaps in the stele, which is usually a siphonostele. The sporangia are on distinctive sporangiophores.

EQUITANT. A condition when a plant member is folded inwards longitudinally, and covers the edges of another member which is similarly folded. This may cover a third, fourth etc.

EQUITORIAL PLATE. The area across the centre of the spindle where the chromosomes become attached at metaphase.

ERECT. (1) Set at right-angles to the organ from which it arises.
(2) Of an ovule, which is upright, with its stalk at the base.

EREMACACEAE. A family of the Saccaromycetales. It is frequently included in the Endomycetaceae.

EREMIUM. A desert formation.

EREMOSYNACEAE (H). A family of the Saxifragales. The leaves are alternate, or all radical. The flowers are rarely solitary. The petals are free, or absent. There are no staminodes, and the anthers are bilocular. The styles are usually free or absent, and the stigmas are dorsal to the carpels. There is 1 erect ovule in each loculus. The fruit is a capsule.

ERGASTIC SUBSTANCES. (1) Non-living inclusions in cells, *e.g.* crystals, starch grains etc.
(2) Non-protoplasmic cell-inclusions, taking some part in a metabolic process.

ERGASTOPLASM = ENDOPLASMIC RETICULUM.

ERGODIC HYPOTHESIS. The succession of vegetation in different places at one time will show the same sequence that one place will show in successive times.

ERGOT. (1) A disease of grasses, and cereals caused by the fungus *Claviceps purpurea*.
(2) The sclerotium of this fungus, which replaces the seeds of the host, and contains alkaloids which cause ergotism in animals and man.

ERICACEAE (EP, BH, H). A family of the Ericales (EP, BH, H). (BH) exclude the Vacciniaceae, but include the Pyrolaceae, and part of the Clethraceae. These are shrubs, trees or herbs, some are epiphytic. The leaves are leathery, often hairy, alternate, opposite or whorled, and without stipules. The flowers are solitary or in racemes, each with a bract and two bracteoles. They are bisexual and regular, or may be slightly zygomorphic. There are 5-4 free sepals which are persistent. The 5-4 petals are fused to form a bell (free in one genus). There are 10-8, or 5-4 free stamens, bearing introrse anthers which open by apical pores. The anthers often have projecting appendages. A nectar-secreting disk is present. The ovary is superior or inferior, consisting of 5-4 fused carpels opposite the petals. The number of loculi varies from 2-7, but is usually 5-4, with axile placentation, with many to 1 anatropous ovules in each. The fruit is a capsule, drupe or berry, usually with many small seeds. The embryo is cylindrical, in copious endosperm.

ERICACEOUS. Heather-like.

ERICALES (EP, BH, H). An order of the Sympetalae (EP), Gamopetalae (BH), Lignosae (H). These are shrubs, rarely herbs or trees, with simple leaves. The flowers are bisexual, and usually regular, with the parts in fours or fives. The petals are usually fused. The stamens are hypogynous, epigynous, or rarely united to the petals at the base. The ovary is superior or inferior with one integument, and having many to 2 carpels, these being opposite the petals when of the same number.

ERICETAL. Growing on moss.

ERICOID. Having narrow, needle-like, rolled leaves, like heather.

ERINOSE. The abnormal development of hairs, usually due to mite damage, or other pests and diseases.

ERIO-. A Greek prefix meaning 'woolly'.

ERIOCAULACEAE (EP) (H) = ERIOCAULEAE (BH). A family of the Farinosae (EP), Glumaceae (BH). The single family of the Eriocaulales (H). These are perennial herbs, often with grass-like leaves. The flowers are in inconspicuous heads. They are unisexual, regular or zygomorphic, with the parts in threes or twos. The perianth is usually of 2 whorls of different texture. The male flower usually has fused perianth segments, with 6-4 or 3-2 stamens with bilocular or unilocular anthers. The female flower has a superior ovary of 3-2 fused carpels, each forming a loculus with 1 orthotropous, pendulous ovule in each. The fruit is a capsule, and the endosperm is floury.

ERIOCAULALES (H). An order of the Calyciferae.

ERIOPHOROUS. Having a thick covering of cottony hairs.

ERODED, EROSE. Appearing as if gnawed, or worn irregularly.

EROSION. The removal of soil from the surface by water, wind, or gravity.

ERRATIC. Of lichens, not fixed to the substratum.

ERROR VARIANCE. The variance arising from agents unrecognized or uncontrolled in an experiment, with which the apparent effect of any recognized agent, controlled, or uncontrolled, is to be compared.

ERUBESCENS. Latin meaning 'bluish-red'.

ERUMPENT. Developing beneath the surface of the substratum, and bursting through it, becoming somewhat spreading.

ERYSIPHACEAE. A family of the Eryisiphales. The mycelium is usually white, with hyaline conidia formed singly or in chains. The perithecia are formed outside the host, and lack an ostiole, being free, or embedded in the cottony mycelium. The external layer of the periderm is dark, and brittle at maturity. The ascospores are hyaline.

ERYISIPHALES. An order of the Euascomycetae. They are parasites, superficial on the host. The ascocarp is closed, with a pseudoparenchymatous peridium. Most members have a layer of parallel asci at the base of the peridial cavity, but a few have the layer reduced to one large ascus.

ERYTHRO-. A Greek prefix meaning 'red'.

ERYTHROPALACEAE (H). A family of the Celastrales (H). The leaves are not lepidote, and the flowers are not enclosed by a bracteole. The calyx is enlarged, and encloses the fruit. The petals are valvate. The carpels and ovary are not enclosed by the disk, which is annular, or of separate glands. The ovary is unilocular, with the ovules pendulous from the top of the loculus.

ERYTHROPHYLL = ANTHOCYANIN.

ERYTHOSE. D-erythose is an aldodextrose sugar.

ERYTHROXYLACEAE (EP, H). A family of the Geraniales (EP), Malpighiales (H). (BH) include it in the Linaceae. These are trees or shrubs with usually alternate, entire, stipulate leaves. The flowers are bisexual and regular. The 5 sepals are free or fused, imbricate or valvate (rarely (H)). The 5 free petals often have appendages on the upper side. There are 2 whorls of stamens with 5 in each, united at the base. The superior ovary consists of 4-3 fused carpels, usually only 1 of which is fertile, containing 2-1 pendulous ovules. The fruit is a drupe (or nut (H)). The seeds may, or may not, contain endosperm.

ERYTHRULOSE. A ketose, tetrose sugar.

ESCALLONIACEAE (H). A family of the Cunoniales. The leaves are often glandular-serrate, and the petiole is not sheathing at the base. Stipules are absent, or adnate to the petiole. Petals are present. The disk is entire, or lobed, with the lobes alternating with the stamens. There are usually 5 stamens, but may be 6-4. The ovules are axile or parietal, and the fruit is a capsule, berry, or drupe.

ESCAPE. An originally cultivated plant, growing wild.

-ESCENS. A Latin suffix meaning '-ish', 'becoming'.

ESCULENT, ESCULENTUS. Edible.

ESKERS. Stratified layers of sand or gravel, formed by streams running under, or through glaciers.

ESPINAL FORMATION. A woodland of spiny trees.

ESSENTIAL ELEMENT. A chemical element, without which a plant cannot develop, and will ultimately die.

ESSENTIAL ORGANS. Sex organs.

ESTABLISHMENT. The successful germination and subsequent growth of a plant, especially in a new area.

ESTIVAL. Referring to the summer.

ESTIVATION = AESTIVATION.

ETAERIO. Aggregate (of a fruit).

ETHANOL, ETHYL ALCOHOL. One of the end-products of an alcoholic fermentation.

$$\begin{array}{ccc} & H & H \\ & | & | \\ H-&C-&C-OH \\ & | & | \\ & H & H \end{array}$$

ETHANOLAMINE. A precursor of choline.

$$CH_2OH$$
$$|$$
$$CH_2—NH_2$$

ETIOLATION. An abnormal condition developing in vascular plants in suboptimal light. The leaves are small and yellowed, and the internodes are abnormally long.

ETIOLOGY. The study of causes.

EU-. A Greek prefix meaning 'well, good, true, typical'.

EUAPOGAMY. The development of a sporophyte from the tissues of a gametophyte, and not from a zygote resulting from sexual fusion.

EUAPOSPORY. The complete failure to form spores.

EUASCOMYCETEAE. The subclass of the Ascomyceteae in which the asci are produced on ascogenous hyphae arising from a zygote, or parthenogenetically from the ascogonium.

EUBACTERIA. The true bacteria. They have rigid cell-walls, the motile species move by flagella, and the cells divide by binary transverse division.

EUBASIDII. A sub-class of the Basidiomycetae, in which the basidia develop directly from the diplophase mycelium, and are usually in a hymenium.

EUBRYA. A sub-class of the Musci. The gametophyte usually has 'leaves' with a mid-rib more than one cell thick. The sporogenous tissue does not overarch the collumella, and develops from the exterior portion of the endothecium. The capsule has a complex organization of the opercular region, and is elevated by the elongation of the seta.

EUBRYALES. An order of the Eubrya. The gametophore is perennial and erect, and the axis bears many rows of 'leaves'. The sporophytes are usually acrocarpous with a drooping or pendant capsule. The peristome is double, with well-developed teeth.

EUCARPIC. (1) Having only part of the thallus functioning as a fruit-body.
(2) Having both sexual, and asexual reproductive organs separate, and functioning at the same time.

EUCHROMATIN. (1) The lightly staining part of a chromosome, probably containing most of the hereditary material.
(2) The part of the chromatin having its maximum nucleic acid attachment on the mitotic spindle.

EUCHROMOCENTRE. A part of a chromosome which stains very deeply, and does not loosen out to form part of the reticulum.

EUCHROMOSOME. A typical chromosome, in contrast to a sex-chromosome.

EUCOMMIACEAE (EP, H). A family of the Rosales (EP), Urticales (H). (BH) include it in the Magnoliaceae. These are trees with alternate, exstipulate leaves. The flowers are unisexual and regular, lacking both sepals and petals. There are 10-6 stamens. The superior ovary is of 2 fused carpels, one of which is abortive. There are 2 anatropous, pendulous ovules. The seeds contain endosperm, and a straight embryo.

EUCRYPHIACEAE (EP, H). A family of the Parietales (EP), Guttiferales (H). (BH) include it in the Rosaceae. These are trees, shrubs, or climbers, with evergreen, opposite, stipulate leaves. The regular, bisexual flowers are solitary. There are 4 free sepals and petals, and many free stamens. The superior ovary consists of 18-5 fused carpels, each with many pendulous ovules. The styles and ripe carpels are free, but joined to the axis by threads. The winged seeds contain endosperm.

EUCYCLIC. Whorled, with the same number of organs in each whorl.

EUGENIC. That which tends to increase the fitness of the race.

EUGEOGENOUS. Weathering readily.

EUGLENALES. An order of the Euglenophyceae, characterized by the flagellate motile cells being the dominant phase in the life-cycle.

EUGLENOCAPSALES = COLACIALES.

EUGLENOPHYTA. A division of the Algae. They have a colourless protoplast, or chloroplasts, and are holophytic, saprophytic, or holozooic. The food reserves are paramylum or fat. They are mostly naked, unicellular flagellates, with 3-1 flagella at the anterior end, usually inserted at the base of an interior chamber, which is connected to the outside by a gullet. Reproduction is by cell-division (rarely sexual), and resting spores are formed frequently.

EUGONIDIUM. A unicellular member of the Chlorophyceae, forming part of the thallus of a lichen.

EUHYMENIAL. Said of a hymenium in which all the basidia are formed nearly at the same time.

EUMITOSIS. A normal mitosis, in which the chromosomes separate distinctly, and clearly divide longitudinally.

EUMORPHIC. Well-formed.

EUMYCETAE. (1) The true fungi. There is no photosynthetic pigment, and the food is stored as glycogen. Most have a definite cell-wall containing cellulose, and/or chitin, and have a branching filamentous thallus (mycelium) consisting of individual hyphae.

(2) Sometimes used to include the ascomycetes, basidiomycetes, and the Fungi Imperfecti.

EUPHORBIACEAE (EP, BH, H). A family of the Geraniales (EP), Unisexuales (BH), Euphorbiales (H). These are trees, shrubs, or herbs, most of which contain latex. The leaves are usually alternate, sometimes opposite, or opposite above, and alternate below. Stipules are usually present, but may be modified as hair-like structures, glands, or thorns. The inflorescence is usually complex, and of great variety. The flowers are always unisexual and regular. The perianth may have 2 whorls, more usually 1, of 5 parts, or may be absent. There are many to 1 stamens, which may be free, or united. The superior ovary consists of 3 fused carpels, with axile placentas and 3 loculi. There are 2-1 ovules in each loculus. The ovules are collateral, pendulous, anatropous, with a ventral raphe. The micropyle is usually covered by a caruncle. The fruit is almost invariably a 'schizocarp-capsule'. It splits into carpels, and at the same time, each carpel opens ventrally releasing the seeds. The seed contains endosperm.

EUPHORBIALES (H). An order of the Lignosae. See *Euphorbiaceae*.

EUPHOTIC ZONE. The top layer of a sea or lake through which light can penetrate, allowing photosynthesis to take place. It is usually about 100m. deep.

EUPHOTOMETRIC. Said of a leaf which is permanently placed in a plane perpendicular to the direction of the strongest diffuse light reaching it.

EUPLOID. Having all the chromosomes of the set present in the same member.

EUPOMATIACEAE (EP, H). A family of the Ranales (EP), Anonales (H). (BH) include it in the Anonaceae. These are shrubs, lacking stipules. The flowers are bisexual, solitary and perigynous. The sepals and petals are not differentiated, being borne on a concave torus. The many stamens are perigynous, with the inner ones sterile and petalloid. There are many carpels immersed in a conical receptacle. The styles are joined. Each carpel contains many ovules. The fruit is a berry with 2-1 seeds in each loculus. The endosperm is ruminate, and the embryo small.

EUPOTAMOUS. Living in rivers and streams.

EUROTIALES. An order of the Ascomycetae. The asci are commonly small and rounded, irregularly distributed within a more-or-less well-marked peridium.

EURYBATHIC. Tolerant of a wide range of depths.

EURYHALINE. Normally living in salt-water, but tolerant of a wide range of salinity.

EURYTHERMOUS. Tolerant of a wide range of temperatures.

EUSPORANGIATAE. A sub-class of the Filicinae. The sporangia are borne, either on outgrowths of the leaf, or in sori on the abaxial side. They have many spores and a jacket-layer more than one cell thick.

EUSPORANGIATE. Having the sporangia each developed from a group of cells.

EUTROPHIC. Said of a type of lake-habitat with gently sloping shores and a wide belt of littoral vegetation.

EVAGINATE. Not having a sheath.

EVEN. Having a smooth surface.

EVER-BEARER, EVER-BLOOMER. A plant that produces leaves and flowers for a long period of the growing season, and often bears flowers and fruits at the same time.

EVERGREEN. Bearing, and losing leaves continuously throughout the year.

EVER-SPORTING. Producing frequent sports; especially a heterozygote which segregates homozygous recessives, but not homozygous dominants in every generation.

EVERTED. Turned outward abruptly.

EVOLUTE. Having the margins rolled outwards.

EVOLUTION. The gradual change in the characteristics of successive generations of a species or race, ultimately giving rise to species or types different from the common ancestor.

EVOLVATE. Lacking a volva.

EXAGGERATION. The expression of a hypomorphic gene placed opposite a deficiency.

EXALBUMINOUS. Lacking endosperm.

EXANNULATE. Lacking an annulus.

EXARCH. Said of a xylem strand which has the protoxylem farthest from the centre of the axis.

EXASPERATE. Having a hard, roughened surface, with short points projecting from it.

EXCEEDING. Projecting beyond a neighbouring member.

EXCELSUS. Latin meaning 'lofty'.

EXCENTRIC. (1) = ECCENTRIC.
(2) Of a pileus in which the stipe is not central.

EXCIPLE = EXCIPULUM. (1) A 'true' exiple is produced as a rim of apothecial tissue from a lichen, and does not contain algal cells. A 'false' exciple is produced by the thallus around the apothecium, and does contain algal cells.
(2) The outer layer of the wall of an apothecium or perithecium.

EXCIPULACEAE. A family of the Sphaeropsidoles. They are mostly sapro-phytes on twigs or stems. The pycnidia open out early to form a more-or-less deep saucer-shaped structure which is tough or hard and black, either sub-epidermally or sub-cortically, breaking through to the outside, or sometimes external from the first.

EXCIPULIFORM. Cup-shaped.

EXCIPULUM. Structure on lichens. The 'thalloid' excipulum is a marginal wall around the apothecium, not arising from it but from the thallus, while the 'proper' or 'true' excipulum is similar, but arises from the apothecium.

EXCITATION. The action of a stimulus on a plant or plant-organ.

EXCLUSIVE SPECIES. A species which is confined to a definite area.

EXCRETION. The removal of the waste products of metabolism. In plants, this is often done by the formation of insoluble salts in the cells.

EXCURRENT. Said of a vein which runs out beyond the lamina of the leaf.

EXENDOSPERMOUS. Said of a seed that lacks an endosperm.

EXERGONIC REACTION. A chemical reaction which releases energy.

EXFOLIATION. The falling away in flakes, layers or scales, as of some bark.

EXINDUSIATE. Having no indusium.

EXINE. The outermost wall of a pollen grain or moss spore.

EXIT PALILLA, EXIT TUBE. (1) A small outgrowth formed by some of the Phycomycetes, which functions to invade the host cell.
(2) An outgrowth from a zoosporangium, through which the zoospores escape.

EXO-. A Greek prefix meaning 'without'.

EXOASCALES. An order of the Euascomycetae. The asci are parallel in a palisade-like layer, and there is no peridium. All are parasitic.

EXOBASIDIACEAE. A family of the Polyporales. These are parasitic on leaves, young stems, and fruits of angiosperms. The basidia are club-shaped, and produced externally in a continuous or interrupted layer. The basidio-spores are septate, and on germination each cell produces a few spindle-shaped conidia.

EXOBASIDIAL. (1) Having the basidia uncovered.
(2) Separated by a wall from the basidium.
(3) Of a lichen sporophore, having no secondary sporing branch.

EXOCARP = EPICARP.

EXOCHITE. The firm outer wall of the macrosporangium of the Fucales.

EXODERMIS. The outer layer of suberized cells on a root. It replaces the piliferous layer in the older parts of the root.

EXOGAMY. Out-breeding.

EXOGENOUS. (1) Produced on the outside of another plant member.
(2) Developed from superficial tissue.
(3) Increasing in thickness by the addition of new layers on the outside.

EXOGENOUS SPORE. A spore formed at the end of a hypha, not inside a sporangium.

EXOGENOUS THALLUS. The thallus of a lichen in which the fungus predominates.

EXOGYNOUS. Having the style projecting beyond the corolla.

EXOLETE. Of a fungus fruit-body, when it is long over-mature, but has nothing inside it.

EXOPERIDIUM. The pseudoparenchymatous outer layer of the peridium of the Lycoperdales.

EXOPHELLODERM. Phelloderm cells cut off on the outside of the phellogen.

EXPONENTIAL GROWTH EQUATION. A compound interest equation, which assumes that the absolute rate of growth of an organism or population is dependent on the original size, or number, and that the relative growth rate is constant.

$$L_nX = L_nX_o + rt,$$

where X = the final size, X_o = the initial size, r = the relative growth rate, and t = time.

EXOSCOPIC. Of a plant embryo, when the first division is at right-angles to the archegonial neck; the apex of the embryo growing towards the archegonial neck, and the basal region is in contiguity with the gametophyte.

EXOSPORE. The outer layer of a zygote (cyst) of some algae and fungi.

EXOSPOREAE. A sub-class (sub-order) of the Myxothallophyta. The spores are borne externally on erect branching fruit-bodies.

EXOSPORIUM. The outer layer of a spore-wall. It is usually thickened in some way.

EXOTHECIUM. (1) The outer layer in the wall of a moss-capsule.
(2) The outer layer of the microsporangium of the gymnosperms.

EXOTIC. Not native.

EXPANDED. Flattened out, and becoming less concave as development proceeds.

EXPANSION THEORY. The theory that the pith of a siphonostele has originated by the non-specialization of the central tissue, which remains parenchymatous, *i.e.* that the pith is part of the stele.

EXPRESSIVITY. Said of a gene. Its influence on the phenotype, as affected by the environment.

EXSERTED. Protruding.

EXSICCATUS. A dried plant, especially one in a herbarium.

EXSTIPULATE. Lacking stipules.

EXTENSION FACTORS. Genes which increase the expressivity of other genes.

EXTERNAL PLASMA MEMBRANE = ECTOPLASM.

EXTRA-. A Latin prefix meaning 'beyond'.

EXTRA-AXILLARY. Said of a bud which is formed elsewhere than in the axil of a leaf.

EXTRACELLULAR. Outside the cell.

EXTRAFLORAL NECTARY. A nectary that is found in some place other than the flower.

EXTRAMATRICAL. Said of a fungus which has the greater part of the thallus, especially the reproductive bodies, outside the host, or on the surface of the substrate.

EXTRORSE. Of anthers which open away from the centre of the flower.

EXUDATION. The liberation of liquid water, or sap from special pores on the plant.

EYE. A bud.

EYESPOT. A light-sensitive organelle found in some of the motile green algae. It consists of a pigment-cup, covered by a light sensitive substance, and sometimes with a lens external to this.

F

F. A class of flowers pollinated by butterflies and moths.

F_1. The first generation of a cross between two individuals.

F_2. The second filial generation, produced by crossing or self-fertilizing individuals of the F_1 generation.

FABACEAE (H) = PAPILIONACEAE.

FACE. The upper surface of an organ that has two distinct sides.

FACIES. The general form and appearance of a plant.

FACILITATED DIFFUSION. Diffusion which is assisted by the presence in the membrane of substances with which the penetrating particle can combine reversibly to form a more soluble complex.

FACTOR. That which is responsible for the independent inheritance of a Mendelian difference. See *Gene*.

FACTORIAL EXPERIMENT. An experiment in which all the treatments or agents under investigation are varied simultaneously, and combined in such a way that any derived effect of one or group of them may be isolated, and evaluated separately.

FACTOR OF THE HABITAT. Anything in the environment which affects, directly or indirectly, the life of a plant.

FACULTATIVE. Incidental; not necessarily.

FACULTATIVE ANAEROBE. A plant that can respire under aerobic and anaerobic conditions.

FACULTATIVE GAMETE. A zoospore that can function as a gamete.

FACULTATIVE PARASITE. A saprophyte, which may become a parasite under certain conditions

FACULTATIVE SAPROPHYTE. A parasite which can live as a saprophyte under special conditions.

FAGACEAE (EP, H). A family of the Fagales (EP, H). (BH) include it in the Cupuliferae. These are trees with simple leaves and scaly stipules that fall off as the leaves expand. The flowers arise in the axils of the leaves of the current year. They are in catkins, or small spikes, and are wind-pollinated, (except *Fagus* male). The perianth consists of 7-4 fused scales. The male flower has 7-4, 14-8, or many undivided stamens. The female flowers are in groups of 3, 2, or 1. The inferior ovary is of 3 fused carpels, with usually 3 styles. There are 3 loculi, with 2 pendulous anatropous ovules in each, borne on axile placentas. The fruit is a 1-seeded nut, and the seeds have no endosperm. The nut, or group of nuts is enclosed in a cupule.

FAGALES (EP, H). An order of the Archichlamydaea (EP), Lignosae (H). These are trees with alternate, stipulate leaves, and the flowers in simple or cymose spikes. They are cyclic rarely naked, and usually monoecious. The stamens are opposite the perianth lobes. The inferior ovary has 6-2 fused carpels, with 2-1 ovules in each. The fruit is a 1-seeded nut. The seed has no endosperm.

FALCATE, FALCIFORM. Sickle-shaped.

FALCIPHORE. See *Falx*.

FALLING STARCH. See *Statolith*.

FALLOW. Farm-land, unused during a growing season.

FALSE ANNUAL RING. A second ring of xylem formed in one season following abnormal defoliation, especially by insects.

FALSE AXIS. A monochasium which looks like one axis, but is really a number of successive lateral branches running more or less in a line.

FALSE BERRY. A fleshy fruit which looks like a berry, but that has some of the flesh derived from the receptacle.

FALSE DICHOTOMY. Branching in which two lateral branches arise on opposite sides of the main axis, and overtop it.

FALSE DISSEPIMENT. A wall which divides the loculus of an ovary into two compartments, but is an ingrowth of the carpel wall, and not a wall between one carpel and the next.

FALSE FRUIT. A fruit formed from other parts of the flower as well as the gynecium.

FALSE GERMINATION. An appearance of germination by a dead seed, due to the swelling of the embryo, as it takes up water.

FALSE HYBRID. A plant developing after cross-fertilization, but having the characteristic of only one parent.

FALSE SEPTUM. See *Spurious dissepiment*.

FALSE TISSUE = PSEUDOPARENCHYMA.

FALX. A 'fertile hypha', or conidiophore of *Zygosporium*, looking like a sickle. Falces may be sessile, or on special hyphae, (falciphores).

FAMILY. The taxonomic division between an order and a genus, but may be a sub-division of a sub-order, or super-family. It contains similar genera. The names of botanical families usually end in -aceae.

FARCTATE. (1) = STUFFED.
(2) Having the centre softer than the outer layer.

FARINACEOUS. (1) Having a mealy surface.
(2) Of a mealy character.
(3) Of an endosperm that is starchy.
(4) Smelling of meal.

FARINOSAE (EP). An order of the Monocotyledoneae. These are usually herbs, with the perianth lobes free, typically in 2 whorls of 3. The whorls may be similar or different in appearance. Typically there are 2 whorls of 3 free stamens, but 1 whorl may be absent, or only 1 stamen. The ovary has three fused carpels, containing orthotropous (usually) ovules. The endosperm is mealy.

FARINOSE, FARINOSUS. Covered with whitish, very short hairs, which are easily detached as a whitish dust.

FASCIATION. Coming together to form a bundle.

FASCICLE. (1) A tuft of branches all arising from about the same place.
(2) A tuft of leaves crowded on a short stem.

FASCICULAR CAMBIUM. A flat strand of cambium between the xylem and phloem of a vascular bundle.

FASCICULATE. In bundles consisting of members all of the same kind.

FASCICULATE BASIDIUM. One of a group of basidia.

FASTIGATE, FASTIGIATE. Having many branches parallel to the main stem, and usually upright.

FASTIGIATE CORTEX. Of lichens made up of parallel hyphae, at right-angles to the axis of the thallus.

FAT. An ester formed by the reaction between a fatty acid and glycerol, *e.g.*

$$\begin{array}{ll}
\text{CH}_2\text{OH} & \text{CH}_2\text{O}(\text{C}_{15}\text{H}_{31}\text{CO}) \\
| & | \\
\text{CHOH} + 3\text{C}_{15}\text{H}_{31}\text{COOH} = \text{CHO}(\text{C}_{15}\text{H}_{31}\text{CO}) + 3\text{H}_2\text{O} \\
| & | \\
\text{CH}_2\text{OH} & \text{CH}_2\text{O}(\text{C}_{15}\text{H}_{21}\text{CO})
\end{array}$$

glycarol   palmitic acid   palmitin   water

FATISCENT. Cracking and falling apart.

FATTY ACID. An acid which forms a fat when it reacts with glycerol. A saturated fatty acid has the basic formula $\text{C}_n\text{H}_{2n}\text{O}_2$, *e.g.* CH_3COOH, while that of an unsaturated fatty acid is $\text{C}_n\text{H}_{2n-x}\text{O}_2$ (where x = 2, 4, 6, or 8). Those of biological importance usually have an even number of carbon atoms, and are in straight chains.

FAULSCHLANIN. A type of lake-bottom deposit, composed of organic detritus covered by mineral deposits, which is characterized by the absence of sediment transporters, and in which anaerobic decomposition takes place.

FAVEOLATE, FAVOSE. Looking like a honey-comb.

FEATHERY. Covered with long, branched hairs.

FEHLING'S SOLUTION. Fehling's A contains 34·66 gm. of copper sulphate crystals, dissolved in 500 ml. of water, and Fehling's B, 173 gm. potassium sodium tartrate, and 50 gm. of sodium hydroxide dissolved in 500 ml. of water. They are used in equal parts, and give a red-orange precipitate when boiled with a reducing sugar. This is due to the reduction of the cupric ions to cuprous.

FELLING SUBSERE. The developmental series of communities, started by the felling of a wood.

FENESTRATE. (1) Having openings.
(2) Of spores, muriform.

-FER. A Latin suffix meaning 'bearing'.

FERAL. Wild; not cultivated.

FERMENT = ENZYME.

FERMENTATION. The enzymatic destruction of a substrate by microorganisms. Basically it is a respiratory process, usually anaerobic. More specifically it refers to the production of ethanol by yeasts, from monosaccharide sugars.

FERNS = FILICALES.

FERRALLITE. A lateritic soil with a low kaolin content.

FERRUGINOUS, FERRUGINEOUS. Rust-coloured.

FERTILE HYPHA = CONIDIOPHORE.

FERTILIZATION. The fusion of gametes, resulting in the formation of a zygote. Strictly it should be confined to the fusion of nuclei, but sometimes it refers to a condition where the cells fuse, but not their nuclei, *e.g.* pseudogamy.

FERTILIZATION TUBE = CONJUGATION TUBE.

FEULGEN STAIN. A stain which stains DNA purple, *e.g.* on chromosomes.

FIBRE. (1) A narrow, elongated, lignified cell, tapering to a wedge-shape at both ends.
(2) A very fine root.

FIBRE TRACHEID(E). An elongated cell found in wood, which has thicker walls and fewer pits than a tracheid, but thinner walls and more pits than a fibre.

FIBRIL, FIBRILLA. (1) A small fibre.
(2) Any fibre-like structure.

FIBRILLOSE. Having fibres.

FIBROUS CORTEX. Of lichens, having a cortex made up of hyphae lying parallel with the longitudinal axis of the thallus.

FIBROUS ROOT SYSTEM. A mass of fine adventitious roots, of more-or-less equal thickness, and bearing finer lateral roots. They are borne on stems or the hypocotyl, *e.g.* grasses.

FIBRO-VASCULAR BUNDLE. A vascular bundle of sclerenchyma, usually on its outer side.

FIBROUS LAYER. A layer of cells found in the wall of an anther. The cell-walls are thickened irregularly by bands of material. They set-up uneven strains in the anther-wall, causing it to rupture.

FICK'S LAW. A law of diffusion. $dm = -D.A.(dc/dx).dt$, where, $dm =$ the amount of substance diffusing in the time dt, $dc/dx =$ the concentration gradient, $A =$ the area of cross-section through which diffusion occurs, $D =$ diffusion coefficient.

FICOIDACEAE (H). A family of the Caryophyllales (H). = AIZOACEAE (EP).

FICOIDALES (BH). An order of the Polypetalae (BH). The flowers are regular, or more-or-less so. The carpels are fused, and the ovary inferior to superior. The ovary is either unilocular with parietal placentation, or many to 2 locular with axile or basal placentation. The embryo is either curved, with endosperm, or cyclical or oblique with no endosperm.

FICOIDEAE (BH). A family of the Ficoidales (BH). = AIZOACEAE (EP).

-FID, FIDUS. Latin for a 'cleft'.

FILAMENT. (1) A chain of cells, set end-to-end.
(2) The stalk of a stamen.
(3) In the Cyanophyta, a trichome(s) and its enclosing sheath.

FILAMENTOUS. Thread-like.

FILAR. Elongated and thin.

FILIAL GENERATIONS. See F_1, F_2.

FILICALES. The single order of the Leptosporangiatae.

FILICES. The homosporous, leptosporangiate Filicales.

FILICINAE. A class of the Pteridophyta. The sporophyte is differentiated into stem, root, and leaf, although the root may be absent. If the vascular cylinder is a siphonostele, there are leaf-gaps. The leaves are macrophyllous, and spirally arranged on the stem. The sporangia are on the margin, or abaxial surface of the leaf.

FILICINEAN. Relating to the ferns.

FILICOPSIDA = FILICINAE.

FILIFORM. (1) Like a thread.
(2) Of bacterial stab-cultures, having equal growth on either side of the line of inoculation.

FILIPENDULOUS. Having swellings of considerable size, along, or at the ends of thin roots.

FILTERABLE VIRUSES = VIRUSES.

FILOPLASMODIUM. A net-like pseudoplasmodium.

FIMBRIATE. Having a fringed margin.

FIMBRICOLOUS. Growing on, or in dung.

FIMETARIACEAE. A family of the Sphaeriales. There are 4, 8, or more asci in a perithecium. The perithecia are separate on the surface of the substrate, and naked or nearly so.

FIMICOLOUS = FIMBRICOLOUS.

FISSIDENTALES. An order of the Eubrya. All the aerial branches have two-sided apical cells, so that the 'leaves' develop in two rows.

FISSILE. Split, or tending to split.

FISSION. Reproduction, by splitting of the nucleus and cytoplasm into equal parts.

FISSIPAROUS. Splitting.

FISTULAR, FISTULOSE. Hollow, like a pipe, and herbaceous.

FISTULINACEAE. A family of the Polyporales. The fruit-bodies are fleshy, resupinate, or reflexed, or laterally stalked. The lower surface grows out to form many elongated separate tubes which are lined by the hymenium. The basidia are chiastic.

FIXATION. The treatment of a specimen with a reagent which will fix its structure and appearance in a life-like condition.

FIXATION OF NITROGEN. The formation of nitrogenous compounds from gaseous nitrogen, by soil bacteria.

FIXATIVE. A reagent which will bring about fixation.

FIXED LIGHT POSITION. The position of a fully developed leaf in respect of the direction of the strongest diffuse light that reaches it.

FLABELLATE, FLABELLIFORM. Fan-shaped.

FLACCID. Limp and flabby.

FLACOURTIACEAE (EP, H). A family of the Parietales (EP), Bixales (H). (BH) put its members partly in the Bixineae, and their family Samydaceae are included in the Flacourtiaceae by (EP). These are trees and shrubs with mostly alternate, stipulate, leathery leaves. The flowers are solitary, or in cymes, or racemes. There is a disk, glands, or scales between the petals and stamens. There are 15-2 free or fused sepals, which are imbricate or valvate. The 15-0 petals are free. The many free stamens (may be united into groups) have anthers which open by longitudinal slits. The superior ovary is of 10-2 fused carpels, with 1 (rarely many) loculus, and parietal placentation. There are many anatropous ovules. The fruit is usually a berry or capsule, containing 1 or many seeds which are often arillate. The embryo is straight, and there is endosperm.

FLAGELLARIACEAE (EP, BH, H). A family of the Farinosae (EP), Calycinae (BH), Commelinales (H). These are strong plants, sometimes climbing, with long, many-nerved sheathing leaves and panicles. The flowers are regular, bisexual, or unisexual, with a perianth of similar lobes. The perianth has 2 whorls of 3, free lobes. The free stamens are also in 2 whorls of 3. The superior ovary has 3 fused carpels with 3 loculi, each of which has 1 axile anatropous ovule. The fruit is trilocular, or with 3-1 stones. The seeds contain endosperm.

FLAGELLATE. (1) Having flagella.
(2) Bearing a long, thread-like appendage.

FLAGELLATE = MASTIGOPHORA.

FLAGELLIFORM. Like a whip-lash.

FLAGELLUM. A whip-like locomotory organ. Typically it consists of an axoneme and a sheath. The axoneme consists of nine long fibres surrounding two central ones. The sheath has an outside membrane.

FLAVESCENT. (1) Becoming yellow, or yellowish.
(2) Having yellow, or yellow-green spots mingled with the normal surface green.

FLAVICIN. An antibiotic produced by some *Aspergillus* spp. and closely similar to or identical with penicillin. It is active against bacteria, but not fungi.

FLAVINS. Mono- or dinucleotides, which occur as components of reductases.

FLAVOGLAUCIN. A yellow pigment in *Aspergillus glaucus*.

FLAVONE. The basis of a number of yellow pigments.

FLAVO-PROTEINS. Conjugated proteins in which the prosthetic group is riboflavin-phosphoric acid, or a compound of this with another nucleotide. They function as hydrogen-carriers in oxidations.

FLAVUS. Latin meaning 'yellow'.

FLESHY. Thick and soft, but not necessarily juicy.

FLESHY DISSEMINULE. A seed or fruit consisting of a large amount of fleshy material.

FLEXUOSE, FLEXUOUS. Said of a stem which is zig-zag, usually changing direction at the nodes.

FLEXUOUS HYPHA. Of the Uredinales. A branched, or unbranched hypha growing from a pycnidium, which may be diplodized by a pycnidium of the opposite 'sex'.

FLOATING. (1) Of structural, or gene changes, for which a mating group is not uniform.

(2) Of a tissue in water-dispersed fruits and seeds hat contains air, andnot easily waterlogged.

FLOCCOSE. (1) Cottony

(2) Having a dense a dense covering of tangled hairs, looking like wool, that is easily detached from the plant.

FLOCCULENT. Of bacterial liquid cultures, when they have small masses of bacteria throughout, or as a deposit.

FLORA. (1) The plants of a particular area.

(2) A descriptive list of the plants of an area, including a key for identification.

FLORAL AXIS = RECEPTACLE.

FLORAL DIAGRAM. A composite diagram of a transverse section through a flower to show the position of the various parts, and their relation with other parts.

FLORAL ENVELOPE. The calyx and corolla, or perianth.

FLORAL FORMULA. A summary of the characteristics of a flower. K = calyx, C = corolla, A = andrecium (stamens), G = gynecium (carpels). The number following represents the number of parts. ( ) means that the parts are fused, *e.g.* C(5) means 5 fused petals. ⌒ indicates 1 whorl fused to another, *e.g.* C(5) A5 would mean a corolla of 5 fused parts to which are joined 5 free stamens. A line below the figure for the gynecium, means that it is superior, *e.g.* G5, beside it, that it is perigynous, *e.g.* G(5)–, and above it inferior, *e.g.* G(3).

FLORAL LEAF. (1) A bract or bracteole.

(2) A petal or sepal.

FLORET. A small flower, usually found in heads.

FLORIBUNDUS. Latin meaning 'producing many flowers'.

FLORIDEAE. A sub-class of the Rhodophyta, in which the growth of the thallus is terminal, and in which the carpospores are formed indirectly from a zygote.

FLORIDEAN STARCH. An insoluble carbohydrate storage product found in red algae. It stains red, rather than blue-black with iodine solution.

FLORIDIOPHYCIDAE = FLORIDEAE.

FLORIDUS. Latin meaning 'showy'.

FLORIGEN. The flower-producing hormone, which may be synthesized in the leaves during a photo-period.

FLORISTIC COMPOSITION. A complete list of the plants forming a plant community.

-FLORUS. A Latin suffix meaning 'flowered'.

FLOS. Latin meaning 'flower'.

FLOWER. The reproductive stem of the angiosperms. Typically it is made up of a calyx of sepals, a corolla of petals. (these two being the perianth), an andrecium of stamens, and a gynecium of carpels. Any of these parts may be missing in a particular flower. The floral axis is the receptacle.

FLOWERING GLUME. A glume which subtends a flower in a grass spikelet.

FLUITANS. Latin meaning 'floating'.

FLUORESCENT. Giving out light when placed in ultra-violet (or other) radiation.

FLUSH. (1) A period of renewed growth in a woody plant.
(2) A limited area watered by a spring etc. distinguished by the luxuriant vegetation.

FLUVALES = HELOBIEAE.

FLUVIATILE, FLUVIATILIS (Latin). Occurring in rivers and streams.

FOLDED VERNATION. The condition in which the leaf is folded about the mid-rib, with the two faces brought together.

FOLIACEOUS, FOLIOSE. (1) Flat, and leaf-like.
(2) Bearing leaves.

FOLIAGE LEAF. An ordinary green leaf.

FOLIAR GAP, FOLIAR TRACE = LEAF GAP, LEAF TRACE.

FOLICOLE. A plant living on leaves, either as a saprophyte or parasite.

FOLICOLOUS. Living on leaves.

FOLIOLOSE. Consisting of minute flattened lobes.

FOLIOSE = FOLIACEOUS.

-FOLIUM, -FOLIUS (suffix). Latin meaning 'leaved'.

FOLLICE. (1) A many-seeded dry fruit, derived from a single carpel, and splitting longitudinally down one side at dehiscence.
(2) A small bladder on the leaves of some mosses.

FONTINALACEAE. A family containing the floating members of the Eubrya.

FOOD BODY. A mass of cells on the outside of a seed-coat, attracting ants which aid in dispersal. The attraction is due to stored food, usually oils.

FOOD POLLEN. Infertile pollen, sometimes produced in special anthers. It attracts insects, which bring about the dispersal of the fertile pollen.

FOOT. (1) The lower part of the embryo sporophyte of the Bryophyta and Pteridophyta. It remains embedded in the gametophyte.
(2) A small thick-walled segment of the hyphae, from which a conidiophore of *Aspergillus* arises.
(3) The lower of two cells derived from a binucleate basal cell in the early stages of the development of a uredospore.

FORAMEN. (1) An opening.
(2) = MICROPYLE.

FORB. Any herbaceous plant, excluding the grasses.

FORCIPATE. Shaped like a pair of forceps.

FORE RUNNER TIP. A leaf tip which becomes active while the rest of the leaf is developing.

FOREST CLIMAX. A climax community composed of trees.

FORFICULATE. Shaped like a scissors.

FORKED. Divided into two or more distinct branches which diverge as they elongate.

FORM GENUS, FORM SPECIES. A group of species, which have similar morphological characters, but are not known certainly to be related by descent. The species are form species.

FORMALDEHYDE HCHO. The aldehyde derivative of methyl alcohol (methanol).

FORMATION = ASSOCIATION. Sometimes referred to by the habitat rather than the dominant plant(s).

FORMATIVE REGION. The region just behind the dividing region of the growing point of stem or root, where the tissues begin to differentiate.

FORMATIVE STAGE OF GROWTH. The stage in development when a cell is formed from a pre-existing cell.

FORMIC ACID. H.COOH.

FORMIC DEHYDROGENASE. An enzyme that breaks-down formic acid to carbon dioxide and water in the presence of co-enzyme 1.

FORMIC HYDROGENLYASE. An enzyme in a few groups of bacteria which catalyses the combination of carbon dioxide and hydrogen, or bicarbonate ions and hydrogen to form formates.

$$CO_2 + H_2 = H^+ + HCO_3$$
$$HCO_3^- + H_2 = H.COO^- + H_2O$$

FORNICATE. Arched and hood-like.

FOUNTAIN-TYPE THALLUS. A thallus of the Rhodophyta, having a central core of axial filaments, each giving rise to lateral filaments.

FOUQUIERIACEAE (EP, H). A family of the Parietales (EP), Tamaricales (H). (BH) include it in the Tamaricaceae. They are shrubs with deciduous leaves, which have persistent, thorny mid-ribs. The bisexual flowers are regular. The 5 sepals are free and imbricate. The 5 petals are fused. There are 15-10 free stamens. The superior ovary is of 3 fused carpels with 6-4 ovules on placentas in the middle of the ventral side. The fruit is spherical with 3 loculi. The winged or hairy seeds contain endosperm.

FOVEOLA. A small pit.

FOVILLA. The material inside a pollen grain.

FRAGMENT. (1) A new acentric product of chromosome breakage.
(2) A small supernumerary chromosome.

FRAGMENTATION SPORES. Conidia produced by hyphae breaking into separate cells.

FRANCOACEAE (H). A family of the Saxifragales (H). The flowers are in lax, elongated racemes. The petals are free, or absent. The anthers are bilocular, on 8-4 (rarely 5) stamens. There are no staminodes. The carpels are fused, and the ovary inferior or superior. The stigmas are lateral to the carpels. The fruit is a capsule.

FRANKENIACEAE (EP, BH, H). A family of the Parietales (EP), Caryophyllinae (BH), Tamaricales (H). These are herbs with jointed stems, and opposite inrolled leaves. The flowers are solitary or in cymes, and are bisexual and regular. The 7-4 sepals are fused, induplicate-valvate. The 7-4 petals are free. There are usually 6 stamens, in 2 whorls, free or slightly joined at the base. The superior ovary is usually of 3 fused carpels, with 1 loculus. There are many anatropous ovules borne on parietal placentas at the base of the ovary. The fruit is a capsule. The seeds contain a mealy endosperm, and a straight embryo.

FREE. (1) Not joined laterally to another member of the same type.

(2) Said of the gills of agarics that reach the stipe, but are not joined to it.

FREE BASAL PLACENTATION. Having ovules arranged on a basal placenta in a unilocular ovary.

FREE CELL-FORMATION. The formation of daughter cells that do not remain united.

FREE CENTRAL PLACENTATION. Having ovules on a central placenta derived from the bottom of a unilocular ovary.

FREE NUCLEAR DIVISION. Nuclear division, not accompanied by the formation of cell-walls.

FREQUENCY FACTOR. The percentage occurrence of a species in a plant community.

FREQUENCY DISTRIBUTION. The distribution of the frequencies of observations with respect to the classes into which the observations are divided.

FREUNDLICH'S ADSORPTION ISOTHERM. An equation for the equilibrium between the concentration of a solute in solution, and on the surface of an adsorbent in the solution;

$$\log a/m \; = \; \log K \; + \; n \log C, \text{ where}$$

m = amount of adsorbent, C = external concentration, K and n are constants for the particular solutes and adsorbent, a = mKCn. If log a/m is plotted against log C, a straight line results, with a slope n, and the intercept on the ordinate = log K.

FRIABLE. Readily powdered.

FRILL. A thin sheet of hyphae forming a horizontal, circular flange around the stipe of an agaric.

FROND. (1) The leaf of a fern.

(2) The leaf of a palm or cycad.

(3) The blade-like thallus of a sea-weed.

(4) A lichen thallus.

FRONDESCENT. (1) Leaf-like.

(2) Having a lot of leaves.

FRONDOSE, FRONDOUS = FRONDESCENT.

FRONT CAVITY. The opening of a stomata nearest the exterior.

FRUTICOLE. Of parasitic fungi, living off fruit.

FRUCTIFICATION. (1) A general term for the body which develops after fertilization, and contains spores or seeds.

(2) Any spore-bearing structure, whether developed after fertilization or vegetatively.

FRUCTOFURANOSE. The furanose form of fructose. The form in which it exists in disaccharides, polysaccharides, and in its phosphates.

D—fructofuranose

FRUCTOPYRANOSE.   The pyranose form of fructose. The form in which it exists in solution.

FRUCTOSANS.   Plant storage-products, produced by the condensation of fructose.

FRUCTOSE.   A keto-hexose sugar, which rotates a beam of polarized light to the left.

FRUCTOSE 1:6 PHOSPHATE.   A diphosphate of fructofuranose. The initial molecule, which is ultimately changed to two molecules of pyruvic acid during respiration.

FRUIT.   The ripened ovary and seeds of an angiosperm.

FRUIT-BODY.   A well-defined group of fungal spores and the hyphae which bear, and surround them.

FRUSTULE.   The wall, or the wall and its contents, of a diatom.

FRUTEX.   An adjective meaning 'shrubby'.

FRUTESCENT.   Shrubby.

FRUTICOLOUS.   Living on shrubs.

FRUTICOSE.   (1) Bushy.
     (2) Said of a lichen thallus, which is attached by its base, and stands out from the substrate, branching and having a bushy appearance.

FUCACEAE.   The single family of the Fucales.

FUCALES.   The single order of the Cyclosporae.

FUCAN, FUCOSAN.   A polysaccharide found in brown algae. It is built up of fucopyranose units, joined in the 1:4 position.

FUCOPYRANOSE.   The pyranose form of fucose.

α-L-Fucopyranose

153

FUCOSE. A methyl sugar found in brown algae.

$$HO-C-\underset{\underset{OH}{|}}{\overset{\overset{H}{|}}{C}}-\underset{\underset{H}{|}}{\overset{\overset{OH}{|}}{C}}-\underset{\underset{H}{|}}{\overset{\overset{OH}{|}}{C}}-\underset{\underset{OH}{|}}{\overset{\overset{H}{|}}{C}}-C-H_3$$

FUCOXANTHIN, FUCOXANTHOL. $C_{40}H_{56}O_6$. The carotinoid that gives the brown colour to the Phaeophyceae. The light-absorbing power is nearly as good as that of chlorophyll.

FUGACEIOUS. (1) Lasting for a short time.
(2) Soon falling from the parent plant.

FULCRUM. (1) An outgrowth from the wall of the zygospore of some fungi.
(2) A sporophore.

FULIGINOUS. Soot-coloured.

FULVIC ACID. An acid in humus. It is soluble in cold alkali, from which it is precipitated by dilute acid, and is then soluble in water. It also forms the yellow pigment of *Penicillium fulvus*.

FULVOUS. Tawny.

FUMAGINOUS. Smoky-coloured.

FUMARASE. The enzyme catalysing the reciprocal change from malic acid to fumaric acid and water.

FUMARIACEAE (H). A family of the Rhoeadales. The flowers are often zygomorphic. The petals may or may not be connivent, the outer 2 often being saccate or spurred at the base. There are 4 or 6 stamens, sometimes united into bundles. The fruit is a capsule or nut.

FUMARIC ACID. The acid produced from succinic acid in the Citric Acid Cycle.

$$\begin{array}{c} COOH \\ | \\ CH \\ | \\ CH \\ | \\ COOH \end{array}$$

FUMIGACIN. An antibiotic produced by some species of *Aspergillus*.

FUMIGATIN. An antibiotic produced by *Aspergillus fumigatus*. It is active against bacteria, but not fungi.

FUNARIALES. An order of the Eubrya. The gametophores are annual or biennial, and erect, having a rosette of 'leaves' at the apex. The sporophores are mostly acrocarpous, and never have cylindrical capsules. The capsules are frequently bent or pendant at maturity. The peristome is sometimes absent, but when present it is double, but may appear to be simple.

FUNDAMENTAL STAGE OF GROWTH. The formation of new protoplasm.

FUNDAMENTAL TISSUE = GROUND TISSUE.

FUNGALES. An obsolete term for the Fungi and lichens.

FUNGAL CELLULOSE. A carbohydrate, like cellulose in fungal cell walls. Sometimes called 'chitin' (*q.v.*).

FUNGI. The Thallophyta lacking photosynthetic pigments. They are saprophytic or parasitic. The plant body is made up of simple filaments (in the true fungi), hyphae, and asexual (and usually sexual) reproduction is by spores. There is no fission. They accumulate glycogen rather than starch.

FUNGICIDE. A substance which kills fungi.

FUNGICOLOUS. Living on fungi.

FUNGIFORM. Looking like a mushroom.

FUNGI FOSSILES. Fossilized fruit-bodies, on or with fossilized stems or leaves.

FUNGI IMPERFECTI. The fungi with septate hyphae, but producing no known sexual stage.

FUNGISTATIC. A substance, or concentration of a substance, which stops the growth of a fungus without killing it.

FUNGISTEROL. A sterol in the fruiting body of the fungus *Polyporus sulphuens*.

FUNGOID. Fungus-like.

FUNICULAR. Cord-like.

FUNICLE. The stalk of an ovule (and later the seed), by which it is attached to the placenta.

FUNICULUS. (1) = FUNICLE.
(2) Of the Nidulariaceae, the cord of hyphae by which the peridioles are first fixed to the inner wall of the peridium.

FUNICULOSE. Forming ropes of intertwined hyphae.

FUNIFORM. Rope-like.

FUNNEL CELL. A cell in the palisade-layer of a leaf, which is widest just below the epidermis, and narrows below.

FUNNEL CLEFT. The termination of the inner fissure of the raphe of the Chrysophyta.

FURAN.

FURANOSE FORM. Said of a compound built up on a 'five-sided' ring.

FURANOSE SUGAR. An hexose sugar, in its furan form.

FURCATE. Divided into two branches which diverge; forked.

FURFURACEOUS. Covered with bran-like particles; scurfy.

6-FURFURYLAMINOPURINE. A nucleic acid derivative which causes the acceleration of cell-division in tissue culture, assists in shoot initiation, and retards the senscence of leaves. It may assist in the mobilization of amino acids.

FURROWING. The formation of a septum, by the development of a ring of thickening on the inside of a cell-wall. The gradual closing of this ring, by the ingrowing of the wall cuts the cell-cavity in two.

FUSCOUS. Dingy-brown.

FUSEAU. A fusoid macroconidium of dermatophytes.

FUSIFORM. Elongated, and tapering towards each end.

FUSEL OILS. Substances produced by yeasts during alcoholic fermentation, from anaerobic nitrogen metabolism.

FUSION NUCLEUS. The product of fusion of nuclei in the embryo-sac. The endosperm initial.

FUSOID. Rounded in section, and tapering at each end: not markedly elongate.

G

GALACTANS. Hemicelluloses, mucilages, pectins and gums, which yield galactose on hydrolysis.

GALACTOARABAN. A polysaccharide which is a polymer containing galactose and arabinose.

GALACTOPYRANOSE. The pyranose form of galactose, which polymerizes to form the hemicellulose galactan.

GALACTOSE. A hexose sugar.

d-galactose

GALACTOSIDASE. An enzyme which catalyses the hydrolysis of galactosides.

GALACTURONIC ACID. The acid formed from galactose. These units are condensed to form pectic acid, which reacts with calcium or magnesium to form insoluble pectates.

α-D-Galacturonic acid

GALBALUS. A strobilus with fleshy cone-scales.

GALEATE, GALEIFORM. Shaped like a helmet or hood.

GALERCULATE. Covered by a cup-like lid.

GALERIFORM. Cup-shaped.

GALL. An abnormal growth, caused by the attack of a pest or disease organism.

GALLIC ACID. $C_6H_2(OH)_3.COOH$. 3,4,5-tgrihydroxybenzoic acid. It is found in nut galls, tea, etc., and is obtained from tannins by hydrolysis.

GALLO-TANNIN = TANNINS.

GALTON'S LAWS. (1) Of ancestral inheritance, 'Any offspring of bisexual parentage derives $\frac{1}{2}$ of its inherited characters from its parents, ($\frac{1}{4}$ from each), $\frac{1}{4}$ from its grandparents, $\frac{1}{8}$ from its great-grandparents etc. These fractions whose numerators are 1, and denominators are successive powers of 2, add up to make 1, *i.e.* $\frac{1}{2}+\frac{1}{4}+\frac{1}{8}$ —— = 1'.

(2) Of filial regression, 'The offspring of exceptional parents tend to regress to mediocrity, in proportion to the degree of the parents exceptionalness'.

GALVANOTROPISM. A tropism in response to an electrical stimulus.

GAMETANGIUM. Any organ in which gametes are produced. Especially of the Thallophyta. They may fuse directly.

GAMETE. A haploid cell taking place in sexual fusion. The two gamete nuclei, and frequently the cytoplasm, fuse to form a diploid zygote. Gametes may develop directly without fusion.

GAMETIC NUMBER. The number of chromosomes in a gamete nucleus, usually the haploid number.

GAMETOGENESIS. The formation of gametes, including meiosis, if it immediately precedes the formation of gametes.

GAMETOPHORE. A stalk bearing sex-organs.

GAMETOPHYTE. An individual of the haploid generation in the life-cycle. Typically it is produced from a haploid spore, and produces the gametes.

GAMETOPHYTIC BUDDING. The formation of gemmae on a prothallus.

GAMETROPHIC. The movement of organs before fertilization.

GAMOGASTROUS. Said of a syncarpous gynecium in which the ovaries are fused, but the styles and stigmas are free.

GAMOGENESIS = SEXUAL REPRODUCTION.

GAMOGENIC. Resulting from sexual fusion.

GAMOPETALAE (BH) = SYMPETALAE.

GAMOPETALOUS = MONOPETALOUS = SYMPETALOUS. Said of a flower with fused, or partially fused petals.

GAMOPHASE. The haploid phase of the life-cycle.

GAMOPHYLLOUS. Having the perianth members fused.

GAMOSEPALOUS. Said of a flower which has the sepals fused, or partly so.

GAMOSTELY. The fusion of steles.

GAMOTROPIC. Movement before fertilization.

GANGLIFORM. Having knots; knotted.

GARRYACEAE (EP, H). A family of the Garryales (EP), Araliales (H). (BH) include it in the Cornaceae. These are shrubs with stems that are quadriangular in cross-section, and bear opposite, evergreen leaves which have the petioles united at the base. The flowers are unisexual, and are in catkins. The male flower has a perianth of 4 free segments, and 4 free stamens. The female flower has no perianth, and has a simple unilocular ovary, made up of 3 fused superion carpels. It contains 2 pendulous, anatropous ovules, with dorsal raphe, or parietal placentas. The fruit is berry-like, with a thin pericarp, and 2-1 seeds. An endosperm is present in the seeds.

GARRYALES (EP). An order of the Archichlamydeae. These are woody plants with opposite evergreen leaves and the flowers in catkin-like panicles. They are unisexual. The male flower has 4 free petals, and 4 free stamens. The female flower has no perianth, consisting only of a superior, unilocular ovary containing 2 ovules. It is made up of 3-2 fused carpels. The seeds contain endosperm.

GASEOUS EXCHANGE. The diffusion of gases in and out through a cell-wall.

GASTERELLACEAE. A family of the Hymenogastrales. It contains one unilocular genus *Gasterella*.

GASTEROMYCETEAE, GASTROMYCETEAE. An order of the Eubasidiae with 4-2 spored basidia (rarely more), with the fruit-body closed at maturity. May be considered as a subdivision of the Autobasidiomycetes.

GASTERPHORE. A thick-walled, globose, asexual spore in *Ganoderma*.

GASTROSPORIACEAE. A family of the Hymenogastrales, having a double peridium.

GAS VACUOLE. A large cavity in the cells of the Cyanophyta. They may contain gas or a viscous substance, and may help to increase buoyancy.

GEASTRACEAE. A family of the Lycoperdales. The outer peridium has a fibrous layer, which splits, and spreads-out like a star.

GEGENPOL. The pole of a resting nucleus, which lies farthest from the centrosome.

GEISSOLOMATACEAE (EP, H). A family of the Myrtiflorae (EP), Thymelaeales (H). (BH) include it in the Penaeaceae. These are small xerophytic shrubs, with opposite evergreen leaves and solitary axillary flowers, which are bisexual. There are 4 free sepals, no petals, and 8 free stamens arranged in 2 whorls of 4. The superior ovary has 4 fused carpels and 4 loculi, each with 2 pendulous ovules. The fruit is a four-loculate capsule, and the seeds contain endosperm.

GEITONOGAMY. The cross-pollination between two flowers on the same plant.

GEL. The apparently solid, often jelly-like, material formed from a colloidal solution. It offers little resistance to liquid diffusion, and may have as little as 0·5% solid matter.

GEL WATER. The water held in the interstices of the gel-structure of a soil.

GELIDIALES. The tetrasporophytic Florideae, in which the carposporophyte develops directly from the carpogonium.

GEMINATE. (1) Paired.
(2) Of two branches from the same node on the same side of the stem.

GEMINUS = BIVALENT.

GEMMA. A cell, or group of cells which serve for asexual reproduction. Found in some algae, fungi, and bryophytes.

GEMMA CUP. A cup- or crescent-shaped outgrowth from the gametophyte of some liverworts. It contains the gemmae.

GEMMATION. The production of gemmae, or new members of a colony which are attached.

GENE. One of the units of inherited material carried on a chromosome. They are arranged in a linear fashion, and are indivisible, but capable of self-replication. Each represents a unit character, which is recognized by its effect on the individual bearing the gene in its cells. There are many 1000 in each nucleus.

GENE DOSAGE. The number of times a gene is present in a nucleus.

GENERALIZED ROOT-SYSTEM. A root-system with a tap-root and laterals well-developed.

GENERATION. (1) Origin.
(2) Production.
(3) The individuals of a species which are separated from a common ancestor by the same number of generations in direct descent.

GENERATION TIME. The time taken by a unicellular organism between one cell-division and the next.

GENERATIVE APOGAMY = REDUCED APOGAMY.

GENERATIVE CELL. A cell in the pollen grain of the gymnosperms which divides to give a stalk cell and a body cell.

GENETIC NAME. The name of a genus.

GENESIS. The origin, formation or development of a group, species, individual, organ, tissue, or cell.

GENETIC. Pertaining to, or analogous with heredity.

GENETIC COMPLEX. The sum total of the hereditary factors contained in the chromosomes and cytoplasm.

GENETIC EQUILIBRIUM. The condition of a population in which successive generations consist of the same genotype with the same frequencies, in respect of particular genes, or arrangement of genes.

GENETICS. The science of heredity, including the study of its chemical foundation, its developmental expression and its bearings on variation, selection, adaptation, evolution, breeding and the activities of man.

GENETIC SPIRAL. A hypothetical line drawn on a stem, passing by the shortest path through the points of insertion of successive leaves.

GENETIC SYSTEM. The reproductive and hereditary organization of a mating group.

GENETIC VARIATION. Variation due to differences in the gametes.

GENIC BALANCE. The hypothesis that the characters of an organism are each determined by the interaction of a large, but unknown number of genes, some affecting development in one direction and some in another, so that the ultimate result is a balance struck between the total effects.

GENICULATE. Elbowed.

GENISTELLACEAE. A family of the Zoopagales. They are parasites (or commensals) in insect intestines. They have branched coenocytic hyphae.

GENOCENTRE = REPRODUCTOCENTRE.

GENOM, GENOME. The chromosome set, especially when considered genetically.

GENOTYPE. The genetic constitution of an organism.

GENOTYPIC CONTROL. The control of chromosome behaviour by the genotype, in contrast to structural control, especially at meiosis through the effects of dissimilarity of the pairing chromosomes in hybrids.

GENOTYPIC ENVIRONMENT. The aggregate of all the genes, considered as acting on one or more of them.

GENTIANACEAE (EP, BH, H). A family of the Contortae (EP), Gentianales (BH) (H). They are mostly herbaceous, with opposite, exstipulate leaves. The inflorescence is usually some form of cyme. Bracts or bracteoles may or may not be present. The flowers are bisexual, and regular. The calyx have usually 5 imbricate fused petals. The 5 petals are fused, to form a bell of some form. The stamens are free, and the same number of the petals, to which they are fused. The superior ovary has 2 fused carpels, with a glandular disk at the base. The placenta is usually parietal, but commonly project far into the cavity; occasionally the ovary is bilocular with axile placentas. There are many anatropous ovules. The fruit is usually a capsule with 7 cavities, with many seeds, (rarely it is a berry). The seeds are small, with a small embryo in abundant endosperm.

GENTIANALES (BH, H). An order of the Gamopetalae (BH), Lignosae (H). The corolla is regular, and the stamens fused to the petals. The ovary is superior. The leaves are usually opposite.

GENTIANOSE. A trisaccharide, consisting of gentiobiose, and fructose, or sucrose and glucose. It is found in gentian roots.

GENTIOBIOSE. A disaccharide reducing sugar, with a 1:6 linkage.

GENUINE TISSUE. Tissue derived from the division of a mass of related cells, with the subsequent differentation of the daughter cells.

GENUS. A taxonomic rank containing related species. Similar genera are collected into a family.

GEO-. A Greek prefix meaning 'earth'.

GEOCALYACEAE. An obsolete family of the Acrogynae, including all members producing a marsupium.

GEOCARPY. The ripening of fruits underground. The young fruits are pushed into the soil by a post-fertilization curvature of the stalk.

GEOGLOSSACEAE. A family of the Inoperculatae. They are found on rotten wood, moss, or soil. The apothecium is stalked. The ascospores are elliptical, 1-2 celled and filiform.

GEONASTY. Curvature towards the ground.

GEOPHILIC. Growing on soil.

GEOPHILLOUS. Having a short, stout stem, with large leaves borne near the ground.

GEOPHYTE. A plant which perennates by subterranean buds.

GEOPLAGIOTROPIC. Growing in a direction at an angle to the ground surface.

GEOTAXIS. The movement of an entire plant in response to gravity.

GEOTOME. An instrument for taking soil-samples, without disturbing the surrounding soil.

GEOTROPISM. The growth of part of a plant due to the influence of gravity.

GERANIACEAE (EP, BH, H). A family of the Geraniales (EP, BH, H). (BH) include the Oxalidaceae, Limnanthaceae, Tropaeolaceae, Balsaminaceae. They are mostly herbs, and are often hairy. The leaves are alternate, or opposite, often with stipules. The flowers are bisexual, and usually regular. The 5 calyx lobes are free or fused, imbricate, with valvate tips, and is persistent. The 5 free petals are imbricate, or convivient. There are 2 or 3 times as many stamens as petals. They are united at the base. The superior ovary is of 5, 3-2, or 5-3 fused carpels with 2-1, or 2-many ovules in each, on axile placentas. The ovules are usually pendulous, with a ventral raphe, and the micropyle facing upwards. The style is long with 5 stigmas. The fruits are usually a schizocarp, the carpels spitting from a central beak. (BH) characters are slightly different.

GERANIAL. An open-chain olefinic aldehyde. $C_9H_{15}.CHO$.

GERANIALES (EP, BH, H). An order of the Archichlamydeae (EP), Polypetalae (BH), Herbaceae (H). The flowers are regular, with sepals and petals, without petals, or without a perianth. The sepals and petals (where present) are usually in fives. The stamens vary in number. The superior ovary is of 5-2 fused carpels (rarely more), often separating when ripe. They usually contain 2-1 (rarely many) pendulous ovules which have a ventral raphe, with the micropyle upwards, or when many are present, some have a dorsal raphe with the micropyle pointing downwards.

GERANIOL. An open-chain olefinic alcohol, $C_9H_{15}CH_2OH$.

GERM. An embryo plant in the seed.

GERM-CELL = GAMETE.

GERMEN. The ovary.

GERMINABLE. Capable of germination in the right conditions.

161

GERMINAL DISK. A flattened plate of cells developing at the end of the filamentous young thallus of some Hepaticae, and ultimately producing the adult thallus.

GERMINAL TUBE. The tube produced by the elongation of the spore-contents during the germination of a spore of the Hepaticae.

GERMINATION. The intake of water by a seed, spore etc. leading to an increase in metabolism and elongation, finally resulting in the formation of new tissue.

GERMINATION BY REPETITION. Producing secondary spores in place of a germ-tube.

GERM PLASM. The hereditary material.

GERM PORE. A thin-walled area on a spore-wall, or pollen grain through which a germ-tube is produced.

GERM SPORANGIUM. A sporangium formed at the end of a germ-tube produced by a zygospore.

GERM TUBE. (1) A tubular outgrowth, put out by a germinating spore, from which the thallus develops by branching, or a germ sporangium is produced.
   (2) A tube put out by a germinating pollen grain, carrying the male nuclei, and growing down through the style.

GESNERIACEAE (EP, PH, H). A family of the Tubiflorae (EP), Personales (BH, H). These are herbaceous or woody with opposite, simple leaves. The flowers are solitary, or cymosely umbelled. They are bisexual, and zygomorphic. The corolla has 5 fused petals, and is two-lipped. There are 4 or 2 free stamens, sometimes with 3-1 staminodes. The ovary is superior or inferior with two fused carpels, and one loculus. There are many ovules on 2 bilobed parietal placentas. The fruit is a capsule with 4 valves. The seeds contain endosperm.

G-HORIZON = GLEY HORIZON.

GIBBERILLINS. A group of growth-promoting substances found widely in plants, first isolated from the fungus *Gibberella*. They increase general physiological activity, *e.g.* increase cell-elongation, promote germination and flowering.

GIBBOSE, GIBBOUS. (1) Swollen, especially at one side.
   (2) Pouched.
   (3) Convex above, and flattened below.

GIBBSITE. A clay mineral, aluminium hydroxide. It is found in octahedral units locked together in plates.

GIGANTISM. An abnormal increase in size, often associated with polyploidy.

GIGARTINALES. An order of the Florideae, in which the axillary cell is a vegetative cell of the gametophyte.

GILL. One of the vertical plates of mycelium that bear the hymenium of an agaric.

GILL CAVITY. The ring-shaped hollow in the young fruit-body of an agaric. In it the early stages of organization of the gills is completed.

GILL-FUNGI = AGARICACEAE.

GILVOUS. Brownish.

GIMPED = CRENATE.

GINKGOACEAE. The single family of the Ginkgoales.

GINKGOALES. A class of the Gymnospermae. There are long and short shoots. The leaves are fan-shaped with open, dichotomous venation. The microsporangiate strobilus is catkin-like, and spirally arranged, while the macrosporangiate (ovuliferous) strobilus is a stalk, with 2 terminal, naked ovules.

GIRDER. An arrangement of mechanical tissue in a stem, or leaf, to give it support.

GIRDLE. A transverse groove on the dinoflagellates, and on the zoospores of the immobile genera of the Dinophyceae.

GIRDLE STRUCTURE. A type of leaf-structure in which the palisade cells are arranged radially, or converge towards central strands of vascular tissue.

GIRDLING. The condition when a leaf trace arises on the opposite side of the stele from the leaf which it serves, and in reaching the leaf curves widely through the cortex.

GLABRESCENT. (1) Very thinly covered with hairs.
(2) Becoming nearly hairless as it matures.

GLABROUS. Hairless.

GLADIATE. Shaped like a sword-blade.

GLAND. An organ, or cell, secreting a specific substance, sometimes surrounding a cavity.

GLANDULAR. Clothed with glands.

GLANDULAR SERRATE. Having a toothed margin, with glands at the tips of the teeth.

GLANS. A hard, dry, indehiscent fruit, with one or more seeds, derived from an inferior ovary, and more or less enclosed in a cupule.

GLAREAL. Growing on dry, exposed ground.

GLAUCESCENT. (1) Slightly glaucous.
(2) Becoming sea-green.

GLAUCOUS. (1) Covered with a waxy bloom.
(2) Sea-green.

GLEBA. The inner, fertile tissue in the fruit-body of the Gasteromycetes and Tuberales.

GLEBULA. A rounded process from a lichen thallus.

GLEBULOSE. Bearing rounded lumps on the surface of the thallus.

GLEICHENIACEAE. A family of the Leptosporangiatae. The sporangia develop simultaneously. There is no indusium, and the annulus is obliquely transverse.

GLEY HORIZON. A horizon in a soil, characterized by the deposition of secondary hydrated ferric oxide.

GLEY SOIL. A soil developed under impeded drainage, and, in consequence, has a horizon of secondary hydrated ferric oxide. More generally, with a greenish-grey shiny horizon, which may contain little iron.

GLIADIN. A protein, found particularly in wheat grains.

GLIDING GROWTH. The slipping, during growth, of a cell-wall over that of its neighbours, over the surface of contact, so that new areas of contact are made with a contiguous cell, and with neighbouring ones with which no previous contact had been established.

GLIOTOXIN. An antibiotic produced by *Aspergillus fumigatus*.

GLOBOID. A rounded inclusion in an aleurone grain, consisting of a double phosphate of calcium and magnesium, combined with globulins.

GLOBULAR, GLOBOSE, GLOBULOSE. Nearly spherical.

GLOBULARIACEAE (EP, H). A family of the Tubiflorae (EP), Lamiales (H). (BH) include it in the Selagineae. These are herbs or shrubs with alternate, exstipulate, simple leaves. The flowers are in heads on spikes, with, or without an involucre of bracts. The flowers are bisexual. The calyx is persistent, consisting of 5 fused sepals. The corolla is zygomorphic, consisting of 5 fused petals, forming a small upper lip of 2 petals, and a lower, larger 1 of 3 petals. There are 2 long, and 2 short stamens, which are free and borne on the petals. The superior ovary has one loculus, with 1 pendulous anatropous ovule. The fruit is a 1-seeded nut. The seed contains a straight embryo and endosperm.

GLOBULINS. Simple proteins, which are insoluble in water, but dissolve in certain salt solutions, from which they can be salted-out with magnesium sulphate.

GLOCHIDIATE. Having barbed bristles.

GLOCHIDIUM. Hair-like processes on the masses of microsporangia produced by *Azolla*. They serve for attachment to the macrosporangium.

GLOEOCYSTIDIUM. A horny or gelatinous cystidium.

GLOMERATE. Collected in heads.

GLOMERULATE. Bearing small clusters of spores.

GLOMERULE. (1) A small bell-like cluster of spores.
(2) A cluster of short-stalked flowers.

GLOMERULUS. A cymose inflorescence in the form of a crowded head of small flowers.

GLUCAN = GLUCOSAN.

GLUCOMANNAN. A polysaccharide made up of glucose and mannose units.

GLUCOPROTEINS. Compounds formed by a protein with a substance containing a carbohydrate group other than a nucleic acid.

GLUCOPYRANOSE. The pyranose form of glucose.

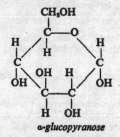

α-glucopyranose

GLUCOSAMINE. $CH_2OH.(CHOH)_3.CHNH_2.CHO$. An amino-sugar. It represents the link between the carbohydrates and proteins.

GLUCOSANS. Carbohydrates, which give only glucose on hydrolysis, and are made up of β-D-glucose only, *e.g.* starch, dextrin, cellulose.

GLUCOSE. A hexose sugar. Only D-glucose exists in plants.

D-glucose

GLUCOSE 1-PHOSPHATE.

GLUCOSE 6-PHOSPHATE.

GLUCOSIDE. A glycoside in which the sugar is glucose, and which is formed from it on hydrolysis.

GLUCOXYLOSES. A group of disaccharide sugars formed between hexose and pentoses.

GLUMACEAE (BH). A series of the Monocotyledons. The flowers are solitary, and sessile in the axils of bracts and arranged in heads of spikelets with bracts. The perianth is of scales, or absent. The ovary is usually unilocular, with 1 ovule. The seeds contain endosperm.

GLUMACEOUS. Thin, brown and papery in texture.

GLUMALES = GRAMINAE.

GLUME. One of a pair of dry bracts, at the base of and enclosing the spikelet of grasses.

GLUMELLA = PALEA.

GLUMIFEROUS. Having the flowers enclosed in glumes.

GLUMIFLORAE (EP, H). A division (order) of the Monocotyledons. They are usually herbs. The flowers lack a perianth (rarely with a hair-like, or true perianth), and are enclosed in bracts (glumes). The superior ovary is unilocular with one ovule.

165

GLUTAMIC ACID. An amino acid. $COOH.CH_2.CH_2CH(NH_2).COOH$ obtained by the hydrolysis of albumenous substances.

GLUTAMIC ACID DEHYDROGENASE. The enzyme that catalyses the reversible reaction α-ketoglutaric acid to glutamic acid.

$$
\begin{array}{ll}
COOH & COOH \\
| & | \\
CH_2 & CH_2 \\
| & | \\
CH_2 + NH_3 + DPNH + H^+ \rightleftharpoons & CH_2 + DPN^+ + H_2O \\
| & | \\
CO & CHNH_2 \\
| & | \\
COOH & COOH \\
\text{α-ketoglutaric acid} & \text{glutamic acid}
\end{array}
$$

GLUTAMINE. An amide. $COOH.CH_2.CH_2CH(NH_2).CONH_2$.

GLUTAMINE SYNTHETASE. The enzyme which catalyses the synthesis o glutamine from glutamic acid and ammonia.

GLUTAMYL PEPTIDES. Peptides which play a part in carrying anions and cations from the cytoplasm to the nucleic acid template.

GLUTATHIONE. A tripeptide, formed from the union of glutamic acid, cystine and glycine. It plays a part in cellular oxidations.

$$
\begin{array}{l}
CH_2.SH \\
| \\
CH.NH.CO.CH_2.CH_2.CH(NH_2).COOH \\
| \\
CO.NH.CH_2COOH.
\end{array}
$$

GLUTELINS. A group of simple proteins, insoluble in water, dilute salts solutions, 70% ethanol, but dissolves in dilute alkalis or acids.

GLUTEN. (1) A reserve protein found in plants.
(2) The sticky coating on the pilei of some agarics.

GLUTINOUS. Covered by sticky substances.

GLYCERALDEHYDE. A triose aldose sugar.

$$
\begin{array}{l}
CHO \\
| \\
HCOH \\
| \\
CH_2OH
\end{array}
$$

GLYCERALDEHYDE PHOSPHATE = GLYCERIC ALDEHYDE PHOSPHATE = TRIOSEPHOSPHATE.

$$
\begin{array}{l}
CH_2O.PO_3H_2 \\
| \\
CHOH \\
| \\
CHO
\end{array}
$$

GLYCERIC DEHYDROGENASE. An enzyme that catalyses the reversible reaction, changing glyceric acid to hydroxypyruvic acid and hydrogen.

GLYCERIDES. The glycerol esters; the most important ones are fats.

GLYCEROL. A trihydric alcohol, which combines with fatty acids to form fats.

$$CH_2OH$$
$$|$$
$$CHOH$$
$$|$$
$$CH_2OH$$

GLYCERYL PHOSPHORYL-CHLORIDE. A substance which is responsible for the transport of some of the phosphorus in the xylem.

GLYCINE. A simple amino acid (α-amino acetic acid) CH_2NH_2COOH.

GLYCININ. A protein in soya bean.

GLYCOCOL. A fungal protein.

GLYCOGEN. A glucosan.

GLYCOL. A diose sugar. $CH_2OH.CH_2OH$.

GLYCOLIC ACID OXIDASE. An enzyme which oxidizes glycolic acid.

GLYCOLYSIS. The anaerobic break-down of a sugar, *e.g.* in an alcoholic fermentation.

GLYCOPHYTES. Plants that are sensitive to a high salt-concentration in the soil.

GLYCOSIDE. The combination of another molecule (ROH) with the H^+ of an OH radicle of a sugar forms the glycoside and water.

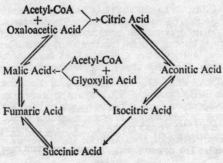

GLYOXALIC ACID = GLYOXYLIC ACID. $CH(OH)_2.COOH$. It occurs in unripe fruit.

GLYOXYLIC ACID CYCLE. A modification of the Citric Acid Cycle. It seems to be limited to tissues where fats are being rapidly consumed, *e.g.* the endosperm of germinating fatty seeds.

Acetyl-CoA
+
Oxaloacetic Acid →→ Citric Acid

Malic Acid ←
Acetyl-CoA
+
Glyoxylic Acid

Aconitic Acid

Fumaric Acid   Isocitric Acid

Succinic Acid

GNETACEAE. The single family of the Gnetales.

GNETALES. An order of the Gymnospermae. It is distinguished by the absence of resin-ducts, having vessels in the secondary wood, and by the presence of a perianth.

GNOMONIACEAE. A family of the Sphaeriales. They are parasites or saprophytes on the stems and leaves of vascular plants. The asci are usually thickened above, with a distinct pore. The perithecia are sunken, with the neck well above the surface. The conidia (when produced) are in gummy acervuli.

GOLGI APPARATUS. A cytoplasmic organelle, found in all cells except bacteria. It consists of a cluster of flattened, parallel, smooth-surfaced sacs, and many smaller vesicles. Its function is obscure.

GOMORTEGACEAE (EP, H). A family of the Ranales (EP), Laurales (H). (BH) include it in the Lauraceae. They are shrubs with opposite, evergreen leaves, with racemes of flowers. The flowers have no distinct sepals and petals, and the perianth, which consists of 7 free segments is spirally arranged. The flowers are bisexual, with 3-2 free stamens, and an ovary of 3-2 fused carpels, each with 1 pendulous ovule. The fruit is a drupe, and the seeds contain endosperm.

GONES. The group of four nuclei, or cells which are the immediate result of meiosis.

GONIDIAL LAYER. The algal layer in a lichen thallus.

GONIDIOPHORE = CONIDIOPHORE.

GONIDIUM. (1) An algal cell in a lichen thallus.
(2) = CONIDIUM.
(3) A non-motile spore formed by some Myxophyceae.
(4) A gemma of some liverworts.

GONIMIUM. A cell of one of the Myxophyceae, when it occurs as part of a lichen thallus.

GONIMOBLAST FILAMENT. In the Florideae, a filament developing from the zygote, and ultimately producing carposporangia.

GONIOAUTOICIOUS. Bearing the antheridia as a bud-like outgrowth from the branch bearing the archegonium.

GONOPLASM. In some Oömycetes, the protoplasm that passes through the fertilization-tube and unites with that of the oosphere.

GONOSPHERE. A zoospore of the Chytridiales.

GONOTOCONT. An organ in which meiosis takes place.

GONYSTILACEAE (EP), GONYSTYLACEAE (H). A family of the Malvales (EP), Thymelaeales (H). (BH) include it in the Thymelaeaceae. They are shrubs with alternate, entire exstipulate leaves, and the flowers in cymes. The flowers are regular and bisexual with 5-4 free sepals and petals. There are many free stamens. The superior ovary consists of 5-3 fused carpels, each with 1 pendulous ovule. The fruit is a berry, and there is no endosperm in the seeds.

GOODENIACEAE (EP, BH, H). A family of the Campanulatae (EP), Campanales (BH), Goodeniales (H). They are herbs or shrubs with radical, or alternate (rarely opposite) exstipulate leaves, and no latex. The bisexual flowers are zygomorphic, and solitary or in cymes, racemes or spikes. The 5 free sepals are usually small. The 5 petals are fused, with the free stamens alternating with them. The ovary is inferior, or semi-inferior, and of 2 fused carpels, forming 2-1 loculi. There are 1, 2, or many ovules in each, and they are generally ascending and anatropous. There is a pollen-cup beneath the stigma. The fruit is usually a capsule, sometimes a nut or drupe. The embryo is straight in a fleshy endosperm.

GOODENIALES (H). An order of the Herbaceae. They are herbs or shrubs with alternate (rarely opposite), simple leaves which may be radical. There are no stipules. The flowers are bisexual and regular to zygomorphic. The calyx tubes are fused to the ovary (rarely free). The corolla lobes are fused, and 2-1 lipped, with the lobes imbricate or valvate. The 5 stamens alternate with the corolla lobes, or there are 2. The filaments are free, or joined around the style. The anthers open lengthwise. The ovary is mostly inferior and of 4-1 loculi. The stigma is indusate.

GOODENOVIEAE (BH) = GOODENACEAE (EP).

GOSSYPINE. Cottony.

GOSSYPOL. A pigment found in cotton seeds.

GOUPIACEAE (H). A family of the Celastrales (H). The leaves are stipulate, and without scales. The flowers are in umbel-like racemes on axillary stalks. The flowers are not enclosed by a bracteole, and the calyx is not enlarged. The petals are valvate. The carpels or ovary are not enclosed by the disk, which is annular or of separate glands. The ovary has 5 loculi, with ovules ascending from the inner angles of the loculi.

GOURMANDISER. A strong, coarsely growing sucker, especially from the stock of a grafted plant.

GRACILIS. Latin meaning 'slender'.

GRADATAE. The megaphyllous pteridophytes, with a basipetal sorus.

GRADATE SORUS. A fern sorus, in which the sporangia develop from the apex of the receptacle downwards.

GRADIENT. The condition when the intensity of a stimulus acting on a plant increases or decreases towards the plant.

GRAFT. A small piece of meristematic tissue, *e.g.* a bud or growing shoot, called the scion, is made to unite with a larger established plant, called the stock.

GRAFT HYBRID. A chimaera produced by grafting dissimilar plants.

GRAMINACEOUS = GRAMINEOUS.

GRAMINALES (H). An order of the Glumiflorae (H). See *Gramineae*.

GRAMINEAE (EP, BH, H). A family of the Glumiflorae (EP), Glumaceae (BH), Graminales (H). The grasses. These are mainly herbs (rarely woody), with joined stems and alternate, two-ranked leaves, with a split sheath and a ligule. The flowers are in panicles, or spike-like inflorescences. The individual flowers are small and bisexual (rarely unisexual), and without a perianth. They are borne in spikelets, each beginning with one or more empty glumes, then paleae with axillary flowers. There are usually 3 free stamens. The superior ovary has 1 ovule, with the micropyle facing downwards. There are 3-1 stigmas. The fruit is a caryopsis with rich endosperm.

GRAMINEOUS. (1) Relating to grasses.
    (2) Grass-like.

GRAMINICOLOUS. Living on grasses, especially of parasitic fungi.

GRANUM. Plate-like structures in chloroplasts. They are stacked together in groups, and have the chlorophyll distributed over the surface.

GRAND PERIOD OF GROWTH. The period of enlargement of a cell, tissue, organ, or organism. This starts slowly immediately after differentiation, increases to a maximum rapidly, and finally falls-off to zero.

GRANATACEAE (H). See *Punicaceae*.

GRANDIFLORUS. Large-flowered.

GRANDIFOLIATE. The leaves are more conspicuous than usual, usually on short stems.

GRANDIS. Latin meaning 'large'.

GRANULAR, GRANULATE, GRANULOSE. (1) Having the surface covered with tiny projecting points, or small warts.
(2) Composed of, or filled with, minute granules.

GRAPHIOLACEAE. A family of the Ustilaginales. They are parasitic on palms. The sori, which are under the epidermis, rupture on spore-dispersal. The spores are in parallel chains, and bud laterally to form 2-4 (or more) sporidia.

GRASSLAND CLIMAX. A climax community consisting of grassland.

GRAVEOLENT. Having a strong, rank smell.

GRAVIPERCEPTION. The perception of gravity by plants.

GRAVITATIONAL INDUCTION. The development of a structure from the underside of a plant-member.

GRAVITATIONAL WATER. Free water, freezing at not less than −1·4°C and available to plants.

GREEN CELL. A cell of the alga *Chlorella*, living inside certain simple animals.

GREGARIOUS. Growing close together, but not matted.

GREYIACEAE (H). A family of the Cunoniales (H). The leaves are simple, with no stipules, or if present they are joined to the petiole. The petiole is sheathing at the base. The flowers have petals. There is a disk, which is crowned by 10 glandular processes. There are 10 stamens. The style is simple, and the ovules are on parietal placentas.

GRIMMIALES. An order of the Eubrya. They are found usually on rocks. The blackish-green gametophores have branched stems with crowded leaves. The sporophyte is acrocarpous, with symmetrical to ovoid capsules. The peristome is simple, with 16 teeth.

GRISEOFULVIN. An antibiotic produced by *Penicillium griseofulvum*.

GRIT CELL. A stone-cell in a leaf or fruit.

GROSSULARIACEAE (H). A family of the Cunoniales (H). The leaves are alternate or in groups, with no stipules, or if they are present, they are joined to the petiole. The flowers, which have petals, are racemose, or sub-solitary. There is no disk. There are 5-4 stamens, and the ovary is unilocular. The fruit is a berry.

GROUND MERISTEM. Those parts of an apical meristem which give rise to the ground-tissue.

GROUND RESPIRATION. The respiration of roots, which is unrelated to salt-absorption.

GROUND TISSUE. The general mass of parenchyma outside and between the vascular strands of a young stem or root.

GROUND WATER. Water naturally contained in, and saturating the sub-soil.

GROWING POINT. The apical meristem of an axis, where cell-division occurs, and differentiation begins.

GROWING ZONE. The portion of an organ in which elongation proceeds.

GROWTH. The increase in size and dry-weight of a plant, or cell, which cannot be reversed. It does not include the increase in size due to the uptake of water.

GROWTH CURVATURE. The curving of an elongating part of a plant organ, due to one side growing faster than the other.

GROWTH FORM = LIFE FORM.

GROWTH INHIBITING SUBSTANCE. A substance formed inside the cells, which slows down or stops growth, often in some other part of the plant. Many of these substances have been identified and synthesized.

GROWTH PROMOTING SUBSTANCE. A substance which promotes, or accelerates growth. It may be formed inside the plant, or may be obtained from an external source.

GROWTH RING. A cylinder of secondary wood laid down in a single season in a stem or root. It is seen as a circle in transverse section, and is not necessarily the same as an annual ring.

GROWTH WATER = AVAILABLE WATER.

GRANULAR VIRUSES. Viruses causing insect-diseases. They are characterized by clumps of small protein-particles inside the host cells. These particles contain the virus.

GRUBBIACEAE (EP, H). A family of the Santalales (EP, H). (BH) include them in the Santalaceae. These are woody plants with opposite, leathery leaves, and small, bisexual, regular flowers. The perianth has 4 free lobes. There are 2 whorls of 4 free stamens. The inferior ovary is of 2 fused carpels, with 2 loculi when young, and 1 when older. There are 2 pendulous orthotropous ovules on central placentas. The fruit is a drupe, and the seeds contain an oily endosperm.

GRUMOUS. Having a flesh composed of little grains.

GUANIDINE. Imino-urea, $NH:C(NH_3)_2$.

GUANINE. A purine found in DNA.

GUANOSINE. A nucleoside of guanine, formed by a ribose, or deoxyribose joining the guanine in the N9 position.

GUARD CELLS. A pair of reniform cells bounding a stoma. The walls are differentially thickened, and the cells contain chloroplasts.

GULLYING. Erosion of soil by running water, when it is confined to narrow channels.

GUMMOSIS. A pathological condition, usually physiological, or caused by bacteria. The cells dry-out causing an enzymatic break-down of the cell-walls to form a gum.

GUMS. These are colloidal plant products, which either dissolve, or swell in water. On hydrolysis they give complex organic acids, pentoses, and hexoses.

GUTTA. (1) An oil drop in a spore or fungal hypha.
(2) A general term for a small vacuole.

GUTTATE. Containing small drops of material.

GUTTATION. The secretion of water from hydathodes, which are commonly at the end of the main veins of leaves.

GUTTER-POINTED. Pointed with the point channelled above to form a spout.

GUTTIFERAE (EP, BH). A family of the Parietales (EP). (BH) exclude the Hypericaceae, but include the Quiinaceae. For (H) see *Clusiaceae*. They are nearly all trees or shrubs, with simple, entire, opposite, exstipulate leaves. Oil-glands or passages are always present. The inflorescence is cymose, or often umbellate. The flowers show considerable variety. Bracteoles are present, and often very close to the calyx. The flower-parts may be in whorls, or partially spirally arranged. The flowers are bisexual and regular. The 5 free calyx lobes are imbricate, and the 5 free petals are imbricate or convolute. There are many stamens, which are free, or united in various ways. The superior ovary is of 5 or 3 fused carpels, with many to 1 loculi. There are many, few, or 1 anatropous ovules. The fruit is a capsule, berry or drupe, and there is no endosperm in the seeds.

GUTTIFERALES (H). An order of the Lignosae. They are mostly woody, with a few herbs. The leaves are opposite, and frequently dotted with glands. Stipules are rare. The flowers are bisexual or unisexual, with imbricate sepals. The stamens are often united into separate bundles, and the seeds rarely contain endosperm.

GUTTULATE. Containing a gutta, or guttae.

GYMNO-. A Greek prefix meaning 'naked'.

GYMNOASCACEAE. A family of the Aspergillales, the members of which lack a firm-walled perithecium.

GYMNOCARPACEAE. A series of the Ascolichens, in which the ascogonium is an apothecium.

GYMNOCARPOUS. Having the hymenium exposed at an early stage of development.

GYMNOCYTE. A cell without a cell-wall.

GYMNODINALES. An order of the Dinophyceae. The protoplasts do not have a cell-wall, or if one is present, it is not divided into plates. The vegetative cells are always motile, with typical dinophycean flagella.

GYMNOGRAMMOID FERNS. The ferns which lack a sorus, and have the sporangia distributed over the lower surface of the leaf.

GYMNOPLASM. An amorphous mass of naked protoplasm.

GYMNOSPERMAE. A class of vascular plants. The seeds are exposed on the sporophyll, *i.e.* the sporophyll is not infolded to form an ovary; and the endosperm is formed before fertilization.

GYMNOSPERMOUS. Having the seeds exposed; not enclosed in an ovary.

GYMNOSTROMOUS. Lacking a peristome.

GYNAECIUM. (1) The group of archegonia in mosses.
(2) The carpel, or carpels in a flower.

GYNANDROUS. Having the stamens and styles united to form a column.

GYNAPHORE. A special structure on the fruit of the peanut (*Arachis hypogaea*), for absorbing ions.

GYNECIUM = GYNAECIUM.

GYNOBASIC. Said of a style which arise near the base of the carpels or ovary-lobes.

GYNOBASIS, GYNOPHORE. An elongation of the receptacle of a flower, forming a short stalk to the ovary.

GYNODIOECY. The condition in a plant population or species in which female and hermaphrodite individuals are formed.

GYNOMONOECY. The condition of plant individuals or species having individuals which bear both female and bisexual flowers.

GYNOPHORE. (1) = GYNOBASIS.
(2) Of the Pyronemaceae, the developing multinucleate female structure.

GYNOSPORE = MEGASPORE.

GYRATE. (1) = CIRCINATE.
(2) Curved into a circle.

GYROSE. Having a folded surface, marked with sinuous lines or ridges.

GYROSTEMONACEAE (H). A family of the Chenopodiales (H). The flowers have no petals, and the anthers open by longitudinal slits. The ovary has 2 or more loculi, with the carpels joined around a central column, which is often dilated into a flat, disk-like top. The fruits are free and dehiscent.

H

H. A flower-class, containing flowers which are usually zygomorphic with a corolla-tube 6-15 mm. long, and suited to pollination by bees.

H-PIECE. A short lateral hypha, joining 2 longer hyphae, and probably aiding in the flow of food-materials through the mycelium.

HABIT. The general external appearance of a plant.

HABITAT. The immediate environment occupied by an organism.

HABITAT FORM. A plant showing features, which are abnormal, but can be related to the place where it is growing, *e.g.* dwarfing under poor conditions.

HABITAT GROUP. A set of unrelated plants which occupy the same kind of situation.

HADROCENTRIC VASCULAR BUNDLE. A concentric vascular bundle in which the xylem is surrounded by phloem.

HADROMASE, HADROMAL. An enzyme present in some fungi, which enables them to decompose wood.

HADROME. The conducting tissues of the xylem.

HAEM. One of a group of iron-porphyrins, which are conjugated with proteins to form peroxidase, catalase, and all the cytochromes.

HAEMATOCHROME. An orange-red pigment, probably a carotinoid, found in some Chlorophyceae.

HAEMATIN. See *Haem*. Its derivatives are present in living cells, and play an important part in cellular oxidations. They exist in combination with nitrogenous organic substances, which are called *haemochromogens*.

HAEMOCHROMOGENS. See *Haematin*.

HAEMODORACEAE (EP, BH, H). A family of the Lilliflorae (EP), Epigynae (BH), Haemodorales (H). (BH) include part of the Liliaceae, and part of the Amaryllidaceae. These are herbs with a panicled inflorescence, of a number of cymes arranged in a racemous way. The flowers are regular or zygomorphic, bisexual, with the flowers parts in threes. The 3 free stamens are joined to the inner perianth-lobes, and have introrse anthers. The ovary is of 3 fused carpels, and is superior, or inferior. There are a few ovules in each loculus, and they are semianatropous. The stigma is capitate, and the fruit is a capsule.

HAEMODORALES (H). An order of the Corolliferae (H). The root stock is a rhizome, rarely a corm. The leaves, which are often very hairy, are entire, often all radical (rarely lobed). The bisexual flowers are solitary, in panicles, or sub-umbels. The perianth mostly has distinct tube-segments, or lobes in 1 or 2 whorls. There are many to 6 (rarely 3) stamens, which are free or in bundles. The anthers are bilocular, and open lengthwise. The ovary is superior to inferior, either with 3 loculi and axile placentas, or one loculus and parietal placentas. The seeds contain endosperm.

HAERANGIOMYCETES. A 'class' of the Ascomycetales, including some of the members in which the ascus does not have a definite cell-wall.

HAERANGIUM. A funnel-like extension formed by filaments from the edge of an ostiole.

HAFT. (1) A leaf stalk, with a strip of photosynthetic tissue running along each side of it to form a wing.
    (2) The stalk of a spatulate leaf.
    (3) The claw of a petal.

HAIR. A uni- or multicellular epidermal outgrowth. It may be absorptive, *e.g.* root-hair, secretory, protective, or reducing the rate of transpiration.

HAIR-PIT = CRYPTOSTOMA.

HALF-INFERIOR. Said of a flower in which the receptacle forms a cup which is joined to the base of the ovary and partly up its side.

HALF-RACE. A race of plants in which only a few of the seedlings show the characteristic of the race, the others have the ordinary characteristic of the species, and in which selection does not lead to the fixing of a pure race.

HALICYSTACEAE. A family of the Siphonales, in which there is only a plasma-membrane between the gametangium, and the vegetative parts of the thallus.

HALIPLANKTON. The plankton of the seas.

HALLOYSITE. A montmorillonite clay, $Al_2O_3.2SiO_2.O.3H_2O$.

HALLOYSITIC ACID. A possible intermediary in the degradation of orthoclase felspar.

$$Si(OH)_2$$

$$Al(OH)$$

HALOBIONTIC. Strictly confined to salt-water.

HALOGENIC SOILS. Soils developing in the presence of sodium salts.

HALONATE. (1) Of a leaf-spot fungus whose symptoms are concentric rings.
(2) Of a spore, having a ring around it.

HALOPHILE. A fresh-water species, capable of surviving in salt-water.

HALOPHOBE. A plant which will not grow in a soil containing an appreciable amount of salt-water.

HALOPHYTE. A plant living in soil containing salt-water in appreciable amounts, where in consequence, there is a physiological drought.

HALORRHAGACEAE (EP, H), HALORAGEAE (BH). A family of the Myrtiflorae (EP), Rosales (BH), Lyrthales (H). (BH) include the Callitrichaceae, and Hippuridaceae. These are hydrophytes of various habit, with a great development of adventitious roots. The leaves are opposite, alternate or whorled, and usually without stipules. The flowers are solitary or in an inflorescence and are inconspicuous. The regular flowers are bisexual or unisexual, with bracts. The perianth has 1 or 2 whorls of 4 lobes, or the perianth is absent. There are 2 whorls of 4 free stamens, or less. The inferior ovary consists of 4-1 fused carpels. It has many loculi, with usually 1 pendulous, anatropous ovule in each. The fruit is a nut or a drupe. The seeds have a straight embryo, and endosperm.

HAMAMELIDACEAE (EP, H), HAMAMELIDEAE (BH). A family of the Rosales (EP), Hamamelidales (H). (BH) include it in the Myrothamnaceae. These are trees with usually alternate, simple or palmate, exstipulate leaves. The inflorescence is a raceme, often a spike or head, frequently with an involucre of coloured bracts. The flowers are bisexual or unisexual, often lacking petals, but rarely without both calyx and corolla. The 5-4 free sepals are usually imbricate. The 5-4 free petals are open or valvate, often rolled like a watch-spring in the bud. There are 5-4 free stamens (rarely fewer). The ovary is superior to inferior with 2 fused carpels, and 2 loculi. There are 1 or more pendulous, anatropous ovules in each loculus. A ventral or lateral raphe is present. The fruit is a capsule. The embryo is straight, and there is endosperm.

HAMAMELIDALES (H). An order of the Lignosae (H). These are trees or shrubs, with simple, alternate (rarely opposite), leaves, which are mostly stipulate. The flowers are usually bisexual and regular, collected into heads or catkins. Petals may or may not, be present. The stamens are perigynous or sub-epigynous. The ovary is inferior to semi-inferior (rarely superior) often with 2 carpels. The pendulous ovules are borne on axile placentas. The seeds have a rather thin endosperm and a straight embryo.

HAMATE, HAMULOSE. Said of (1) a narrow leaf hooked at the tip.
(2) A bent trichome.

HANDLE CELL. One of a group of cells supporting the outer cover of the globule of the Charales. This leaves spaces in which the antheridial filaments develop.

HANNA SOIL. A Czechoslovakian soil of the chernozem type.

HAPANTHOUS, HAPAXANTHIC. Flowering once, then dying.

HAPLOBIONT. (1) A plant which has only one type of individual in its life-cycle.
(2) Having one type of thallus which is typically haploid.

HAPLOBIONTICAE. A sub-class of the Ascomycetes in which the haploid generation is more important than the diploid.

HAPLOCAULESCENT. Having a single axis.

HAPLOCHLAMYDEOUS CHIMAERA. A periclinal chimaera in which one component is present as a single-celled layer, forming an epidermis.

HAPLODIOECIOUS = HETEROTHALLISM.

HAPLODIPLONT. A sporophyte in which the cells contain the haploid number of chromosomes.

HAPLOGONIDIUM, HAPLOGONIUM. A gonidium, produced singly, not in a group.

HAPLOHETEROECIOUS = HETEROTHALLIC.

HAPLOID. Having a single set of unpaired chromosomes.

HAPLOMONOECIOUS = HOMOTHALLIC.

HAPLONT. A plant which reproduces sexually, having a diploid zygote, while all the other cells are haploid.

HAPLOPHASE. The haploid stage in the life-cycle of plants with an alternation of generations.

HAPLOPOLYPLOID. A plant derived from a polyploid by haploid partheno-genesis, and therefore having half the number of chromosomes of the original polyploid.

HAPLOSIS. The halving of the number of chromosomes at meiosis.

HAPLOSTELE. A protostele, with a smooth circular outline in transverse section.

HAPLOSTEMONOUS. Having a single whorl of stamens.

HAPLOSTICHINEAE. A sub-class of the Heterogeneratae. The sporophyte is trichothallic, made up of free to closely packed filaments. The sporophyte produces zoospores or neutral spores. The gametophytes are microscopic, and are isogamous or oogamous.

HAPLOSTROMATIC. Having only an ectostroma or endostroma.

HAPLOSYNOECIOUS = HOMOTHALLIC.

HAPLOXYLIC. Said of a leaf having one vascular strand.

HAPTERES. (1) The root-like branches of the hold-fast of the brown algae. (2) Organs of attachment of some angiosperms living in fast-running water.

HAPTERON. A cell, or cellular organ attaching a plant to the substrate. Used especially of the Thallophyta.

HAPTONASTY. A movement induced by touch, but not influenced by the direction of the stimulus.

HAPTOTROPISM. A tropism in response to contact, *e.g.* a tendril twining around the object it has touched.

HARD BAST. Sclerenchyma produced in the phloem.

HARD PAN. An inpenetrable layer formed in the B-horizon of a soil.

HARDY-WEINBERG LAW. That the random mating of individuals is equivalent to the random union of gametes.

HARPELLACEAE. A family of the Zoopagales. They are parasites (or commensals) in the alimentary canal of aquatic insects, and are distinguished by having unbranched, flexuous, coenocytic hyphae, and spherical zoospores.

HARTIG NET. In an ectotrophic mycorrhiza, the mycellium inside the cells of the root disintegrates, leaving a network of hyphae in the inter cellular spaces (the Hartig net). From this the mycellium grows out to form a sheath over the root.

HASTATE. (1) Having two somewhat out-turned lobes at the base.
(2) Halbert-shaped.

HAULM. A stem of a grass.

HAUSTORIUM. A specialized branch of organ of a parasite, which penetrates the host tissue, and absorbs nutrients and water.

HEAD. (1) A group of conidia and sterigmata crowded into a dense mass, which is rounded in outline.
(2) A dense inflorescence, of small, crowded, usually sessile flowers, usually surrounded by an involucre.

HEARTWOOD. The central xylem of tree-trunks, frequently impregnated with tannins, resins etc. which can be considered to be excretory products. It contains no living cells.

HEBETATE. Having a blunt or soft point.

HELICOID. Coiled like a spring.

HELICOID CYME. A sympodium in which the branches all develop on the same side of a relatively main axis, but not in the same plane.

HELICOID DICHOTOMY. A branch system of repeated dichotomies, giving a weak and strong branch, the latter always being on the same side.

HELICOSPORE. A coiled, cork-screw-shaped spore, which is generally septate.

HELIOPHYTE. A plant able to live in a very sunny situation.

HELIOSCIPHYTE. A plant which will tolerate shade, but grows better in the sun.

HELIOTROPISM = PHOTOTROPISM.

HELIOZOOID. Amoeboid, but having definite ray-like pseudopodia.

HELLEBORACEAE (H). A family of the Ranales (H). The flowers have no disk, but the torus is sometimes much enlarged with sunken carpels. The stamens are centripetal, or few, with introrse anthers. The ovary consists of more than one carpel, and each carpel contains more than one ovule. The ovules are basal or axile, inserted on the adaxial suture of the carpels. The fruit is a follicle, rarely berry-like, or connate, and dehiscing at the apex.

HELMINTHOID. Worm-shaped.

HELOBIEAE (EP). An order of the Monocotyledoneae. These are marsh or water plants, with scales in the axils. The flowers are cyclic or hemicyclic, with the perianth in 1 or 2 whorls, or absent. The perianth lobes may all be similar, or in a distinct calyx and corolla. The flowers are hypogynous or epigynous. There are many to 1 free stamens, and many to 1 carpels, which are free or united. The seeds contain little or no endosperm.

HELODRIUM. A thicket formation.

HELOPHYTE. A bog plant.

HELOTIACEAE. A family of the Inoperculatae. Most are parasitic or saprophytic on plant tissue. The apothecia do not originate in sclerotia. The apothecia are mostly fleshy, and disk- or cup-shaped. They are closed at first, and often stalked. The excipulum consists of filamentous hyphae, sometimes grading into an outer layer of shorter, thicker cells.

HELOTIALES. An order of the Ascomyceteae (doubtful), and includes certain Pezizales. Most are saprophytic, but some are parasites. The apothecium is disk- to cup-shaped, dark or brightly coloured, and sessile or stalked. The paraphyses generally make an epithecium. The ascospores are usually 1-celled, though frequently septate.

HELOTISM. A form of symbiosis, in which one partner benefits more than the other, *e.g.* the fungus in a lichen thallus benefits more than the alga.

HELVELLACEAE. A family of the Operculatae. The apothecia are stalked and convex, and attached to the apex of the stalk, or grow fast to its upper portion.

HELVELLALES. An order of the Euacsomycetae. The ascocarp is sessile or stalked, with a freely exposed hymenium, which is smooth or wrinkled with an everted ascogenous layer. The asci are parallel with one another.

HEMI-. A Greek prefix meaning 'half; partial'.

HEMIANGIOCARPIC. (1) Said of the fruit-body of a fungus, when the hymenium begins its development enclosed, but is exposed at maturity.
(2) Said of a sporocarp, which opens just before it is mature.

HEMIASCUS. An atypical, multispored ascus.

HEMIAUTOPHYTE. A parasite that contains chlorophyll, so that it can manufacture some carbohydrate.

HEMIBASIDAE. A sub-class of the Basidiomycetae, in which the basidia develop from a special resting-spore, rather than directly from a hymenium.

HEMICARP = MERICARP.

HEMICELLULOSES. These are carbohydrates related to cellulose, but are mixed polysaccharides, containing other sugars besides glucose. They are easily hydrolysed by dilute acids.

HEMICRYPTOPHYTE. A plant which develops its buds just above, or below the soil-surface, where they are protected.

HEMICYCLIC. Said of a flower when some parts are in spirals, and others in whorls.

HEMIFORM. A form of life-cycle in the Uredinales, which includes only uredospores and teleutospores.

HEMIKARYON. A cell with the haploid number of chromosomes.

HEMIPARASITE = FACULTATIVE PARASITE.

HEMISAPROPHYTE = FACULTATIVE PARASITE.

HEMISPHAERIACEAE. A family of the Hemisphaeriales. A mycelium is lacking, or superfical and reticulate. The cover of the stroma is not radial in structure, and under it may be a single hymenium, with, or without paraphysis-like threads, or several smaller hymenia may be produced under one cover.

HEMISPHAERIALES. An order of the Ascomyceteae. Its members are entirely superfical, or sub-cuticular, or have a hypodermal stroma connected to an epiphyllous stroma by strands of hyphae emerging through the stomata or other openings.

HEMISPORE. Used especially for the fungi attacking skin: a cell at the end of a filament, which later becomes a deuteroconidium by division.

HEPATICAE. A class of the Bryophyta. The gametophytes are dorsiventrally differentiated, and develop the sex-organs terminally, or from superficial layers of the dorsal surface of the thallus. The sporophytes are all alike, in that they are strictly limited in their growth.

HEPATOPHYTA = HEPATICAE.

HEPTAMEROUS. Having the parts in sevens.

HEPTANDROUS. Having seven stamens.

HEPTOSE. A monosaccharide with seven carbon atoms. $C_7H_{14}O_7$.

HERB. A plant having no persistent parts above the ground.

HERBACEAE (H). A division of the dicotyledons, including the herbs, and the shrubs derived from them. All its members are basically not woody.

HERBACEOUS. Soft and green, containing little woody tissue.

HERBARIUM. A collection of preserved plant material.

HERBICOLOUS. Living on herbs.

HERCOGAMY, HERKOGAMY. The condition of a flower, when the stamens and stigmas are so placed, that self-pollination is impossible.

HEREDITY. The process whereby the qualities of the parents are passed on to the offspring.

HERKOGAMOUS. Bisexual, but incapable of self-fertilization.

HERMAPHRODITE = BISEXUAL.

HERNANDIACEAE (EP, H). A family of the Ranales (EP), Laurales(H). (BH) include it in the Lauracae. They are trees or shrubs with alternate, exstipulate leaves. The regular flowers are bisexual or unisexual. The perianth consists of 10-4 free lobes. The stamens are free in a whorl before the outer perianth. The carpels are united to form an inferior, unilocular ovary, with 1 pendulous, anatropous ovule. The fruit is often winged. The embryo is straight, and there is no endosperm.

HESPERIDIUM. A fleshy fruit formed from a superior, syncarpous gynecium. The flesh is formed from fluid-filled hairs projecting into the loculi, *e.g.* an orange.

HETERANDROUS. Having stamens that are not all the same size.

HETERO-. A Greek prefix meaning 'different; not normal'.

HETEROAUXETIC COEFFICIENT = ALLOMETRIC COEFFICIENT.

HETERAUXIN = β-INDOLYLACETIC ACID.

HETEROBASIDIAE. A sub-class of the Basidiomyceteae, in which the basidia are variously septate.

HETEROBASIDIUM. A septate basidium.

HETEROBRACHIAL. A chromosome bent into two parts of unequal length.

HETEROCAPSALES. An order of the Xanthophyceae. They are palmelloid. The immobile vegetative cells can return directly to the motile phase, or the zoospores have the ability to divide directly into new zoospores.

HETEROCARPOUS. Having more than one kind of fruit.

HETEROCARYOSIS. Of fungi, the property of cells with two or more dissimilar nuclei.

HETEROCHLAMYDEOUS. Having a distinct calyx and corolla.

HETEROCHLORIDALES. An order of the Xanthophyceae, including all those with flagellated vegetative cells.

HETEROCHROMATIN. Part of a chromosome containing no known genes, differing in nucleic acid content from euchromatin, and staining darker than it. It probably controls the nucleic acid metabolism of the nucleus.

179

HETEROCHRONOGENOUS SOIL. A secondary soil.

HETEROCOCCALES. An order of the Xanthophyceae. They are non-filamentous. The vegetative cells are immobile and surrounded by a wall. They are incapable of returning directly to a motile phase.

HETEROCYCLIC COMPOUNDS. Cyclic, or ring, compounds containing carbon and other elements as part of the ring.

HETEROCYST. A special type of cell produced in most filamentous Myxophyceae. They differ from the other cells in the structure of the walls and their transparent contents. They function as spores.

HETERODYNAMIC. Of unequal potentiality.

HETERODYNAMIC HYBRID. A hybrid which resembles one parent more than the other.

HETEROECIUS. Of parasitic fungi which produce different forms of spores on different and unrelated hosts.

HETEROECY. (1) = DIOECIOUS.
(2) = HETEROECIOUS.

HETEROFERTILIZATION. The fertilization of the endosperm nucleus and egg-nucleus by gametes of different genetic constitution.

HETEROGAMETANGIC. Having gametangia of more than one type.

HETEROGAMY. (1) = ANISOGAMY.
(2) Having male, female, hermaphrodite, and neuter (or any 2 or 3 of these types) in one inflorescence.

HETEROGENEITY. (1) Of a population containing several different species.
(2) Lack of agreement between different bodies of data in regard to the value of one or more parameters of which they are all capable of yielding estimates.

HETEROGENERATAE. A class of the Phaeophyta in which the two alternating generations are unlike in their vegetative structure.

HETEROGENIC. Of a population, or gamete, containing more than one allelomorph of a particular gene or genes.

HETEROGONIC COEFFICIENT = ALLOMETRIC COEFFICIENT.

HETEROICOUS. Said of the Bryophyta which have more than one kind of arrangement of the antheridia and archegonia on the same plant.

HETEROKARYOSIS. Of a cell having more than one genetically different nucleus.

HETEROKARYOTISE. The fusion of haploid structures of opposite sex, which does not give a conjugate arrangement of the nuclei.

HETEROKINESIS. Differential division of chromosomes.

HETEROKONTAE = XANTHOPHYCEAE.

HETEROMERAE (BH). A series of the Gamopetalae (BH). The ovary is usually superior, with more than two carpels. The stamens are epipetalous, or free from the corolla, and opposite or alternate with its segments, or twice as many.

HETEROMEROUS. Said of a lichen thallus in which the layer of algal cells lies between two layers of fungus hyphae.

HETEROMORPHIC CHROMOSOMES. Homologous chromosomes which are different in size or form, especially at meiosis.

HETEROMORPHIC INCOMPATIBILITY. Incompatibility associated with, or dependent on morphological variations.

HETEROMORPHOUS. (1) Existing in more than one form.
(2) Having more than one kind of flower on the same plant.

HETEROPHYLLY. Bearing leaves of two different forms.

HETEROPLOID = ANEUPLOID.

HETEROPUCNOSIS. Excessive charging of heterochromatin with nucleic acid at meiosis, and premeitotic divisions.

HETEROPYXIDACEAE (H). A family of the Rhamnales. The leaves are dotted with glands. The calyx lobes are imbricate, and there are many ovules of axile placentas.

HETEROSIPHONALES. An order of the Xanthophyceae. It contains all the multinucleate siphonaceoes members.

HETEROSIS. Hybrid vigour. The increase in vigour, size, and fertility of a hybrid as compared with its parents, resulting from the union of genetically different gametes, and assumed to be due to special recombinations of dominant and recessive genes.

HETEROSPORANGY. The formation of more than one kind of sporangium, containing more than one kind of spore.

HETEROSPOROUS. Producing large megaspores which give rise to the female gametophyte, and small microspores which give rise to male gametophytes.

HETEROSTYLACEAE (H) = LILAEACEAE (H).

HETEROSTYLY. The division of a species into two or three kinds of individual by the relative positions of the stigma(ta) and anthers.

HETEROSYNAPSIS. The pairing of two dissimilar chromosomes.

HETEROTHALISM. Referred to Algae and Fungi having separate, physiologically different, male and female thalli; or in the fungi, both types of sex-organ may be present on the same thallus, but self-fertilization is impossible. Sexual reproduction only occurs when two appropriate strains come together.

HETEROTROPH. A true saprophyte.

HETEROTROPHIC. Needing an organic source of food. Unable to synthesize food from simple inorganic substances, *e.g.* fungi.

HETEROTRICHALES. An order of the Xanthophyceae. All the members have cells joined end-to-end to form a simple or branched filament.

HETEROXENY = HETEROECIOUS.

HETEROZYGOTE. A zygote derived from the union of gametes dissimilar in respect for the quality, quantity, or arrangement of their genes, so that it is heterozygous for one or more allelomorphs.

HEX-. A Greek prefix meaning 'six'.

HEXANDROUS. Having six stamens.

HEXARCH. Having six protoxylem strands.

HEXAPLOID. An individual having six sets of chromosomes.

HEXASOMIC. An otherwise normal diploid, but having one chromosome represented six times.

HEXOKINASE. The enzyme which combines the phosphate in ATP with glucose to give glucose 6-phosphate.

HEXOSANS. Hemicelluloses formed from hexose sugars.

HEXOSE. A monosaccharide sugar, having six carbon atoms. The general formula is $C_6H_{12}O_6$.

HEXOSE PHOSPHATES. Hexose sugars in which one or more of the $-$ OH groups is replaced by an H_3PO_4 group, *e.g.* $C_6H_{11}O_5(PO_4H_2)$.

HEXOPHOSPHATE DEHYDROGENASES. An enzyme catalysing the reaction:— Glucose-6-phosphate $+$ TPN$^+$ = 6-phosphogluconate $+$ TPNH $+$ H$^+$.

HIANS. Latin meaning 'gaping'.

HIASCENT. Becoming wide open.

HIBERNATION. Remaining quiescent during the winter.

HIEMAL ASPECT. The appearance and condition of a plant community during the winter.

HIEMATIS. Latin meaning 'winter'.

HIGH ENERGY PHOSPHATE = ADENOSINE TRIPHOSPHATE.

HIGHER FUNGI. The ascomycetes and basidiomycetes.

HILL REACTION. Isolated chloroplasts, in suspension, when illuminated reduce ferric ions, and produce oxygen at the same time. They cannot utilize carbon dioxide, so that it was assumed that the illuminated chloroplasts, on their own, reduced water, the hydrogen being removed by an oxidizing agent.

$$4Fe^{+++} = 2H_2O \quad \overset{\text{light}}{\underset{\text{chloroplasts}}{=}} \quad 4Fe^{++} + 4(H)^+ + O_2.$$

HILL WASH = COLLUVIUM.

HILUM. (1) The scar on the testa of a seed where it was attached to the funicle.

(2) The lateral depression in which the flagella are inserted in reniform zoospores.

(3) A small granule at the centre of a starch grain.

(4) The scar on a spore at its point of attachment to a conidiophore or sterigma.

HIMANTANDRACEAE (EP, H). A family of the Ranales (EP), Magnoliales (H). These are trees with alternate, exstipulate leaves, with shield-like hairs. The flowers have bracts, with the perianth lobes, free and in spirals. The flowers are bisexual. The many free stamens are perigynous, and there are staminodes. The carpels are united at the base.

HIMANTOID. Said of a mycelium which spreads out in fan-like cords.

HINGE. A thin strip in the wall of a guard-cell, about which movement can occur.

HIPPOCASTANACEAE (EP, H). A family of the Sapindales (EP, H). (BH) include it in the Sapindaceae. They are trees or shrubs. The leaves are opposite, and exstipulate. There are more than 2 sepals. A disk is present. There are few ovules. The fruit is a trilocular, or 2-1 locular capsule, which is not compressed contrary to the septum. The leaves are digitate.

HIPPOCRATEACEAE (EP, H). A family of the Sapindales (EP), Celastrales (H). (BH) include it in the Celastraceae. These are shrubs, mostly lianes, with opposite or alternate simple leaves. The regular, bisexual flowers are in cymes, and have a disk. There are 5 free calyx lobes, and a similar number of free petals. There are 3 free stamens (rarely 5, 4, or 2). The superior ovary has 3 fused carpels, with 10-2 anatropous ovules in each loculus. The fruit is a berry or schizocarp. The seeds have no endosperm.

HIPPOCREPIFORM. Shaped like a horse-shoe.

HIPPURIDACEAE (EP). A family of the Myrtiflorae (EP). (BH) include it in the Haloragidaceae. These are water-plants, with creeping rhizomes and erect shoots, whose upper parts usually project above the water. The leaves are linear, in whorls, the submerged ones are longer and more flaccid than the aerial. The flowers are sessile in the axil of the leaves. They are sometimes bisexual (or sometimes female on some stocks). There is 1 stamen, and 1 carpel, with 1 pendulous ovule, and no integuments. They are wind-pollinated.

HIRCINUS. Latin meaning 'with a goaty smell'.

HIRSUTE. Having long, but not stiff, hairs.

HIRSUTIDIN. An anthocyanin.

HIRTOSE, HIRTOUS. Having hairs.

HIRTUS. Latin meaning 'hirsute'.

HISPID. With rough bristly hairs.

HISPIDULOUS. Somewhat or delicately hispid.

HISTIDINE. An amino-acid.

$$\underset{HC\!=\!\!=\!\!=\!C.CH_2.CH(NH_2).COOH}{\overset{\overset{\displaystyle CH}{HN\quad N}}{|\qquad\quad|}}$$

HISTOGEN. See *Histogen Theory*.

HISTOGENESIS. The differentiation of tissues from undifferentiated cells derived from a meristem.

HISTOGENOUS. (1) Produced from tissues.
(2) Of spores, produced from hyphae or cells, without conidiophores.

HISTOGEN THEORY. The theory that the tissues of a stem or root form definite regions called histogens.

HISTOLOGY. The study of individual tissues.

HISTOLYSIS. The disintegration of tissues, due to the disappearance or solution of the walls or tissues.

HISTONES. Simple proteins with a relatively high molecular weight, and having a large variety of amino-acids; but they are highly basic. Histones often occur combined with nucleic acids.

HOARY. Covered with a short greyish-white down.

HOLARD. The whole of the water in the soil.

HOLDFAST. An organ, usually branching, attaching a thallophyte to its substrate.

HOLO-. A Greek prefix meaning 'complete'.

HOLOBASIDIAE. A division of the Heterobasidiae, in which the basidia are not actually septate, but forked, or the hypobasidia are separated from the epibasidia by septae.

HOLOBASIDIUM. A basidium which is not septate.

HOLOCARPIC. (1) Having the whole thallus formed at maturity into a sporangium, or a sorus of sporangia.

(2) Said of a fungus, which is completely enclosed in the host cell.

HOLOENZYME. A complete enzyme, made up of an apoenzyme and coenzyme.

HOLOGAMY. The fusion of two mature cells, each of which has been completely changed into a gametangium.

HOLOPHYTIC. Capable of synthesizing complex molecules from simple ones, thus storing chemical energy, and building new tissues from them. This refers especially to photosynthesis.

HOLOSAPROPHYTE. A true saprophyte.

HOLOTYPE = ALLOTYPE.

HOMEOSTASIS. The maintenance of the constancy of the internal environment of an organism.

HOMO-, HOMIO-. A Greek prefix meaning 'alike, similar'.

HOMOBASIDIOMYCETAE. A sub-class of the Basidiomycetae. The basidia are aseptate, and typically club-shaped.

HOMOBASIDIUM. A non-septate basidium.

HOMOBIUM. A self-supporting association of a fungus and an alga, as in lichens.

HOMOCHLAMYDEOUS. Having a perianth consisting of members all of the same kind, and distinguished into separate petals and sepals.

HOMOCYCLIC COMPOUNDS. Compounds containing a ring composed entirely of atoms of the same kind.

HOMODROMOUS. (1) Having leaves arranged in spirals, all running in a uniform direction.

(2) Having all leaves turned the same way.

HOMODYNAMIC HYBRID. A hybrid having an equal grouping of characters derived from each parent, and so differing in appearance from both of them.

HOMOEOSIS. A type of variation in which a plant member takes on the characters of an unlike member, *e.g.* when a petal changes into a stamen.

HOMOGAMOUS. (1) Having all the flowers in an inflorescence the same.

(2) Having anthers and stigmas ripening at the same time.

HOMOGENEOUS. Uniform.

HOMOGENIC. Of a population or gamete, containing only one allelomorph of a particular gene or genes.

HOMOIMEROUS. Said of a lichen-thallus in which the algal and fungal components are mixed, not in layers.

HOMOKARYOTIC. Having the same sort of nuclei, as in a line of isolates without variation.

HOMOLOGOUS. Said of structures which are similar, or may be structurally or functionally different, but which have descended from a common type, *e.g.* stamens and staminodes.

HOMOLOGOUS ALTERNATION OF GENERATIONS. A theory that the sporophyte has originated as a modification of the gametophyte, and not as a new phase introduced into the life-cycle.

HOMOLOGOUS CHROMOSOMES. Chromosomes having identical loci. They are paired in a diploid nucleus, and single in a haploid one.

HOMOLOGOUS VARIATION. The occurrence of similar variations in related species.

HOMOMORPHIC. Said of chromosome pairs which have the same form and size.

HOMOMORPHIC INCOMPATIBILITY. Incompatibility, not dependent for its action on morphological variation, *e.g.* the incompatibility between two strains of a fungus, which look identical.

HOMOMORPHOUS. Alike in form.

HOMONEMEAE. An obsolete term for the algae and fungi.

HOMONYM (1) The same specific name for the same plant when it is placed in another genus.
(2) A name which has been used earlier in a different sense, and so should not be used again.

HOMOPHYLLOUS. Having foliage-leaves all of the same kind.

HOMOPHYTIC. Having the diploid (sporophytic) thallus bisexual.

HOMOPLASMIC = HOMOKARYOTIC.

HOMOPLASTIC. Of the same structure and manner of development, but not descending from a common source.

HOMOSPORANGIC. Having only one kind of sporangium.

HOMOSPOROUS. Producing spores, all of the same size.

HOMOSTYLY. Having styles all of the same length.

HOMOSYNAPSIS. The pairing of two similar chromosomes.

HOMOTHALLISM. Said of the Thallophyta which bear male and female sex-organs on the same thallus, and are capable of self-fertilization. If both male and female organs are present, but there is no self-fertilization, the plant is physiologically heterothallic.

HOMOTROPOUS. (1) Curved, or turned in one direction.
(2) Of an anatropous ovule with the radical next to the hilum.

HOMOZYGOUS. Having identical genes at the same locus on each member of a pair of chromosomes which are homologous.

HONEY GUIDE. Lines, dots etc. on perianth lobes showing insects the way to honey.

HOOKED DISSEMINULE. A fruit, seed or spore, which bears hooks. These aid in dispersal by becoming attached to animals.

HOOKERIALES. An order of the Eubrya. The gametophytes are generally prostrate and freely branching, with the 'leaves' tending to lie in one plane. The leaves are asymmetric, and frequently have two mid-ribs. The sporophyte is generally pleurocarpous, with a double peristome.

HOPLESTIGMATACEAE (H). A family of the Bixales (H). The leaves are alternate. The flowers are bisexual with fused petals, and an ovary with parietal, forked placentas.

HORDENINE. A glutelin protein.

HORIZON. A stratum of the soil. They start with the A horizon at the surface, and usually end with the C horizon, which is the parent material.

185

HORMESIS. The stimulus given to an organism by a non-toxic concentration of a normally toxic substance.

HORMOCYST. A short hormogonium enclosed in a thick stratified sheath.

HORMOGONALES. An order of the Myxophyceae. The cells are united in definite trichomes that can always form hormones.

HORMOGONE. Short sections of the trichome, which are organs of vegetative reproduction in some of the Myxophyceae.

HORMONE. An organic substance produced in small amounts at one site in an organism, and transported to another site where it has a great effect.

HORMOSPORE. A thick-walled multicellular body, which is spore-like, and produced by some Myxophyceae.

HOROLOGICAL. Said of a flower which opens and shuts at a definite time of day.

HORTENSIS. Latin meaning 'of gardens'.

HORTUS SICCUS. A herbarium, or collection of dried plants.

HOST. The plant or animal on which a parasite lives.

HUACEAE (H). A family of the Malpighiales. The leaves are simple, alternate, and never unifoliate. The sepals are imbricate (rarely valvate). The anthers have 4 loculi. The ovary has 1 loculus, with 1 basal ovule.

HUMIC ACID. A fraction of the soil organic matter, which is soluble in cold alkali, then precipitated in dilute acid, and then insoluble in ethanol. Its possible formula is $C_{60}H_{52}O_{24}(COOH)_4$.

HUMICOLE, HUMICOLOUS. Growing on soil, or humus.

HUMIFUSUS. Latin meaning 'spreading on the surface'.

HUMILIS. Latin meaning 'dwarf'.

HUMIN. The fraction of the soil organic matter which is insoluble in cold alkali.

HUMIFICATION. The transformation of organic material in the soil, into humus.

HUMIRIACEAE (EP, BH, H). A family of the Geraniales (EP, BH), Malpighiales (H). These are shrubs, with alternate, simple leaves, and regular, bisexual flowers. The 5 free sepals are imbricate (rarely valvate). The 5 petals are free. There are many to 10 free stamens, and a cup-like disk. The ovary has 7-5 loculi, with 1-2 (1-3, H) ovules pendulous from the apex of the axis. The fruit is a drupe, and the seeds contain copious endosperm.

HUMP. A tiny outgrowth on the side of a growing point, the rudiment of a future lateral member.

HUMUS. The organic matter in the soil, formed by the decomposition of plant and animal remains. It contains a large number of elements necessary for plant-growth, and its colloidal mature improves the texture of the soil, and its water-retaining capacity.

HUMUS NUCLEUS. A combination of lignin and protein, which forms the basis of humus.

HUMUS PLANT. A flowering plant, often poorly provided with chlorophyll, and which grows into deep humus. Its roots then form a mycorrhiza with a fungus.

HUMUS THEORY. First stated by Aristotle, 'that plants absorb food from the soil in an elaborated form'.

HYALINE. Thin and translucent.

HYALINE CELL. A colourless cell, lying between those containing chlorophyll, in a 'leaf' of *Sphagnum*.

HYALO-. A prefix referring to spores which are hyaline, or brightly-coloured.

HYALOGEN. A particle formed by the secretory processes of a cell.

HYALOGENESIS. The secretory processes of a cell.

HYALOPLASM. Clear, non-granular protoplasm.

HYALORIACEAE. A family of the Tremellales. The fruit-bodies are stalked and gelatinous, or sessile forming a flimsy layer on the substrate. The basidia are 2-4-celled with tubular extensions which may taper to a long fine thread. The basidiospores are borne symmetrically at the apex, and break off with part of the supporting thread attached. They are not discharged from the sterigma.

HYALOSPOROUS. Having hyaline, one-celled spores.

HYBRID. The offspring of two related individuals of the same species, race, or variety.

HYBRID VIGOUR = HETEROSIS.

HYBRIDIZATION. The production of a hybrid by crossing two individuals of unlike genetic constitution.

HYDATHODE. A water-secreting gland found on the edges and tips of leaves of many plants.

HYDNACEAE. A family of the Agaricales. They are saprophytic or parasitic, mainly on wood or earth. The fruit-body is stalked or resupinate, and leathery or fleshy. The hymenium is borne on teeth or papillae, and the spores are unicellular.

HYDNANGIACEAE. A family of the Hymenogastrales. The gleba develops as a convoluted layer over a central columella. The palisade layer of the underside of the curved pileus is formed into folds which anastomose with each other, and the columella to form a multilocular gelba. The columella may be much reduced, and the stipe is a very small projection below the pileus.

HYDNORACEAE (EP, H). A family of the Aristolochiales (EP, H). (BH) include it in the Cytinaceae. These are parasites on the roots of trees and shrubs. They have no leaves. The flowers are bisexual and regular. The perianth has 4-3 fleshy fused lobes. There are 4-3 free anthers. The ovary is inferior, and consists of 3 fused carpels with parietal placentation and many ovules. The fruit is a berry. The seeds contain endosperm, and perisperm.

HYDRANGEACEAE (H). A family of the Cunoniales (H). There are no stipules, or if present, they are joined to the petiole. The flowers are cymose or corymbose, and have petals. There is no disk. There is no indumentum, or it is of simple hairs. There are 8 or more stamens which have untoothed filaments. The fruit is a capsule, rarely a berry.

HYDRARGILLITE = GIBBSITE.

HYDRASE. An enzyme which can add or remove water, without hydrolysis.

HYDRATION. The addition of water, either physically, or in chemical combination.

HYDROCARBON. A compound of hydrogen and carbon only.

HYDROCARPIC. Said of aquatic plants which ripen their fruits under water, after pollination has taken place in the air above.

HYDROCARYACEAE (H) = TRAPACEAE (H).

HYDROCHARIDEAE (BH) = HYDROCHARITACEAE.

HYDROCHARITACEAE (EP, BH, H). A family of the Helobieae (EP), Microspermae (BH), Butomales (H). These are aquatic plants, usually with ribbon-like submerged leaves. The inflorescence is axillary, and unisexual or bisexual. The female inflorescence usually has 1 flower, and the male more than 1 flower, enclosed in a spathe. The flowers are usually regular with the parts in threes. The perianth has two unlike whorls, and there are 5-1 stamens, the inner ones often staminodes. The inferior ovary consists of 15-2 fused carpels, and has 1 loculus with parietal placentas. The many ovules are orthotropous, to anatropous, and erect to pendulous. There are the same number of stigmas as carpels. The fruit dehiscence is irregular. It contains many non-endospermic seeds.

HYDROCHORIC. Dispersed by water.

HYDRODICTYACEAE. A family of the Chlorococcales. The cells are in free-floating colonies, in which the cell-number is a multiple of 2. Sexual reproduction is by the fusion of biflagellate gametes.

HYDROGEN ACCEPTOR. In the equation $AH + B = BH + A$, AH is the hydrogen donor, and B is the hydrogen acceptor.

HYDROGEN DONOR. See *Hydrogen acceptor*.

HYDROGENIC SOIL. A soil formed under water-logged conditions, chiefly in cold climates.

HYDROGEN ION CONCENTRATION = pH.

HYDROIDS. Empty thin-walled cells in the central cylinder of certain mosses.

HYDROLASES. Enzymes which bring about hydrolysis, or condensation.

HYDROLYSIS. The addition of water to a large molecule, to produce smaller ones, *e.g.* $C_{12}H_{22}O_{11} + H_2O = 2C_6H_{12}O_6$.

HYDROMORPHIC SOIL. A soil whose profile has developed under impeded drainage.

HYDROPHYLIC. Becoming attached to water molecules, readily.

HYDROPHILOUS. (1) Living in water.
(2) Pollinated by water.

HYDROPHYLLACEAE (EP, BH, H). A family of the Tubiflorae (EP), Polemoniales (BH, H). These are herbs or shrubs, with various forms and arrangement of the exstipulate leaves. The regular, bisexual flowers are scattered, or in cincinni, which are usually without bracteoles. The 5 calyx lobes are fused and imbricate with the odd sepal posterior. The corolla lobes are 5 in number and fused to form a bell. There are usually 5 free epipetalous stamens, alternating with the petals, often with scale-like appendages at the base. ((H) states no scales). There may, or may not be a disk. The superior ovary consists of 2 fused carpels, with 2-1 loculi, and 2-1 styles. There are many to 2 ovules in each carpel. They are sessile or pendulous, anatropous, and often parietal. The fruits are usually loculicar capsules. The embryo is small, and there is a rich endosperm.

HYDROPHYTE. A plant adapted to living in water, especially a vascular plant.

HYDROPHYTIUM. A water-plant formation.

HYDROPONICS. The large scale culture of plants, using solutions as the sole source of nutrient salts.

HYDROPOTE. A cell, or group of cells easily permeable to water and dissolved salts; found in submerged leaves.

HYDROPTERIDINEAE = SALVINIACEAE.

HYDROQUINONE.

HYDROSERE. A sere beginning in a wet habitat.

HYDROSTACHYACEAE (EP, H). A family of the Rosales (EP), Podostemales (H). (BH) include it in the Podostemaceae.
These are aquatic herbs with spikes of unisexual flowers. There is no perianth. The male flower consists of 1 stamen, and the female of 2 fused carpels, with many ovules. The fruit is a capsule.

HYDROTROPISM. A tropism in response to water.

HYENIALES. An order of the Equisetinae. The sporophytes have jointed stems, but the leaves are not in pronounced whorls. The leaves are long and narrow, and forked once or twice at the tip. The sporangiophores are cylindrical, and once or twice forked. The tip of each branch recurves and ends in a single sporangium. The sporangiophores are in lax strobili, and do not have sterile appendages.

HYGROCHASTIC. Of a fruit which opens by the absorption of water.

HYGROMETRIC. Said of movement due to changes in the atmospheric humidity.

HYGROPHANOUS. Darkening in colour, due to the entry of water; looking soaked.

HYGROPHOBE. Living in dry conditions.

HYGROPHYTIC. Living in a plentiful supply of water.

HYGROSCOPIC. (1) Readily absorbing water, and changing shape in consequence.
(2) Moving as a result of intake, or loss, of water.

HYGROSCOPIC MOISTURE. The water that can be driven-off from a soil by heating at 100°C–105°C. It is probably held in monomolecular layers.

HYLAEA. The upper region of the Amazon valley.

HYLIUM. A forest formation.

HYLODIUM. An open dry woodland.

HYLOPHYTE. A plant growing in damp woods.

HYMATOMELANIC ACID. A constituent of soil organic matter. It is soluble in cold alkali, from which it is precipitated by dilute acid, and is then soluble in ethanol.

HYMENIUM. The palisade-like layer of basidia.

HYMENOGASTRALES. A class of the Gasteromyceteae. The fruit-bodies are subterranean, or rarely superficial while young, but in most cases growing to the surface at maturity. The gleba retains its structure until maturity, not being digested to a sticky mass. Dispersal is not by the production of a mass of wind-dispersed spores.

HYMENOGASTRACEAE. A family of the Hymenogastrales. The columella is frequently lacking, with the cavity containing the hymenium typical, and lined by the hymenium. The gleba is not clearly wrinkled.

HYMENOMYCETAE = AGARICALES.

HYMENOPHORE. Any structure bearing a hymenium.

HYMENOPODIUM, HYMENOPODE. The tissue under the hymenium.

HYMENOPHYLLACEAE. A family of the Leptosporangiatae. The sporangia are in basipetalous, gradate succession on an elongated receptacle, which is always surrounded by a cup-shaped, or two-lipped indusium. Most have semi-transparent leaves which are one-cell thick.

HYPANTHODIUM. The deeply hollowed receptacle of the fig.

HYPERBASAL CELL. One of the two lowermost cells of the developing sporophyte of bryophytes and pteridophytes. It usually develops into part of the seta and foot.

HYPERCHIMAERA. A chimaera in which the components are intimately mixed.

HYPERDIPLOIDY. The condition where the full diploid complement of chromosomes is present, as well as a portion of one chromosome which has become translocated.

HYPERICACEAE (H). A family of the Guttiferales (H). The leaves are often dotted with glands, and there are no stipules. They are trees, herbs, or shrubs, with a resinous juice. If hairs are present, they are often stellate. The flowers are bisexual, and the stamens are often in bundles. The seeds have no aril, and lack endosperm.

HYPERICINEAE (BH). A family of the Guttiferales (BH). (EP) include it in the Guttiferae (*q.v.*).

HYPERMORPH. A mutant gene, having a similar, but greater effect than the non-mutant allelomorph.

HYPERPARASITE. A parasite, parasitic on another parasite.

HYPERPLASIA. The abnormal increase in the size of a tissue by cell-division.

HYPERPLOID. Having a chromosome number slightly exceeding an exact multiple of the haploid number.

HYPERSENSITIVITY. A method of resistance to invasion by an obligate parasite. The first few host cells that are invaded first die rapidly, thus removing the food-supply for the parasite, which consequently dies.

HYPERSTROMATIC. Having the stroma on the upper surface of a leaf.

HYPERTONIC. Of a solution, having a greater osmotic pressure than the one with which it is being compared.

HYPERTROPHY. The abnormal increase in the size of a tissue by enlargement of the original cells.

HYPERTROPHYTE. A parasite causing hypertrophy.

HYPHA. An individual filament of a fungus thallus.

HYPHAL BODY. A thin-walled, multinucleate segment of a hypha, serving for the propagation of some fungi which are parasites on fungi.

HYPHOCHYTRIDIACEAE. A family of the Hyphochytridiales. Its members are hypha-like and without rhizoids, having terminal and intercalary zoosporangia. They have a number of centres of growth, and more than one reproductive organ, but only part of the thallus produces reproductive organs.

HYPHOCHYTRIDIALES. An order of the Phycomyceteae. They closely resemble the chytridiales, but the zoospores have one anterior flagellum.

HYPHOMYCETEAE = MONILIALES.

HYPHOMYCETOUS. (1) Mould-like.
(2) Cobwebby.

HYPHOPODIUM. A more or less lobed outgrowth from a hypha, often serving to attach an epiphytic fungus to a leaf.

HYPNOCYST. A thick-walled resting spore of some algae.

HYPNOSPORE. A thick-walled aplanospore.

HYPNOZYGOUS. Said of a zygote which remains inert for some time after formation.

HYPNOASCIDIUM. (1) An abnormal cup-shaped outgrowth from a leaf.
(2) A transformation of a leaf, the inner surface, corresponding to the lower surface of the leaf.

HYPOBASAL HALF. The posterior portion of an embryo.

HYPOBASIDIUM. An enlarged cell of a hyphae, in which a nuclear fusion takes place before the formation of the basidium.

HYPOCARPOGENOUS. Flowering and fruiting underground.

HYPOCHNOID. Having effused, resupinate, dry, rather loosely intertwined hyphae.

HYPOCOTYL. The part of a seedling axis between the radical and the cotyledon(s).

HYPOCRATERIFORM. Trumpet-shaped, possibly saucer-shaped.

HYOCREACEAE. A family of the Hypocreales. The perithecia are buried in a stroma, and the ascospores are ellipsoid to cylindrical, with 1 to many cells.

HYPOCREALES. An order of the Euacsomycetae. The ascocarp is a perithecium, with a light-coloured, soft periderm, and distinct from the rest of the mycelium.

HYPODERM, HYPODERMIS. A layer of one, or more cells thick, of thickened cells lying immediately below the epidermis.

HYPODERMAL. Beneath the epidermis.

HYPODERMATACEAE. A family of the Inoperculatae, here included in the Phacidiaceae.

HYPODERMII = UREDINALES and USTILAGINALES.

HYPOGEAL GERMINATION. When germination takes place, the cotyledons stay below the ground.

HYPOGENOUS. Produced lower down.

HYPOGYNOUS. (1) Said of a flower in which the other parts arise below the gynecium.
(2) Of an antheridium developing on a branch of the oögonial stalk.

HYPOMORPH. Of a mutant gene having an effect similar to, but less than that of the non-mutant allelomorph.

HYPONASTY. The lower side of an organ growing more quickly than the upper, resulting in an upward bending.

HYPOPHLOEODAL. Growing just within the surface of the bark.

HYPOPHYLLOUS. Attached to, or growing on the under-surface of a leaf.

HYPOPLASIA. A state of having growth less than others. Developmental deficiency.

HYPOPLOIDS. Diploids lacking a piece or pieces of chromosomes from their complement.

HYPOPOLYPLOID. A polyploid in which one or more of the chromosomes are lost from the complete set.

HYPOSTATIC. Recessive, when relating to one of two characters which are not allelomorphic.

HYPOSTOMATIC. Having stomata on the lower surface of the leaf.

HYPOSTROMA. A stroma found beneath the epidermis of the host.

HYPOTHALLUS. (1) Of the Myxomycetae; a discarded remnant of the plasmodium at the base of a sessile or stalked sporangium.
(2) The first formed weft of hyphae in the development of the thallus of a lichen, often remaining at the base or edge of the thallus.

HYPOTHECA. The inner of the two half-walls of the cell-wall of diatoms, and dinoflagellates.

HYPOTHECIUM. The lower layer of an ascocarp of the Hysteriales. It contains the ascogenous hyphae.

HYPOTONIC. Said of a solution having a smaller osmotic pressure than that of the solution with which it is being compared.

HYPOTROPHY. Eccentric thickening of the underside of an approximately horizontal shoot, or root.

HYPOXYLOID. Forming a cushion-shaped, or crust-like stroma.

HYPOXIDACEAE (H). A family of the Haemodorales (H). The leaves are radical, or towards the base of the stem. The flowers are regular, or slightly zygomorphic, with 6 or many (rarely 3) stamens which are free or in bundles. The ovary is inferior and trilocular.

HYPSOPHYLLARY LEAF. A bract.

HYSTERANGIACEAE. A family of the Hymenogastrales. The columella is mostly absent, and the hymenial cavities are typical; lined by the hymenium. The gleba is cartilagenous or gelatinous, and clearly ridged.

HYSTERANTHOUS. Developing leaves after flowering.

HYSTERESIS. A lag in movement at one level, in response to a stress at another level.

HYSTERIACEAE. A family of the Hysteriales (*q.v.*). It is the only important family.

HYSTERIACEOUS. Long and cleft.

HYSTERIALES. An order of the Euascomycetae. The ascocarps are small and elongate, developing longitudinal slit-like openings. The asci are in a palisade-like layer at the base of the ascocarp.

HYSTERIFORM, HYSTERIAEFORM. Having the shape of a long, narrow ridge, with a longitudinal opening along the top.

HYSTEROTHECIUM. An elongated perithecium, remaining closed as it develops, and opening when ripe, by a slit-like cleft at the top.

I

IAA = INDOLYLACETIC ACID.

IBA = INDOLYBUTYRIC ACID.

ICACINACEAE (EP, H). A family of the Sapindales (EP), Celastrales (H). (BH) include it in the Olacineae. These are mainly trees or shrubs, often lianes. The leaves are alternate, usually entire, often leathery, and lack stipules. The regular flowers are usually bisexual, in compound panicles. The 5-4 calyx lobes are fused. The 5-4 petals are rarely united, valvate or imbricate. The 5-4 free stamens alternate with the petals, and usually have introrse anthers. A disk is rarely developed (absent, (H)). The superior ovary is of 3 fused carpels, rarely 5 or 2. It is usually unilocular, but may be multilocular. There are 2 pendulous anatropous ovules per loculus. There is a dorsal raphe, with the micropyle facing upwards, and the funicle usually thickened above it. The fruit is unilocular, and 1-seeded. It is usually a drupe, but sometimes a samara. The endosperm is usually present, and the embryo is straight or curved.

IDIOBLAST. (1) A supporting cell, found among chlorophyllaceous tissue. It lacks chlorophyll, has thick walls, and is usually elongated.

(2) Any cell that varies in cell-content, and wall thickness from its neighbours.

IDIOCHROMATIN. A substance within the nucleus of a cell, which controls the reproduction of the cell.

IDIOPLASM Germ-plasm.

IDOSES. Monosaccharides belonging to the aldohexoses.

IGNITION LOSS. The loss of weight from an oven-dried soil sample, on burning.

ILICACEAE = AQUIFOLIACEAE.

ILICINEAE (BH) = AQUIFOLIACEAE.

ILLECEBRACEAE (BH, H). A family of the Curvembryae (BH), Polygonales (H). (EP) include it in the Caryophyllaceae. These are herbs, which are rarely shrubby, with leaves which are usually opposite, entire and stipulate. The inconspicuous bisexual flowers are usually in cymes. The persistent perianth is herbaceous, or leathery, and is of 5-4 fused lobes. The 5-4 free stamens (rarely more or less) are opposite the perianth lobes. The ovary is unilocular containing 1 ovule (rarely 2), which is amphitropous or anatropous. The seeds contain endosperm.

ILLEGITIMATE POLLINATION. Self-pollination, which takes place, in spite of the flower appearing to be adapted for cross-pollination.

ILLICIACEAE (H). A family of the Magnoliales (H). These are trees or small shrubs, with simple, alternate leaves which lack stipules. The bisexual flowers are solitary. The many to 7 perianth lobes are in whorls, grading from being bracteole-like at the outside, to petaloid on the inside. The many to 4 stamens are free. The gynecium consists of 21-5 free carpels in a single whorl, each with 1 ovule. The fruit is a follicle. The seed contains copious endosperm, and a small embryo.

IMBIBITION. This takes place when a solvent enters a colloid, between the free capillary spaces, and the inter-micellar spaces. It causes the colloid to swell, (and ultimately to be dispersed). The swelling causes considerable pressure—the *imbibitional pressure*.

IMBIBITIONAL WATER. Water held within the lattice of the colloidal matter of the soil.

IMBIBITIONAL MECHANISM. A mechanism that functions hygroscopically.

IMBRICATE. Overlapping, like the tile of a roof.

IMINO ACID. A secondary amine acid, with two hydrogen atoms from the ammonia replaced by alkyl radicals.

IMMACULATE. Not spotted.

IMMARGINATE. Lacking a distinct edge.

IMMEDIATE GERMINATION. Germination, without undergoing a period of inactivity.

IMMERSED. (1) Embedded in the tissues of the plant.
(2) Arising beneath the surface of the substratum.

IMMOBILIS. Latin meaning 'immovable'.

IMMUNITY. The ability to prevent invasion by a pathogen. This is of three types:
(*a*) Structural, *e.g.* a thick cuticle may prevent the penetration of the fungal hyphae.
(*b*) Physiological, *e.g.* the protoplasm is an unfavourable environment for the development of the parasite.
(*c*) Acquired, *e.g.* plants attacked by a mild strain of a virus, which appears to have no ill-effects, may be immune to the attack of a more virulent strain.

IMPARIPINNATE. Said of a pinnate leaf, which has a terminal leaflet.

IMPERFECT FLOWER. A flower which lacks either anthers or carpels, or if they are present, they are non-functional.

IMPERFECT FUNGI = FUNGI IMPERFECTI.

IMPERFECT HYBRIDIZATION. An abortive attempt to form zygospores between the hyphae of two distinct species of *Zygomycetes*.

IMPERFECT STAGE. The asexual stage of a fungus.

IMPERFORATE. Having no opening.

IMPRESSED. Having the surface marked by slight depressions.

INAEQUALE. Latin meaning 'unequal'.

INARTICULATE. Not joined.

INBREEDING. The raising of progeny of mating of two or more closely related gametes or zygotes.

INBREEDING COEFFICIENT. A measure of the intensity of inbreeding.

INCANUS. Latin meaning 'hoary-white'.

INCEPT. The rudiment of an organ.

INCERTAE SEDIS. Of uncertain taxonomic position.

INCIPIENT PLASMOLYSIS. The condition when about half the cells of a tissue are plasmolysed. It is in this condition that the osmotic pressure of cell-sap is usually measured.

INCISE, INCISED. Cut sharply, and deeply at the margin.

INCLUDED. Not projecting beyond surrounding members.

INCLUSION. A body occurring in the cytoplasm of a cell.

INCOMPATIBILITY. (1) The failure of self- or cross-fertilization by reason of genetic similarity, within an otherwise freely inbreeding group.
(2) Differences in the physiology of a host and parasite which inhibits or stops the growth of the latter.

INCOMPLETAE (BH) = MONOCHLAMYDEAE (EP).

INCOMPLETE FLOWER. A flower lacking sepals or petals, or both.

INCRASSATE. Made thick.

INCRUSTATION. A coating of iron carbonate, or iron compounds on the surface of some algae.

INCUBATION PERIOD. The time between the inoculation with a pathogen, and the development of symptoms.

INCUBOUS. Said of the 'leaf' of a liverwort, when its upper border, (*i.e.* the border towards the apex of the 'stem') overlaps the lower border of the next leaf above it, and on the same side of the stem.

INCUMBENT. (1) Lying on.
(2) Said of a radical which is bent back to lie on one of the cotyledons.

INCURVED. (1) Curved inwards.
(2) = CAMPYLOTROPOUS.

INDEFINITE. (1) Having a large, but indefinite number.
(2) Not ending in a flower, and so theoretically, capable of further elongation.
(3) Racemose.

INDEHISCENT. Said of fruits, fruit-bodies etc. which do not open to disperse their contents.

INDEPENDENCE. The relation between two variates, the variation of one being uninfluenced by variation in the other.

INDEPENDENT ASSORTMENT OF GENES. See *Mendel's Second Law.*

INDEPENDENT COMPARISONS. Comparisons between observations whose values are uninfluenced by changes in each other.

INDETERMINATE. (1) Without a distinctive edge.
(2) = INDEFINITE.

INDICATOR PLANT. A plant which grows under special conditions of climate, or in a particular soil, or in a particular community, and thus, by its presence, indicates the general nature of the habitat.

INDIFFERENT SPECIES. A species which occurs in two or more distinct communities.

INDIGENOUS. Native to an area; not introduced.

INDIVIDUAL. (1) A unit of life, considered either physiologically, or genetically.

INDOLE-3-ACENTONITRILE. A precursor of IAA.

INDOLYLACETIC ACID. A plant hormone which causes elongation of cells, when it is present in suitable concentrations.

β-indoleylacetic acid

INDOLYLBUTYRIC ACID. A plant hormone.

INDUCED MUTATION. A mutation determined by external conditions.

INDUMENTUM. The hairy covering.

INDUPLICATE. Folded inwards.

INDURATE. Hardened.

INDUSIATE. Like a small cup.

INDUSIUM. The cover growing over the sporangia of some ferns.

INERMIS. Latin meaning 'unarmed, thornless'.

INFECTION TUBE. The germ-tube which penetrates the host from the germinating spore of a parasitic fungus.

INFERAE (BH). A series of the Gamopetalae (BH). The ovary is inferior, with the stamens, usually as many as the corolla lobes.

INFERIOR. (1) Said of an ovary where the receptacle encloses it, so that the other floral parts arise above the ovary. The flower is then epigynous.
(2) Said of the annulus of an agaric, when it is placed low on the stipe.

INFERIOR PALEA = LEMMA.

INFLEXED. Curved, or bent inwards.

INFLORESCENCE. A flowering shoot, bearing more than one flower. See *Raceme, cyme, panicle, corymb, spadix, umbel, capitulum, monochasium, dichasium, catkin.*

INFLUX. The gross uptake of ions by a plant root.

INFRUCTESCENCE. The inflorescence after the flowers are fallen, and the fruits are formed.

INFUNDIBULIFORM. Tubular below, gradually opening upwards, *i.e.* funnel-shaped.

INHERITANCE = HEREDITY.

INHIBITOR. A substance which limits, or destroys the catalytic activity of an enzyme.

INITIAL CELL. A cell which remains meristematic, divides repeatedly, and gives rise to many daughter-cells, from which, after further division, the permanent tissues of the plant are differentiated.

INITIAL SPINDLE = NETRUM.

INJECTED. Having the inter-cellular spaces filled with water.

INNATE. (1) Sunken into the thallus.
(2) Originated in the thallus.
(3) Said of an anther which is joined to the filament only by its base.

INNER ENDODERMIS. The endodermis internal to the vascular tissues in a solenostele.

INNER FISSURE. The inner layer of the raphe (wall) of the Bacillariophyceae.

INNER GLUME = PALE.

INNOVATION. (1) An accessory branch, produced in some mosses, after the sporangium is developed.
(2) A newly-formed shoot, formed during one season.

INOCULATION. (1) The conveyance of an infection to a host plant by any means of transmission.
(2) The entry of the germ-tube of a parasitic fungus into a host.
(3) The placing of spores, or piece of mycelium of growing fungi, or bacteria, into a culture medium.

INOCULUM POTENTIAL. The energy of growth of a fungus, or other microorganism, available for the colonization of a substrate.

INOPERCULATAE. A sub-order of the Pezizales, the members of which lack an operculum.

INOPERCULATE. Said of a sporangium, which lacks a lid.

INORDINATE. Not arranged in any special order.

INSCULPT, INSCULPTATE. Bearing holes, or depressions in the surface.

INSERTED. Growing out of another member.

INSERTION. (1) The manner of attachment.
(2) The place where one plant member grows out of another, or is attached to another.

INSIGNIS. Latin meaning 'notable'.

INSOLATION. Exposure to the sun.

INSPERSED. Having granules penetrating the substance of the thallus.

INTEGRIFOLIUS. Latin meaning 'simple-leafed'.

INTEGUMENT. (1) The membrane(s), enclosing the nucellus, finally forming the testa.
(2) The membrane enclosing the female gameteophyte in the Lepidocarpales.

INTERACTION OF GENES. The process by which one gene-difference affects the expression of another gene-difference.

INTERBREEDING. Experimental hybridization of different species or varieties.

INTERCALARY. (1) Describing a meristem occurring between non-dividing tissue.
(2) Lying between other bodies of a row, or placed somewhere along the length of a stem, filament, or hypha.

INTERCALARY CELL. A small cell between two aecidiospores, which disintegrates as the spores ripen, and breaks down as they are set free.

INTERCALARY MERISTEM. A meristem located somewhere along the length of a plant member, and by its activity giving *intercalary growth*.

INTERCALARY PLATES. A layer of plates between the apical and precingualar plates of the cell-wall of the Peridiniales.

INTERCELLULAR. Occurring between cells.

INTERCHANGE. An exchange of non-homologous terminal segments of chromosomes.

INTERCHROMOCENTRE. The areas of a chromatid composed of euchromatin.

INTERFASCICULAR CAMBIUM. A strand of cambium between two adjacent vascular bundles. The formation of interfascicular cambium is the first stage in the normal secondary thickening of a stem.

INTERFERENCE. The property by which one cross-over interferes with the occurrence of another cross-over in its neighbourhood.

INTERFIBRILLAR SPACES. The spaces between the microfibrils of the cell-wall, filled mostly with water and pectic substances.

INTERGRANUM AREA. The zone between the grana of a chloroplast. Called the stroma in older literature.

INTERGRANUM LAMELLAE. The fine plate-like structure of a chloroplast, made up of a layer of chlorophyll and lipid, bounded on either side by a layer of protein.

INTERKINESIS = INTERPHASE.

INTERMITOSIS. The period between two mitotic divisions of a cell.

INTERNAL FACTOR. Any factor which depends on the genetic constitution of the plant and which influences its growth and development.

INTERNAL PHLOEM. A phloem lying between the xylem and the centre of the stem.

INTERNODE. (1) The stem between two successive nodes.
(2) The part of the thallus of the Charales, where the branch-like filaments arise.

INTERPETIOLAR. Between the petioles.

INTERPHASE. The resting stage that may occur between the first and second meiotic divisions.

INTERRUPTED. Said of organs, not evenly spaced-out along an axis.

INTERRUPTEDLY PINNATE. Said of a pinnate leaf where pairs of small leaflets alternate with pairs of larger ones.

INTERSEX. An individual which exhibits characters intermediate between those of the male and those of the female of the same species.

INTERSPECIES. Said of a cross between two separate species.

INTERTEXTIC. A type of soil-aggregate, of uncoated minerals bound by friable colloidal bridges.

INTERVARIETAL. Said of a cross between two individuals of the same variety of the same species.

INTERXYLARY PHLOEM. A strand of secondary phloem surrounded by secondary xylem.

INTINE. (1) The inner layer of the wall of a pollen grain.
(2) The endospore in the spores of bryophytes.

INTRACELLULAR. Occurring within the cell.

INTRAMATRICAL. Said of a parasitic fungus which lives inside the host cell, or in the matrix.

INTRAMOLECULAR RESPIRATION = ANAEROBIC RESPIRATION.

INTRAPETIOLAR. Within the petiole.

INTRASPECIFIC. Within the same species.

INTRAXYLARY PHLOEM = INTERNAL PHLOEM.

INTRICATE. Intertwined, entangled.

INTRORSE. Said of an anther which dehisces inwards.

INTRUDED. Projecting forwards.

INTUMESCENCE. A localized pathological swelling, consisting mainly of parenchyma.

INTUSSUSCEPTION. The depositing of material between the microfibrils of a cell-wall.

INULASE. The enzyme that hydrolyses inulin to fructose.

INULIN. See *Fructosans*.

INVADER. A plant occurring in a community, to which it does not belong.

INVARIANCE. The reciprocal of variance.

INVASION. The movement of plants from one area to another, and their establishment in the latter.

INVASION THEORY. The theory that the siphonostele originated through the invasion of the vascular tissue by the cortex to form a pith, *i.e.* that the pith is non-vascular.

INVERSE. Said of the condition of an embryo in which the radicle is turned towards a point in the seeds at the opposite end to the hilum.

INVERSICATENALES = PRIMOFILICES.

INVERSION. (1) The breaking and re-uniting of parts of a chromosome so that the genes are lying in the reverse order.
(2) The turning inside-out of the colony of some of the Volvocales.

INVERTASE = SUCRASE.

INVOLUCEL. The group of bracts, sometimes at the base of a partial umbel.

INVOLUCRAL BRACT. One of the leafy appendages forming an involucre.

INVOLUCRE. (1) Any leaf-like structure, protecting the reproductive structures.
(2) A ring of bracts or bristles.
(3) A shield-like structure growing over the calyptra and enclosed embryo of some Hepaticae, and Anthocerotae.

INVOLUTE. Having inrolled margins.

IONIC ANTAGONISM. The effect of adding one element as its ions to the soil, causing deficiency of another in a crop, *e.g.* adding potassium may cause symptoms of magnesium deficiency.

IONIC LINKAGE. The linkage between an acid and a base, *e.g.*

$$\text{HC}-\text{CH}_2-\text{NH}_3^+ - - - - \text{OOC}-\text{CH}_2-\text{CH}$$

IRIDACEAE (EP, BH, H). A family of the Liliiflorae (EP), Epigynae (BH), Iridales (H), (*q.v.*).

IRIDALES (H). An order of the Corolliferae. It is similar to the Liliales (*q.v.*), but the ovary is inferior (superior in *Isophysis*), and has 3 stamens. The style-arms is often divided, and sometimes petaloid.

IRIDEAE (BH) = IRIDACEAE (EP).

IRREGULAR. (1) Asymmetric, not arranged on an even line or circle.
    (2) Not divisible into halves by an indefinite number of longitudinal planes.
    (3) Having members of a whorl which are not all alike.

IRRITABILITY. Sensitiveness to stimulus.

IRRORATE. As if covered with dew.

IRVINGIACEAE (H). A family of the Malpighiales (H). The leaves are simple, and alternate, with large stipules folded around the terminal bud. The sepals are imbricate (rarely valvate). The ovary is bilocular, with 1 pendulous ovule from near the top of the axis in each loculus.

ISIDIUM. A small lump-like outgrowth on the free surface of a lichen thallus.

ISOBILATERAL. (1) Referred to leaves having the same structure on both surfaces, *e.g.* the leaves of monocotyledons.
    (2) Divisible into symmetrical halves by two distinct planes.

ISOBRACHIAL. Said of a chromosome which is bent into two equal arms.

ISOBRYALES. An order of the Eubrya. The gametophyte is generally perennial or creeping, with the leaves so twisted on the freely branched stem, that they appear to be in two rows. The sporophyte is generally plerocarpous, and the capsules generally erect. The peristome is double.

ISOCHROMOSOME. A chromosome with two homologous arms, derived from sister chromatids by sister-union within a terminal centromere.

ISOCHRYSIDINEAE. A sub-order of the Chrysomonadales. The cells are biflagellate, with flagelle of equal size, and are solitary or in colonies of definite form.

ISOCITRIC ACID. An acid in the Citric Acid Cycle; it is formed from aconitic acid, and is oxidized to oxalo-succinic acid.

$$
\begin{array}{l}
COOH \\
| \\
CH_2 \\
| \\
CH{-}COOH \\
| \\
CHOH \\
| \\
COOH
\end{array}
$$

ISOCITRIC ENZYME. The enzyme that catalyses the formation of isocitric acid from α-ketoglutaric acid.

$$
\begin{array}{l}
COOH \\
| \\
CH_2 \\
| \\
H^+ + CH_2 + CO_2 + TPNH \\
| \\
CO \\
| \\
COOH
\end{array}
\overset{(Mn^{++})}{\rightleftharpoons}
\begin{array}{l}
COOH \\
| \\
CH_2 \\
| \\
CH{-}COOH + TPN^+ \\
| \\
CHOH \\
| \\
COOH
\end{array}
$$

ISODIAMETRIC, ISODIAMETRICAL. Of the same length, vertically and horizontally.

ISOELECTRIC POINT. See *Amphoteric electrolyte.*

ISOETACEAE. A family of the Isoetales. The sporophyte has a corm-like stem not externally differentiated into axis and rhizophore. The sporophylls are not in strobili. The sporophyte is heterosporous. The antherozoids are multiflagellate.

ISOETALES. An order of the Lycopodinae. These are herbaceous with secondary thickening of the stem. The leaves are microphyllous and ligulate. There are large rhizophores, and these plants are heterosporous. The living genera have multiflagellate antherozoids.

ISOGAMETANGIC. Having gametangia of the same size, and form.

ISOGAMETE. One of a pair of uniting gametes of similar size and form, but may be physiologically different.

ISOGENERATAE. A class of the Phaeophyta, having the alternating generations of identical structure.

ISOGENIC. Propagating entirely by means of apogamy.

ISOGENOMATIC. Said of chromosome complements which are composed of similar genoms.

ISOKONTAN. Bearing two (or more) flagella of equal length.

ISOLATION. (1) The condition in which individuals of common ancestry are separated into two mating groups. It may be geographic, by space, or genetic, by genotype.

(2) The condition in which chromosomes or parts of chromosomes of common ancestry are prevented from undergoing effective recombination.

ISOLATION TRANSECT. A belt of land to which grazing animals are admitted under observation, so that the effect of grazing on the vegetation may be studied.

ISOLEUCINE. An amino-acid found in protein.

$$CH_3—CH—CH_2—CH_3$$
$$|$$
$$CH—NH_2$$
$$|$$
$$COOH$$

ISOMER. A compound having the same percentage composition, and the same molecular weight as another, but of different constitution.

ISOMERASES. Enzymes which catalyse isomeric changes, *e.g.* glucose 6-phosphate to fructose 6-phosphate.

ISOMORPHISM. The apparent likeness between individuals belonging to different species or races.

ISOPHASE. The condition when one hereditary factor influences the development of several characteristics.

ISOPRENE. One of the substances from which terpenes and ultimately essential oils are formed.

$$\begin{array}{ccc} CH_2 & & H \\ & C{=}C & \\ CH_3 & & CH_2 \end{array}$$

ISOREAGENT. A variety of microspecies.

ISOMOTIC SOLUTION. A solution that has the same osmotic pressure as the cellular contents.

ISOSPOROUS. Having asexually produced spores of one kind only.

ISOSTEMONOUS. Having as many stamens as petals, and in a whorl.

ISOTONIC. Said of a solution having the same osmotic pressure as the one with which it is being compared.

ISOTOPE. An element which has two or more atomic weights. The ones with the higher atomic weights are often radio-active.

ISTHMUS. The narrow connecting zone between the semi-cells of the Desmidiaceae.

IXONANTHACEAE (H). A family of the Malpighiales (H). The leaves are simple and never unifoliate. The sepals and petals are contorted. The fruit is a septicidal capsule, and the seeds contain endosperm.

J

JACKET. Any layer of tissue which covers an archegonium, capsule, or sporogonium.

JACKET CELL. An outer cell, usually of a reproductive organ.

JAMIN'S CHAIN, JAMINIAN CHAIN. A series of short threads of water separated by bubbles of air, in the vessels of plants.

JOINT, JOINING. A node.

JOINTED. Said of an elongated plant member which is constricted at intervals, and ultimately separates into a number of portions by breaking across the constrictions.

JORDANIAN SPECIES. A species distinguished by very small differences.

JUGLANDACEAE (EP, BH, H). a family of the Juglandales (EP, H), Unisexales (BH). These are trees with alternate stipulate leaves, (no stipules, (H)). The inflorescences are monoecious, the male flowers being as catkins on the twigs of the previous year, and the female as sessile flowers on the stems of the current year. The perianth, typically has 4 free lobes. The male flower has 40-3 free stamens, and the female has an inferior ovary of 2 fused carpels, each with 1 erect orthotropous ovule. Pollination is by wind. The fruit is a drupe or nut. The testa is thin, and there is no endosperm.

JUGLANDALES. An order of the Archichlamydeae (EP), Lignosae (H). See *Juglandaceae.*

JUGLANDEAE (BH) = JUGLANDACEAE (EP).

JUGUM. A pair of opposite leaves.

JULACEOUS. (1) Cylindrical and smooth.
(2) Resembling a catkin.

JULIANIACEAE (EP, H). A family of the Julianiales (EP), Sapindales (H). (BH) include it in the Anacardiaceae. These are trees or shrubs with alternate, usually pinnate exstipulate leaves and dioecious flowers. The male flowers are numerous, in panicles, and the females in fours at the end of downward directed spikes. The male flower has a perianth of 8-6 lobes which are free, and 8-6 free stamens. The female flower is naked with a superior unilocular ovary, with 1 ovule on a cup-like funicle. The seed has no albumen.

JULIANIALES (EP). An order of the Archichlamydeae (EP). See *Julianiaceae.*

JUNCACEAE (EP, BH, H). A family of the Liliiflorae (EP), Calycinae (BH), Juncales (H). These are usually creeping sympodial rhizome, one joint of the sympodium appearing above ground each year as a leafy shoot. The stem does not often lengthen above the ground, except to bear the inflorescence. The leaves are usually narrow, and sometimes centric. The flowers are usually in cymes of various kinds, and are usually bisexual and regular. The perianth lobes are free in 2 whorls of 3 members each, and usually sepaloid. There are 3 or 6 free stamens. The superior ovary has 3 fused carpels, with axile or parietal placentas with many or few anatropous ovules. The fruits are loculicar capsules. The seeds have a straight embryo, and a starchy endosperm.

JUNCAGINACEAE (H). A family of the Juncaginales (H). There are 6 or 4 perianth segments. The style is short, or absent from all flowers. The female flowers are all of one kind.

JUNCAGINALES (H). An order of the Calyciferae. These are perennial or annual herbs, with fibrous or tuberous roots from a rhizome. The leaves are mostly radical with an open sheath at the base. The flowers are bisexual or unisexual, in racemes or spikes. There are no bracts. The perianth has 6 or 3 lobes, or if only 1 it is bract-like. There are 6-1 stamens with extrorse anthers. The ovary is superior with 6-1 free or fused carpels. The ovules are more-or-less basal, and the seeds have no endosperm.

JUNCALES (H). An order of the Glumiflorae. These are perennial or annual herbs, with linear often grass-like leaves which have open or closed sheaths at the base. The leaf-blade may be reduced, so that the sheath embraces the stem. The flowers are mostly wind-pollinated, very small, and in heads of various forms. The perianth is scaly, in 2 whorls, much-reduced, or absent. There are 6, 3, 2, or 1 stamens. The anthers have 2 or 1 loculi. The ovary is superior, and the fruit is a capsule or nut.

JUNCEOUS. Latin meaning 'rush-like'.

JUNGERMANNIALES. An order of the Hepaticae. The gametophyte is simple, foliose or differentiated into 'stem' and 'leaves'. There is little internal differentiation. The antheridia are globose, or sub-globose, usually on long stalks. The archegonia have necks of five vertical rows of cells. The jacket-layer of the capsule is more than one cell thick, and usually dehisces longitudinally into 4 parts.

JUNGLE. A low or thin forest.

JURASSIC. A geological period extending from approximately 140 to 170 million years ago.

JUVENILE FORM. A young plant that has leaves and other features different from those of a mature plant of the same species.

JUVENILE LEAF. The form of leaf found on a sporeling or seedling, when it differs markedly from the leaf of the adult plant.

JUVENILE STAGE. A special stage in the life-history of some algae from which the ordinary plant develops as an outgrowth.

K

KANKAR = CALICHE.

KAOLIN. A group of clay minerals, having the general formula, $Al_2O_3.SiO_2.xH_2O$.

KAOLINITE. A kaolin type of mineral, with the formula, $Al_2O_3.2SiO_2.2H_2O$.

KARYASTER. A group of chromosomes arranged like the spokes of a wheel.

KARYO-. A Greek suffix meaning 'nucleus'.

KARYOGAMY. (1) The dividing of a nucleus into two, passing to the opposite sides of the cell.
　(2) The union of two nuclei, especially gametic nuclei.

KARYOKINESIS = MITOSIS.

KARYOLOGY. Nuclear cytology.

KARYOLYMPH. The matrix lying in the reticulum of the nucleoplasm.

KARYOMERE. A swollen condition sometimes seen in chromosomes towards the end of a nuclear division.

KARYOMICROSOME. A nuclear granule.

KARYOMITE = CHROMOSOME.

KARYON = NUCLEUS.

KARYOPLASM = NUCLEOPLASM.

KARYOPLASMATIC RATIO. The ratio of the volume of the nucleus and that of the cytoplasm of the same cell.

KARYORHEXIS. The disintegration of the chromatin of the nucleus into darkly staining granules, during the necrosis of the cell.

KARYOSOME. (1) A nucleus.
　(2) A chromosome.
　(3) An aggregation of chromatin in a resting nucleus.
　(4) A type of nucleolus well-shown by many of the lower plants, which stain with basic dyes, and furnishes material for the chromosomes during mitosis.

KARYOTHECA = NUCLEAR MEMBRANE.

KARYOTIN = CHROMATIN.

KARYOTYPE. The character of a nucleus as defined by the size, shape, and number of the mitotic chromosomes.

KATA-. A Greek prefix meaning 'down'.

KATABIONS. Organisms in which katabolism processes predominate over anabolic processes, as in animals.

KATABOLISM = CATABOLISM.

KATADROMOUS. Said of the venation in the ferns when the first nerves in each leaf-segment come off on the basal side of the mid-rib.

KATAKINETIC. Tending to discharge energy.

KATAKINETOMERES. Energy-poor, stable protoplasm molecules.

KATAPHASE. The stages of mitosis from the formation of the chromosomes up to the division of the cell.

KEEL. (1) = CARINA.
(2) A longitudinal narrow outgrowth from the underside of a leaf, or leaf-like structure.
(3) Any prominent ridge.

KEIMPLASM = GERMPLASM.

KELP. A general term for large sea-weeds, especially *Laminaria* spp.

KERNEL. (1) The seed inside the stony endocarp of a drupe.
(2) An obsolete term for the nutritive tissue and asci in a perithecium.

KETOGLUTARIC ACID. An acid in the Citric Acid Cycle, produced from oxalosuccinic acid, with the evolution of carbon dioxide, and oxidized to succinic acid.

$$
\begin{array}{l}
COOH \\
| \\
CO \\
| \\
CH_2 \\
| \\
CH_2 \\
| \\
COOH \qquad \text{α-ketoglutaric acid}
\end{array}
$$

KETONE. The oxidation products of secondary alcohols having the general formula $C_nH_{2n}O$, but having a $\rangle C = O$ group in the molecule.

KETOSE. A hexose sugar having a = O group on the second carbon atom, and no HO group on the first one, *e.g.* fructose.

$$
\begin{array}{l}
CH_2OH \\
| \\
C=O \\
| \\
H—C—OH \\
| \\
HO—C—H \\
| \\
HO—C—H \\
| \\
CH_2OH
\end{array}
$$

KETO-URACIL. One of the precursors of nucleosides.

KICKXELLACEAE. A family of the Mucorales, in which the sporangia are reduced to 1-celled indehiscent sporangioles, borne singly on sterigmata arranged on one side of a branch, rather like the teeth of a comb.

KINASE. The enzyme which catalyses the transfer of phosphate to ADP, forming ATP.

KINETIC BODY. A small granular body lying where a chromosome is attached to the spindle.

KINETIC CONSTRICTION, KINETOCHORE. The portion of a chromosome where the attachment is made to a spindle fibre.

KINETIN = 6-FURFURYLAMINOPURINE.

KINETOCHORE = CENTROMERE = KINETIC CONSTRICTION.

KINETOMERES. Molecules of protoplasm which may be energy-rich and reactive, or energy-poor and stable.

KININS. A group of plant hormones, which seem to be particularly important in influencing the rate of mitosis. Found in young developing tissue and embryos.

KINO. A resin-like substance, soluble in water, astringent, used medicinally and in tanning.

KINOPLASM. Protoplasm which appears to be composed of fibrils and which in cell-division composes the spindle-fibres, attraction sphere, and astral rays.

KLINOSTAT = CLINOSTAT.

KLINOTROPHIC. Placed at a slant to the direction of a given stimulus.

KOEBERLINACEAE (H). A family of the Celastales (H). (EP) place it in the Capparidaceae. These are woody trees, but never climbers, with astipulate leaves. The sepals and petals are imbricate (petals rarely contorted), and the petals are free. There is no disk, and the fruit is rarely enclosed by an enlarged calyx. The ovules spread from a central axis. The fruit is a capsule with winged seeds, or a berry.

KOJIC ACID. An acid produced exclusively by the white-spored *Aspergillus* spp.

KRAMERIACEAE (H). A family of the Polygalales (H). The leaves lack stipules. The lower sepal is not spurred. There are 8-4 fertile stamens, with anthers mostly opening by apical pores. The ovary is unilocular, forming into bristly fruits. The seeds are often pilose, with conspicuous strophiole.

KREB'S CYCLE = CITRIC ACID CYCLE.

KREMNOPHYTE. A plant which grows alongside a steep wall.

KURTOSIS. The departure of a symmetrical frequency-distribution from the normal by excess (platykurtosis) or deficiency (leptokurtosis) in its shoulders, as opposed to tails and centre.

L

LABIATAE (EP, BH, H). A family of the Tubiflorae (EP, Lamiales (BH, H). These are herbs or shrubs with decussate, or whorled leaves without stipules. The inflorescence is a cyme, but may be compressed to appear as a whorl. The flowers are bisexual, zygomorphic. The 5 calyx lobes are fused, as are the 5 petals, usually forming 2 lips. There are 4 free stamens, with one pair longer than the other, or with 1 pair of staminodes. The superior ovary has 2 fused carpels, each with 2 erect ovules, with folds between them. The fruit is 4 nutlets, and the seeds have little or no endosperm.

LABIATE. Said of a corolla with one or more petals formed into a lip.

LABILE = PLASTIC MATERIAL.

LABILE CHEMICAL BINDING. The temporary holding of cations and anions in more labile complexes.

LABIUM. A lip.

LABOULBENIALES. An order of the Euascomycetae. These are ectoparasites on the cutinous integuments of living insects. The ascogonium has a trichogyne and fertilization is by a spermatia. The mature asci are in small perithecia.

LABYRINTHIFORM. (1) Marked by sinuous lines.
(2) Maze-like.

LABYRINTHUALES. An order of the mycetozoa, whose members are parasitic in the cells of algae and of submerged aquatic plants, forming a net-plasmodium. There are no serial sporangia. The germinating spores produce myxamoebae, or anteriorly uniflagellate zoospores.

LACCATE. Having a shining surface.

LACERATE, LACINIATE. Irregularly cut, as if torn.

LACINIA. (1) An incision in a leaf, petal etc.
(2) A slender lobe, projecting from the margin of a thallus.

LACINIATE. Said of an edge, which appears as if cut into bands.

LACISTEMACEAE. A family of the Piperales (EP), Bixales (H). (BH) place it as an anomolous family of the Monochlamydeae. These are shrubs with lanceolate, exstipulate leaves arranged in 2 ranks. The minute bisexual flowers are in spikes. The flowers are naked, or with a sepaloid perianth. There is 1 stamen. The superior ovary is of 2-3 fused carpels, with parietal placentas with 1-2 pendulous ovules on each. The fruit is a one-seeded capsule. The embryo is straight, and the seeds contain endosperm.

LACRIMIFORM, LACRIMOID. Like a tear, a watery secretion.

LACYEOUS. Milky.

LACTIC ACID. An organic acid produced by many bacteria, *e.g. Lactobacter*, during the fermentation of glucose or lactose. It can be utilized by some fungi. The most important form is $CH_3.CH(OH).COOH$.

LACTIC DEHYDROGENASE. The enzyme which reduces pyruvic acid.

LACTOFLAVIN. A yellow pigment, produced by *Eremothecium ashbyii*.

LACTORIDACEAE (EP, H). A family of the Ranales (EP), Magnoliales (H). (BH) include it in the Piperaceae. These are shrubs with alternate stipulate leaves. There are 3 free perianth lobes, 2 whorls of 3 free stamens, and a gynecium of 3 free carpels, with pendulous anatropous ovules. The seeds have endosperm.

LACTIFEROUS. Containing latex.

LACTIFEROUS HYPHAE, LACTIFEROUS TUBES. These are tubes occurring in some fungi. They are filled with latex which coagulates on a wound.

LACTIFIC. Producing latex.

LACUNA. (1) A large multicellular cavity.
(2) A depression in the surface of a plant.

(3) A cavity formed in rapidly elongating stems by the breakdown of the protoxylem.

(4) Any cavity in a plant.

LACUNOSE. Having a pitted surface.

LACUSTRINE, LACUSTRIS (Latin). (1) Pertaining to a lake, or lakes.

(2) Living in, or on the shores of a lake.

LACUSTRINE PEAT. The decomposition of vegetation to form peat, under anaerobic conditions.

LAEVIGATE, LAEVIGATUS, LAEVIS = LEVIGATE.

LAGENIDIACEAE. A family of the Lagenidiales. Its members are parasitic (rarely saprophytic) in the cells of algae, microscopic animals, or their eggs, and in one genus, on grasses. The contents of the zoosporangium are released into a vesicle, in which the zoospores attain their final form and are released. The zoospore encysts on the surface of the host, and germinates, penetrating the host by means of a variously shaped infection tube. The plant-body inside the host may be coenocytic, or variously septate, with each segment becoming a zoosporangium, or gametangium. Separate cells in the same hypha, or inseparate hyphae become male or female gametangia, and fertilization takes place through a conjugation-tube. The zygote is thick-walled, and there is no periplasm.

LAGENIDIALES. An order of the Phycomycetes. Most of these are parasitic, developing in a single host cell. The mycelium is sparingly branched, and is cellular, at least at reproduction, each cell forming a sporangium or gametangium. Asexual reproduction is by biflagellate zoospores. Sexual reproduction is oögamous by fusion of aplanogametes.

LAGENIFORM. Flask-shaped.

LAGGING. The slow movement towards the poles of the spindle by one or more chromosomes in a dividing nucleus, with the result that these chromosomes do not become incorporated into a daughter nucleus.

LAG-PERIOD. The period between the initial inoculation of a culture medium with an organism, and the resumption of normal growth.

LAMARCKISM. The theory that acquired characters are inherited during sexual reproduction.

LAMELLA. (1) A plate-like structure.

(2) A gill of an agaric.

(3) The plate-like structure in the grana and stroma in the chloroplast.

LAMELLATE. Made up of thin plates.

LAMELLOSE. Stratified.

LAMIALES (BH, H). An order of the Gamopetalae (BH), Herbaceae (H). See *Labiatae*.

LAMINA. (1) The flattened part of a leaf.

(2) Any flattened part of a thallus.

LAMINARIALES. An order of the Polystichineae. The sporophyte has a holdfast, stipe and blade. There is an intercalary meristem between the stipe and the blade. There are only unilocular sporangia on the sporophyte. The sori are spread extensively on the blade, some of which are specialized to bear sori.

LAMINARIN. A polysaccharide storage-product found in the brown-algae. It is constructed of β-D-glucopyranose units, linked in the 1:3 position.

LAMINATED BULB. A bulb composed of a number of swollen leaf-bases, each of which completely encloses all parts of the bulb inside it, *e.g.* an onion.

LANATE, LANATUS (Latin), LANOSE, LANGUINOSE. Covered with long, and loosely tangled hairs.

LANCEOLATE. Flattened, two or three times as long as broad, widest in the middle and tapering to a pointed apex.

LANCET-SHAPED. Flattened, and shortly lanceolate, with a bluntish apex.

LANGMUIR EQUATION. $a/m. = (k_1C) (1 + k_2C)$, where k_1, k_2 are constants, (see *Freundlich's equation*). If the reciprocal of a/m is plotted against the reciprocal of C, a straight line is formed.

LANGUID. (1) Feeble.
(2) Hanging down.

LANUGINOSE, LANUGINOUS = **LANATE.**

LANATE. Woolly.

LARDIZABALACEAE (EP, H). A family of the Ranales (EP). Berberidales (H). (BH) include it in the Berberidaceae. These are climbing shrubs with compound leaves, and bisexual or unisexual flowers, solitary or in racemes; they are regular. The perianth-lobes are free, and in 2 whorls of 3, with 2 whorls of honey-leaves. The stamens are free, with 2 whorls of 3. The superior ovary is of 3 free carpels, each with many ovules. The fruit is a berry, and the seeds contain endosperm.

LATENT TIME. The period of time between the beginning of stimulation and the first signs of a response.

LATERAL. (1) Arising from the side of the parent axis.
(2) Attached to the side of another member.
(3) On or near the edge of a thallus or fruit.

LATERAL MERISTEM. A meristem located at the side of a plant member.

LATERAL PLANE. A plane passing through a flower, parallel to the earth's surface.

LATERITE. A term widely applied to tropical red soils, with various definitions, but basically the two following conditions are necessary:
(1) in deep horizons, the parent rocks are permanently saturated with water, and have iron compounds which dissolve.
(2) In higher zones, where ferrous substances are oxidized as they are in contact with air. The top layer is usually brick-hard, and of ferruginous material.

LATERITOUS. Brick-red.

LATEX. A milky fluid containing sugars, proteins, alkaloids, oils, etc. Produced in latex vessels of many plants.

LATEX CELL. A simple or branched cell, derived from the enlargement of a single cell, and containing latex.

LATEX DUCT. An elongated, branched, aseptate system of anastomosing hyphae, containing latex, and present in some of the larger agarics.

LATEXOSIS. An abnormal secretion of latex due to a pathological condition.

LATEX TUBE = **LATEX VESSEL or LATEX CELL.**

LATEX VESSEL. A simple, or branched tube, usually anastomosing with other similar tubes, derived from the enlargement and union of a chain of cells, and containing latex.

LATI-. A Latin prefix meaning 'broad'.

LATIFEROUS DUCT. A cavity into which latex is secreted.

LATIFOLIUS. Latin meaning 'broad-leaved'.

LATISEPTATE. Having wide septa or dissepiments.

LATTICE. A vestigial sieve-plate on the side-wall of a phloem sieve-tube, having vaguely defined edges and very minute pores.

LATTICED. (1) Cross-barred.
(2) Like a net-work.

LAURACEAE (EP, H, BH). A family of the ranales (EP), Laurales (H), Daphnales (BH). (BH) include the Hernandiaceae. These are woody plants with leathery, alternate exstipulate leaves, and oil-cavities in the tissues. The flowers are bisexual or unisexual, in various forms of inflorescence. The perianth is of 2 similar whorls of 3 free segments. The free stamens are in 3 or 4 whorls, 1 of which is sometimes staminodes. The anthers open from the base by valves. The superior ovary consists of 3 fused carpels, with 1 loculus containing 1 pendulous ovule. The fruit is a fleshy berry, enclosed in a fleshy axis. The seeds have no endosperm.

LAURALES (H). An order of the Lignosae (H). All members are woody, with bisexual or unisexual flowers wich are hypogynous to perigynous, and cyclic. The flowers have no petals, with a definite number of free stamens. The ovary is apocarpous, to having 1 carpel. The endosperm is uniform, occasionally ruminant, or absent. The leaves are alternate, or opposite, simple and lack stipules.

LAURINEAE (BH). = LAURACEAE.

LAURIUM. A drain formation.

LAX, LAXUS (Latin). Arranged loosely.

LAYER. A stratum of vegetation, as the shrubs in a wood.

LAYERING. (1) The banding seen in thick cell-walls, due to the presence of wall-layers differing in water-content, chemical composition, and physical structure.
(2) The grouping of vegetation in a wood into 2 or more well-defined layers differing in height, as trees, shrubs, and ground vegetation.
(3) A method of artificial propagation in which stems are pegged down until they root, and then detached from the parent plant.

LEACHING. The removal of mineral salts from the soil by percolating water.

LEADER. One of the main shoots of a tree.

LEAF. An outgrowth from the stem of a vascular plant, usually green and mainly concerned with photosynthesis, and transpiration. It consists of a leaf-base, petiole, and flattened lamina, which is usually conspicuously veined.

LEAF BASE. The base of the leaf-stalk, where it joins the stem.

LEAF BUD. A bud containing vegetative leaves only.

LEAF CUSHION. A swollen leaf-base.

LEAF DIVERGENCE. The angle of the intersection of the planes passing longitudinally through the middles of two successive leaves.

LEAF-FALL. The organized shedding of leaves by deciduous trees in autumn.

LEAF GAP. A space in the stele of a plant where the leaf-trace emerges.

LEAF INCEPT. The earliest recognizable rudiment of a leaf.

LEAFLET. One separate portion of the lamina of a compound leaf.

LEAF MOSAIC. The arrangement of the leaves on a shoot or a plant in such a way that as much leaf as possible is exposed to light, and as little as possible is shaded by other leaves.

LEAF PRIMODIUM. A localized pass of meristematic tissue near the apical meristem. It ultimately divides to form a leaf.

LEAF SCAR. The scar on a stem left after the leaf has fallen. It is commonly covered with a layer of cork.

LEAF SCAR PERIDERM. A periderm developing across a leaf-scar beneath the initial protective layer.

LEAF SHEATH. (1) An expansion of the petiole or leaf-blade to surround the stem.
(2) Scale-like leaves growing over the apical cell of a stem of the Equisetaceae.

LEAF TRACE. The vascular bundle(s) extending from the stem to the leaf.

LEAFY RACEME. A raceme in which the bracts differ little, or not at all, from the ordinary foliage leaves of the plant.

LECANORALES. An order of the Ascomycetae. They produce apothecia, and show a specialized form of parasitism on land species of the Chlorophyceae and Myxophyceae.

LECANORINE. Said of a lichen apothecium, having a thalline margin.

LECIDEINE. Said of a lichen apothecium, which is dark-coloured or carbonaceous, and generally having no thalline margin.

LECITHINS. These are phosphatides, which are constituents of protoplasm, especially membranes. On hydrolysis they yield one molecule of glycerol, two molecules of fatty acids, R_1 and R_2, one of orthophosphoric acid, and one of choline.

$$CH_2{-}OOC{-}R_1$$
$$CH{-}OOC{-}R_2 \quad OH \qquad\qquad CH_2$$
$$CH_2{-}\!\!-\!O{-}\!\!-\!P{-}\!\!-\!O{-}CH_2{-}CH_2{-}N{-}CH_3$$
$$\overset{\|}{O} \qquad\qquad\qquad CH_3$$

LECITHOPROTEIN = LIPOPROTEIN.

LECTOTYPE = ALLOTYPE.

LECYTHIDACEAE (EP, H). A family of the Myrtiflorae (EP), Myrtales (H). (BH) include it in the Myrtaceae. These are trees with alternate, simple exstipulate leaves bunched at the end of the twigs. The bisexual flowers are solitary or in racemes. They are either regular, or petals and stamens are zygomorphic. The receptacle and ovary are completely fused. There are 6-4 free calyx lobes, 6-4 imbricate, free (rarely fused) petals, and many stamens in several whorls, more or less united at the base. The anthers are versatile, and bent inward in the bud. The ovary consists of many-6-4 fused carpels, with many loculi, containing many to 1 anatropous ovules. The fruit is a berry or capsule, and the seeds have no endosperm.

LEDOCARPACEAE (H). A family of the Malpighiales (H). The opposite leaves are simple, and never unifoliate. The sepals are imbricate (rarely valvate). There are 10 stamens. The style is very short, or the stigma sessile. There are many ovules, or there are two collateral, pendulous ones. The fruit is a capsule.

LEGUME. (1) A member of the Leguminosae.

(2) A dry fruit consisting of 1 carpel, splitting by 2 longitudinal sutures at dehiscence, and having a row of seeds on the inner side of the ventral suture, *e.g.* a pea-pod.

LEGUMINALES (H). An order of the Lignosae (H). See *Leguminosae*.

LEGUMINOSAE (EP, BH). A family of the Rosales (EP, BH). These are trees, shrubs, or herbs, usually with alternate, stipulate leaves. The flowers are in racemes, and are usually bisexual, regular or zygomorphic. There are 5 free calyx and corolla lobes. There are usually 2 whorls of 5 stamens, but may be more; they may be variously fused. The ovary is superior, and is mainly of one carpel, but may be 2-5-15, with many ovules. The fruit is a pod, or is indehiscent. There is usually no endosperm.

LEIOSPOROUS. Having smooth spores.

LEITNERIACEAE (EP, BH, H). A family of the Leitneriales (EP), (*q.v.*), Unisexales (BH), Leitneriales (H).

LEITNERIALES (EP, H). An order of the Archichlamydeae (EP), Lignosae (H). These are woody plants with simple alternate leaves, and spikes of dioecious flowers. The male flowers consist only of 12-3 free stamens. The female flowers are haplochlamydeous with a perianth of small, scaly united leaves. The single carpel is superior, with 1 amphitropous ovule. The fruit is a drupe, and there is a thin endosperm.

LEITNERIEAE (BH) = LEINTNERIACEAE (EP).

LEJEUNEACEAE. A family of the Acrogynae. The elaters have a single spiral thickening, and each elater extends from top to bottom of the capsule, and has the upper end fixed to the capsule wall.

LEMMA. The outer bract of a grass floret, *i.e.* has the grass floret in its axile.

LEMNACEAE (EP, BH, H). A family of the Spathiflorae (EP), Nudiflorae (BH), Arales (H). These are free-floating water-plants, usually with no leaves, (no stems, (H)). The flowers are unisexual, and lack a perianth. The male consists of 1 stamen, and the female of 1 carpel, with 6-1 basal erect ovules. The seeds have a thin endosperm.

LENNOACEAE (EP, BH, H). A family of the Ericales (EP, BH, H). These are root-parasites, with many bisexual, regular flowers. There are many to 5 flower-parts, with as many stamens as petals. The superior ovary consists of 6-14 (10-15 (H)) fused carpels, each containing 2 ovules. The fruit is a drupe, with 12-24 stones. The seeds contain endosperm.

LENTIBULARIACEAE (EP, BH, H). A family of the Tubiflorae (EP), Personales (BH, H). These are insectivorous plants of damp ground and water. The flowers are bisexual, and zygomorphic. The 5 calyx lobes are fused, as are the petals which form 2 lips. The 2 stamens are free, and borne on the petals. The superior ovary consists of 2 fused carpels, and contains many ovules on 2 parietal bilobed placentas. The fruits are 4-valved capsules. The seeds contain endosperm.

LENTICEL. A pore in the periderm of a woody stem. It is packed with a loose aggregate of cells derived from the phelloderm, and acts as an organ of gaseous exchange.

LENTICULAR. Shaped like a double convex lens.

LENTIGINOSE, LENTIGINOUS. Minutely dotted.

LENTO-CAPILLARY POINT. The point when the capillary movement of water in a soil becomes sluggish, and ineffective.

LEPIDOBOTRYACEAE (H). A family of the Malpighiales (H). The alternate leaves are compound, or unifoliate, with caducous stipules.

LEPIDOCARPACEAE. A family of the Lepidocarpales. The sporophylls are heterosporous, and in strobili. The older macrosporangia are completely surrounded by an integument which opens at the apex by a slit. Initially there are 4 macrospores but only 1 develops to give a seed-like structure, consisting of gametophyte, sporangium, and integument.

LEPIDOCARPALES. An order of the Lycopodinae. The sporangia are heterosporous, and the macrogametophyte is permanently retained within the macrosporangium.

LEPIDODENDRACEAE. A family of the Lepidodendrales, which is distinguished primarily by the spirally arranged leaf-scars on the trunks.

LEPIDODENDRALES. An order of the Lycopodonae. These were tree-like, with secondary thickening of the stems and roots. The leaves were microphyllous and ligulate. The roots were borne on rhizophores. The sporophytes were heterosporous, and the sporophylls in strobili.

LEPIDOTE. Covered with scale-like leaves.

LEPRARIOID, LEPROSE. Having a whitish, mealy, or scurfy surface.

LEPTO-. A Greek prefix meaning 'slender'.

LEPTOCENTRIC VASCULAR BUNDLE. A concentric vascular bundle, in which a central strand of phloem is surrounded by xylem.

LEPTODERMOUS. Having a thin wall, especially of a bryophyte capsule which is soft.

LEPTO-FORM. A form of life-cycle of the rusts in which there are only teleuto-spores, which germinate without rest.

LEPTOME. The conducting elements in the phloem.

LEPTOMITACEAE. A family of the Saprolegnilaes. The mycelium is constricted at regular intervals, and there may, or may not be a stout axis and slender branches. The sporangia are elongated and club-shaped, or short, and pear-shaped. It is permanently attached to the thallus, and produce many biflagellate zoospores. Sexual reproduction is oögamous.

LEPTOMITALES. A doubtful order of the Phycomycetae, differentiated due to the formation of a periplasm around the egg. This has been disproved in one genus.

LEPTONEMA = LEPTOTENE.

LEPTOSPORANGIATAE. A sub-class of the Filicinae. The sporangial jacket is one cell-thick. The tapetal layer is differentiated early from a single internal cell of a developing sporangium. The antheridia are small, and are more or less emergent, with relatively few antherozoids. The archegonial neck protrudes considerably from the gametophyte. The foot, primary root, cotyledon, and stem are each referable to a cell in the four-celled embryo.

LEPTOSPORANGIATE. Said of ferns in which the sporangium originates from a single cell.

LEPTOSTROMATACEAE. A family of the Sphaeropsidales, distinguished by the pycnidia having a well-developed roof, but poorly developed basal portions.

LEPTOTENE. The initial stage of meiosis. The chromosomes appear longitudinally, single rather than double, and the structure is more definite than in mitosis. There is a series of dense chromomeres.

LEPTOTICHOUS. Said of a tissue which is thin-walled.

LESION. (1) A wound.
(2) A well-marked, but limited diseased area.

LETHAL. Causing death; this may apply to a normal or abnormal environmental factor, or to a hereditary factor.

LETHAL GENE. A dominant or recessive gene, which, when substituted for its normal allelomorph converts a viable to an inviable gamete or zygote.

LEUC-, LEUCO-. A Greek prefix meaning 'white'.

LEUCANTHOUS, LEUCOANTHUS. White-flowered.

LEUCINE. An amino acid.

$$CH_3{-}CH{-}CH_2{-}CH_3$$
$$\overset{|}{C}H{-}NH_2$$
$$\overset{|}{C}OOH$$
Isoleucine

LEUCOANTHOCYANINS. See *Anthocyanins*.

LEUCOPLAST. A starch-storing organelle in the cytoplasm. It is a colourless plastid made up of concentric layers of starch around a central body.

LEUCOSPOROUS. Having white spores.

LEVANS. A polysaccharide; a polymer of fructose.

LEVIGATE, LEVIGATUS (Latin). Having a smooth, polished surface.

LEY. A short-term pasture, sown to last from one to a few years, after which it is ploughed-up, and another crop sown.

LIANA, LIANE. A woody, climbing plant, found in tropical forests. They usually have anomalous secondary thickening.

LIBER = PHLOEM.

LIBRIFORM FIBRE. An elongated thick-walled element of the xylem, formed from a single cell.

LICHENES, LICHENS. A group of composite plants, consisting of an alga and a fungus in intimate association.

LICHEN ACIDS. Organic acids, special to the lichens.

LICHENICOLE. Living on lichens, said especially of parasitic fungi.

LICHENIN. A glucosan.

LID. The covering of a sporangium of the mosses, usually detachable by a horizontal slit.

LEIBIG'S LAW OF MINIMUM. The amount of plant growth is regulated by the factor present in minimum amount, and rises or falls accordingly, as this is increased or decreased in amount.

LIFE CYCLE. The changes taking place between the production of gametes by one generation, and the production of gametes by the next generation. It may be synonymous with the life-history, but may involve a number of individuals, as in the alternation of generations in the bryophytes and pteridophytes.

LIFE FORM. The form of a plant determined by the position of its resting buds (if any), in respect to the surface of the soil.

LIFE HISTORY. The change in form and/or habits of a plant from its earliest stages to death.

LIGASES. A general term for enzymes which catalyse many biological substances.

LIGHT SEED. A seed which will not germinate unless it has been exposed to light.

LIGHT STAGE. The stage in photosynthesis when triosephosphate is formed from a 3-carbon acid (Phosphoroglyceric acid). The energy absorbed by the chloroplasts is used to split the water molecule into hydrogen and oxygen.

LIGNEOUS, LIGNOSE. Woody.

LIGNICOLE, LIGNICOLOUS. Growing on, or in wood, or on trees.

LIGNIFICATION. The deposition of lignin on and in a cell-wall.

LIGNIN. A complex carbohydrate deposited in the cellulose micella of the cell-walls of woody tissue.

LIGNOCELLULOSE. A compound of lignin and cellulose found in the wood and other fibrous materials.

LIGNOSAE (H). A division of the dicotyledons. This is basically a woody group, including trees and shrubs, but also herbs derived from them.

LIGULATE, LIGULIFORM. (1) Strap-shaped.
(2) Said of a corolla which has a very short tube and is prolonged above into a flattened group of united petals.
(3) Said of a capitulum in which all the flowers have a ligulate corolla.

LIGULE. (1) A flattened membrane arising from the base of the leaves of some lycopods.
(2) A membrane at the junction of the leaf-sheath and leaf-base of many grasses.

LIGULIFLORATE. Having ligulate flowers.

LIKELIHOOD FUNCTION. The function relating an unknown parameter to observations, from which it can be estimated.

LILAEACEAE (H). A family of the Juncaginales. There is only one, bract-like perianth segment. The style is of one kind in the female flower, it is much elongated, and whip-like. There are two kinds of female flower, one type at the base of the spike, and sessile within the leaf-sheath, and the other in a spike.

LILIACEAE (EP, BH, H). A family of the Liliiflorae (EP), Coronarieae (BH), Liliales (H). These are herbs, with rhizomes or bulbs, shrubs, or trees. The flowers are usually homochlamydeous and in racemes. They are usually regular and bisexual. The perianth lobes may be free or fused, and are in 2 whorls of 3, and petaloid. There are 6 free stamens, in 2 whorls of 3. The ovary is superior to inferior, of 3 fused carpels, with 3-1 loculi. The fruit is of various forms. The seed contains a straight or curved embryo, and a fleshy or cartilaginous endosperm.

LILIALES (H). An order of the Corolliferae. See *Liliaceae*.

LILIIFLORAE (EP). An order of the Monocotyledoneae. These are usually herbs, rarely with stout stems. The flowers are cyclic with the perianth usually of 2 whorls of 3 free segments, the whorls being alike or different. There are 2 whorls of 3 free stamens (1 whorl may be absent), or there may be only 1 stamen. The ovary is of 3 fused carpels. The ovules are usually anatropous, and the endosperm fleshy or oily.

LIMB. (1) The lamina of a leaf.
(2) The widened upper part of a petal.
(3) The upper, often spreading, part of a sympetalous corolla.

LIMBATE. Edged with another colour.

LIME KNOT. A widening in the threads of the capillitium of Myxomycetes, containing calcium carbonate.

LIMITING FACTOR. A factor which stops, or reduces the speed of a reaction, when all other factors are present in abundance.

LIMIT OF TREES. The line, north or south, or upwards on mountains, beyond which trees do not naturally occur.

LIMNANTHACEAE (EP, H). A family of the Sapindales (EP), Geraniales (H). (BH) include it in the Geraniaceae. These are annuals with alternate exstipulate leaves, with solitary, axillary, bisexual, regular flowers. Both the calyx and corolla consist of 5-3 free segments. There are 10-6 free stamens. The ovary consists of 5-3 fused carpels, and is superior. Each carpel contains 1 ovule, and they separate when ripe. There is no endosperm.

LIMNIUM. A lake formation.

LIMNODIUM. A salt-marsh formation.

LIMONIFORM. Lemon-shaped.

LINACEAE (EP, BH, H). A family of the Geraniales (EP, BH), Malpighiales (H). (BH) include the Erythroxyaceae. These are herbs or woody, with alternate leaves, with, or without stipules. The flowers are bisexual and regular, lacking a disk. There are 5-4 free sepals and petals. The 20-5 free stamens are united at the base. The ovary consists of 5-4 (or less) fused carpels, and is superior. It is multilocular, often with extra partitions, containing 2-1 pendulous anatropous ovules each. The fruit is a capsule or drupe. The embryo is usually straight, in a fleshy endosperm.

LINAMARIN. A cyanogenetic glucoside, present in flax roots.

LINEAE (BH) = LINACEAE (EP).

LINEAR. Having parallel edges, and at least 4-5 times as long as broad.

LINEAR ORDER. See *Chromosome map*.

LINEATE. Marked with lines.

LINE BREEDING. The mating in successive generations of individuals having the same common ancestor.

LINEOLATE. Marked with fine lines.

LINE SURVEY. A record of plants occurring along a line taken across a piece of country.

LINE TRANSECT. A chart showing the position and names of the plants occurring on a line drawn across a piece of country.

LINEAGE. In evolution, the development of a character, or individual, through time, giving rise to a complex of lines of descent.

LINEAR TETRAD. A row of four megaspores as is usual in flowering plants.

LINGUIFORM, LINGULATE. Tongue-shaped; shorter and wider than ligulate, and somewhat fleshy with a blunt apex.

LININ. The more solid, form-conserving part of the nucleus, which holds the chromioles in definite relationship with each other.

LINKAGE. When two genes are relatively close together on a chromosome, so as not to be separated during crossing-over in meiosis, *i.e.* they are inherited together, and are said to be linked.

LINKAGE GROUP. A group of hereditary characteristics which remain associated with one another through a number of generations.

LINKAGE MAP. A chromosome map, determined by recombination relations.

LINKED REACTION. A chain reaction, *e.g.* A results in B, which results in C —— X.

LINNAEAN (LINNEAN) CLASSIFICATION. The system of classification and binomial nomenclature established by Linnaeus.

LINNAEAN SPECIES. A wide conception of a species, in which many varieties are included.

LINNAEON. A Linnaean species.

LIP. A large projecting lobe of a corolla.

LIPASE. An enzyme breaking-down a true fat into its component fatty acid(s) and glycerol.

LIP-CELL = STOMIUM.

LIPID (LIPIDE). An ester formed from an alcohol and one or more fatty acids.

LIPID HYPOTHESIS. The concept that solubility in a lipid is important in determining the rate at which substances are transferred across membranes.

LIPIN (LIPINE). A fat containing nitrogen, phosphorus, or sulphur.

LIPOCHONDRIA = GOLGI APPARATUS.

LIPOIC ACID. It may be concerned with the transfer of the acetyl radical to coenzyme A, and hydrogen to coenzyme 1 in the pyruvate oxidation system.

$$H_2C \underset{S\text{——}S}{\overset{\overset{\displaystyle CH_3}{|}}{\diagdown}} CH\text{—}CH_2\text{—}CH_2\text{—}CH_2\text{—}CH_2\text{—}COOH$$

LIPOID. A fat, or a substance resembling a fat in its solubility.

LIPOPHILIC. Attracted by hydrocarbons, and repelled by hydroxy- (OH) substances.

LIPOPLAST. A fatty globule.

LIPO-PROTEIN. Compounds of protein and a fat, which are most commonly found in membranes.

LIPOSOME. A fatty or oily globule in the cytoplasm.

LIPOXIDASE. An enzyme which plays a part in fatty acid oxidation by breaking the chain at the double bond.

LIRELLA. A long, narrow apothecium with a ridge in the middle, found in some lichens.

LIRELLIFORM. Like a furrow.

LISSOCARPACEAE (H). A family of the Styacales (H). The corolla lobes are contorted, and the 4 calyx-lobes are imbricate. There are 8 stamens, with linear anthers having the connective produced at the apex.

LITHOPHYTE. A plant growing on rocks and stones.

LITTER. The more or less undecomposed plant residues on the surface of soil in a wood.

LITTORAL, LITTORALIS (Latin). Living on, or pertaining to the shore, especially the sea-shore.

LITTORAL ZONE. The part of the sea-shore inhabited by plants, below the average low-water level.

LITUATE. Forked, and having the tips turned out a little.

LIVERWORTS = HEPATICAE.

LIVEUS. Latin meaning 'pale lead-coloured'.

LOASACEAE (EP, BH, H). A family of the Parietales (EP), Passiflorales (BH), Loasales (H). These are herbs (rarely shrubs) which are sometimes twining. The leaves are opposite or alternate, with no stipules, and often with stinging hairs. The flowers are bisexual. The 5 (rarely 7-4) calyx lobes are free. The 5 petals are free, (rarely united) and are often boat-shaped. There are many free stamens, with those opposite the calyx lobes often modified as nectaries. The inferior ovary consists of 7-3 fused carpels each with many-to-1 ovules usually on parietal placentas. The fruits are capsules, sometimes spirally twisted. The seeds have endosperm.

LOASALES (H). An order of the Lignosae. These are woody or herbaceous plants with astipulate leaves. The flowers are hypogenous to perigynous. The petals are contorted or valvate. There are many-to-few stamens which are sometimes in bundles. The carpels are fused with parietal placentation. The seeds have copious endosperm and are often arillate. The embryo is straight.

LOASEAE (BH) = LOASACEAE (EP).

LOBATE, LOBED, LOBOSE, LOBULATE. Having lobes.

LOBE. (1) One of the parts into which a flattened plant member is cut, when the parts are too large and distinct to be called teeth, but not wholly separate from one another.

(2) A portion of a divided (not compound) stigma.

LOBELIACEAE (H). A family of the Campanales. The corolla is zygomorphic, and the anthers are joined into a tube around the style (rarely free).

LOBING. Formed into lobes.

LOCALIZATION. The restriction of crossing-over and chiasma formation to certain corresponding parts of all the chromosomes. This is genetically determined. It may be procentric or protermal, according to whether the contact points are near the centromere or near the ends.

LOCELLATE. Subdivided into smaller loculi.

LOCHMIUM. A thicket formation.

LOCK. The cavity of the ovary in plants; a locule.

LOCULAR, LOCULATUS (Latin). Divided into compartments by septa.

LOCULAMENT, LOCULE, LOCULUS. (1) A chamber of an ovary or an anther.

(2) One portion of a septate spore.

LOCULICIDAL. Said of a fruit which splits open along the midribs of the carpels.

LOCULUS = LOCULAMENT.

LOCUS. (1) The position of a particular gene on its chromosome.

(2) = *hilum* of starch grains.

LODICULE. A scale below the ovary of a grass flower representing the reduced perianth. There are usually 1 or 2 (rarely 3), and they become distended with water and assist in the separation of the glumes.

LOESS. A wind-dispersed soil.

LOGANIACEAE (EP, BH, H). A family of the Contortae (EP), Gentianales (BH), Loganiales (H). These are woody (rarely herbs) with opposite, or whorled often stipulate leaves. The flowers are bisexual, unisexual, and regular, being borne in cymose umbels. The calyx is usually imbricate. The many-5-4 corolla lobes are fused, and are valvate, imbricate, or convient. There are as many stamens as corolla lobes, or only 1. The superior ovary is of 2 fused carpels (rarely more), with many to 1 ovules on axile placentas. The fruits are capsules, and the seeds contain endosperm.

LOGANIALES (H). An order of the Lignosae. The leaves are opposite, simple (rarely compound) and lack stipules, or have stipules. The sepals are mostly valvate. The corolla lobes are contorted, imbricate, or valvate. They are fused, or rarely free. The stamens are borne on the corolla-tube, alternating with the corolla-lobes, or fewer in number. The ovary is superior and has 2-4 loculi. There are usually many ovules. The seeds have a straight embryo and endosperm.

LOMENTUM. A fruit, usually elongated, which develops constrictions as it matures, finally breaking across these into one-seeded portions.

LONG-DAY PLANT. A plant that needs more than 12 hours of daylight, followed successively by shorter periods of darkness before it will flower.

LONGI-. A Latin prefix meaning 'long'.

LONGICOLLOUS. Having a long beak or neck.

LOPHIUM. A hill formation.

LOPHOTRICHOUS. Said of an organism which has the flagella in one group arising at one point on the surface of the cell.

LOPHOZIACEAE. A family of the Acrogynae, distinguished by the triangular perianth having its single angle towards the dorsal surface.

LORANTHACEAE (EP, BH, H). A family of the Santalales (EP, H), Achlamydosporae (BH). These are woody semiparasites, usually on trees. The flowers are bisexual or unisexual, and homochlamydeous. They are regular, with the perianth of 2 whorls of 2-3 lobes. There are as many stamens as perianth lobes. The inferior ovary is unilocular, usually without differentiation of ovule and placenta, and a layer of viscin around the seed. The seeds contain endosperm.

LORATE, LORIFORM. Shaped like a strap.

LORICA. An obsolete term for the testa.

LOWIACEAE (H). A family of the Zingiberales (H). The leaves and bracts are distichous. The sepals are united into a tube, and the median petal is large, forming a labellum. There are 6-5 stamens, with bilocular anthers. The fruit is a capsule.

LUBRICOUS. Having a slippery surface.

LUCENS, LUCIDUS. Latin meaning 'with a shining surface'.

LUCIFUGOUS. Shunning light.

LUCIPHILOUS. Seeking light.

LUNATE. Half-moon shaped.

LUMEN. The space enclosed by a cell-wall, especially after the contents have disappeared.

LUNDEGÅRDTH HYPOTHESIS. That anion respiration is mediated through cytochrome oxidase, and that cytochrome may be a carrier of anions.

LURID, LURIDUS (Latin). Dingy yellowish-brown.

LUTEIC ACID. A metabolic product of *Penicillium luteum.*

LUTEIN, LUTEOL. $C_{40}H_{56}O_2$. A dihydroxycarotene, leaf xanthophyll. One of the most common carotenoids in leaves. In autumn an isomer occurs, namely zeaxanthol. Lutein also occurs in some petals.

LUTEOLEERSIN. A metabolic product of *Helminthosporium leersi.*

LUTEOLUS. Latin meaning 'pale-yellow'.

LUTEOUS, LUTEUS. Latin meaning 'a good yellow colour'.

LUTESCENS (Latin). LUTESCENT. Yellowish.

LYASES. Enzymes which break-down complex molecules to simpler ones, without hydrolysis, *e.g.* carbonic anhydrase breaks down carbonic acid to water and carbon dioxide, $H_2CO_3 = H_2O + CO_2$.

LYCOMARASMIN. A toxic product of *Fusarium lycopersici*. It causes wilting of tomatoes, by affecting the permeability of the membranes.

LYCOPERDACEAE. A family of the Lycoperdales. There is no stipe. The outer periderm is shed in patches or granules, and the inner periderm is mostly thin, opening by an ostiole or breaking away in pieces. A columella is sometimes present.

LYCOPERDALES. An order of the Gasteromyceteae. The fruit-body is medium-sized to large, and, with few exceptions, epigeous, at least at maturity. The gleba usually has definite hymenial cavities, or they sometimes become obliterated. When it is mature it becomes a powdery mass of spores and capitulum, or the basidia may remain intact.

LYCOPODIACEAE. A family of the Lycopodiales. These are herbaceous, with the stem bearing many small leaves which have simple apices. The sporophylls and foliage leaves may be similar or dissimilar. The sporophylls may or may not be in strobili.

LYCOPODIALES. An order of the Lycopodinae. These are herbaceous, with no secondary thickening. The leaves are microphyllous and lack ligules. The sporangia are homosporous, and the sporophylls are usually in strobili. The gametophytes are more or less subterranean, with the antheridia embedded in them. The antherozoids are biflagellate.

LYCOPODINAE. A class of the Pteridophyta. The sporophyte has a stem, root, and leaf. The leaves are microphyllous, and spirally arranged. There are no leaf-gaps in the vascular cylinder. The sporangia are on the adaxial side of the leaf and towards the base. A leaf bears only one sporangium.

LYCOPSIDA. The Pteridophyta lacking leaf-gaps in the stele. It includes the lycopods, horsetails, and the psilotaceous series.

LYRATE. (1) Lyre-shaped.
(2) Said of a leaf which is pinnately lobed, and has a terminal lobe which is much larger than the lateral lobes.

LYSIGENIC, LYSIGENETIC, LYSIGENOUS. Said of a space formed by the disintegration of cells, especially of secretory cells leaving a cavity containing the secretion.

LYSINE. An amino acid. $CH_2(NH_2).(CH_2)_3).CH(NH_2).COOH$.

LYSIS. The bursting of a bacterial cell when attacked by a bacteriophage.

LYSOGENIC 'PHAGE. A bacteriophage which does not kill its host.

LYSOZYME. A substance, present in some plants, which has the power to kill bacteria. It resembles an enzyme in some respects, but cannot reproduce itself.

LYTHRACEAE (EP, BH, H). A family of the Myrtiflorae (EP), Myrtales (BH), Lythrales (H). These are herbs or shrubs, with simple, entire, usually opposite, and stipulate leaves. The flowers are bisexual, regular or zygomorphic, and are borne in various forms of inflorescence. The sepals and petals are 4 or 6 in number, and the sepals are fused to form a tube. The calyx is valvate, and the petals may be absent, but when present they are sometimes crumpled in the bud. There are twice as many stamens as petals, many, or only 1. The superior ovary consists of 6-2 fused carpels, with 6-2, rarely 1 loculi, each with many to 2 ovules. The fruit is a capsule, and there is no endosperm.

LYTHRALES (H). An order of the Herbaceae (H). These are herbaceous to woody, with reduced forms which are aquatic. The leaves are simple, usually opposite, and lacking stipules. The flowers are regular and perigynous to epigynous. The calyx (rarely absent) is tubular with valvate lobes. Petals are usually present and often clawed. There are as many, or twice as many stamens as petals, and they are sometimes in two distinct whorls. The ovary has axile placentation. The seeds do not usually contain endosperm.

LYTIC 'PHAGE. A bacteriophage which causes lysis.

M

MACCHIE, MAQUI. A copse association of Mediterranean coasts.

MACRO-. A prefix meaning 'long'. It is sometimes used in the sense of 'Mega-'.

MACROCONIDIUM. (1) A long, or large conidium.
(2) The larger, generally more diagnostic conidium of a fungus, which has microconidia in addition.

MACROCYCLIC. Said of rusts which have 2 alternative hosts, and all the stages in the life-cycle.

MACROCYST. The resting form, of the young plasmodium of the myxomycetes.

MACROFUNGI. Fungi having large fruit-bodies.

MACROGAMETE. A large gamete containing food reserves, *i.e.* the female gamete.

MACROGAMETOPHYTE. The larger of the gametophytes of the heterosporous pteridophytes. It produces the female gametes.

MACROGONIDIUM. A large conidium.

MACROMOLECULE. A molecule of very large molecular weight.

MACROMYCETES = MACROFUNGI.

MACRONEMEAE. A group of the Moniales having conidia unlike the hyphae and the conidiophores.

MACRONUTRIENTS. Minerals that are required in relatively large amounts for the healthy growth of plants.

MACROPHYLLINE. Divided into, or having, large lobes.

MACROPHYLLOUS. Said of leaves, which have a branched vascular system.

MACROPODOUS. Said of an embryo, without cotyledons.

MACROSCOPIC. Able to be seen without a lens.

MACROSPORANGIUM = MEGASPORANGIUM.

MACROSPORE = MEGASPORE.

MACROSPOROPHYLL = MEGASPOROPHYLL.

MACULA, MACULE. (1) A blotch, or spot of colour.
(2) A small tubercule.
(3) A small, shallow pit.

MACULATUS. Latin meaning 'spotted'.

MACULICOLE, MACULICOLOUS. Growing on spots, *e.g.* leaf-spots.

MAGNOLIACEAE (EP, BH, H). A family of the Ranales (EP, BH), Magnoliales (H). These are woody plants, with alternate, simple leaves. The flowers are solitary, regular, bisexual or unisexual. The perianth is usually petaloid. There are many free anthers. The gynecium is superior, and of many carpels, which are rarely united. The seeds contain endosperm.

MAGNOLIALES (H). An order of the Lignosae (H). This is an entirely woody group, with leaves that are usually alternate, and sometimes stipulate. The flowers are hypogynous, and usually bisexual. The petals, if present are free. The numerous stamens are free, or massed and joined. The carpels are numerous and free, or reduced to one. The endosperm is copious, not ruminant, and the embryo is small.

MAJUS. Latin meaning 'greater'.

MALACOID. Like mucilage.

MALACOPHILY. Pollinated by snails.

MALACOPHYLLOUS. Said of xerophytic plants, which have fleshy leaves containing much water-storage tissue.

MALATE DEHYDROGENASE. See *Malic acid*.

MALE FLOWER. A flower containing stamens, but no carpels.

MALESHERBIACEAE (EP, H). A family of the Parietales (EP), Passiflorales (H). (BH) include it in the Passiflorales. These are herbs or undershrubs, with alternate, exstipulate, usually hairy leaves. The regular bisexual flowers are in racemes or cymes. There are 5 free sepals, and petals. The axis is tubular with a central androphore bearing 5 stamens. The ovary is superior consisting of 3 fused carpels with parietal placentas, and many anatropous ovules. There are 3-4 styles below the apex of the ovary. The fruit is a capsule, and the seed has no aril.

MALIC ACID. An acid in the Citric Acid Cycle. It is formed from fumaric acid by hydrolysis (enzyme fumarase) and is changed to oxalo-acetic acid by dehydrogenation by malic dehydrogenase. It is also fairly widely distributed in plants, *e.g.* in apple fruits.

$$\begin{array}{c} \text{COOH} \\ | \\ \text{CH}_2 \\ | \\ \text{CHOH} \\ | \\ \text{COOH} \end{array}$$

Malic acid

MALONIC ACID. $KOOC.CH_2.COOH$.

MALOL. An alcohol, $C_{30}H_{42}O_3$. A constituent of apple wax.

MALPIGHIACEAE (EP, BH, H). A family of the Geraniales (EP, BH), Malpighiales (H). These are woody usually climbing plants with opposite, stipulate leaves. The flowers are bisexual, with a convex, or flat axis, sometimes with a gynophore. The calyx has 5 fused lobes, often with nectaries. The 5 free petals are often clawed. The free stamens are in 2 whorls of 5, with the outer whorl opposite the petals; some are sometimes aborted. The anthers are bilocular, rarely 4-locular. The superior ovary is of 3 fused carpels, each with a single ovule. The fruit is a schizocarp, nut, or drupe. There is no endosperm in the seeds.

MALPIGHIAN CELL. One cell of a layer of closely packed, radially directed thick-walled cells occurring in the testas of some seeds.

MALTASE. An enzyme breaking maltose into its two component glucose molecules.

MALTHUSIAN PARAMETER. A measure of the relative rate of population increase or decrease when in the appropriate steady state.

MALTOSE. A disaccharide sugar ($C_{12}H_{22}O_{11}$), made up from the condensing of two D-glucose molecules, in the 1:4 and α position.

MALTOTRIOSE. A trisaccharide sugar.

MALVACEAE (EP, BH, H). A family of the Malvales (EP, BH, H). (BH) include the Bombacaceae. This is the single family of the Malvales according to (H) (*q.v.*) for his characters. These are herbs, shrubs, or trees, with simple, lobed leaves which have stipules. The flowers are bisexual solitary or in inflorescences. The 5 free calyx lobes are free, often with an epicalyx. The 5 petals are free and convient. There are usually many stamens in 2 whorls, and they are united at the base. Each anther represents a half-anther, and has thorny pollen. The ovary is superior and of 2-5 fused carpels, with many to 2 ovules. The fruit is a capsule or schizocarp. The seeds have endosperm.

MALVALES (H). An order of the Lignosae. The characters are similar to the above, but exclude the tree forms.

MALVALES (EP, BH). An order of the Archichlamydeae (EP), Polypetalae (BH). The bisexual (usually) and regular flowers have their parts in whorls (sometimes except the stamens). There are usually 5 sepals and petals, the latter being valvate (rarely absent). The stamens are free, and many, or in 2 whorls with the inner branched. The ovary consists of 2 to many fused carpels, each with many to 1 anatropous ovules, with 2 integuments.

223

MALVIDIN. A blue anthocyanin pigment.

Malvidin

MAMILLAR, MAMILLATE. Having a rounded outgrowth, ending in a papilla, or point.

MAMILLIFORM. Shaped like a papilla.

MANDELONITRILE. Its glucoside prulaurasin occurs in leaves of cherry laurel.

Mandelonitrile   *Prulaurasin*

MANGROVE. An association of plants of the muddy swamps at the mouth of rivers and elsewhere in the tropics, over which the tide flows daily, leaving the mud bare at low tide.

MANNON. A polymer of D-mannose.

MANNITOL. A sugar-alcohol of mannose.

D—Mannitol

MANNOSE. A hexose sugar.

L-Mannose

MANOCYST. The receptive papilla of some Oömycetes.

MANOXYLIC WOOD. Wood of a somewhat loose texture, and contains a large amount of parenchyma.

MANUBRIUM = HANDLE-CELL.

MAR = MOR.

MARANTACEAE (EP, H). A family of the Scitamineae (EP), Zingiberales (H). (BH) include it in the Scitamineae. These are perennial herbs with two-ranked leaves with a pulvinus at the end of the stalk. The bisexual flowers are asymmetrical and heterochlamydeous. The perianth is of 2 whorls of 3 free lobes. There are 5-4 free stamens. The inner one has but 1 anther lobe, and all the other stamens are petaloid. The inferior ovary is of 3 fused carpels, and has 3 or 1 loculi, each with 2 ovules. The fruit is a capsule. The seed is arillate, and has endosperm and perisperm.

MARATTIACEAE. The single family of the Marattiales.

MARATTIALES. An order of the Eusporangiatae. The sporangia are in sori on the abaxial side of the leaf-blade. There are stipules at the leaf bases.

MARBLED. Marked with irregular streaks of colour.

MARCESCENT. Withered, but remaining attached to the plant.

MARCGRAVIACEAE (EP, H). A family of the Parietales (EP), Theales (H). (BH) include it in the Ternstroemales. These are woody plants, often climbing or epiphytic, with simple exstipulate leaves, and racemes of bisexual regular flowers. The bracts are modified to form nectaries. There are 5-4 free sepals, 5-4 fused petals. The many-6-3 stamens are free. The superior ovary is of 5, many-8-2 fused carpels, with many ovules or originally parietal placentas, which later meet at the centre. The fruit is a capsule, and the seeds have no endosperm.

MARCHANTIACEAE. A family of the Marchantiales. It includes all the genera with archegonia on special stalked, vertical branches (archegoniophore).

MARCHANTIALES. An order of the Hepaticae. The gametophyte is internally differentiated into various tissues, and the jacket-layer of the sporophyte is one cell thick.

MARGIN. (1) The edge of a growing fungal mycelium.
(2) The edge of a leaf or other flattened plant member.

MARGINAL. (1) Situated on, or arising from, the edge of a member.
(2) Of a placenta, in single carpels, or on the edges of the carpels.

MARGINAL COMMUNITY. A plant community bordering on another community of slightly different character.

MARGINALES. An order of the ferns, including the members that bear the spores on the leaf margins.

MARGINAL RAY CELL. A more or less specialized cell, occurring with others of the same kind at the edge of a vascular ray.

MARGINAL SPECIES. A plant that grows along the edge of a woodland.

MARGINAL VEIL = PARTIAL VEIL.

MARGINATE. Having a well-marked border, which is often composed of cells or elements differing in form or colour from those making up the rest of the member.

MARITIME, MARITIMUS (Latin). Living by the sea.

MARMORATE, MARMORATUS (Latin). Marked, or coloured like marble.

MARSILIACEAE. A family of the Leptosporangiatae. They are aquatic and heterosporous. The sporocarp contains both microspores and macrospores.

MARSUPIUM. Upgrowths around the 'stem' apex of the Acrogynae.

MARTYNIACEAE (EP, H). A family of the Tubiflorae (EP), Bignoniales (H). (BH) include it in the Pedaliaceae. This family is similar to the Pedaliaceae (*q.v.*) but the anther thecae are spurred. The superior ovary has two bilobed parietal placentas. The fruits are capsules, and the seeds have a thin endosperm. The flowers are in racemes.

MASS ACTION, LAW OF. The rate of a chemical reaction is proportional to the concentration of the reacting substances. In the equation $k_1[A] + [B] \rightleftharpoons k_2[C] + [D]$ where the bracketed figures are the concentrations of the particular substances,

$$\frac{[A] \times [B]}{[C] \times [D]} = \frac{k_2}{k_1} = K.$$

K is the equilibrium concentration for a particular reaction, and is unaffected by the initial concentrations of the substances.

MASSULA. One of the four alveolar from the plasmodial tapetum of *Azolla*. One contains the functional macrospore.

MAST. The fruit of the beech and related trees.

MASTER FACTOR. Any powerfully acting ecological factor, which plays the main part in determining the occurrence in a given area of a plant community of major rank.

MASTIGOPHORA. A class of Protozoa characterized by the presence of 1 or more flagella, and 1 nucleus (if more than one is present, the nuclei are of the same kind). Some authorities include the unicellular Volvocales in this class.

MASTIGOPOD. An obsolete term for a swarm-cell of the Myxomycetes.

MASTOID. Like a nipple.

MATIÈRE NOIRE. The fraction of the soil organic matter which is soluble in a 4% ammonia solution. It is used by some as the definition of humus.

MATONIACEAE. A family of the Leptosporangiatae. The sporangia develop simultaneously in sori which are protected by an umbrella-like indusium. The sporangia dehisce transversely by approximately vertical annuli.

MATERNAL INHERITANCE. The occurrence of hereditable differences referable to materials transmitted by the egg, but not through the male gamete. This is due to *cytoplasmic inheritance*.

MATING CONTINUUM. An aggregate of individuals whose genes systematically recombine.

MATING GROUP. A group of individuals, haploid or diploid, within which mating is favoured at the expense of mating outside the group, by genetic or environmental conditions characteristic of the group.

MATORRAL. A xerophilous shrub community, found on leached acid soils in Spain and Portugal.

MATRIX. (1) An outer layer of stainable material in a chromosome.
(2) Any substratum, living or dead, in which a fungus grows.

MATROCLINOUS. Exhibiting the characters of the female parent more prominently than the male.

MATROMORPHIC. Resembling the female parent.

MATURATION. The formation of gametes or spores by meiosis.

MATURATION DIVISION = MEIOSIS.

MAXIMUS. Latin meaning 'very large'.

MAYACACEAE (EP, BH, H). A family of the Farinosae (EP), Coronarieae (BH), Commelinales (H). These are marsh herbs with alternate, linear leaves. The bisexual, regular flowers are solitary or in umbels. The perianth segments are in threes. There are 3 free stamens, and the superior ovary has 3 fused carpels and 1 loculus containing a few orthotropous ovules. The fruit is a capsule.

MAZAEDIUM. A fruit-body, as in some lichens and fungi, in which the spores, generally with sterile elements, become free from the asci as a mass of powder.

MEALY. Covered by a scurfy powder.

MEAN (ARITHMETICAL). The arithmetic average of a series of observations or quantities.

MECHANICAL TISSUE. Tissues, usually made up of thick-walled cells, which give support to the plant body.

MEDIAL. (1) Central, middle.
(2) The value of a variate on each side of which lie equal numbers of observations.

MEDIAN PLANE. The plane passing through the middle of an organ.

MEDIUM. A nutritive substance on, or in which, tissues or cultures of microorganisms may be grown.

MEDULLA. (1) The central part of an organ, *e.g.* the pith of a stem.
(2) A tangle of loose hyphae in a sclerotium, rhizomorph, or any other large fungal structure.
(3) A loose hyphal layer in a thallus of a lichen.

MEDULLARY. Relating to or belonging to the pith.

MEDULLARY BUNDLE. A vascular bundle running in the pith.

MEDULLARY RAY. A sheet of parenchyma running radially through the vascular tissue of a stem or root. They are concerned with food storage and lateral conduction of food materials, water etc. They may be primary, or secondary, *i.e.* produced from the cambium, when they are called vascular rays.

MEDULLARY SHEATH. The peripheral cells, surrounding the pith. They are smaller cells than those of the pith, and usually have thicker walls and denser contents.

MEDULLARY SPOTS. Small patches of thick-walled parenchyma in the medulla.

MEDULLARY STELE. A meristele lying in the centre of a fern stem.

MEDULLATE. (1) Having pith.
(2) = STUFFED.

MEDUSAGYNACEAE (H). A family of the Theales (H). The leaves are opposite, and the styles are in a ring on the shoulders of the carpels.

MEDUSANDRACEAE (H). A family of the Olacales (H). The leaves have small stipules which are sometimes very caducous. The flowers are in catkin-like racemes. The sepals and petals are distinct, with the latter imbricate. There are 5 staminodes, and the ovary has a slender free-central placenta.

MEGA-. A Greek prefix meaning 'of great size, large'.

MEGAGAMETE = MACROGAMETE.

MEGALOGONIDIUM = MACROGONIDIUM.

MEGAPHANEROPHYTE. A tree over 30 m. high.

MEGAPHYLLOUS. Having very large leaves, especially when supplied with a vascular bundle(s).

MEGASPORIUM. The sporangium in which megaspores are produced by division of sporogenous tissue, *e.g.* the ovule of flowering plants.

MEGASPORE. The larger of 2 spores of heterosporous plants. It ultimately develops to produce the female gametophyte.

MEGASPOROCYTE. The female gametophyte of angiosperms. The nucleus divides to give the egg-cell and the accessory nuclei.

MEGASPOROPHYLL. A leaf bearing megaspores. The leaf can be highly modified, *e.g.* the carpels of a flowering plant.

MEIOCYTE. Any cell in which meiosis is begun.

MEIOMEROUS. Having a small number of parts.

MEIOPHASE. The part of a life cycle in which a diploid nucleus undergoes reduction.

MEIOSIS. The reduction division, ultimately resulting in the production of gametes. The chromosome number is halved by two successive divisions. The nucleus divides twice, but the chromosomes only once.

MEIOTAXY. The failure of a whorl, or whorls to develop.

MEIOTIC EUAPOGAMY. The development of a sporophyte from a cell or cells of a gametophyte, without any fusion of gametes, giving a plant whose nuclei have the gametic number of chromosomes.

MELA-. A Greek prefix meaning 'black'.

MELAMPSORACEAE. A family of the Uredinales. The teliospores lack stalks, and are produced singly or in groups of 2 or 4 in the mesophyll, or just below, or within the epidermal cells, or are united laterally into a crust below the epidermis or cuticle. They may be united into separate chains, or into chains which are joined into a waxy column that emerges through the epidermis. The aecia are mostly on species of the Pinaceae.

MELANCONIACEAE. The single family of the Melanconiales.

MELANCONIALES. An order of the Fungi Imperfecti. The asexual spores are produced in acervuli.

MELANOGASTRACEAE. A family of the Hymenogastrales, distinguished by the hymenial cavity being filled with a gelatinous substance, or with irregular masses of basidia.

MELANOSPORACEAE. A family of the Sphaeriales. The perithecial walls are thin and usually light-brown. They are ostiolate, with the opening on a papilla or well-developed neck. There are no paraphyses. The ascospores are dark-coloured, and are discharged at maturity.

MELANOSPOROUS. Having black spores.

MELASTOMATACEAE (EP, BH, H). A family of the Myrtiflorae (EP), Myrtales (BH, H). These are herbs or woody plants, with opposite or whorled exstipulate leaves. The flowers are bisexual, and regular, having a hollow axis. There are many to 3 flower parts in each whorl, with an equal number of calyx and corolla lobes. There are usually twice as many stamens as corolla lobes, with the anthers opening by apical pores, and with connectives usually with appendages. The ovary has fused carpels, having the same number as the corolla lobes. There are many seeds in capsules or berries. They have no endosperm.

MELIACEAE (EP, BH, H). A family of the Geraniales (EP, BH), Meliales (H). These are woody usually with pinnate exstipulate leaves, and usually regular bisexual flowers in cymose panicles. The calyx and corolla are sometimes united. The stamens are usually in a tube, obdiplostemonous, or 5. The superior ovary consists of 5 or less fused carpels, with many loculi, each with 2-1 (rarely more) ovules. The fruit is of various form, and the seeds may, or may not, have endosperm.

MELIALES (H). An order of the Lignosae (H). See *Meliaceae*.

MELIANTHACEAE (EP, H). A family of the Sapindales (EP,H). (BH) include it in the Sapindaceae. These are woody plants, with alternate, usually pinnate leaves, with or without stipules. The bisexual, zygomorphic flowers are in racemes. There are 5 free sepals and petals, and 5-4 stamens (rarely 10), which are unequal or partially united. The ovary is superior, and of 4 or 5 fused carpels, each with 2 ovules. The fruit is a capsule. An aril may or may not be present, and there is endosperm.

MELIBIOSE. A disaccharide of D-glucose and α D-galactose linked in the 6:1 position.

MELIOLACEAE. A family of the Erysiphales. The aerial mycelium (if present) is dark, and the outer peridium layer is not brittle. The peridium is parenchymatous, and not slimy. The mycelial hyphae are cylindrical, and not slimy.

MELIOLALES = MELIOLACEAE.

MELIEUS. Latin meaning 'honey-coloured, or -tasting'.

MEMBER. Any part of a plant considered from the standpoint of morphology.

MEMBRANE. A thin sheet-like structure, usually fibrous, connecting other structures or covering or lining a part or organ.

MEMBRANACEOUS, MEMBRANOUS. Thin, dry, not green, flexible.

MENDELISM. See *Mendel's Laws*.

MENDELIAN CHARACTER = ALLELOMORPH.

MENDELIAN INHERITANCE. Inheritance obeying Mendel's laws.

MENDEL'S LAWS OF INHERITANCE. (1) The Law of Segregation—that the gametes produced by a hybrid or heterozygote contain unchanged, either one or the other of any two factors determining alternative unit characters in respect of which its parental gametes differed.
(2) The Law of Recombination—that the factors determining different unit characters are recombined at random in the gametes of an individual heterozygous in respect of these factors.

MENISPERMACEAE (EP, BH, H). A family of the Ranales (EP, BH), Berberidales (H). These are climbing shrubs usually with alternate simple leaves, and small usually regular unisexual flowers. The sepals, petals, and stamens are usually in 2 whorls. The superior ovary is of many, 3 or 1 free carpels, each with 1 ovule. The fruit is a drupe. An endosperm may or may not be present.

MENTUM. (1) A chin.
(2) An axial growth in the orchids, carrying the sepals forward.

MENYANTHACEAE (H). A family of the Gentianales (H). The leaves are alternate, entire, or trifoliate, sometimes peltate. The petiole is sheathing at the base. The corolla lobes are valvate, or induplicate-valvate.

MERENCHYMA = PLECTENCHYMA.

MERICARP. A one-seeded portion of a fruit which splits-up at maturity.

MERICLINAL CHIMAERA. A chimaera in which one component does not completely surround the other.

MERIDONAL. Southern.

MERISM. The development of more than one member of the same kind, usually in such a way that a symmetrical arrangement or pattern is formed.

MERISMATOID. Said of a pileus, made up of smaller pelei.

MERISPORE. One segment of a multiple spore.

MERISTELE. The vascular tissue of a dictyostele, running between two overlapping leaf-gaps.

MERISTEM. A localized group of cells, which are actively dividing and undifferentiated, but ultimately giving rise to permanent tissue.

MERISTEM SPORE = PHIALO-SPORE.

MERISTIC VARIATION. Variation in the number of organs or parts.

MERISTOGENETIC. Formed from, or by, a meristem.

MERISTOGENOUS. Said of pycnidia etc. formed by growth and division of one hypha.

MEROGAMY. The union of two individualized gametes.

MERONT. One of the daughter myxamoebae, cut-off in turn by a parent myxamoeba.

MEROSPORANGIUM. Of the Mucorales, a cylindrical outgrowth from the swollen end of a sporangiophore, in which a chain-like series of sporangio-spores is generally produced.

MERULIACEAE. A family of the Polyporales. The fruit-bodies are flat on the surface, or slightly uplifted at the edges, with the hymenium flat, then thrown up into shallow ridges, which may join to form pits. The fruit-body is usually more-or-less gelatinous, spreading in all directions. Many species destroy wood.

MESARCH. Having the protostele surrounded by metaxylem.

MESOCARP. The middle layer of the periderm.

MESOCHIL. Of a lip.

MESOCHITE. The firm inner wall of the macrosporangium of the Fucales.

MESOGELATIN. A gelatinous layer separating the exochite from the mesochite of the Fucales.

MESOMITOSIS. Mitosis which takes place within the nuclear membrane without any cooperation from cytoplasmic elements.

MESOPHANEROPHYTE. A tree having a height of 8-30 m.

MESOPHILIC BACTERIA. Bacteria which grows best at a temperature of 10-40°C.

MESOPHYLL. The parenchyma of a leaf, differentiated into the cyhe longy palisade cells with a large number of chloroplasts and arranged with lindrica axis at right-angles to the epidermis; and the spongy mesophyll of looselt packed cells with fewer chloroplasts, and large air-spaces.

MESOPHYTE. A plant occurring in places where the water-supply is neither scanty nor abundant.

MESOPHYTIC ENVIRONMENT. An environment in which the water-supply is neither very scanty nor abundant.

MESOPOD. Said of the fruit-body of a fungus which has a central stipe.

MESOSAPROBE. A plant living in somewhat foul water.

MESOSPORE. (1) A teliospore, usually one-celled.

(2) The middle layer of the wall of the zygote in the Volvocales.

(3) A layer of a spore-wall developed inside the first formed outer layer, and sometimes bearing ridges, pointed outgrowths or other ornamentations.

MESOSPORIUM. The thick middle layer of a spore-wall of the Ricciaceae.

MESOTAENIACEAE. A family of the Zygnematales. The Saccoderm desmids. The cells are uninucleate and of various shapes. They may be solitary or in filaments. The cell-walls lack pores and dividing cells do not regenerate new 'half-cells'. Conjugation is usually by definite conjugation tubes.

META-. A prefix meaning 'changed in form or position', 'between', 'with', 'after'.

METABASIDIUM = HETEROBASIDIUM.

METABIOSIS. The association of two organisms acting, or living one after the other.

METABOLIC INHIBITOR. A substance which is closely related chemically to the normal metabolite, and in consequence can occupy the same place on an enzyme preventing the normal reaction.

METABOLIC NUCLEUS. A nucleus when it is not dividing, and when the chromatin is in the form of a net-work.

METABOLIC STAGE. The resting stage between two successive cell divisions.

METABOLISM. The nett result of the biochemical processes of a living organism, or cells. The balance between anabolism and catabolism.

METABOLITE. A substance used during metabolism, *e.g.* oxygen during respiration, or carbon dioxide during photosynthesis.

METABOLY. The power possessed by some cells of altering their external form.

METACELLULOSE. The cellulose of fungi and lichens.

METACHROMATIC CORPUSCLE, METACHROMATIC GRANULE. An inclusion in the cytoplasm consisting of metachromatin.

METACHROMATIN. A complicated substance, probably a compound of nucleic acid occurring in granules in cytoplasm.

METAKINESIS = METAPHASE.

METAMITOSIS. Mitosis in which the nuclear membrane disappears and the karyokinetic figure lies free in the cytoplasm.

METAMORPHOSIS. A change in form or structure, during the development of an individual or species etc.

METAPHASE. A stage in mitosis or meiosis when the chromosomes become arranged on the equator of the spindle.

METAPHASE PLATE. The group of chromosomes arranged in the equitorial plane of the spindle.

METAPHLOEM. Completely developed primary phloem, consisting of sieve-tubes, fibres and parenchyma.

METAPHYSIS = PARAPHYSIS.

METAPLASM. Any substance within the body of a cell which is not protoplasm, especially food material.

METASPERMAE = ANGIOSPERMAE.

METASYNDESIS. End-to-end union of the elements of a pair of chromosomes.

METAXENIA. Any effect that may be exerted by pollen on the tissues of the female organs.

METAXYLEM. Primary xylem which is derived from the procambium. The cells are heavily lignified, and have reticulate thickening or pitted walls, in consequence, they cannot be stretched.

METHIONINE. An amino acid.

$$
\begin{array}{c}
S\!-\!CH_3 \\
| \\
CH_2 \\
| \\
CH_2 \\
| \\
CH.HH_2 \\
| \\
COOH
\end{array}
$$

METHOD OF MAXIMUM LIKELIHOOD. The method of estimation depending on the maximization of the logarithm of the likelihood (and hence of the likelihood) function.

METHOXONE. A synthesized substance with the properties of auxin.

METHYL ALCOHOL, METHANOL. The simplest alcohol, many of whose esters are found in plants.

$$
\begin{array}{c}
H \\
| \\
H\!-\!C\!-\!OH \\
| \\
H
\end{array}
$$

METHYL GLYOXAL. $CH_3.CO.CHO$. It is hydrolysed to lactic acid by the enzyme glyoxalase, in the presence of glutathione.

$CH_3.CO.CHO + H_2O = CH_3CHOH.COOH$ lactic acid.

METHYL SUGAR. A hexose sugar in which the $-CH_2OH$ of the last carbon atom is reduced to CH_3.

METOECIOUS = HETEROECIOUS.

METOXENOUS = HETEROECIOUS.

METROMORPHIC. Resembling the female parent.

METULA. A sporophore branch, bearing further branches, and chains of conidia.

METULIFORM. Resembling a pyramid.

METULOID. An encrusted cystidium.

MIADESMIACEAE. A family of the Lepidocarpales. The sporophyte resembled *Selaginella*. The sporophylls were in lax strobili. It is not known whether the micro- and megasporophylls were borne in the same strobilus. The macrosporangium had one functional macrospore.

MICACEOUS. Said of a pileus surface, covered with bright particles.

MICELLA. A crystalline structure, which may, with many other similar structures, form the foundation of cell-walls, starch-grains etc.

MICELLAR REGION. An area of crystallinity in a cellulose molecule.

MICELLE = MICELLAR REGION.

MICRANTHOUS. Latin meaning 'small-flowered'.

MICRO-. Prefix meaning 'small'.

MICROAEROPHILE. An organism which cannot grow well in the atmospheric concentration of oxygen, but only when it is lower than normal.

MICROCONIDIUM. A small conidium produced by some species of fungi, differing in form as well as in size from the larger conidia characteristic of the species.

MICRO-CULTURE. A culture of an organism in very small amounts of medium, usually so that the whole can be viewed under the microscope.

MICROCYCLIC. Of the Uredinales = SHORT-CYCLIC.

MICROCYST. An encysted myxamoeba, or swarm-spore.

MICROFIBRIL. A fine thread of cellulose in a cellulose cell-wall.

MICROFLORA. The microscopic plants, *e.g.* fungi and bacteria of an organ or area.

MICROFUNGI. (1) The fungi which need microscopic examination for their adequate study.
  (2) Fungi having microscopic fruit-bodies.

MICROGAMETANGIUM. A gemetangium which produces microgametes.

MICROGAMETE. The smaller, male gamete.

MICROGAMETOPHYTE. The gametophyte which produces microgametes.

MICROGONIDIUM. A very small green body in a lichen thallus.

MICRON (μ). A thousandth of a millimetre.

MICRONEMEAE. A group of the Moniliales, having conidiophores or conidia like the hyphae, or having no hyphae.

MICRONEMOUS. Having small hyphae.

MICROPHANEROPHYTE. A woody plant from 2-8 m. high.

MICROPHYLLINE. Composed of small scales or lobes.

MICROPHYLLOUS. (1) Said of leaves, which are small, with a single central vein, especially of the Pteridophyta.
  (2) Said of leaves of xerophytic plants.

MICROPYLE. (1) A canal into the nucellus formed by an upgrowth of the integument(s). It remains as a minute pore in the testa of the seed through which water passes prior to germination.
  (2) The opening in the integument of the Lepidocarpales.

MICROSOME. A minute particle free, or fused, to the endoplasmic reticulum, and containing RNA. They are responsible for metabolism, especially respiration, and protein synthesis.

MICROSPECIES. A variety of a species.

MICROSPERMAE (EP, BH). An order of the Monocotyledonae. The flowers are cyclic, with the perianth lobes in 2 whorls of 3. Typically the flowers are diplostemonous, but there is commonly great reduction in the number of stamens. The ovary is inferior with 1 or 3 loculi, with many small ovules. Endosperm may or may not be present.

MICROSPORACEAE. A family of the Ulotrichales. The filaments are unbranched, and are of cells with single, variously lobed chloroplasts. The cell-wall is of two pieces that are H-shaped in optical section.

MICROSPORANGIUM. A sporangium in which microspores are produced.

MICROSPORE. (1) The smaller of two kinds of spores of heterosporous plants, ultimately producing the male gametophyte.
(2) A small spore of a species which produces spores of two sizes.

MICROSPOROCYTE. (1) A male gametophyte in the Angiosperms. The pollen mother-cell, which divides to give 4 pollen grains.
(2) A cell which divides to give microspores.

MICROSPOROPHYLL. A leaf bearing microsporangia. They may be highly modified, *e.g.* the stamens of flowering plants.

MICROTHYRIACEAE. A family of the Hemisphaeriales. The ascocarps have a more or less radial structure. The vegetative mycelium and ascocarps are entirely superficial. The ascocarps are round or laterally compressed.

MICROTOME. An instrument for cutting thin sections of specimens.

MIDDLE LAMELLA. A layer of pectin, running between adjoining primary cell-walls.

MIDRIB. The largest vein in a leaf, running through the middle of the lamella longitudinally.

MID-SUMMER GROWTH. The second period of active growth shown by some trees.

MIHI. Quoted as an authority to species, accepted by the author as the correct form.

MIKTOHAPLONT. A haplont made up of cells having genotypically different nuclei.

MILDEW. (1) A general term for a superficial growth of fungus.
(2) A plant disease caused by a powdery or downy mildew.

MINIOLUTEIC ACID. A metabolic product of *Penicillium minioluteum*.

MIOCENE. A geological period, a subdivision of the Tertiary period, from 35-15 million years ago.

MIMOSACEAE (H). A family of the Leguminales (H). The flowers are regular. The petals are free, or united in a tube, and are valvate (rarely imbricate). The anthers open lengthwise, and sometimes have a deciduous gland at the apex.

MINIMAL MEDIUM. Of culture media, which include the minimum amount of materials for complete metabolism to occur.

MINERALIZATION. The deposition of inorganic substances in a cell-wall.

MINIMAL SURFACES, LAW OF. A free-floating cell will always assume a spherical form, *i.e.* exposing the smallest surface area for a given mass.

MINUS STRAIN. One of the two distinct strains of a heterothallic fungus.

MISDIVISION. Spontaneous cross-wise (instead of length-wise) division of the centromere on the spindle, especially of univalents.

MISTUS. A crossbreed between two forms of a species.

MITOCHONDRIA. Small bodies in spaces of the ground cytoplasm. They are spherical, long rods, or threads, and are the sites of many important enzymatic processes. The inner layer of the wall is infolded into finger-like processes.

MITOGENETIC RAY. A form of radiant energy, possibly ultra-violet radiations of low intensity, emitted by some actively growing parts of plants, and said to influence development.

MITOSIS. The normal process of cell-division in which the chromosomes duplicate themselves into two separate nuclei, and ultimately into two new cells.

MITOSOMES = MITOCHONDRIA.

MITOTIC INDEX. The proportion of dividing cells in any tissue.

MITRATE, MITRIFORM. (1) Rounded, bonnet-shaped, and folded inwards at the top.
(2) Of a calyptra when not split down one side.
(3) Split on two or more sides at the base in a symmetrical manner.

MITSCHERLICH'S LAW. Assuming that a plant produces its maximum yield if all conditions are ideal, but if an essential factor is absent, then there is a corresponding reduction in yield, and assuming an increase in yield by unit increment of the lacking factor is proportional to the decrement from the maximum, then:—

$$\frac{dy}{dx} = (A-y)\,C,$$

where y is the yield when x is the amount of the factor present, A is the maximum yield if x is in excess, and C is a constant.

MIXED BUD. A bud containing young foliage leaves, and the rudiments of flowers or inflorescences.

MIXED PITH. A pith consisting chiefly of parenchyma, but with isolated tracheids scattered in it.

MIXOCHIMAERA. A chimaera produced experimentally in fungi by mixing the contents of two hyphae of different strains.

MIXOCHROMOSOME. A new chromosome formed by the fusion of a pair of normal chromosomes.

MIXOPLOID. A diploid-polyploid chimaera.

MIXTUS (Latin) = MISTUS.

MOCARRERO = MURRAM.

MODE. The value of a variate shown by the class most frequently observed.

MODER. May mean *mor* or *raw humus*. Material in which biological decomposition has been arrested, but still shows cellular structure, sometimes more or less felted together with fungal hyphae.

MODIFICATION. A non-heritable change (caused by a difference in the environment) lasting only as long as the operative conditions last.

MODIFYING GENE. A gene whose differences are revealed by their effect on the expression of another gene.

MOLAR SOLUTION. The molecular weight, in grammes, of a substance dissolved to make 1 litre of solution.

MOLECULAR FRAMEWORK THEORY. An old theory of cytoplasmic structure. The protein forms a loose framework, random in form, but held together by intermolecular bonds. Fats and phosphatides may also play a part in the framework, which was thought to be constantly reforming.

MOLECULAR SIEVE HYPOTHESIS. The theory that membranes have pores of fixed size through which particles can pass if they are not too big.

MOLKENBÖDEN. Basically a meadow soil developed on sandstone, under impeded drainage.

MOLLIS. Latin meaning 'soft, and hairy'.

MOLLUGINACEAE (H). A family of the Caryophyllales (H). There are 3 or more sepals. The stamens are opposite the sepals, if of the same number, but may be twice as many. They are hypogynous. The ovary is superior with 3-5 loculi and axile placentas. The fruit is a capsule which opens by individual loculi, or a transverse slit (rarely indehiscent). There is no endosperm.

MONADELPHOUS. Said of stamens whose filaments are united in bundles or to form a tube.

MONANDROUS. (1) Having one antheridium.
(2) Having one stamen.

MONANGIAL. Said of a sorus consisting of a single sporangium.

MONARCH. Having a single strand of protoxylem in the stele.

MONILIACEAE. A family of the Moniliales. The hymenium and conidiophores are hyaline, or brightly coloured, as are the conidia, which are borne on conidiophores which are not distinguishable from the other branches of the mycelium, or are terminal or lateral on distinct, branched or unbranched separate conidiophores.

MONILIALES. An order of the Fungi Imperfecti. The conidia are formed on conidiophores which are separate, at least at their apical portions, or the vegetative mycelium breaks up into conidia.

MONILOID, MONILIFORM. Like a string of beads.

MONIMIACEAE (EP, BH, H). A family of the Ranales (EP), Micrembryae (BH), Laurales (H). These are woody plants, usually with opposite leaves, and solitary flowers, or cymes. The flowers are regular or actinomorphic, bisexual or unisexual. There are many, 4 or no perianth lobes. There are many or few free stamens. The superior gynecium is of many carpels, each with 1 ovule. The fruit is an achene. There is no endosperm.

MONO-. A prefix meaning 'one'.

MONOAXIAL THALLUS. Said of the Florideae in which there is a single axial filament that gives off filaments laterally on all sides.

MONOBLEPHARIDACEAE. The single family of the Monoblepharidales.

MONOBLEPHARIDALES. An order of the Phycomyceteae. They are oögamous, a large non-motile egg being fertilized by a small motile antherozoid. The zoospores and antherozoids are posteriorly uniflagellate.

MONOCARPELLARY. Having, or consisting of, a single carpel.

MONOCARPIC. (1) Forming a single fruit and then dying.
(2) Dying after one flowering season.

MONOCENTRIC. Said of the thallus of the Chytridiales, having one centre of growth and development.

MONOCEPHALIC, MONOCEPHALOUS. One-headed.

MONOCHASIUM. A cyme in which each flowering branch bears one other flowering bud in its turn.

MONOCHLAMYDEAE (BH). A division of the Dicotyledons. The flowers have a perianth of one whorl, commonly sepaloid, or it is absent.

MONOCHLAMYDEOUS. Said of a flower with only one whorl in the perianth.

MONOCHLAMYDEOUS CHIMAERA = HAPLOCHLAMYDOUS CHIMAERA.

MONOCHROMOGENOUS SOIL. A soil derived from the direct decomposition of crystalline rocks.

MONOCLINOUS. (1) Having the antheridium on the oogonial stalk.
(2) Having stamens and carpels in the same flower.

MONOCLEACEAE. A family of the Marchantiales. They have a hood-like sheath posterior to the female receptacle. The capsule of the sporophyte is elongated.

MONOCOTYLEDONEAE. A class of the Angiospermae. The seeds have a single cotyledon. The floral parts are in 3, or multiples of 3. The leaves have parallel veins, and the vascular bundles of the stem are scattered and closed, (*i.e.* without a cambium).

MONOCOTYLEDONOUS. (1) Belonging to the Monocotyledonae.
(2) An embryo having a single cotyledon.
(3) A seedling having a single cotyledon.

MONOCYCLIC. (1) An annual plant.
(2) Having each kind of member in one whorl.

MONODESMIC. Said of a petiole which contains one vascular strand.

MONOECIOUS. (1) Bearing unisexual flowers on the same plant.
(2) Having male and female sex-organs on the same thallus.

MONOGENIC. Of an hereditary difference determined by one gene difference, as opposed to 2, 3, or many.

MONOGENOCENTRIC. Said of the Chytridiales, having the development of one reproductive structure at the centre of gravity of the thallus.

MONOHYBRID. Heterozygous in respect of one gene.

MONOICOUS. Said of Mosses which have the antheridia and archegonia borne on the same plant, but in separate groups.

MONOKARYON. A nucleus with only one centriole.

MONOLITH. A carefully prepared column of the natural soil-profile, taken to preserve records of soil-structure.

MONOMITIC. Having one kind of hypha.

MONOMYCELIAL. Of an isolate of a fungus grown from one spore, or hyphal end.

MONOPETALAE = SYMPETALAE.

MONOPETALOUS = GAMOPETALOUS.

MONOPHAGOUS. Said of a fungal parasite which attacks one host-cell only.

MONOPHYLETIC. Descended from a single parent form.

MONOPLANETIC. Having one period of locomotion.

MONOPLOID. Having the basic number of chromosomes.

MONOPODIUM. A stem increasing in length by the division of an apical meristem, and branching by similar lateral branches occurring in acropetal succession.

MONOREPRODUCTO-CENTRIC = MONOGENOCENTRIC.

MONOSPERMOUS, MONOSPORIC, MONOSPOROUS. One-spored.

MONOSACCHARID. A sugar which cannot be further broken-down and still retain the properties of a sugar.

MONOSEPALOUS = GAMOSEPALOUS.

MONOSIPHONOUS THALLUS. A thallus of a single row of cells, joined end-to-end.

MONOSOMIC. Of an otherwise diploid organism, lacking one chromosome of its proper complement.

MONOSPORANGIUM. A sporangium, producing monospores, borne terminally on short lateral branches.

MONOSPORE. A large 4-nucleate aplanospore produced by some brown algae.

MONOSPOROUS. (1) Containing one spore.
(2) Derived from one spore.

MONOSTELE. A single central stele in a shoot.

MONOSTICHOUS. Forming one row, line or series.

MONOSY. The separation of parts that are usually fused.

MONOSYMMETRICAL = ZYGOMORPHIC.

MONOTRICHIATE, MONOTRICHOUS. Having a single flagellum.

MONOTROPACEAE (H). A family of the Ericales (H). These are leafless parasites devoid of chlorophyll. The petals are united into a tube, rarely free or absent. The superior ovary has 6-1 loculi and contain many ovules on axile or parietal placentas.

MONOTROPEAE (BH). A family of the Ericales (BH). (EP) include it in the Pyrolaceae.

MONOTYPIC. Said of a species or genus which is exemplified in only one type.

MONOXEROUS. Said of a parasitic fungus which is restricted to one species of host plant.

MONSTROSITY. A marked aberrant variation suddenly appearing.

MONTANUS. Latin meaning 'mountain'.

MONTMORILLONITE. A clay mineral, $(Ca,MgO).Al_2O_3.5SiO_2. 5H_2O$.

MONTMORILLONITIC ACID. A hypothetical acid, assumed to be a degradation product in the formation of a clay.

MOR. Like mull (*q.v.*) but it is practically pure organic matter, more or less compacted and felted together.

MORACEAE (EP, H). A family of the Urticales (EP, H). (BH) include it in the Urticaceae. These are usually trees or shrubs with stipulate leaves, and containing latex. The flowers are small and unisexual, and are borne in cymes which are usually spike-like. The 4 perianth lobes are free or fused, or sometimes absent. There are 4 free stamens opposite the perianth lobes. The superior ovary is of 2 fused carpels, with 1 loculus, usually with 1 pendulous ovule. The fruit is a nut or drupe. Endosperm may or may not be present.

MOR HUMUS. A humus developing on acid, sandy heaths. The soil is too acid for an abundant microflora to live, so the humus develops on the surface of the soil.

MORIFORM. Like a mulberry fruit in shape.

MORINGACEAE (EP, BH, H). A family of the Rhoeadales (EP), Capparidales (H). (BH) place it as an anomolous family in the Disciflorae. These are trees with pinnate exstipulate leaves, and panicles of bisexual zygomorphic flowers. There are 5 free sepals and petals, 5 free stamens, and 5 staminodes. The superior ovary is on a short gynophore, and is of 3 fused carpels with parietal placentas and many ovules. The fruit is a capsule, and there is no endosperm in the seeds.

MORINGEAE (BH) = MORINGACEAE (EP).

MORPHOGENESIS. (1) The development of structure or form.
(2) The organization of tissues to form the mature organism.

MORPHOLOGICAL ALTERNATION OF GENERATIONS. The alternation of generations, where the two generations do not look alike.

MORPHOLOGY. (1) The study of external shape, in contrast to function.
(2) The actual shape of a member.

MORTAR FRUIT. A structure consisting of a persistent calyx from which the true fruits are thrown-out by wind, or by shaking caused by animals.

MORTIERELLACEAE. A family of the Mucorales. The asexual spores are aerial. The sporangia are spherical, many-spored with a basal septum, but no columella.

MOSAIC. (1) See *Chimaera*.
(2) The arrangement of the leaves on a plant in such a way that the leaves above do not shade those below.
(3) The symptoms of certain virus diseases, resulting in the patchy yellowing of the leaves.

MOSAIC GROWTH THEORY. An old theory that cell-walls grow by the introduction of cellulose into certain areas only. The areas that are not filled become pit-fields.

MOSSES = MUSCI.

MOTHER CELL. A cell with a diploid nucleus, which gives rise to four haploid nuclei by meiosis.

MOTILE. Able to move as a whole by means of flagella or other organs of locomotion.

MOTOR CELL. One of a number of cells which together expand or contract and so cause movement in a plant member.

MOTOR SYSTEMS. Any system which causes or allows movement. These are of three types:
(1) Having non-living tissues which move suddenly, *e.g.* for dehiscence.
(2) Parts capable of differential growth, *e.g.* tendrils.
(3) Movement by alternation of turgor, *e.g.* pulvini.

MOULD. A loose term for any superficial fungus growth.

MUCEDINEOUS, MUCEDINOUS. (1) Mould-like.
(2) White and cottony.

MUCILAGE. A polymer of galactan.

MUCILAGINOUS. (1) Pertaining to, containing, resembling, or composed of mucilage.
(2) Sticky when wet, slimy.

MUCORACEAE. A family of the Mucorales. The asexual spores are produced aerially. All the sporangia are many-spored, with a well-developed columella. The sporangial-wall is relatively thin, and breaking, or deliquescent.

MUCORALES. An order of the Phycomyceteae. The mycelium is very extensive and non-septate, or septate in the older aerial hyphae. Asexual reproduction is typically by aplanospores produced in terminal sporangia. These are sometimes reduced to indehiscent sporangioles which act as conidia. Sexual reproduction is usually present.

MUCOSE. Slimy.

MUCRO. A short sharp point formed by the continuation of the midrib.

MUCRONATE. Said of a leaf tipped with a short sharp point of much the same texture as the leaf.

MULL. A humus layer with a marked crumb-structure and containing mineral matter.

MULTI-. A prefix meaning 'a great number, many, much'.

MULTIALLELE. This condition arises when more than one allele occupies the same position on a chromosome. These genes do not usually exhibit any dominance.

MULTIAXIAL THALLUS. Said of the Florideae that have a central core of axial filaments, each giving off lateral filaments.

MULTICELLULAR. Consisting of a number of cells.

MULTIENZYME SYSTEM. Such a condition exists when several enzymes can catalyse a similar number of reactions in the same place at the same time, without mutual interference.

MULTIFID. Divided into a number of lobes.

MULTIFOLIATE. Having many leaflets.

MULTIFORM. Diverse in shape.

MULTILAYERED. Consisting of several layers of cells.

MULTILOCULAR, MULTILOCULATE. Having a number of compartments.

MULTI-NET THEORY. A theory of the method of growth of the primary cell-wall. It states that the cell-wall is laid-down from the inside with micro-fibrils transverse to the cell-axis. The space between the fibrils become larger from the inside outwards, as the successive layers are stretched.

MULTINOMIAL SERIES. The series obtained by expanding to any power, the sum of three or more quantities, *i.e.* $(a + b + c \text{------} + x)^n$.

MULTINUCLEATE. With many nuclei.

MULTIOVULATAE AQUATICEAE (BH). A series of the Incompletae (BH). These are aquatic plants with a syncarpous ovary and many ovules.

MULTIOVULATAE TERRESTRES (BH). A series of the Incompletae (BH). These are terrestrial plants with a syncarpous ovary and many ovules.

MULTIPLE ALLELOMORPH. A group of two or more allelomorphs of one gene, only two of which can be present in a diploid cell. They originate by mutation.

MULTIPLE CHROMOSOME. The product of the fusion of two chromosomes.

MULTIPLE FACTORS = POLYMERIC GENES.

MULTIPLE FRUIT. A fruit formed from the flowers of an inflorescence, and not from one flower.

MULTIPLICATION. Increase by vegetative means.

MULTISEPTATE. Having a number of septa.

MULTISERIATE. (1) Said of a vascular ray, which is several-to-many cells wide.

(2) Said of ascospores arranged in several rows in the ascus.

MULTISPOROUS. Having a number of spores.

MURALIS. Latin meaning 'growing on walls'.

MURICATE. Having a surface roughened by short, sharp points.

MURICULATE. Delicately muricate.

MURIFORM. Said of a spore made up of a mass of cells formed by divisions in three intersecting planes.

MURRAN. A highly indurated concretionary material with deposits of iron oxide. It develops in certain tropical soils under impeded drainage.

MUSACEAE (EP, H). A family of the Scitamineae (EP), Zingiberales (H). (BH) include it in the Scitamineae. These are very large herbs with a 'false' stem, or trees, with a compound inflorescence with large often petaloid bracts. The flowers are bisexual or unisexual, zygomorphic, and homochlamydeous or heterchlamydeous. There are two whorls of 3 perianth lobes which are petaloid, and free or united. There are 2 whorls of free stamens, there are 3 in the outer one, and 2, with 1 staminode in the inner. The inferior ovary is of 3 fused carpels, with 3 loculi, each with 1 to many ovules. The fruit is a berry or capsule. The seeds have endosperm and perisperm.

MUSCARIN. One of the poisons in poisonous mushrooms.

MUSCI. A series of the Bryophyta. There are two phases in the development of the gametophyte. The spore grows into a filamentous protonema from which the gametophore develops. The gametophore grows by a pyridimal apical cell, usually. The rhizoids are multicellular with diagonal cross-walls. There is greater sterilization of the tissues of the capsule than in the other Bryophyta. There are no elaters.

MUSHROOM. A general term including the edible Agaricales.

MUTABLE GENE. A gene that is liable to frequent mutation.

MUTAFACIENT. One gene or genetic element which determines or increases the chance of mutation of another.

MUTAROTATION. Said of a compound, which, if it crystallizes under different conditions, has different physical properties, and if dissolved in water rotates a beam of polarized light through different angles.

MUTATION. The sudden alteration of the chemical structure of a gene, or the alteration of its position on the chromosome by breaking and rejoining of the chromosome. Mutations rarely occur naturally, but may be caused artificially by irradiation, or by chemicals, *e.g.* mustard gas. They may occur in somatic cell or gametes, in which case they can be inherited.

MUTATION PRESSURE. The measure of the action of mutation in tending to alter the frequency of a gene in a given population.

MUTATION RATE. A proportion of individuals or cells in a given group, which show mutation for a given gene under given conditions in one generation, or other stated unit of time.

MUTATION THEORY. That new species arise by single mutations.

MUTICATE, MUTICOUS. Pointless; blunt.

MYC-, MYCET-, MYCETO-, MYCO-. A prefix meaning 'fungus'.

MYCELIANAMIDE. A metabolic product of *Penicillium griseofulvum*.

MYCELIA STERILIA. A heterogeneous group of fungi, which produces no known spores at any stage of its development.

MYCELIUM. A collective term for the vegetative hyphae of a fungus.

MYCELIOID. Like a mycelium.

MYCELOCONIDIUM = STYLOSPORE.

MYCETISM. Mushroom poisoning.

MYCETOME. A cellular organ inside an aphid, containing symbiotic yeasts.

MYCETOZOA = MYXOMYCETES = MYXOTHALLOPHYTA.

MYCOBIOTA. The mycological flora.

MYCOCECIDIUM. A gall caused by a fungus.

MYCOCRINY. The decomposition of plant material by fungi.

MYCODERMA. A name sometimes applied to the yeasts.

MYCOGENOUS. Coming from, or living on, fungi.

MYCOLOGY. The study of fungi.

MYCOPHAGY. Using fungi as food.

MYCOPHTHOROUS. Said of a fungus which is parasitic on another fungus.

MYCOPLASM. (1) Legume bacteria.
(2) Erroneously; a symbiotic phase of a rust and host protoplasm.

MYCOPLASMATALES. An order of the Bacteria. These are small organisms or irregular morphology, and without a rigid cell-wall. They develop into filamentous 'large bodies' that can break-down and liberate viable elements.

MYCORRHIZA. The association between a fungus and the roots of a plant. It is probably a symbiosis, but may be a weak parasitism. Ectotrophic mycorrhiza have the fungus forming an envelop over the small roots. Endotrophic mycorrhiza have the fungus in the cells of the root-cortex.

MYCOSIN. A nitrogenous material, like animal chitin in the cell-wall of fungi.

MYCOSE = TREHALOSE.

MYCOSIS. A disease of an animal caused by invasion by a fungus.

MYCOSPHAERELLACEAE. A family of the Pseudoaphaeriales. The stromata are small and perithecia-like, under the epidermis often eventually external. At maturity only a single large cavity with a spreading cluster of asci are left, with only, at most, the remnants of the stromatic tissue between the basal portions of the asci.

MYCOSTATIC = FUNGISTATIC.

MYCOTIC. Said of a disease caused by fungi.

MYCOTROPHIC. Having mycorrhiza.

MYIOPHILLOUS. Having inconspicuous, evil-smelling flowers, which are pollinated by flies.

MYLLITA. A large sclerotium.

MYOPORACEAE (EP, BH, H). A family of the Tubiflorae (EP), Lamiales (BH, H). These are mostly trees or shrubs with alternate or opposite entire exstipulate leaves, which are often covered with woolly or glandular hairs, frequently reduced in size. The bisexual, regular or zygomorphic flowers, are axillary, solitary or in cymose groups. The 5 calyx lobes are fused, as are the 5 petals. There are 4 free didynamous stamens, with confluent anther-loculi. The ovary consists of 2 fused carpels and is superior. There are 2 loculi, or 3-10 by segmentation. In the former case there are 1-8 and in the latter 1, pendulous, anatropous ovules in each loculus. The fruit is a drupe, and the seeds contain endosperm.

MYRIANGIACEAE. The single family of the Myriangiales.

MYRIANGIALES. An order of the Ascomyceteae. These are parasitic on scale-insects, but may eventually parasitize the plant host. The basal stroma is well-developed bearing one or more ascigerous portions throughout whose tissues the asci are scattered in no definite order or layers.

MYRICACEAE (EP, BH, H). A family of the Myricales (EP, H), Unisexuales (BH). These are woody, usually with simple leaves and flowers in simple, rarely compound spikes. The unisexual flowers lack a perianth, but may have bracts at the base. There are 2-16 (usually 4) free stamens. The superior ovary consists of 2 fused carpels, with 1 loculus containing 1 basal orthotropous ovule. Fertilization is through the micropyle. The fruit is a drupe with a waxy exocarp. The seed has no endosperm.

MYRICALES (EP, H). An order of the Archichlamydeae (EP), Lignosae (H). See *Myricaceae*.

MYRIOSPOROUS. Having a great number of spores.

MYRISTICACEAE (EP, BH, H). A family of the Ranales (EP), Micembryae (BH), Laurales (H). These are woody plants with evergreen simple leaves and axillary racemes of unisexual regular cyclic flowers. The 3 perianth segments are fused. The 18-3 stamens are fused. The ovary is superior and of 1 carpel with 1 basal ovule. The fruit is fleshy and indehiscent, an aril is present. The seeds have a ruminate endosperm.

MYRMECOPHILOUS. (1) Pollinated by ants.
(2) Said of fungi being a covering or food for ants.

MYRMECOPHILY. A symbiotic association between plants and ants.

MYROTHAMNACEAE (EP, H). A family of the Rosales (EP), Hamamelidales (H). (BH) include it in the Hamamelidaceae. These are xerophytic shrubs with opposite stipulate leaves and spikes of unisexual, regular, achlamydeous flowers. The male flower has 8-4 stamens which may be free or united. The female has an ovary of 4-3 carpels which are fused. It is superior and mutilocular, with a short thick style and flattened stigma. There are many anatropous ovules. The fruit is a capsule, and the seeds contain endosperm.

MYRSINACEAE (EP, BH, H). A family of the Primulales (EP, BH), Myrsinales (H). (BH) include the Theophrastaceae. These are woody often with evergreen entire alternate, exstipulate leaves. The regular flowers are bisexual or unisexual. The 5 calyx lobes are free and the 5 petals are fused. There are 5 free stamens, rarely with 5 staminodes. The ovary is superior to inferior, with 1 loculus and many ovules on basal or free-central placentas. The fruit is a drupe with 1 or a few seeds. The seed contains endosperm.

MYRSINALES (H). An order of the Lignosae. The leaves are mostly gland-dotted and have no stipules. The flowers are small with fused petals (rarely free), which are usually contorted or imbricate. The stamens are of the same number, and opposite the corolla-lobes, to which they are usually joined. The anthers open lengthwise, or by apical pores. The ovary is superior to half-inferior, and contain many ovules on free basal placentas.

MYRTACEAE (EP, BH, H). A family of the Myrtiflorae (EP), Myrtales (BH, H). (BH) include the Lecythicaceae. These are woody, with opposite or alternate, entire exstipulate leaves. The flowers are bisexual and regular. The sepals and petals are free usually 5-4. There are many stamens, usually in bundles. The inferior ovary has 2 to 5 to many loculi, each with many to 1 ovules. The fruit is of various types, and the seeds have no endosperm.

MYRTALES (BH, H), MYRTIFLORAE (EP). An order of the Archichlamydeae(EP), Calyciflorae (BH), Lignosae (H). These are herbs or woody plants with regular heterochlamydeous (rarely apetalous) flowers. The flowers may be zygomorphic, but they are always on a concave axis. The stamens are in one or two whorls, and are sometimes branched and in bundles. The ovary is of 2-to-many fused carpels, usually united to the axis (rarely of 1 free carpel).

MYTILIFORM. Resembling a mussel shell in shape.

MYXAMOEBA. A naked, non-flagellate amoeboid cell, produced on the germination of a spore of the Myxothallophyta.

MYXOBACTERALES. An order of the Bacteria. These are long bacillus-like cells massed in a slime to form a colony. The cells at the edge of the colony are non-flagellate, but motile, so that the colony is amoeboid. The fruit-bodies are stalked or sessile, and often coloured. They may be aggregations of cysts in which resting-spores are enclosed, or masses of slime surrounding rod-shaped or rounded spores.

MYXOGASTRALES. An order of the Mycetozoa. These are saprophytes, or surround and ingest fungi, bacteria etc. The aerial sporangia have a thick or thin periderm, and most have a capillitium. The spores germinate to form an anteriorly uni- or bi-flagellate swarm cell (rarely non-flagellate myxamoeba).

MYXOMYCETAE = MYXOGASTRALES.

MYXOPHYCEAE = CYANOPHYCEAE.

MYXOPHYTA = MYXOPHALLOPHYTA.

MYXOPOD = MYXAMOEBA.

MYXOSPORE. An obsolete term for a myxomycete spore.

MYXOTHALLOPHYTA. A division of the Thallophyta. The slime moulds. The vegetative body is a naked mass of protoplasm which may be single and multinucleate, or an aggregate of many uninucleate units. Reproduction is by walled uninucleate spores in most genera produced in sporangia or definite shapes.

MYZODENDRACEAE (EP, H). A family of the Santalales (EP, H). (BH) include it in the Santalaceae. These are semiparasites, undershrubs with alternate leaves and minute unisexual flowers without a perianth. The male has 3-2 or 1 free stamens, each with 1 anther. The inferior ovary is of 3 fused carpels with axile placentas and 3 ovules. The fruit has 3 feathery bristles in the angles.

N

N. The haploid, or gametic number of chromosomes.

NACREOUS. Like mother-of-pearl.

NAD = NICOTINAMIDE ADENINE DINUCLEOTIDE.

NADP = NICOTINAMIDE ADENINE DINUCLEOTIDE PHOSPHATE.

NÄHRHUMUS. The material in humus, consisting of carbohydrate, organic acids, protein and soluble or hydrolysable substances, which is readily decomposed by microorganisms.

NAIADACEAE (EP, BH). A family of the Helobieae (EP), Apocarpae (BH). These are submerged herbs with opposite linear, toothed leaves, and unisexual flowers. The male has 2 free perianth lobes and 1 free terminal stamen. The female has 1 (or no) perianth lobe. The superior ovary has 1 carpel, with 1 basal anatropous ovule.

NAIADEAE (BH) = NIADACEAE (EP).

NAJADACEAE (H) = NAIADACEAE. A family of the Najadales (H).

NAJADALES (H). An order of the Calyciferae (H). These are submerged aquatic annuals or perennials, with alternate, or opposite, sheathing leaves. The flowers are minute and unisexual, with a perianth of small scales, or it is absent. There are 3-1 stamens, having the anthers mostly sessile, with 4-1 loculi. The ovary has 1-9 free carpels with 1 ovule. The fruit is indehiscent, and the seeds have no endosperm.

NAKED. (1) Lacking a perianth.
(2) Without any appendages.
(3) Not enclosed in a pericarp.
(4) Bractless.

NAMATIUM. A brook formation.

NANDINACEAE (H). A family of the Berberidales (H). These are shrubs with 2-3 times pinnately compound leaves, with large panicles of small bisexual flowers. There are 6 stamens with anthers opening by longitudinal slits. There is 1 carpel with pendulous ovules. There are many sepals which are spirally arranged.

NANNANDRIUM. A dwarf male-filament produced by some Oedogoniales.

NANNOCYTE. A small, spore-like cell, produced by some non-filamentous Myxophyceae. They are not true spores.

NANNOPLANKTON. Plankton which will pass through the finest nets.

NANOPHANEROPHYTE. A plant 25 cm. to 2 m. in height, with resting buds above the ground surface.

NANUS. Latin meaning 'dwarf'.

NARIFORM. Shaped like a turnip.

NARROWED. Tapering, especially downward.

NASTIC MOVEMENT. A response to a stimulus, independent of its direction, *e.g.* opening of buds under different light intensities.

NATANS. Latin meaning 'swimming (under water)'.

NATIVE PROTEIN. Protein in its naturally occurring state.

NATURALIZED. Introduced from another region, reproducing freely by seed and maintaining its position in competition with indigenous plants.

NATURAL SELECTION. The mechanism of evolution propounded by Darwin (1859). Within a population, individuals vary slightly. Certain variations are more favoured by the environment, and consequently survive, and reproduce, thus propagating the variation.

NAVICULAR, NAVICULATE. Shaped like a boat.

NEBULOUS. Clouded, dark.

NECK. (1) The upper tubular part on an archegonium, and of a perianth.
(2) The lower part of the capsule of a moss, just above the junction with the seta.

NECK CANAL CELL, NECK CELL. One of the central cells of the central canal in the neck of an archegonium.

NECK INITIAL. A cell derived from a jacket-cell of the bryophyte and pteridophyte archegonium. It ultimately divides to give the neck of the archegonium.

NECROGENIC ABORTION. The speedy death of the tissues of a plant just under the point of attack by a parasite, thus checking the spread of the latter.

NECRON. Dead plant material, not rotted into humus.

NECROPHAGOUS = SAPROPHYTE.

NECROSIS. Death of a cell or group of cells, while still part of the living plant.

NECTAR. (1) A sugar fluid produced in the nectaries of some insect-pollinated flowers.
(2) A similar solution produced by fungi, to attract insects to disperse spores.

NECTAROMYCETACEAE. A family of the Saccharomycetales. It contains one genus *Nectaromyces* which grows in the nectar of flowers. It produces yeast-like budding cells, but sometimes they are in cross-like groups of four. Long, branched hyphae may occur, which bear conidia at the tips.

NECTARY. A gland in some flowers, or other part of the plant, secreting nectar.

NECTON = NEKTON.

NECTONIC BENTHOS. Small organisms, floating at the bottom of the water.

NECTRIACEAE. A family of the Hypocreales. The perithecia are without stroma, or external to them. The ascospores are ellipsoidal to cylindrical, and 1 to many celled.

NECTRIOIDACEAE (ZYTHIACEAE). A family of the Sphaeropsidales. The pycnidia (or stroma) is bright-coloured, usually fleshy or leathery.

NEEDLE. A long, narrow, stiff leaf, from which water-loss is greatly reduced.

NEGATIVE HETEROPYCNOSIS. A subnormal amount of nucleic acid in the heterochromatin in mitotic and premeiotic divisions.

NEGATIVE REACTION. A taxis or tropism in which an organism moves or a member grows from a region where the stimulus is stronger, to one where it is weaker.

NEKTON. Actively swimming aquatic organisms, in contrast to those which are primarily drifting, *i.e.* plankton.

NEMALIONALES. An order of the Florideae. They lack a tetrasporophytic generation, and in almost all genera the carposporophyte develops from the carpogonium.

NEMATHECIUM. A sorus-like, fertile layer in the Rhodophyceae. It contains carpogonia, spermatongia, or sporangia.

NEMATOPARENCHYMATOUS THALLUS. An algal thallus composed of united threads which are, however, still recognizable as individuals.

NEMATOTHECIUM. A cushion-like projection on the thallus of a sea-weed, and bearing the reproductive organs.

NEMORAL, NEMORALIS (Latin). Living in woods.

NEO-. A Greek prefix meaning 'new'.

NEOMORPH. An amorph having an effect apparently unrelated to that of the non-mutant allelomorph.

NEOTROPICAL. Of the New World tropics.

NEPENTHACEAE (EP, BH, H). A family of the Sarraceniales (EP), Multiovulatae Terrestres (BH), Aristolochiales (H). These are climbers with alternate leaves, the lower ones with pitchers, and the upper ones tendrils. The flowers are bisexual and regular, being borne in racemes or panicles. The perianth is in 2 whorls of 2 like members. There are 16-4 fused stamens. The superior ovary is of 4 fused carpels and 4 loculi with many ovules. The fruit is a capsule, and the seeds contain endosperm.

NEPHROID = RENIFORM.

NERITIC. Living in the sea at a depth of less than 200 m.

NERVATION, NERVATURE = VENATION.

NERVE. The midrib, or large vein in a leaf.

NERVICOLOUS. Said of a parasitic fungus which lives on the veins of leaves.

NET ASSIMILATION RATE. The rate of increase in dry weight per unit area of a leaf. This is approximately equivalent to the rate of photosynthesis.

NET KNOT. A small accumulation of chromatin, particularly at the intersection of the nuclear reticulum.

NEOPLASM. (1) A newly formed tissue.
(2) A tumour.

NET-PLASMODIUM. The plasmodium of the Labyrinthulales, which develops from the 4 naked cells that penetrate the cell-walls of the host plant.

NETRUM. A minute spindle which arises within the centrosome during the division of the centromere.

NETTED. (1) Covered with lines which form a net-work.
(2) Forming a net-work.

NET-VEINED. Having veins running irregular courses, and forming a net-work.

NEUROMOTOR APPARATUS. An organelle found in the motile cells, probably including gametes and zoospores, of the Chlorophyta. It is intimately associated with the nucleus, and is responsible for the production of flagella.

NEUTER. (1) Apparently sexless, especially of strains of fungi which usually show sexuality.

(2) Of flowers in which the andrecium and gynecium are functionless or absent.

NEUTRAL SPORE. An asexual spore produced by some algae. They are produced in various ways, but are never in sporangia.

NICHE. A term used to describe the status of a plant or animal in its community, *i.e.* its biotic and trophic relationship.

NICOTINAMIDE-ADENINE-DINUCLEOTIDE = NAD = DPN.

NICOTINAMIDE ADENINE DINUCLEOTIDE PHOSPHATE = NADP = TRI-PHOSPHOPYRIDINE NUCLEOTIDE.

NICOTINAMIDE MONONUCLEOTIDE (NMN). A constituent of diphosphopyridine nucleotide.

NICOTINE. An alkaloid, *1*-1-methyl - 2 (3-pyridyk) - pyrrolidine, derived from tobacco (*Nicotiana tabacum*) leaves. Use as an insecticide.

NICOTINIC ACID.

It promotes the growth of some bacteria.

NIDOSE, NIDOROSE. Having an unpleasant smell.

NIDULARIACEAE. A family of the Nidulariales. The peridioles are few, with thick hard walls. The peridium is beaker-like, opening by the rupture of the diaphragm-like top, leaving the peridioles like eggs in a nest.

NIDULARIALES. An order of the Gasteromyceteae. These are medium-sized to small, not hypogeous at maturity. The hymenial cavities have definite linings of basidia. Towards maturity, the tramal tissue surrounding each cavity encloses it in a thin or thick wall, producing separate structures called peridioles.

NIEDERUNGSMOOR. A fen-peat developed by the decomposition of vegetation which encroaches on, and ultimately obliterates a lake.

NIGER. Latin meaning 'black'.

NIGRESCENT, NIGRESENS, NIGRICANS (Latin). Blackish, or becoming so.

NIGRO-PUNCTATE. Marked with black dots.

NIPE CLAY. A red earth, which has developed practically to the stage of ferrallite. It has a relatively large quantity of ferric oxide.

NITID, NITIDOUS. Smooth and clear, lustrous.

NITRIFICATION. The oxidation of nitrogen-containing ions in the soil, by bacteria, *e.g. Nitromonas* oxidizes ammonium ions to nitrites, and *Nitrobacter* oxidizes nitrites to nitrates.

NITROGEN CYCLE. The circulation of nitrogen in Nature. Dead organic matter is converted to ammonium compounds during decay in the soil. These are converted to nitrates (see *nitrification*), which are utilized again by plants.

Some of the nitrates are converted to gaseous nitrogen by the denitrifying bacteria, but this loss is counterbalanced by the fixation of nitrogen by *Bacillus radicicola* and the tissue of legume root-nodules in which it grows.

NITROGEN FIXATION. The conversion of atmospheric nitrogen into organic nitrogen compounds. This is done by some soil bacteria, or by others living in the nodules on the roots of various plants, *e.g.* leguminous plants.

NITROPHILOUS. Said of plants characteristic of places with a high concentration of nitrogen compounds.

NITROSATION. The conversion of ammonium compounds to nitrites by bacteria.

NIVALIS. Latin meaning 'growing near snow'.

NIVEUS. Latin meaning 'snow-white'.

NMH = NICOTAMIDE MONONUCLEOTIDE.

NODE. (1) The part of a stem where the leaf or leaves emerge.
  (2) The zone on the thallus of the Charales, which gives rise to the lateral branches and the cortex.

NODE-CELL = HYPHOPODIUM.

NODOSE. Bearing knot-like swellings.

NODOSE-SEPTUM = CLAMP-CONNECTION.

NODULAR, NODULOSE. Bearing local thickenings, especially of an elongated plant member.

NOLANACEAE (EP, H). A family of the Tubiflorae (EP), Solanales (H). (BH) include it in the Convolvulaceae. These are herbs or undershrubs with alternating leaves. The bisexual regular flowers are solitary or in racemes. The 5 calyx and corolla lobes are fused in 2 whorls. The 5 stamens are free. The 5 carpels are free (usually) with many ovules. The fruit is divided longitudinal, or transverse constrictions, each of which contains 7-1 seeds. The seeds have no endosperm.

NOMEN. Latin meaning 'name'.

NOMEN AMBIGUUM. One name having different senses.

NOMEN CONFUSUM. The name of a taxonomic group based on two or more different elements.

NOMEN CONSERVANDUM. A name made valid by a decision of the International Botanical Congress.

NOMEN CONSERVANDUM PROPOSITUM. A name put up for conservation.

NOMEN DUBIUM. A name of uncertain sense.

NOMEN PROVISORIUM. A name proposed provisionally.

NOMEN NUDUM. The name of a group, having no diagnosis.

NONACOSAN. One of the higher paraffins, found in the wax from Brussel sprout leaves.

NON-ARTICULATE. Not cut off by an absciss-layer.

NON-CONJUNCTION. The complete failure of synapsis.

NON-DISJUNCTION. The complete failure of two chromosomes to disjoin at meiosis.

NON-ESSENTIAL ORGANS. Sepals and petals.

NON-EXCHANGEABLE IONS. Ions in the cell, or surrounding water which cannot be exchanged for ions in the surrounding water, or cell, respectively.

NON-FREE SPACE. The volume of a cell, made up of vacuoles, and probably part of the cytoplasm, from which ion-exchange takes place slowly.

NON-HOMOLOGOUS PAIRING. The association of non-homologous segments of one or two chromosomes at pachytene.

NON-REDUCTION = AMEIOSIS.

NON-SENSIBILITY. The ability of a plant to support a parasite without showing marked signs of disease.

NONTRONITE. An octahedral clay mineral, in which the normal aluminium ions are replaced by ferric iron. $Fe_2O_3. 3SiO_2.O.4H_2O$.

NON-VIABLE. Incapable of survival.

NORM. The value of a quantity or a state which is statistically most frequent.

NORMAL DEVIATE. The ratio of an observed deviation to the appropriate, or corresponding standard deviation, as fixed by hypothesis.

NORMAL DISTRIBUTION, NORMAL CURVE OF ERRORS. The limit which is reached either by the binomial or the multinomial series, where the power is large and none of the summed quantities very small in relation to the power and to one another. This frequency distribution is expected from a series of observations on a variate whose magnitude is affected by a large number of agents having small independent effects.

NOTATE. Said of surfaces marked by straight or curved lines.

NS QUOTIENT. The measure of the relative humidity of a climate. N/S where N is the mean annual rainfall in mm. and S is the mean deficit from saturation with water in mm. of mercury.

NUCELLAR EMBRYONY. A form of apomixis where the embryo arises directly from the nucellus.

NUCELLUS. The central tissue of the ovule, containing the embryo-sac and surrounded by the integument(s).

NUCLEAR BUDDING. The production of two daughter-nuclei of unequal size by constriction of the parent nucleus.

NUCLEAR CAP. Of the Blastocladiaceae, a body at one side of the nucleus of a zoospore or gamete.

NUCLEAR DIVISION = MITOSIS, MEIOSIS, AMITOSIS.

NUCLEAR FRAGMENTATION. The formation of two or more portions from a nucleus by direct break-up, and not by mitosis.

NUCLEAR GENES. Genes that occur in the nucleus, in contrast to plasmagenes.

NUCLEAR MEMBRANE. A layer bounding the nucleus. It consists of two layers (each about 75-100Å thick) bounding a cisterna-like space, and the outer one being continuous with the endoplasmic reticulum. It is perforated by round holes (200-500Å in diameter) called nuclear pores.

NUCLEAR PLATE. The aggregation of chromosomes in the equitorial plane during mitosis or meiosis.

NUCLEAR PORE. See *Nuclear membrane*.

NUCLEAR RETICULUM. A mesh-work of delicate threads of chromatin seen in stained preparations of metabolic nuclei.

NUCLEAR SAP. The fluid which is lost by the chromosomes as they contract during prophase, and which fills the space of the nucleus = KARYOLYMPH.

NUCLEAR SPINDLE. The fusiform structure, composed of fine fibrils arranged longitudinally and converging at the poles, which appears in the cytoplasm of a cell surrounding the nucleus during mitosis and meiosis.

NUCLEAR STAIN. A stain which will pick out the nuclei in a tissue or organism a different colour or shade.

NUCLEASES. Enzymes inducing hydrolysis of nucleic acid.

NUCLEATE. (1) Having a nucleus or nuclei.
(2) An obsolete term meaning guttate.

NUCLEIC ACIDS. The non-protein constituents of nucleo-proteins. They have a high molecular weight, and consist of alternate units of phosphate and a pentose sugar, which has a purine and pyrimidine base attached to it.

NUCLEOCHYLEMA = NUCLEAR SAP.

NUCLEO-CYTOPLASMIC RATIO. The ratio of nucleus volume to cytoplasm volume. Alteration of this ratio may be a cause of cell-division.

NUCLEOLAR ORGANIZER. A secondary constriction on a spiralized chromosome, which organizes the production of the nucleolus.

NUCLEOLATE. Said of a spore which contains one or more conspicuous oil-drops.

NUCLEOLINE. Special particles occurring within the nucleosis which do not disappear during mitosis.

NUCLEOLO-CENTROSOME. A prominent deeply staining body found in the nucleus of some lower plants.

NUCLEOLUS. A body, not containing desoxyribose nucleotides and secreted by a specific organizer, gene or super-gene, in the resting nucleus.

NUCLEOME. The whole of the nuclear substance in a protoplast.

NUCLEONEMA. A fibre-like net-work that is sometimes present in the nucleolus.

NUCLEOPLASM. The ground-substance of the nucleus. It may appear granular or fibrils.

NUCLEOPROTEIN. A protein which is combined with, or combinable with, a nucleic acid.

NUCLEOSIDE. A chemical group consisting of a purine or pyrimidine base, and a pentose sugar.

NUCLEOTIDE. A chemical group, consisting of a purine or pyrimidine base, ribose or deoxyribose sugar, and phosphoric acid.

NUCLEUS. (1) The chief organelle of a cell. It is a cavity (usually spheroid) in the cytoplasm, bounded by a nuclear membrane containing, bathed in nuclear-sap, a complicated system of nuclear-proteins, DNA etc. in the form of a network in the stained cell and/or of rounded nucleoli. This constitutes the chromosomes and controls the activities of the cell. It also determines the transmission of inheritable characters when the nucleus divides.
(2) An obsolete term for the *nucellus*.
(3) An oil-drop.

NUCULE. The female fruit-body of the Charales.

NUDATION. The formation of an area bare of plants, by natural or artificial means.

NULL HYPOTHESIS. The hypothesis from which the expectations are formulated for the purpose of a test of significance.

NULLIPLEX. A condition of a polyploid in which all the chromosomes of one homologous type carry the recessive allelomorph of a particular gene.

NUMERICAL HYBRID. A hybrid whose parental gametes differed in respect of the number of chromosomes.

NUMERICAL MUTATION. A change in the number of chromosomes, either balanced to give polyploidy, or unbalanced to give aneuploidy.

NURSE CELL, NURSE TISSUE. (1) Any cell or tissue in contact with developing gametes, and concerned with their nutrition.
(2) In *Scleroderma*, hyphae giving food material to spores which have become detached from the basidia.

NUT. A hard, dry, usually one-seeded indehiscent fruit, derived from a syncarpous ovary.

NUTANT, NUTANS (Latin). Hanging with the apex downwards; nodding.

NUTATION. (1) The spiral growth of the apex of a plant organ, e.g. shoots, tendrils, etc. It is caused by variation in the rate of growth of the meristematic areas.
(2) The lateral swaying of the tip of a growing organ.

NUTLET. A one-seeded portion of a fruit which fragments at maturity.

NUTRILITE. One of a group of substances that are needed in small amounts for the healthy growth of fungi.

NYCTAGINACEAE (EP, BH, H). A family of the Centrosperae (EP), Curvembryatae (BH), Thymelaeales (H). These are herbs or woody plants with opposite exstipulate leaves. The flowers are regular, bisexual or unisexual in cymes. They have bracts at the base, and these may be united and petaloid. The perianth is of 5 fused petaloid lobes. There are typically 5 stamens, but may be 1-30. The ovary consists of 1 superior carpel, with 1 basal erect ovule. The fruit is an achene, and a perisperm is present.

NYCTAGINEAE (BH) = NYCTAGINACEAE.

NYCTANTHOUS. Said of flowers that open at night.

NYCTINASTY. The response of plant organs to the periodic alternation of day and night, *e.g.* opening and closing of flowers.

NYCTITROPIC. A term referring to the position assumed by leaves etc. at night.

NYIROK SOILS. Red soils of Moravia and Hungary, with affinities to a terra-rosa, but may have resulted from Tertiary weathering.

NYMPHAEACEAE (EP, BH, H). A family of the Ranales (EP, BH, H). These are water or marsh plants, with usually submerged or swimming leaves and solitary, regular bisexual flowers. The axis is often hollowed. There are many to 6 free perianth lobes, and many to 6 free stamens. The gynecium consists of many to 3 free or fused carpels, and may be superior or inferior. Each carpel contains many to 1 ovules. Endosperm may, or may not be present.

NYSSACEAE (EP, H). A family of the Myrtiflorae (EP), Araliales (H). (BH) include it in the Cornaceae. These are shrubs with alternate exstipulate leaves, and small bisexual or bisexual flowers. The axis is hollow, and the male flowers are in racemes, with the female solitary. The 5 or more sepals are free, the 5 petals are free, valvate or absent. The stamens are free and twice as many as the corolla lobes. The inferior ovary is usually unilocular (rarely 6-10) with 1 ovule in each. The fruit is a drupe, and the seeds contain endosperm.

O

OB-. A prefix meaning 'reversed, turned out'.

OBCLAVATE. Club-shaped, but widest at the base.

OBCOMPRESSED. Flattened from front to back.

OBCONIC, OBCONICAL. Cone shaped, but attached at the point.

OBDIPLOSTEMONOUS. Having two whorls of stamens, the members of the outer whorl being placed opposite the petals, and not alternating with them.

OBLANCEOLATE. Lanceolate, tapering, but towards the base.

OBLATE. Globose, but noticeably wider than long.

OBLIGATE. Compelled; can live only under one type of circumstance.

OBLIGATE PARASITE. A parasite which is incapable of a free existence.

OBLIGATE SAPROPHYTE. A saprophyte, living on dead material, and incapable of attacking living tissue.

OBLIQUE DIVISION. The development of a septum which is neither parallel to the long axis of the cell, nor across it at right-angles.

OBLIQUE PLANE. Any plane, of a flower, other than the median and lateral plane.

OBLITERATION. The crushing and closing-up of tubular elements within a plant by the pressure set-up by new elements as they develop.

OBLONG. Elliptical, blunt at each end, having nearly parallel sides, and two to four times as long as broad.

OBLONG-ELLIPSOID. Having the long side almost parallel, and the ends almost hemispherical.

OBOVATE. Having the general shape of the longitudinal section of an egg; not exceeding twice as long as broad, and with the greatest width slightly above the middle, hence attached at the narrow end.

OBOVOID. Solid, egg-shaped, and attached at the narrow end.

OBSCURE. Said of a venation which is poorly developed, so that hardly more than the mid-rib can be seen.

OBSOLETE = ABORTION.

OBSUBULATE. Very narrow; pointed at the base, and a little wider at the top.

OBTUSE. Rounded or blunt, or being greater than a right-angle.

OCCASIONAL SPECIES. A species which is found from time to time in a given plant community, but is not a regular member of that community.

OCCIDENTALIS. Latin meaning 'western'.

OCCLUSION. The blocking of a stoma by the ingrowth of parenchymatous cells into the sub-stomatal cavity.

OCEANIC. (1) Living in sea more than 200 m. deep.
(2) An ocean formation.

OCELLATE. Marked by a round patch different in colour from the rest of the organ.

OCELLUS. (1) An enlarged discoloured cell in a leaf.
(2) A swelling on the sporangium of some fungi; it may be a light-receptor.

OCHNACEAE (EP, BH, H). A family of the Parietales (EP), Geraniales (BH), Ochnales (H). These are woody, or under shrubs, with evergreen stipulate leaves, usually with parallel lateral nerves. The bisexual regular (rarely zygomorphic) flowers are in showy panicels, and the floral axis is often enlarged after flowering. There are 10-4 free sepals, and 5 (rarely 10-4) free petals. The many to 5 stamens are free, and staminodes may be present. The superior ovary consists of 10-5-2 fused carpels, with many to 1 erect or pendulous ovules. Endosperm may or may not be present.

OCHNALES (H). An order of the Lignosae (H). They are similar to the Theales, but the leaves have intra- or extra-petiolar stipules. The calyx lobes are often enlarged and wing-like in the fruit, or involucrate by bracts.

OCHRACEOUS, OCHREOUS, OCHERY. Yellowish-brown.

OCHREA, OCREA. A cup-shaped structure around a stem, formed from united stipules or united leaf-bases.

OCHREATE. Sheathing.

OCHROLEUCOUS. Yellowish-white.

OCHROMONADINEAE. A sub-order of the Chrysomonadales. The cells have two flagella of unequal length. The cells are solitary, or united into colonies.

OCHROSPOROUS. Having yellow or yellowish-brown spores.

OCTANOIC ACID. A fatty acid formed in coconut oil, $CH_3CH = CH.COOH$.

OCTANT DIVISION. The division of an embryonic cell by walls at right-angles, giving 8 cells.

OCTOKNEMACEAE (H). A family of the Olacales (H). The leaves have no stipules, and there is a covering of star-shaped, or forked cells. The flowers lack petals and a disk. The stamens are free and opposite the sepals. The ovary is inferior.

OCTOSES. A group of monosaccharides containing eight carbon atoms.

OCTOSPOROUS. Containing 8 spores.

ODONT-, ODONTO-. A Greek prefix meaning 'tooth'.

ODONTOID. Tooth-like.

OECOLOGY = ECOLOGY.

OEDEMA, EDEMA. A large mass of unhealthy parenchyma. An abnormal swelling of large areas of tissue due to excess water.

OEDOCEPHALOID. Having a swelling at the end or tip.

OEDOGONIACEAE. The single family of the Oedogoniales.

OEDOGONIALES. An order of the Chlorophyceae. The cells are uninucleate, in simple or branched filaments. The cell-division is unique in that there is a distinctive annular splitting of the lateral wall. The motile reproductive cells have a transverse whorl of flagella on the anterior end. Asexual reproduction is usually by zoospores, but may be by akinetes. Sexual reproduction is always oögamous.

OENOTHERACEA (H). See *Onagraceae*.

OFFICINALIS. Latin meaning 'medicinal'.

OFF-SET = STOLON. A short runner, bending up at the end.

OGIVE. The integral of a frequency distribution.

-OID. A suffix meaning 'resembling, imitating, like'.

OIDIOSPORE = OIDIUM.

OIDIUM = ARGTHROSPORE. (1) A spore that is formed simultaneously throughout the length of a filament.

(2) Spermatia formed on hyphal branches, especially in heterothallic Hymenomycetes.

(3) Flat-ended asexual spores formed by the breaking of hyphae.

OIL. A fat which is liquid at normal atmospheric temperature. Essential oils have relatively low molecular weights, and are volatile, many having a strong perfume.

OIL BODY. A large single droplet of oil in isolated cells of some of the liverworts, especially the Marchantiaceae.

OIL DROP. Any small droplet of oily material included in the cytoplasm.

ÖKOTYPE = ECOSPECIES.

OLEIC ACID. A fatty acid found in plant and animal fats. $CH_3(CH_2 = CH = CH(CH_2)_7COOH$.

OLACACEAE (EP, H). A family of the Santalales (EP), Olacales (H). (BH) include it in the Olacineae. These are trees and shrubs usually with alternate entire leaves, and small regular, bisexual flowers. There are 6-4 very small calyx lobes, and the same number of free petals. The stamens are free and as many or twice or three times as many as the petals. The ovary is superior and consists of 5-2 fused carpels, with the same number of loculi at the base, with one at the apex. Each loculus contains one pendulous ovule. The seed is a 1-seeded nut or drupe. The seeds contain endosperm.

OLACALES (BH, H). An order of the Polypetalae (BH), Lignosae (H). These are trees or shrubs, with alternate simple, exstipulate leaves. The flowers are regular, bisexual or unisexual. The calyx is small, with or without a disk, which if present, is of various forms. The ovary is entire with many to 1 loculi, with 3-1 ovules in each loculus. The raphe is dorsal, and the integument is confluent with the nucellus. The endosperm is copious and fleshy. The embryo is small.

OLACINEAE (BH). = OLACACEAE + ICACINACEAE (BH).

OLEACEAE (EP, BH, H). A family of the Contortae (EP), Gentianales (BH), Loganiales (H). These are woody, sometimes climbing, rarely herbs. The leaves are opposite or whorled, simple or pinnate, and lack stipules. The regular, bisexual or unisexual flowers are in compound inflorescences. There are 6-5-4 or no petals, which are free or united, imbricate or valvate. The 2 free stamens are epipetalous or hypogynous. The superior ovary consists of 2 fused carpels, each usually with 2 (rarely 1 or 8-4) axile ovules. The fruit is a capsule, berry or drupe. The seeds have or have no endosperm.

OLEIFERUS. Latin meaning 'oil-bearing'.

OLEOSOME. A large fatty inclusion in the cytoplasm of a cell.

OLERACEOUS. Latin meaning 'edible'.

OLIG-, OLIGO-. A Greek prefix meaning 'few, small'.

OLIGOMEROUS. Consisting of but a few parts.

OLIGOSPOROUS. Containing or having only a few spores.

OLIGOTROPHIC. Said of a lake-habitat having steep or rocky shores, and scanty littoral vegetation.

OLIGOTROPHOPHYTE. A plant growing in a soil poor in soluble mineral salts.

·**OLINIACEAE (EP, H).** A family of the Thymelaeales (EP), Cunoniales (H). (BH) include it in the Lythraceae. These are shrubs with opposite entire leaves and panicles of bisexual flowers. The sepals, petals and stamens are all free, and 5-4 in number. The ovary is inferior, consisting of 5-3 fused carpels, with 5-3 loculi, each with 2-3 ovules. The fruit is a drupe, and there is no endosperm in the seeds.

OLIVACEOUS, OLIVE. Greyish-green, with a touch of orange.

OLPIDIACEAE. A family of the Chytridiales. The entire plant-body of the fungus lies within the host-cell, and all of it is fertile. Reproduction is by uniflagellate zoospores, or zoogametes.

OLPIDIOPSIDIACEAE. A family of the Lagenidiales. These are one-celled, free in the host cell and producing a cellulose wall early in development. The zoospores complete their development in the zoosporangium, and resting spores are produced sexually or parthenogenetically. They are parasitic in algae or fungi.

OMBROPHILE. A plant which thrives in a place where rain is abundant.

OMBROPHOBE. A plant which cannot survive in continuous rain.

OMBROPHYTE. A plant inhabiting rainy places.

OMNIVOROUS. Said of a parasitic fungus which attacks several or many species of host-plant.

ONYGENACEAE. A family of the Aspergillales. They grow on animal material such as fur, feathers, horn etc. The ascocarps are 2-3 mm. to 1-2 cm. tall, consisting of a stalk and a somewhat enlarged head in which the asci are scattered. The tissues break-up into a sort of capillitium.

ONAGRACEAE (EP, BH, H). A family of the Myrtiflorae (EP), Myrtales (BH), Lythrales (H). These are usually herbs with opposite or alternate exstipulate leaves. The bisexual, usually regular, flowers have a tubular axis and are axillary or in racemes. The calyx has 4-2 (rarely more) free sepals, and there are 4-2 (more or none) free petals. There are 8-4 free stamens. The ovary is inferior usually of 4 fused carpels, each with many-to-1 ovules. The fruit is a capsule nut, or berry. The seed has little or no endosperm.

ONTOGENESIS, ONTOGENY. The history of the development of an individual.

OÖBLAST. Found in the Florideae. A tubular outgrowth from the carpogonial base, which connects the carpogonium and the auxillary cell.

OÖCYSTACEAE. A family of the Chlorococcales. It contains all the autosporic Chlorococcales in which the cells are solitary or in colonies of indefinite number of cells. Reproduction is only by autospores.

OÖCYTE. A cell with a diploid nucleus, which, by meiosis, gives four haploid nuclei, which ultimately form female gametes, usually by further division.

OOGAMY. The fertilization of a large, non-motile female gamete by a small, motile male gamete.

OÖGENESIS. The development of an egg from an oöcyte.

OÖGONIUM. A cell giving-rise to oöcytes, directly or by mitosis.

OÖLYSIS. The conversion into leafy structures of carpels and ovules.

OÖPHYTE. A gametophyte in the Bryophyta and Pteridophyta.

OÖPLASM. The central plasma in the oögonium of some Oömycetes, representing a more or less undifferentiated egg.

OÖSPHERE. (1) The large non-motile fertile gamete of some algae and fungi.
(2) An ovum.

OÖSPORE. The thick-walled resting zygote formed from a fertilized oösphere.

OPAQUE. Dull, not shining.

OPEN AESTIVATION. Aestivation in which the perianth leaves neither overlap nor meet by their edge.

OPEN BUNDLE. A vascular bundle that has a cambium.

OPEN COMMUNITY. A plant community which does not occupy the ground completely, so that bare spaces are visible.

OPEN VASCULAR BUNDLE = OPEN BUNDLE.

OPEN WOODLAND. Grassy ground with trees here-and-there, often forming groups.

OPERCULAR CELL. A lid cell by means of which some antheridia open.

OPERCULATAE. A sub-order of the Pezizales, in which the ascus opens by an operculam.

OPERCULATE. (1) Possessing a lid.
(2) Opening by means of a lid.

OPERCULUM. A lid.

OPHIO-. A Greek prefix meaning 'snake'.

OPHIOGLOSSACEAE. The single family of the Ophioglossales.

OPHIOGLOSSALES. An order of the Eusporangiatae. The sporangia are borne on an outgrowth (the fertile spike) that projects from the adaxial surface of the leaf, near the junction of the blade and petiole.

OPHIOSTOMATACEAE. A family of the Sphaeriales. The preithecia are superficial or slightly sunken, with a long neck. There are no paraphysis and the asci dissolve at maturity. The hyaline 1-celled ascospores are exuded in a drop at the end of the neck.

OPILIACAEA (EP, H). A family of the Santalales (EP), Olacales (H). (BH) include it in the Olacineae. These are parasitic, with bisexual heterochlamydeous flowers. The calyx is seam-like. The ovary has one ovule and no integument.

OPIOTHIAL APERTURE. The opening between the base of the stomatal pore and the sub-stomatal cavity.

OPPOSITE. (1) Inserted at the same level.
(2) Said of leaves inserted in pairs at each node, with one on each side of the stem.
(3) Said of a stamen which lies next to the middle of a petal.

-OPSIS. A Greek suffix meaning 'like'.

OPSIS-FORM. Said of heteroecious or autoecious Uredinales that produce pycnidia, uredospores and teleutospores.

OPUNTIALES = CACTALES.

ORBICULAR. Flat, with a circular, or almost circular outline.

ORCHIDACEAE. The single family in the Orchidales (H). A family of the Microspermae (EP, BH). These are perennial herbs of various forms, often epiphytic with pseudobulbs, and bisexual zygophytic flowers, which are resupinated and homo- or hetero-chlamydeous. The perianth is of two whorls of 3 free segments. The 1 or 2 free stamens are united with the style to form a column. The ovary is inferior and consists of 3 fused carpels, which form 1 loculus and contains many ovules. The fruit is a capsule, and there is no endosperm in the seeds. The pollen is in tetrads, usually united to form pollina. There are 3 stigmas. The third is usually rudimentary or from a rostellum.

ORCHIDALES (H). An order of the Corolliferae (H). See *Orchidaceae*.

ORCHIDEAE (BH) = ORCHIDACEAE (EP).

ORCULIFORM. Said of a 2-celled spore having a thick septum pierced by a connecting-tube between the two cells.

ORDER. The next taxonomic group below a class, and containing related families.

ORDOVICIAN. A geological period which lasted about 420 to 340 million years ago.

OREO-. A Greek prefix meaning 'mountain'.

ORGADIUM. An open woodland formation.

ORGAN. Part of a plant made up of different tissues making a functional unit.

ORGANELLE = CELL-ORGAN.

ORGANIC AXIS. The principle axis of a cell, passing through the centrosome and nucleus of the resting cell.

ORGANIZED. Showing the characteristics of an organism; having the tissues and organs formed into a unified whole.

ORGANISM. A living animal or plant.

ORGANOGENESIS, ORGANOGENY. The differentiation of organs, or the study thereof.

ORGANOGRAPHY. A descriptive study of the external form of plants, with relation to function.

ORIENTALE. Latin meaning 'eastern'.

ORIENTATION. (1) The movement of the centromeres so that they lie axially with respect to the spindle either as to their potential halves at mitosis or as to members of a pair or higher configuration at meiosis.

(2) The position, or change of position, of a part or organ with relation to the whole, or change of position of an organ under stimulus.

ORNITHINE. A precursor in the manufacture of argenine. An essential amino-acid. $CH_2(NH_2).(CH_2)_2.CH(NH_2).COOH$.

ORNITHO-. A Greek prefix meaning 'bird'.

ORNITHOPHILY. Pollinated by birds.

OROBANCHACEAE (EP, BH, H). A family of the Tubiflorae (EP), Personales (BH, H). These are parasitic herbs with scaly leaves. The bisexual, zygomorphic flowers have the sepals and petals in whorls of 5, and are terminal or in racemes. The corolla is two-lipped, and the 4 stamens are didynamous. The ovary is superior, consisting of 2 or 3 fused carpels, each with 2 parietal placentas, sometimes united in the middle, and many ovules. The fruit is a capsule, and the seeds contain endosperm.

OROPHYTIUM. A sub-alpine plant formation.

ORTHO-. A Greek prefix meaning 'upright, straight'.

ORTHOCLADOUS. Having long, straight branches.

ORTHOCLASE.

This is the simplest formula. It is probably the initial material which forms clay minerals by degradation.

ORTHOGENESIS. The theory of the mechanism of evolution which postulates that variation is determined by the action of the environment on the fixed constitution of the organism, so that the possibilities of variation are limited to certain definite lines.

ORTHOGEOTROPISM. Growth of a stem, vertically upwards, or of a root vertically downwards, in relation to gravity.

ORTHOGONAL. (1) Said of functions from which independent (orthogonal) comparisons are made.

(2) The manner of arrangement of four members of a flower, when two are median, and two lateral.

ORTHOPLOCOUS. Said of cotyledons which are folded longitudinally, with the inner one lying in the groove formed by the outer.

ORTHOSELECTION. Modification resulting from the elimination of all other lines of variation through the selective struggle.

ORTHOSTICHIES, ORTHOSTICHY. A vertical rank of leaves on a stem.

ORTHOTROPISM. A tropism such that the member comes to lie in the line of action of the stimulus.

ORTHOTROPOUS. (1) Said of an ovule which is straight, *i.e.* with the micropyle in a straight line with the funicle.

(2) Said of an organ which shows a sharp positive or negative tropism in respect to a given stimulus.

ORTHOTROPIC. Placing itself in line with the stimulus.

OSCILLATION VOLUME. The volume occupied by an adsorbed particle, as it vibrates around the point of adsorption.

OSCULE. A pore in a rust spore.

-OSIS. A suffix meaning 'condition of, state caused by'.

OSMOSIS. The passage of water (or solvent) from a dilute solution to a more concentrated one through a semi-permeable membrane.

OSMOTIC PRESSURE. The pressure developed due to the passage of water by osmosis. The maximum osmotic pressure of a given solution is developed if it is separated from pure water by a semi-permeable membrane.

OSMUNDACEAE. A family of the Leptosporangiatae. All the sporangia on a leaf develop simultaneously in sori which lack indusia. The sporangia dehisce longitudinally by a shield-shaped annulus at one side of the sporangial jacket.

OSTEOSCLEREIDE. A thick-walled idioblast which is shaped something like a thigh-bone.

OSTERHOUT'S HYPOTHESIS. A hypothesis accounting for the active-transport of cations. An acidic substance HX, located in the outer protoplasmic membrane combines with an entering base, $KOH + HX = KX + H_2O$. The neutral, undissociated complex KX, diffuses across the protoplasm, and is decomposed on the inner side, where the sap is more acid than the external medium.

OSTIOLATE. Having an opening.

OSTIOLE. (1) A general term for an opening.
(2) The opening by means of which spores etc. escape from a conceptacle or a perithecium.

OSTROPACEAE. A family of the Inoperculatae. The apothecia are not buried in a fleshy stroma. The asci are cylindrical, the apical thickening hemispherical, with a long slender canal. The ascopores are hyaline, thread-like, often septate and falling apart into cylindrical cells. The apothecia are sessile, or stalked, or immersed and somewhat perithecium-like. They are saprophytes.

OUTBREEDING. The mating of relatively unrelated gametes.

OUTER FISSUE. The upper arm of the cleft in the raphe (wall) of some Bacillariophyceae.

OUTER SPACE = WATER-FREE SPACE.

OVAL. Flat, rounded at each end, with curved sides, and about twice as long as broad, being widest in the middle.

OVARIICOLOUS. Living in ovaries.

OVARY. (1) The hollow basal region of a carpel, containing one or more ovules. In a flower with 2 or more united carpels, they form a single compound ovary.
(2) Loosely used as meaning the pistil.

OVATE. Flat and thin, shaped like the longitudinal section of an egg, widest below the middle.

OVOID. Solid, like an egg in form, and attached by the broader end.

OVULE. (1) The nucellus containing the embryo-sac and enclosed by 1 or 2 integuments, which after fertilization, and subsequent development, becomes a seed.
(2) A young seed in the course of development.

OVULIFEROUS SCALE. One of the scales of a fertile cone, in the Coniferae. It bears ovules and later seeds.

OVUM. A non-motile female gamete.

OXALIC ACID. A dibasic acid occurring in many plants.

OXALIDACEAE (EP, H). A family of the Geraniales (EP, H). (BH) include it in the Geraniaceae. These are usually herbs, with alternate, compound leaves which may or may not have stipules. The regular, bisexual flowers have the sepals and petals in fives, and have no disk. The 10 stamens are obdiplostemonious, and united at the base. The superior ovary consists of 5 fused carpels with many to 1 ovules. The fruit is a capsule or berry, and the seeds contain endosperm.

OXALOACETIC ACID. COOH.CH$_2$CO.COOH. It is produced in the citric acid cycle and glyoxylic cycle, from malic acid, and produces citric acid.

OXIDASE. An enzyme which brings about the oxidation of a substrate. Several may be involved in oxidizing a substrate to its final form.

OXIDATION. The addition of oxygen to a compound. More generally it is the addition of negative ions, or the removal of positive ions from a molecule.

β-OXIDATION. An oxidation which occurs between No. 2 and No. 3 carbon atoms to form a double bond.

OXIDATIVE DECARBOXYLATION. Oxidation by the removal of carbon dioxide, *e.g.*

pyruvic acid

OXIDATIVE PHOSPHORYLATION. The formation of ATP from DPNH and oxygen, or TPNH and oxygen.

OXIDOREDUCTASES. A group of enzymes which catalyse oxidations and reductions.

OXODIUM. A humus marsh formation.

OXYCHROMATIN. A form of chromatin which stains comparatively lightly, and contains little nucleic acid.

OXYGEN DEBT. Oxygen consumed in excess of normal amounts, or when an organism, or part of an organism, has been respiring with an inadequate oxygen supply.

OXYGENOTAXIS, OXYTAXIS, OXYGENOTACTIC, OXYTACTIC. A response or reaction of an organism to the stimulus of oxygen.

P

P. The parental generation from which a breeding experiment starts. All the female and male parents are each uniform as regards the allelomorphs under observation.

PACHY-. A Greek prefix meaning 'thick'.

PACHYCARPOUS. Having a thick pericarp.

PACHYNEMA = PACHYTENE.

PACHYPHYLLOUS. Having thick leaves.

PACHYPLEUROUS. Thick-walled.

PACHYTENE. The double thread. By extension, the third stage in meiotic prophase when the chromosomes pair, *i.e.* form a double thread.

PACKING CELL. An individual parenchymatous cell.

PAEDOGENESIS. Markedly precocious flowering.

PAEONIACEAE (H). A family of the Ranales (H). The flowers have a disk, and the many stamens are centrifugal. The seeds are arillate.

PAGINA = LAMINA.

PAIRING. Said of chromosomes. It is said to be active when they come together at zygotene, and passive when they stay together during the first metaphase.

PAKIHI SOIL. A soil developed under podsolic conditions and impeded drainage. Found in parts of New Zealand.

PALAEARCTIC. Old-world arctic.

PALAEOBOTANY. A study of fossil plants.

PALAEOCENE. A geological period (a sub-division of the Tertiary). It lasted approximately from 70 to 60 million years ago.

PALAEOPTERIDALES = PRIMOFILICES.

PALAEOTROPICAL. Old-world tropical.

PALATE. The prominent part of the lower lip of a corolla, which closes the opening of the corolla.

PALE, PALEA, PALET. (1) The inner bract of a grass floret.
(2) A general name for the glumes associated with the grass-flower.
(3) The scales which form the ramentum on ferns.

PALEA = PALE.

PALEACEOUS. Chaffy in texture.

PALEOLA = LODICULE.

PALET = PALE.

PALIFORM. Having the form of a stake.

PALISADE CELL. (1) A single cell of a palisade layer.
(2) One of the terminal cells of the hyphae forming the cortex of a lichen thallus.

PALISADE FUNGI = BASIDIOMYCETAE.

PALISADE LAYER. A layer of elongated cells set at right-angles to the surface of a leaf or thallus, and underlying the upper epidermis, or layers of cells. Its cells contain numerous chloroplasts and is concerned with photosynthesis.

PALISADE STEREIDE. A rod-shaped cell found in the testa of a seed. It is thick-walled, elongated, and lies at right-angles to the surface of the seed.

PALISADE TISSUE. One or more layers of palisade cells beneath the epidermis of a leaf.

PALITANTIN. A metabolic product of *Penicillium palitans*.

PALLENS. Latin meaning 'pale coloured'.

PALLESCENT. Becoming lighter in colour with age.

PALLID. Light-coloured.

PALMAE (EP, BH, H). The single family of the Principes (EP). A family of the Calycinae (BH). The single family of the Palmales (H).

PALMALES (H) = PRINCIPES. An order of the Corolliferae.

PALMATE. Having several (5-7) lobes, segments etc. spreading from the same point, like the fingers of a hand. The term is especially applied to leaves.

PALMELLA. The zooglea stage of bacteria, when forming a jelly-like mass.

PALMATISECT. Having the leaf-blade cut nearly to the base, so forming a number of diverging lobes.

PALMIFID. Having a leaf-blade cut about half-way down, forming a number of lobes.

PALMITIC ACID. A fatty acid which occurs as glycerides in vegetable oils and fats.

$$CH_2O(C_{15}H_{31}CO)$$
$$|$$
$$CHO(C_{15}H_{31}CO)$$
$$|$$
$$CH_2O(C_{15}H_{31}CO)$$

PALUDICOLOUS. Living in ponds, streams and marshes.

PALUDAL, PALUDOSE, PALUSTRIS (Latin). Inhabiting wet places.

PALYGORSKITE. A group of clay minerals, based on double chains of tetrahedra, with $(OH_2)_4(OH)_2Mg_5Si_8O_{20}.4H_2O$ as the basic unit.

PALYNOLOGY. The science of identifying plant spores.

PAMPINODY. The change of leaves, or parts of leaves into tendrils.

PAN. A hard layer in a soil. It is impervious to water. The soil particles are cemented together by organic material, or by iron or other compounds, usually under conditions of impeded drainage.

PANDACEAE (EP, H). The single family of the Pandales (EP), a family of the Celastrales (H).

PANDALES (EP). An order of the Archylamydeae. The flowers are cyclic and heterochlamydeous with 4-5 whorls. They are dioecious. The superior ovary is of 3 fused carpels, each with 1 pendulous orthotropous ovule. The fruit is a drupe.

PANDANACEAE (EP, BH, H). A family of the Pandanales (EP), Nudiflorae (BH). The single family of the Pandanales (H). These are woody plants, sometimes climbing, with 3-ranked leaves and terminal or racemed spikes of male and female flowers. The male flowers have numerous stamens as a raceme or umbel on long or short axes. The female flowers are of many to 1 fused carpels with sessile stigmas and many to 1 ovules. The fruits are berries or drupes, and are in heads. The endosperm is oily.

PANDANALES (EP, H). An order of the Monocotyledoneae (EP), Corolliferae (H). For (H) definition see *Pandanaceae*. These are marsh herbs or trees, with linear leaves and compound heads or spikes of naked, haplochlamydeous, or homochlamydeous unisexual flowers. The perianth is bract-like. There are many to 1 stamens, and many to 1 carpels. The seeds have endosperm.

PANDANEAE (BH) = PANDANACEAE (EP).

PANDURATE. Fiddle-shaped.

PANGENESIS. The theory that the heredity of organisms is determined by the summation of influences from an indefinite number of particles (pangenes) derived from all parts of the body-tissues and variably affected by variation in the environment.

PANICLE. (1) A branched raceme, with each branch bearing a raceme of flowers.

(2) More generally any branched inflorescence which is in any way complex.

PANMIXIS. Unrestricted, random mating.

PANNIFORM. Looking like felt.

PANNOSE, PANNOSUS (Latin) = PANNIFORM.

PANPHOTOMETRIC. Said of a narrow leaf which stands nearly or quite erect.

PANTOTHENIC ACID. Part of co-enzyme A, and one of the B group of vitamins. It is needed for the growth of yeasts and some bacteria.

$$HOCH_2-\underset{\underset{CH_3}{|}}{\overset{\overset{CH_2\ OH}{|}}{C}}-CH-CONH-CH_2-CH_2-COOH$$

PAPAIN. An enzyme (or group of enzymes) found in paw-paw leaves. In neutral solution, it hydrolyses proteins to polypeptides.

PAPAVERACEAE (EP, BH, H). A family of the Rhoeadales (EP), Parietales (BH), Rhoeadales (H). These are usually herbs with alternate leaves and latex. The flowers are bisexual and regular or zygomorphic. There are 2 free sepals, and 4 (rarely 6 or more) free petals, or they are absent. There are many, 4 or 2 free stamens, which are branched. The superior ovary consists of 16-2 fused carpels with parietal placentas and many ovules, or 1 basal ovule. The fruit is a capsule, and the seeds have oily endosperm.

PAPAYACEAE = CARICACEAE.

PAPILIONACEAE (H). A family of the Leguminales (H). The flowers are zygomorphic, with imbricate petals. The upper petal is outside the adjacent lateral petals. The stamens are monadelphous or diadelphous.

PAPILIONACEOUS. Said of a flower which looks like a butterfly.

PAPILLA, PAPILLOSE. (1) A minute blunt hair.

(2) A small rounded process.

PAPPOSE, PAPPOUS. Having a pappus.

PAPILLOSE, PAPILLOUS. Having short protuberances.

PAPPUS. A ring of fine, sometimes feathery, hairs, developed from the calyx, and covering the fruit (usually of the Compositae). It acts as a parachute, and aids in wind-dispersal.

PAPYRACEOUS. Papery in texture.

PARA-. A Greek prefix meaning 'beside'.

PARABOLIC. Having a broad base, and gradually narrowing by curved sides to a blunt point.

PARACHUTE DISSEMINULE. A fruit or seed with hairs, or some other mechanism which aids in dispersal by wind.

PARACOROLLA = CORONA.

PARADESMOSE. A small transverse fibril of stainable material which connects the belepharoplasts at the base of the flagella of some Chlorophyta.

PARAFFINIC ACID. One of the acids found in humus. $C_{25}H_{43}O.COOH$.

PARAGYNOUS. Said of antheridium which is applied to the side of an oögonium.

PARALLEL DESCENT. The manner of derivation of structures which are similar, but occur on plants not descended from an obvious common ancestor.

PARALLEL-VEINED. Having the main veins running side-by-side for some distance in the leaf.

PARALLELODROMOUS. Having parallel veins.

PARALLELOTROPIC. Said of a plant member set along the direction of a stimulus.

PARAMETER. A quantity necessary for the specification of a population.

PARAMO. The alpine region of the north Andes.

PARAMYLUM. A substance resembling starch found in granules in the Euglenophyta.

PARANEMA. A sterile hair.

PARAPHOTOTROPIC = DIAPHOTOTROPIC.

PARAPHYLL, PARAPHYLLUM. A leaf-like, or filamentous appendage near a 'leaf' in mosses, but they are not in any specific position.

PARAPHYSIS. A sterile hair growing around, or among reproductive structures. They are found in some algae, ascomycetes, basidiomycetes, and mosses.

PARAPHYSOID. A plate of cells occurring between the asci in some ascomycetes.

PARAPLASM. The inactive vegetative part of the protoplasm.

PARAPLECTENCHYMA = PSEUDOPARENCHYMA.

PARASITE. A plant or animal living on, or in another organism, called the host. The parasite extracts food from the host, and lives at the latter's detriment.

PARASITIC CASTRATION. The condition when a plant is unable to fruit due to damage to the reproductive organs by a parasite.

PARASPORANGIUM. A sporangium containing paraspores.

PARASPORE. A spore produced on the tetrasoprophyte of some Rhodophyta. They are very likely to be diploid, and always germinate to produce a new tetrasporophyte.

PARASTICHY. A spiral line passing once around a stem through the bases of successive leaves.

PARASYMBIOSIS. The condition when two organisms grow together without assisting or harming one another.

PARASYNAPSIS, PARASYNDESIS. The side-by-side association of a pair of chromosomes as in zygotene and pachytene.

PARATHECIUM. A layer of hyphae around the apothecium of a lichen or Discomycete.

PARATONIC. Said of a plant movement induced by external stimulus.

PARATRACHEAL. Said of xylem parenchyma which occurs at the edge of the annual ring, around the vessels, but nowhere else.

PARATYPE = ALLOTYPE.

PARENCHYMA. A tissue of undifferentiated cells, which are more or less spherical, frequently unspecialized, and with cellulose cell-walls. Air-spaces are often present, and the tissue is often for storage.

PARENT MATERIAL. The material from which a soil is formed by physical and chemical weathering.

PARICHNOS. In certain lower vascular plants, a pair of scars, one on each side of the leaf base; each scar marks the end of a strand of parenchyma passing into the stem.

PARIETAL. Joined to the wall.

PARIETAL PLACENTATION. Having ovules attached to the walls of a unilocular syncarpous ovary.

PARIETALES. An order of the Archichlamydeae (EP), Polypetalae (BH). The flowers are spiral or cyclic, often with many stamens and ovules which may or may not be fused, with parietal placentas, very rarely with a basal ovule. The flowers are heterochlamydeous, rarely without petals, hypogeal to epigeal.

PARIPINNATE. Said of a compound pinnate leaf which has no terminal leaflet, *i.e.* has an even number of leaflets.

PARKERIACEAE. A family of the Leptosporangiatae. These are the only aquatic, homosporous leptosporangiate ferns. The sporangia was borne singly on the abaxial side of the leaf-blade.

PARK-LAND. Savannah; open grassy country with patches of forest or copse.

PARNASSIACEAE (H). A family of the Saxifragales (H). The flowers are solitary and the petals are free or absent. The anthers are bilocular, with staminodes alternating with the stamens. The carpels are united into a superior or inferior ovary which is unilocular with parietal placentas. The fruit is a capsule.

PAROICOUS. Said of the Bryophyta in which the antheridia and archegonia occur on the same branches, but are not mixed, the antheridia being lower on the stem than the archegonia.

PARONYCHIACEAE (BH). A family of the Caryophyllinae (BH). (EP) include it in the Caryophyllaceae (*q.v.*).

PARTED. Cleft nearly to the base.

PARTHENOAPOGAMY. The fusion of vegetative nuclei.

PARTHENOCARPY. The development of a fruit without the formation of seeds as a result of (a) lack of pollination, (b) lack of fertilization, (c) lack of embryo development. The condition can be artificially induced by application of hormones.

PARTHENOGAMY. The union of two female gametes, or the structures equivalent to them.

PARTHENOGENETIC. Reproduction by the formation of egg-cells capable of development without fertilization.

PARTHENOSPORE = AZYGOSPORE.

PARTHENOTE. An individual developed from an egg, containing only one nucleus which is haploid.

PARTIAL. (1) Secondary.
(2) Subsidiary.
(3) Not general.

PARTIAL HABITAT. The habitat occupied by a plant during one stage of its life-cycle.

PARTIAL PARASITE. A plant that has at least some power of photosynthesis, but obtains some material, mainly mineral salts and water, from a host.

PARTIAL UMBEL. One of the smaller groups of flowers which altogether make-up a compound umbel.

PARTIAL VEIL. Said of agarics, a layer of tissue joining the pileus edge to the stipe during the development of the hymenium. It may later become the annulus or cortina.

PARTICULATE INHERITANCE. Inheritance, in one individual, of distinctive characteristics from both parents.

PARTIM. Latin meaning 'partly'.

PARTITION. A wall dividing the loculi of a syncarpous ovary.

PART SPORE. One of the one-celled spores resulting from the breaking up of a two-or-more-celled ascospore.

PARVIFLORUS. Latin meaning 'small-flowered'.

PARVIFOLIATE. Having leaves which are small in relation to the size of the stem.

PASCUAL. Inhabiting pastures.

PASSAGE CELL. An unthickened cell in the endodermis of a root, found opposite the protoxylem element, through which solutions can diffuse in a transverse direction.

PASSIFLORACEAE (EP, BH, H). A family of the Parietales (EP), Passiflorales (BH, H). (BH) include the Achariaceae, Caricaceae, and Malesherbiaceae. These are herbs or shrubs often climbing by tendrils with simple, usually palmately lobed leaves. There may or may not be stipules. The regular flowers are bisexual or unisexual, solitary, or in racemes or cymes. The axis is often more or less tubular, ending in effigurations. There are usually 5 free sepals and petals (rarely 3-8). There are 5, 8-4, rarely many free stamens, which are united to an elongation of the axis. The superior ovary consists of 5-3 fused carpels with many ovules on parietal placentas. The fruit is a capsule or berry. The seed usually has an aril, and endosperm.

PASSIFLORALES (BH, H). An order of the Polypetalae (BH), Lignosae (H). The bisexual or unisexual flowers are usually regular. The ovary is usually inferior and syncarpous with one loculus and parietal placentae. Sometimes 3 or more loculi are produced by placentae. The styles are free or connate.

PASSIVE ABSORPTION. The accumulation of a substance inside a cell by purely physical processes.

PATANAS. Acidic mountain-steppe soils found in Ceylon.

PATELLA. A sessile apothecium, which is saucer-like with a distinct margin.

PATELLATE, PATELLIFORM. Shaped like a saucer or dish.

PATENS (Latin), PATENT. Spreading.

PATENT. Said of leaves or branches which spread-out widely from the stem.

PATHFINDER = HONEY-GUIDE.

PATHOGEN. An organism causing disease.

PATHOLOGY. The study of diseases.

PATROCLINAL, PATROCLINOUS. Exhibiting the characteristics of the male parent more prominently than those of the female parent.

PATROMORPHIC. Resembling the male parent.

PATULIN = CLAVICIN = CLAVIFORMIN. An antibiotic produced by some *Penicillium* spp.

PATULOUS. Spreading fairly widely.

PAUCIFLORUS. Latin meaning 'few-flowered'.

PEAT. An accumulation of partly decomposed plant matter, due to the lack of oxygen, *i.e.* waterlogging, not allowing the bacteria that cause decomposition, to live.

PECTIC COMPOUNDS. Acid polysaccharides present in the cell-walls of unlignified tissue. They are soluble in water, and forms gels under certain conditions. A = arabinose; GA = galacturonic acid; G = galactose; MGA = Methylated galacturonic acid.

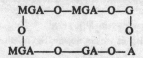

PECTIC SHEATH. A layer of pectose formed on the outside of the cellulose cell-wall of the Chlorophyta.

PECTINASE. An enzyme destroying the pectin of the middle lamella of cell-walls. It is produced by many parasitic fungi.

PECTINATE. Comb-like.

PEDALFER. A soil developed under humid conditions. There is no accumulation of calcium carbonate, but the clay complex differentiates (or tends to differentiate) into horizons, resulting in the accumulation of sesquioxides.

PEDALIACEAE (EP, BH, H). A family of the Tubiflorae (EP), Personales (BH), Bignoniales (H). (BH) include the Martyniaceae. These are herbs, with glandular hairs and opposite leaves, which may be opposite above. The bisexual zygomorphic flowers are axillary or in cymes. The sepals and petals are 5 in number. There are 2 or 4 free stamens. The ovary is superior or inferior, and of 2 fused carpels, (rarely 3-4) with many ovules. The ovary is 2-4 locular, transversely divided, with axile placentas. The fruit is a capsule or nut, and the seeds has a thin endosperm.

PEDALINEAE (BH) = PEDALIACEAE (EP).

PEDALIS. Latin meaning 'a foot long or high'.

PEDATE LEAF. A palmately divided compound leaf, having three main divisions, and having the outer divisions one or more times.

PEDAFID. Having the lamina deeply cut in a pedate manner.

PEDICEL. (1) The stalk of an individual flower of an inflorescence.
(2) A small stalk.

PEDICEL CELL. A large basi-central cell in the globule of the Charales. It bears all the other structures of the globule, which have been derived by division of a cell at its apex.

PEDICELLATE. (1) Said of a flower or a fruit having a stalk.
  (2) Having a stalk.

PEDIGREE. A table of ancestry or of posterity.

PEDOCALS. A group of soils, usually associated with drier climates, and having a freely drained profile with a horizon of calcium carbonate.

PEDOGAMY. Pseudomixis between mature and immature cells, as in certain yeasts.

PEDOGENESIS. Reproduction in young or immature organisms.

PEDOGENIC FACTOR. The factors which affect the final appearance of a soil profile, *i.e.* parent material, climatic factors, topography, vegetation, and time.

PEDOLOGY. The study of soil.

PEDUNCLE. A stalk of a flower or inflorescence.

PEG. An outgrowth from the hypocotyl of seedlings of some Cucurbitaceae. It plays some part in assisting the seedling to emerge from the testa.

PELAGIUM. A surface sea-formation.

PELARGONIDIN. A red anthocyanidin pigment.

PELLICLE. (1) The outer layer of the upper surface of a pileus, when it can be stripped off as a delicate membrane.
  (2) Of bacteria, a growth on the surface of a liquid culture.

PELLICIERACEAE (H). A family of the Theales (H). The leaves are simple and alternate. These are not epiphytes, and there are no leaf-bud scales. The petals are imbricate. There are 5 stamens with elongated anthers which open by longitudinal slits. There is 1 ovule in each loculus.

PELLUCID. (1) Translucent.
  (2) Transparent.

PELLUCID-PUNCTATE. Scattered with translucent dots.

PELLUCID-STRIATE. Said of a pileus which has a translucent top, so that the gills can be seen through it as rays.

PELOPHILE. A plant which occurs in clayey soil.

PELORIA, PELORY. An abnormal condition in which an individual of a species normally producing irregular flowers, produces regular ones.

PELOTON. A type of endophytic mycorrhiza.

PELTATE. More or less flattened, and having the stalk attached to the middle of the lower surface.

PENAEACEAE (EP, BH, H). A family of the Myrtiflorae (EP), Daphnales (BH), Thymelaeales (H). (BH) include the Geissolomaceae. These are shrubs with opposite evergreen leaves with solitary, axillary, bisexual flowers. There are 4 free valvate sepals, and no petals. There are 4 free stamens, and the superior ovary consists of 4 fused carpels, each with 4-2 erect ovules. The fruit is a capsule, and the seeds have no endosperm.

PENDULAR WATER. If the soil particles are considered as touching spheres, the pendular water is that which is held in circles around the point of contact of the spheres, but there are still air-spaces between the circles. At this stage, about 24% of the pore space is occupied by water.

PENDULOUS. (1) Hanging down.
(2) Said of an ovule which is suspended from a point at or near the top of the ovary.

PENETRANCE. The proportion of individuals homozygous for a particular gene which show the phenotype for that gene. Penetrance of many genes is practically 100% but in other cases it is much less, the value being affected by the environment or geneotype.

PENICILLATE. Brush-shaped.

PENICILLIC ACID. An antibiotic produced by *Penicillium puberlum*, and *P. cyclopium*.

PENICILLIN. An antibiotic produced by strains of *Penicillium notatum*, and *P. chrysogenum*.

PENICILLIOPSIN. A yellow pigment produced by *Penicilliopsis clavarii-formis*.

PENNALES. An order of the Bacillariophyceae. The valves are bilaterally symmetrical or asymmetrical in surface view, but the ornamentation is always bilaterally symmetrical about a line, and never a point. A raphe or pseudoraphe is always present. There are no statospores or motile macrospores formed. Many species conjugate to form auxospores.

PENTA-. A Greek prefix meaning 'five'.

PENTADIPLANDRACEAE (H). A family of the Celastales (H). These are trees, shrubs or climbers. The leaves are not lepidote, and the flowers are not enclosed by a bracteole. There is a cup-shaped or annular disk, which does not enclose the carpels. The petals are imbricate, and have a scale at the base. There are 10 stamens, and no staminodes.

PENTACYCLIC. Having parts arranged in whorls of 5 parts each.

PENTAGONAL. Having 5 angles, with convex surfaces between them.

PENTAPHYACACEAE (EP, H). A family of the Sapindales (BH), Theales (H). (BH) include it in the Ternstroemiaceae. These are shrubs with alternate leathery leaves. The small bisexual flowers are regular, and in racemes below the leaves. The flower-parts are in fives. The superior ovary consists of 5 fused carpels, each with 2 pendulous ovules. The fruit is a capsule, and there is slight endosperm.

PENTAPLOID. An individual having 5 complete sets of chromosomes.

PENTOSAN. A gum made up of pentose sugars, *e.g.* arabinose and xylose, by condensation.

PENTOSE. A monosaccharide sugar having five carbon atoms, and the general formula $C_5H_{10}O_5$.

PENTOSE-PHOSPHATE CYCLE. An oxidation cycle similar to the Citric Acid Cycle. This cycle results from the oxidation of glucose-6-phosphate by oxidized NADP, with the formation of pentose phosphates and the release of carbon dioxide. Some glucose-6-phosphate is re-formed thus:—

6-glucose-6-phosphate $+$ 12 NADP $=$ 4-glucose-6-phosphate $+$ 2 triose phosphate $+$ 12 $NADPH_2$. The $NADPH_2$ is only oxidized in the presence of oxygen, and the cycle does not occur anaerobically.

PEONIDIN. A magenta-coloured anthoxanthin pigment.

PEPO. A fleshy or succulent fruit, often of large size, formed from an inferior syncarpous ovary, and containing many seeds. It is a particular type of berry, *e.g.* a cucumber.

PEPTIDE. A compound formed from two or more amino acids by peptide link(s).

PEPTIDE LINKAGE. The bond between two amino acids by the amino (NH_2) group of one joining with the carboxyl (COOH) group of the next to give a $-NH-CO-$ linkage between them, with the elimination of water.

PEPTONE. One of the first derivatives of the digestion of protein.

PERCEPTION. The first changes which must be assumed to occur when a plant is stimulated; they lead to the appropriate reaction in due course.

PERCNOSOME. An inclusion of obscure nature found in the sperm mother-cell cytoplasm of some mosses.

PERCURRENT. Said of a vein which runs through the whole length of a leaf, but does not project beyond the tip.

PERDIGON. A soil concretion, markedly sesquioxidal in character, but containing some silica. Characteristic of Cuba.

PERENNATION. The survival from season to season, generally with a period of reduced activity between each season.

PERENNIAL. A plant living for three or more seasons, and normally flowering and fruiting at least in the second and subsequent seasons.

PERFECT. (1) Said of a flower which has both functional anthers and ovules.
(2) Said of the stage in the life-cycle of a fungus which produces spores by sexual fusion.

PERFOLIATE. Said of a leaf base which surrounds a stem completely so that the latter appears to pass through it.

PERFORATE. (1) Pierced by holes.
(2) Containing small rounded transparent dots so as to appear pierced.

PERFORATION. An interruption in the continuity of a stele, not due to a leaf-gap.

PERGAMENEOUS, PERGAMENOUS, PERGAMENTACEOUS. Like paper or parchment in appearance or texture.

PERIANDRA. The 'leaves' surrounding a group of antheridia in mosses.

PERIANTH. (1) The floral envelope, it includes the calyx and corolla, or any one of them.

(2) The cup-shaped or tubular sheath surrounding the archegonia of some liverworts.

PERIBLEM. The tissue at the growing point of a stem or root lying between the dermatogen and plerome. It gives rise to the cortex.

PERICAMBIUM = PERICYCLE.

PERICARP. (1) A layer of vegetative tissue covering a fruit-body, as in some fungi.

(2) The body of a fruit developed from the ovary-wall, and enclosing the seeds.

PERICENTRAL CELL. A cell cut off in a radial plane, found in the thalli of some Rhodophyta.

PERICENTRAL SIPHON. One of the tubular elements surrounding the central siphon in the thallus of some Rhodophyta.

PERICHAETIAL BRACT. One of the 'leaves' composing the perichaetium in a moss.

PERICHAETIUM. (1) A cup-like sheath surrounding the archegonia in some liverworts.

(2) The group of 'leaves' surrounding the sex-organs of some mosses. They are closely packed, and somewhat different in structure from the normal 'leaves'.

PERICHYLOUS. Having water-storage tissue surrounding the chlorophyll-containing tissue.

PERICLINAL. (1) Said of cell-walls running parallel to the surface of the plant.

(2) Curved in the direction of, or parallel to the surface of a plant member.

PERICLINAL CHIMAERA. A chimaera in which the distinct tissues are arranged concentrically.

PERICYCLE. A cylinder of vascular tissue, 3-6 cells thick, lying immediately inside the endodermis of a root. It consists of parenchyma, and sometimes fibres.

PERICYCLIC FIBRE. A strand of sclerenchyma in the pericycle.

PERICYSTACEAE. A family of the Saccharomycetales. There is one genus *Pericystis* which attacks pollen and larvae in beehives. The mycelium is septate with multinuclear segments, and it is heterothallic. Male and female gametangia are produced, and fertilization is through a conjugation-tube. Several zygotes are formed in each oögonium, and each zygote is considered to be an ascus.

PERIDERM. A secondary protective tissue formed in secondarily thickened stems and roots. It consists of the phellogen (cork cambium), phellem (cork), and phelloderm (secondary cortex).

PERIDERMIUM. A form of aecium with the peridium irregularly split or broken.

PERIDINALES. An order of the Dinophyceae. The walls are composed of a definite number of plates arranged in a specific manner. The entire wall is never vertically separated into two halves or valves.

PERIDININ. An alcohol-soluble, reddish pigment in the chromatophores of the Dinophyceae.

PERIDIOLE, PERIDIOLUM. Especially of the Nidulariaceae, a division of the gleba, having a separate wall, and frequently acting as a unit of distribution.

PERIDISCACEAE (H). A family of the Tiliales (H). The calyx is deeply lobed, or the sepals are free. There is a disk present. The anthers are bilocular, and the ovules are pendulous in a unilocular ovary.

PERIDIUM. The outer wall of the fruit-body of a fungus, when it is constructed of sterile tissue, and organized as a distinct layer.

PERIGAMIUM. Part of a reduced branchlet, containing the achegonia of some mosses.

PERIGONE. A perianth which is not clearly differentiated into calyx and corolla.

PERIGONIAL BRACT. One 'leaf' of the perigonium in mosses.

PERIGONIUM. A group of 'leaves' often forming a flat rosette around the base of the group of antheridia in mosses.

PERIGYNUM. (1) The 'leaves' surrounding the group of archegonia in mosses. (2) A tubular sheath surrounding the archegonia in liverworts.

PERIGYNOUS. Said of a flower in which the receptacle is developed into a flange or concave structure, on which the sepals, petals, and stamens are borne. The receptacle remains distinct from the carpels.

PERIMEDULLARY ZONE = MEDULLARY SHEATH.

PERINIUM = EPISPORE.

PERIODICITY. Rhythmic activity.

PERIPHERAL CELL. (1) One of the derivatives from the division of a periclinal cell in the tetrasporophytic generations of the Ceramiales.

PERIPHERAL CYTOPLASM. The cytoplasm lining the cell-membranes.

PERIPHERAL INITIAL = PERIPHERAL CELL.

PERIPHYSIS. A hair-like extension of the end of a hypha, forming with many others of the same kind, a pile-like lining in the ostiole of a globose structure containing reproductive organs.

PERIPLASM. The plasma lying just within the oögonial wall in some Oömycetes. It encloses the egg-cell and contains degenerating nuclei. It contributes to the formation of the wall of the oösphere.

PERIPLASMODIUM. The material formed by the disintegration of the tapetum in sporangia of Pteridophyta and Phanerogamae. It helps in the nutrition of the developing spores and pollen grains.

PERIPLOCACEAE (H). A family of the Apocynales (H). The stamens have a coronal appendage, and the pollen is granular. The ovules are not arranged parietally in pairs. The styles are united and thickened at the top, but often free below. The fruits are follicular dehiscing ventrally down one side only.

PERIPTERYGIACEAE (H). See *Cardiopteridaceae (H)*.

PERISPERM. A nutritive tissue present in some seeds, derived from the nucellus of the ovule.

PERISPORALES = ERYSIPHALES.

PERISPORE. (1) The remains of the contents of the cells of the tapetum, forming a deposit on the outside of the walls of the spores of ferns.
(2) = EPISPORE.

PERISPORIACEAE = MELIOLACEAE.

273

PERISPORIALES = ERYSIPHALES.

PERISTOMATE. Having a peristome.

PERISTOME. (1) A fringe of elongated teeth around the mouth of the capsule of a moss. The teeth are formed from the persistent remains of unevenly thickened cell-walls.

(2) A fringe of hyphae around the opening of the fruit-body of some Gasteromycetes.

(3) A form of lip arising as an outgrowth in some Protophyta. It assists in the ingesting of solid food.

PERISTROMIUM. The bounding membrane of a chloroplast.

PERITHECIUM. A rounded, or flask-shaped fruit-body of certain ascomycetes and lichens. They have an internal hymenium of asci and paraphyses, and with an apical pore (ostiole) through which the ascospores are discharged.

PERITRICHOUS, PERITRICHIATE. Said of bacteria when there are flagella distributed over the whole surface of the cell.

PERMANENT COLLENCHYMA. Functional collenchyma found in the stems and petioles of herbaceous plants.

PERMANENT HYBRID. A hybrid which breeds true because some types of possible offspring are prevented from developing by the operation of lethal factors.

PERMANENT TISSUE. Tissue consisting of fully differentiated elements.

PERMANENT WILTING. Wilting from which a plant does not recover if placed in a saturated atmosphere.

PERMEABILITY. The extent to which molecules of a given kind can pass through a membrane.

PERMEASE. A possible enzyme responsible for the translocation of ion(s) across a membrane.

PERMIAN. A geological period lasting approximately from 220 until 190 million years ago.

PERONATE. Having a stipe, covered by a thick, felted sheath, particularly at the base.

PERONOSPORACEAE. A family of the Peronsporales. They are strictly parasitic, with the intercellular mycelium producing haustoria into the host cells. All have branching sporangiophores projecting beyond the host, and bearing conidiosporangia singly at the ends of the ultimate branches.

PERONOSPORALES. An order of the Phycomyceteae. All are parasitic on land plants. The conidiosporangia germinate to form biflagellate zoospores, or germinate directly. Reproduction is oögamous. There is a single egg in an oögonium, and it is differentiated from the periplasm. Fertilization is by a male aplanogamete.

PEROXIDASE. An enzyme which catalyses an oxidation of a substrate by the removal of hydrogen. The hydrogen is removed by its combination with hydrogen peroxide.

PERPUSILLUS. Latin meaning 'very small'.

PERRUMPENT. Breaking through.

PERSISTENT PERIANTH. A perianth which remains unwithered, and often enlarged around the fruit.

PERSONALES (BH, H). An order of the Gamopetalae (BH), Herbaceae (H). The flowers are usually very irregular. The corolla is hypogynous and often two-lipped. There are usually 4 stamens, joined to the petals, or there are 2. The ovary is 1-2, or rarely 4 locular, usually with many ovules.

PERSONATE. Said of a two-lipped corolla which has some likeness to a mask or face of an animal.

PERTHOPHYTE. A necrophyte on dead tissues of living hosts.

PERTUSATE. (1) Perforated.
   (2) Pierced by slits.

PETAL. One of the parts forming the corolla of a flower, usually brightly coloured and conspicuous.

PETALODY. The transformation of stamens into petals.

PETALOID. Looking like a petal.

PETALOMANIA. Abnormal in the number of petals.

PETERMANNIACEAE (H). A family of the Alstroemeriales (H). These are woody climbers with reticulately-veined leaves. The inflorescence is leaf-opposed. The unilocular ovary is inferior with parietal placentas.

PETIOLATE. Having a leaf-stalk.

PETIOLE. The stalk of a leaf.

PETIOLULE. The stalk of a leaflet of a compound leaf.

PETIVERIACEAE (H). A family of the Chenopodiales (H). Stipules are often present, but may be modified. The calyx is 3-5 lobed, or partite, and there are no petals. The stamens alternate with the calyx lobes, if they are of the same number, with the anthers opening by longitudinal slits. The ovary is unilocular, and the fruits are free, not united into a fleshy mass.

PETRAEUS. Latin meaning 'growing on rocks'.

PETRIUM. A rock formation.

PETRODIUM. A boulder-field formation.

PETROSAVIACEAE (H). A family of the Alismatales (H). These are saprophytes with the leaves reduced to scales. The flowers are small and in racemes. The perianth segment are colourless. There are 3 carpels.

PEZIZACEAE. A family of the Operculatae. The apothecia are cup-shaped or discoid, sometimes convex. It may or may not have a stalk, and is never cap-shaped.

PEZIZALES. An order of the Ascomyceteae. The apothecia are fleshy or leathery, external from the first or emerging from the substratum. It is rounded in outline, or occasionally narrowed. A stalk may or may not be present. A stroma is not usually present. The ascospores are discharged into the air.

PEZIZOID. Resembling a cup-shaped apothecium.

pF. (1) A measurement of the suction force by which a soil holds water. It is the logarithm of the free-energy difference expressed in cm. In practice, it is a measurement of the availability of water, pF7 is oven-dry soil, pF4 is wilting point, pF2 is found in a well-drained wet soil.
   (2) It is also a measurement of the rate of movement of a dissolved substance in chromatography.

PFITZER'S LAW. That cells become progressively smaller during successive generations.

PFLÜGER'S RULE. The mitotic figure elongates in the direction of least resistance.

PGA = PHOSPHOROGYCERIC ACID.

pH. The negative logarithm of the concentration of hydrogen ions in grams per litre. Thus pH7 is neutral, less than 7 is acid, and more than 7 is alkaline.

PHACIDIACEAE. A family of the Inoperculatae. The apothecia are not buried in a fleshy stroma, which is lens-shaped, black and often brittle. The excipulum is weakly developed, and the ascospores are oblong to needle-shaped.

PHACIDIALES. An order of the Euascomycetae. They have a flattened, rounded ascocarp in which the over-arching peridium opens at maturity by rupturing into a star-shaped split, or by developing a circular pore.

PHAENOGAMOUS. Relating to flowering plants.

PHAENOGAMS = PHANEROGAMAE.

PHAENOLOGY. The study of the periodic phenomena of vegetation.

PHAEO-. A prefix meaning 'dark-coloured, swarthy'.

PHAEOPHYCEAE, PHAEOPHYTA. The brown algae. The chromatophores have the photosynthetic pigments masked by the gold-brown fucoxanthin. The thallus is always multicellular, and generally macroscopic. The motile reproductive cells are pyriform mostly with 2 laterally inserted flagella of unequal length.

PHAEOPHYTIN. A greyish pigment in chlorophyll.

PHAEOPLAST. The chromatophore in the Phaeophyceae.

PHAEOSPOROUS. Having dark-coloured one-celled spores.

'PHAGE = BACTERIOPHAGE.

PHAGOTROPIC = HOLOZOIC. Feeding on elaborated food-material, either protoplasm or its derivatives.

PHAILIDE. A one-celled, flask-like structure from the end of which conidia (phailospores) are abstricted.

PHAILOSPORE. See *phailide*.

PHALANGE. A bundle of stamens.

PHALLACEAE. A family of the Phallales. The receptacle occupies the upper portion of a stout, hollow, stipe, either grown fast to it, or forming a bell-shaped structure attached to the top of it.

PHALLALES. An order of the Gasteromyceteae. The mature gleba deliquesces into a slimy, usually evil-smelling mass, which covers or is supported by a definite framework.

PHANEROGAMAE, PHANEROGAMIA. An old term which includes the Gymnospermae and Angiospermae. It is now replaced by the term Spermatophyta.

PHANEROGAMOUS. Relating to the flowering plants.

PHANEROPHYTE. A tree or shrub with resting buds freely exposed on branches raised above the soil.

PHARYNGEAL ROD. One of a pair of rod-like organs running alongside the gullet of the Euglenales, and probably supports the distended gullet.

PHASEOLIFORM. Shaped like a bean.

PHELLEM. Cork. Suberized cells formed in the outside regions of a stem or root, from a phellogen.

PHELLIUM. A rock-field formation.

PHELLODERM. A cylinder of unthickened cells formed from a phellogen on the inner side. The cells often contain starch, and sometimes chloroplasts. It is a kind of secondary cortex.

PHELLOGEN. The cork cambium. A layer of meristematic cells lying in the cortex of a stem or root. It forms cork on its outer surface, and phelloderm on the inside.

PHELLANDRENE. A terpene found in *Eucalyptus* spp.

PHELLOID. A crust of non-suberized, or slightly suberized cells on the surface of some plants, replacing true cork.

PHENOCOPY. The phenotype of a given genotype changed by external conditions to resemble the phenotype of a different genotype.

PHENOLS. Compounds having a hydroxyl (OH) group substituted for a hydrogen atom in a benzene ring.

Phenol

PHENOLOGY. The study of periodical phenomena in plants, *e.g.* the time of flowering in relation to climate.

PHENOTYPE. The kind or type of organism produced by the reaction of a given genotype with the environment.

PHENOTYPIC. Caused or produced by environmental factors.

PHENYL ANALINE. An amino acid found in proteins.

PHENYLPROPANE. The repetitive unit recurring in lignin.

PHENYL UNIT = PHENOL.

PHIALIDE. A short, flask-shaped sterigma.

PHIALIFORM. Shaped like a saucer or cup.

PHIALOMERISTEM SPORE = PHIALOSPORE.

PHIALOPORE. A small hole in the cynobium of the Volvocaceae, found at the posterior end.

PHIALOSPORE. A spore abstricted from a phialide.

PHILADELPHACEAE (H). A family of the Cunoniales (H). There is an indumentum of stellate hairs. The flowers are in racemes, heads, cymes, or panicles. The stipules, if present, are joined to the petiole. The flower has no disk. There are 8 or more stamens, with the filaments often toothed near the apex. The fruit is a lodicular capsule, rarely a berry.

PHILESIACEAE (H). A family of the Alstroemeriales (H). These are herbaceous or woody, often climbers. The flowers are axillary or terminal. The ovary is superior, and the fruit is a berry.

-PHILOUS. A Greek suffix meaning 'loving, dwelling in'.

PHILYDRACEAE (EP, BH, H). A family of the Farinosae (EP), Coronarieae (BH), Haemodorales (H). These are herbs with 2-ranked narrow leaves and spikes of homochlamydeous flowers. The flowers are bisexual, and strongly zygomorphic with flower-parts in threes (4 perianth segments (H)). There is 1 anterior stamen. The superior ovary consists of 3 fused carpels with 3 or 1 loculi, and many ovules. The fruit is a capsule, and the seeds contain endosperm.

PHLEOGENACEAE. A family of the Auriculariales. They are saprophytic on wood, bark, etc. There is no distinction between the hypobasidium and epibasidium. The spore fruit is stalked with a head of radiating, more or less coiled, hyphae among which the curved basidia are found. These bear 2 or 4 spores without visible sterigmata. Clamp-connections are found in some species.

PHLOBAPHENE. A yellow-brown substance found in cork cells, probably formed from the decomposition of tannins.

PHLOEM. The vascular tissue which conducts synthesized foods in vascular plants. It is characterized by the presence of sieve-tubes, and in some plants companion cells, fibres and parenchyma.

PHLOEM FIBRE. An element of sclerenchyma (or a strand of such elements) present in the phloem. It probably helps to support the sieve-tubes.

PHLOEM ISLAND. A patch of phloem surrounded by secondary wood.

PHLOEM PARENCHYMA. The unspecialized cells found in the phloem.

PHLOEM RAY. The part of a vascular ray that passes through the phloem.

PHOENICEUS. Latin meaning 'scarlet'.

PHLOEOTERMA, PHLOËTERMA. An endodermis in which the radial walls and the inner tangential walls are heavily suberized.

PHOBOTAXIS. The response or reaction of an organism to a nocuous stimulus, the organism withdrawing at an angle not necessarily related to the direction of the stimulus.

-PHOBOUS. A suffix meaning 'avoiding, disliking'.

PHOENOCIN. A dye produced by some fungi.

-PHORUS. A suffix meaning 'stalk, bearing'.

PHOSPHATASE. An enzyme which splits phosphate radicals from their esters.

PHOSPHATIDE. A glycerol tri-ester resembling fats, but only two fatty acids are involved. The third glycerol —OH group is combined with a nitrogen-containing base through a molecule of orthophosphoric acid.

PHOSPHATIDIC ACID. A possible intermediary in a carrier mechanism across a cell-membrane, formed from, and producing licethin in a cyclic mechanism.

PHOSPHILIPINE = PHOSPHATIDE.

PHOSPHOENOLPYRUVIC ACID. A carbon dioxide acceptor during photosynthesis, resulting in the production of pyruvic acid and phosphate.

$CH_2COH_2PO_2COOH + CO_2 + H_2O = COOHCH_2COCOOH = H_3PO_4$.
phosphoenol-   pyruvic acid ·
pyruvic acid

PHOSPHOGLYCERIC ACID (PGA). The product formed from ribulose-diphosphate by combination with carbon dioxide during the dark phase of photosynthesis. It is then reduced by $NADH_2$. It is also an intermediat-product in the anaerobic respiration of hexosediphosphate to pyruvic acid.

$$\begin{array}{c} CH_2OP \\ | \\ CHOH \\ | \\ COOH \end{array}$$

PHOSPHOLIPOID, PHOSPHOLIPID = PHOSPHATIDE.

PHOSPHONUCLEOTIDE. A nucleotide combined with phosphoric acid.

PHOSPHOROGLYCERIC ACID = PHOSPHOGLYCERIC ACID.

PHOSPHORESCENCE = BIOLUMINESCENCE.

PHOSPHORYLASE. An enzyme which catalyses the conversion of glucose 1-phosphate to amylase, or, more generally, it brings about a combination with phosphoric acid.

PHOSPHORYLATION. Generally a combination with phosphoric acid, but, more specifically, the accumulation of energy by ATP, and the 'release' of the energy by the back reaction to ADP.

PHOSPHORLYL-CHOLINE. One of the forms in which phosphorus is transported in the xylem.

PHOSPHOSERINE. The amino acid serine combined with phosphoric acid.

$$\begin{array}{c} \qquad\qquad OH \\ \qquad\qquad | \\ CH_2-O-P=O \\ | \qquad\quad | \\ CH-NH_2 \quad OH \\ | \\ COOH \end{array}$$

PHOTOGENIC. Emitting light, producing light.

PHOTOLYSIS. (1) The grouping of the chloroplasts in relating to the amount of light falling on the plant.

(2) The decomposition of dis-association of a molecule as the result of the absorption of light; this is brought about by the destruction of an O–H bond by its excitation with photons of light.

PHOTOMORPHOSIS. A change in the structure of a plant after exposure to strong light.

PHOTON. A 'packet' of light energy. It may be considered as a particle of mass hv/c^3, where h is Planck's constant, v is the frequency of vibration, and c is the velocity of light.

PHOTONASTY. Response to a general, non-directional illumination stimulus, *e.g.* the opening and closing of flowers at night.

PHOTOPATHY. Negative phototaxis.

PHOTOPERCEPTOR. The part of an eye-spot which is sensitive to light.

PHOTOPERIOD. The time exposed to light.

PHOTOPERIODISM. The effect of the length of alternating light and dark periods on the growth and formation of flowers and fruits. Most plants have their optimum day-length for flower formation.

PHOTOPHILOUS. (1) Light-seeking or light-loving.
(2) Said of plants which inhabit sunny places.

PHOTOPHOSPHORYLATION. The formation of ADP and ATP in the presence of inorganic phosphate during the light-phase of photosynthesis.

PHOTOREDUCTION. Anaerobic reduction in the presence of light, by molecular hydrogen.

PHOTOSTAGE. (1) An early stage in the development of a seedling, during which it needs a supply of light.
(2) The stage in photosynthesis in which light is necessary. In this stage PGA is converted into triose-phosphate.

$$
\begin{array}{ccc}
CH_2OP & & CH_2OP \\
| & & | \\
CHOH & \longrightarrow & CHOH + [O] \\
| & & | \\
COOH & & CHO
\end{array}
$$

PHOTOSYNTHESIS. The building-up, in the green cells of a plant, of simple carbohydrates from carbon-dioxide and water, with the liberation of oxygen. This proceeds only when the plant has sufficient light. The chlorophyll acts as an energy transformer, which enables the plant to use the light as a source of energy. Proteins are synthesized from intermediate products of photosynthesis.

PHOTOSYNTHETIC CAPACITY. The efficiency of a plant cell or a chloroplast in carrying out photosynthesis.

PHOTOSYNTHETIC EFFICIENCY. The ratio of light-energy absorbed by the chloroplasts of a tissue in unit time, to the amount of energy fixed per unit time.

PHOTOSYNTHETIC NUMBER. The ratio between the number of grams of carbon dioxide absorbed per hour by a unit of a leaf to the number of grams of chlorophyll which that unit contains.

PHOTOSYNTHETIC QUOTIENT, PHOTOSYNTHETIC RATIO. The ratio between the volume of carbon dioxide absorbed to the volume of oxygen set free, during a given time, by plant material occupied in photosynthesis.

PHOTOSYNTHETIC UNIT. The minimum unit of chlorophyll necessary to reduce 1 molecule of carbon dioxide, *e.g.* it is estimated that 2,500 chlorophyll molecules are needed.

PHOTOTAXIS. The movement of a whole organism in response to light.

PHOTOTONUS. The condition of a leaf which is able to respond to a stimulus, because it has received an adequate amount of light.

PHOTOTROPIC CONDUCTION. The differential conduction of auxin down either side of a shoot which is unilaterally illuminated, so that growth is towards the light.

PHOTOTROPIC INDUCTION. The effect of light on the apex of the plant affects the growth of the stem below it, *i.e.* the tip, under the influence of light, induces growth lower down.

PHOTOTROPISM. The growth curvature of part of a plant in response to light.

PHRAGMOBASIDIUM. A basidium which becomes septate, and is then divided into four cells.

PHRAGMOPLAST. The cell-plate formed during cell-division.

PHRAGMOSPORE. A spore having two or more transverse septa.

PHRETIUM. A tank formation.

PHRYMACEAE (EP, H). A family of the Tubiflorae (EP), Verbenales (H). (BH) include it in the Verbenaceae. These are herbs with opposite leaves and small zygomorphic, axillary flowers. There is 1 carpel, with 1 erect orthotropous ovule.

PHYCOCHRYSIN = CHRYSOCHROME The golden-brown pigment of the Chrysophyceae.

PHYCOCYANIN. The blue pigment in the chloroplasts of the Rhodophyta, and distributed through the cells of the Cyanophyta.

PHYCOERYTHIN. The red pigment of the Rhodophyta. It is found in the cell-sap, and is protein in nature and soluble in water.

PHYCOLOGY. The study of Algae.

PHYCOMYCETAE, PHYCOMYCETEAE. A class of fungi. The mycelium is usually aseptate, and there is no one distinctive type of spore. There is an indefinite number of spores in a sporangium, and in most members the fusion of gametes results in a zygote that forms a thick-walled resting spore.

PHYCOPYRRIN. A red-brown water-soluble pigment in the chromatophores of the Dinophyceae.

PHYLETIC CLASSIFICATION. A scheme of plant-classification based on the presumed evolutionary descent of organisms.

PHYLLACHORACEAE. A family of the Sphaeriales. These are leaf-parasites with the stroma extending from the upper to the lower surface, or between the cuticle and epidermis, or between the epidermis and the palisade layer. Perithecial walls are present, and true paraphyses are produced.

PHYLLARY. One of the involucral bracts on the outside of a capitulum.

PHYLLOCARPIC MOVEMENT. A curvature of the fruit-stalk bringing the young fruit under the protection of the leaves.

PHYLLOCHLORIN. A possible chloroplast pigment, containing chromoprotein molecules, lipoid molecules, and molecules of carotinoids.

PHYLLOCLADE = CLADODE.

PHYLLODE. A flattened, leaf-like petiole.

PHYLLODY. The transformation of parts of a flower into leaves.

PHYLLOME. A general term for leaves, and all leaf-like organs.

PHYLLOPODIUM. The petiole and rachis of a fern leaf.

PHYLLOSIPHONACEAE. An order of the Siphonales. All are endophytic or endozoic with tubular or vesticular coenocytic thalli. The only known method of reproduction is by aplanospores.

PHYLLOSIPHONIC. Said of a siphonostele which has both leaf-gaps, and branch-gaps.

PHYLLOTAXIS. The arrangement of leaves on the stem.

PHYLOGENESIS, PHYLOGENY. The history of the development of race.

PHYLON. A line of descent.

PHYLUM. A major division of the plant or animal kingdom. The term 'division' used to be used more frequently in the plant kingdom.

PHYSCION. A lichen acid.

PHYSIOLOGICAL BALANCE. The balance between the ions in a culture solution, so that none of them can have an adverse effect on the growth of a plant.

PHYSICAL BARRIER. Any physical object, *e.g.* a sea or mountain range which imposes a barrier on the migration of plants or animals.

PHYSIOGNOMY. The characteristic appearance of a plant community by which it can be recognized at a distance.

PHYSIOGRAPHIC CLIMAX. A plant community maintained in a certain stage of development by some natural features of the habitat.

PHYSIOLOGICAL. Relating to the functions of a plant (or animal) as a living organism.

PHYSIOLOGICAL ANATOMY. The study of the relation between structure and function.

PHYSIOLOGICAL DROUGHT. The condition when a plant is unable to take in water, because of the low ground-temperature, or because it holds substances in solution which hinder the absorption of water by the plant.

PHYSIOLOGICAL SPECIATION. Existence within a particular species of a number of races or forms which are indistinguishable in structure, but show physiological, biochemical, or pathogenic characters.

PHYSIOLOGIC VARIETY = BIOLOGIC FORM.

PHYSIOLOGICAL ZERO. The threshold temperature below which the metabolism of a cell, organ, or organism ceases.

PHYSIOLOGIC FORM. A race showing physiological speciation.

PHYSIOLOGICAL RACE = PHYSIOLOGIC FORM.

PHYSIOLOGY. The study of the physical and chemical processes which go on in living organisms.

PHYSODERMATACEAE. A family of the Chytridiales. They are parasitic on higher plants. The primary infection produces an external sporangium with rhizoids penetrating the epidermal cells of the host.

PHYT-, PHYTO-. A Greek prefix meaning 'plant'.

PHYTASE. An enzyme capable of hydrolysing phytin, releasing phosphate. It may be capable of hydrolysing other phosphate ester linkages.

PHYTIN. An acid, calcium and magnesium salt of inositolhexaphosphoric acid $(C_6H_6(H_2PO_4)_6)$. It is found in certain seeds.

PHYTOBENTHON. Plant benthon.

PHYTOCHROME. A light-receptive pigment.

PHYTOGENIC SOIL. A soil which develops in temperate zones under a wide range of humidity. The dominant factor in their development is the vegetation which they carry.

PHYTOHORMONE. A substance produced by plants (or manufactured) which influences the growth and/or development of all or part of the plant.

PHYTOLACCACEAE (EP, BH, H). A family of the Centrospermae (EP), Curvembryae (BH), Chenopodiales (H). These are herbs or woody. The regular flowers are in racemes or cymes, and usually bisexual. There are usually 5-4 free perianth lobes, and 4-5 free stamens with anthers which open by longitudinal slits. There may be many stamens. The superior ovary (rarely inferior) consists of many to 1 carpels which are free or united. There is 1 ovule in each. The fruit is a drupe or nut, rarely a capsule, and a perisperm is present.

PHYTOL. $C_{20}H_{20}OH$. It is a constituent of chlorophyll.

PHYTOMITOGENS. Materials of vegetable origin which induce mitosis, and possibly causes cancer.

PHTYOMYXMAE. A class of the Myxothallophyta. The vegetative body is a parasitic plasmodium in the tissues of angiosperms. The plasmodium breaks into a mass of spores that have no definite arrangement.

PHYTON. A hypothetical plant unit composed of leaf, blade, and stalk.

PHYTOPATHOLOGY. The study of plant diseases, especially of plant in relation to parasites.

PHYTOPHILOUS, PHYTOPHAGOUS. Plant-eating.

PHYTOPLANKTON. The plants of the plankton.

PHYTOSTEROL. $C_{26}H_{44}O.H_2O$. A substance in the humic fraction of soil. See *Sterol*.

PHYTOTOMY = ANATOMY.

PHYTOTOXIC. Poisonous to plants.

PHYTOTRON. A large group of experimental buildings maintained at a variety of controlled conditions for research on plants.

PICRODENDRACEAE (H). A family of the Juglandales (H). The leaves are trifoliate with minute setiform stipules. The ovary is bilocular.

PICTUS. Latin meaning 'coloured'.

PIGMENTOSA. The pigmented part of an eye-spot.

PILEATE. (1) Shaped like a cup.
(2) Having a pileus.

PILEUS. The cap of an agaric, bearing the hymenium on its lower surface.

PILIFEROUS. (1) Ending in a delicate hair-like point.
(2) Bearing hairs.

PILIFEROUS LAYER. The part of a root epidermis which bears the root-hairs.

283

PILIFORM. Resembling a long zig-zag hair.

PILOBOLACEAE. An order of the Mucorales. The sporangia are aerial and have many spores. Its wall is thickened above, and neither break-up nor is deliquescent. The columella is of moderate size.

PILOCYSTIDIUM. A cystidium on the pileus surface.

PILOSE. Bearing a scattering of simple moderately stiff hairs.

PIMELIC ACID. An acid promoting the growth of various bacteria.

PINACEAE. A family of the Coniferae. These are mostly monoecious having perfect cones. The seeds are concealed between scales, with a woody or leathery testa, and no aril.

PIN-EYED. Having the throat of the corolla more or less closed by a stigma shaped like a pin-head. The term is applied to the primrose and its relatives.

PINNA. (1) The secondary division of a compound fern-leaf.
(2) A branch of a thallus, when they are arranged in opposite rows.

PINNA BAR. A plate-like vascular strand formed by the fusion of two pinna traces.

PINNATE. (1) Said of a compound leaf having leaflets arranged in two ranks on opposite sides of the rachis.
(2) Said of a thallus having branches arranged on each side of a middle axis.

PINNATIFID. Said of a leaf-blade which is cut about half-way towards the mid-rib, into a number of pinnately arranged lobes.

PINNATISECT. Pinnatifid, but with the cuts reaching nearly to the mid-rib.

PINNA TRACE. The strand of vascular tissue running from the main vascular tissue of the rachis to that of a pinna.

PINNULE. (1) One of the lobes or segments when the leaflet of a pinnate leaf is itself divided in a pinnate manner.
(2) One of the fine terminal branches of the Bryopsidaceae. They may break off and serve for vegetative reproduction.

PINOCYTOSIS. The formation of a small vacuole or vesicle in the cytoplasm.

PIONEER COMMUNITY. The first plant community to become prominent on a piece of formerly bare ground.

PIONEER SPECIES. A species whose members tend to be among the first to occupy bare ground. These plants are often intolerant of competition, and may be crowded out as the community develops.

PIONNATE. Having a continuous layer of fungus spores. It is often slimy.

PIPERACEAE (EP, BH, H). A family of the Piperales (EP, H), Micembryae (BH). (BH) include the Saururaceae and Lactoridaceae. These are herbs or shrubs with alternate leaves and spikes etc. of bisexual or unisexual flowers. There is no perianth. There are 10-1 free stamens. The superior ovary consists of 4-1 fused carpels, forming 1 loculus with one basal ovule. Endosperm and perisperm are present.

PIPERALES (EP, H). An order of the Archichlamydeae (EP), Lignosae (H). The leaves are simple with or without stipules. The small flowers are bisexual or unisexual and are in spikes. The perianth is absent or bract-like. There are 10-1 free stamens, and 4-1 carpels which are free or united.

PIPERIDINE. A heterocyclic secondary amine; a reduction product of pyridine.

$$
\begin{array}{c}
\text{H}_2 \\
\text{C} \\
\text{H}_2\text{C} \quad\quad \text{CH}_2 \\
\text{H}_2\text{C} \quad\quad \text{CH}_2 \\
\text{N} \\
\text{H}
\end{array}
$$

PIPTOCEPHALIDACEAE. A family of the Mucorales. The sporangia are aerial and narrow, containing 1 to many spores. There is no columella. The sporangia are more or less in heads, and often break-up into 1-spored segments.

PIRIFORM = PYRIFORM.

PISIFORM. Pea-shaped.

PISTIL. (1) Each separate carpel of an apocarpous gynecium.
(2) The gynecium as a whole, whether it is apocarpous or syncarpous.

PISTILLAR. Club-shaped.

PISTILLATE. Said of a flower which has carpels but no anthers.

PISTILLIDIUM = ARCHEGONIUM.

PISTILLODE. An abortive or non-functional pistil.

PISTON MECHANISM. A method of moving pollen. It is shed into a tube from which it is pushed by the style thus coming in contact with an insect visitor.

PIT. (1) A small sharply-defined area of a plant cell-wall which remains unthickened when the rest of the wall thickens.
(2) The two opposite thin areas in the walls of two cells or vessels in contact.
(3) A local thin spot in the wall of the oögonium of some Oömycetes.

PIT CAVITY. The excavation in a cell-wall where the thinning is apparent.

PITCHER. An urn-shaped or vase-shaped modification of a leaf or part of a leaf. It serves to trap insects or other small animals which are killed or digested.

PIT FIELD. The thin-walled area of a pit in a cell-wall.

PITH. A cylinder of cell, chiefly parenchymatous, lying centrally in an axis surrounded by vascular tissue.

PITH RAY = VASCULAR RAY.

PITH-RAY FLECK = MEDULLARY SPOT. A dark spot found in timber composed of cells which have filled a cavity resulting from attacks by insects in the cambium.

PIT MEMBRANE. The thin sheet of unbroken wall between two opposite pit cavities.

PITTED. (1) Having pits in the wall.
(2) Having the surface marked by small excavations.

PITTED VESSEL. A vessel with pits in the wall.

PITTOSPORACEAE (EP, BH, H). A family of the Rosales (EP), Polygalinae (BH), Pittosporales (H). These are woody, sometimes climbing with alternate leaves and resin passages. The bisexual flowers are regular with the perianth segments in whorls of 5 members. The superior ovary consists of 2 or more fused carpels, with 1-5 loculi. The placentas are parietal or axial and bear many anatropous ovules in 2 ranks. The style is simple. The fruit is a capsule or berry, and the seeds contain endosperm.

PITTOSPORALES (H). An order of the Lignosae. These are trees, shrubs, or climbers with simple, alternate to verticillate leaves which lack stipules. The flowers are hypogynous, mostly bisexual and regular. The sepals and petals are imbricate or valvate. The stamens are free, the same as or twice the number as the petals. The anthers open longitudinally or by pores. The carpels are fused with parietal to axile placentas. The seeds have a copious endosperm and a minute embryo.

PLACENTA. (1) The part of an ovary to which the seeds are attached.
(2) Any mass of tissue to which sporangia or spores are attached.

PLACENTAL CELL. A food-supplying cell for the developing carposporophyte or some red algae. It is produced by the fusion of the carpogonial filament cells with the inferior daughter cell of the carpogonium.

PLACENTAL SCALE = OVULIFEROUS SCALE.

PLACENTATION. The arrangement of the placentas in a syncarpous ovary.
(a) *parietal*—the carpels are fused only by their margins so that the placentas then appear as internal ridged on the ovary wall.
(b) *axile*—the margins of the carpel fold inwards fusing together in the centre of the ovary, forming a single central placenta.
(c) *free central*—the placenta is a central upgrowth from the base of the ovary.

PLACENTIFORM. Like a flat cake or cushion.

PLACODIOID. Said of a lichen thallus which is rounded in outline and is edged with small scales.

PLACODIOMORPH. A polarilocular lichen spore.

PLACODIUM. A hardened hyphal layer surrounding the openings of the perithecial ostiole, when the perithecium is embedded in a stroma.

PLAGIO-. A Greek prefix meaning 'oblique'.

PLAGIOGEOTROPISM. A growth response to gravity, so that the axis of the plant member makes an angle other than 90° with the line of the gravitational field. It may be used to mean that the organ makes any constant angle with the axis, and would thus include *diageotropism* as a special term.

PLAGIOTROPIC, PLAGIOTROPOUS. Said of plant members which become arranged at right-angles across the line of the stimulus.

PLAKEA. A curved plate of eight cells formed during the early stages in the development of a colony of the Volvocaceae.

PLANAR WATER. Water absorbed in soils, but held by strong electric fields on the surface of Si-O-Si or OH-Al-OH planes.

PLANE. Flat.

PLANE OF SECTION. The direction in which a plant member is cut, or assumed to be cut, in ascertaining its structure.

PLANETISM. Of Oömycetes, the condition of having motile spores.

PLANKTON. The more or less free-floating animals and plants living near the surface of a sea or lake.

PLANK ROOT. A root which is very markedly flattened so that it stands out from the base of a stem like a plank set on edge. They give additional support to the plant.

PLANO-. A prefix meaning 'motile'.

PLANO-CONVEX. Convex, but somewhat flat.

PLANOCYTE, PLANONT. A motile cell.

PLANOGAMETE = ZOOGAMETE.

PLANOSOL. A soil with a well-developed hard pan.

PLANOSOME. An odd chromosome resulting from the non-disjunction of a pair during meiosis.

PLANOSPORE = ZOOSPORE.

PLANOZYGOTE. A motile zygote.

PLANT. An organism which takes in all its nutrients in solution. Green plants are capable of synthesizing complex substances from simple ones, while the non-green ones need to be suppied with amino-acids, sugars etc.

PLANTAGINACEAE (EP, BH, H). A family of the Plantaginales (EP, BH, H). These are usually herbs with alternate or opposite leaves. The flowers are unisexual or bisexual and regular. There are 4 fused membranous sepals and petals. The 4 free stamens are joined to the petals. The superior ovary has 2 fused carpels (or 1), with 4-1 loculi which contain few or 1 anatropous ovules. The fruit is a capsule or nut, and the seeds contain endosperm.

PLANTAGINALES (EP, BH, H). An order of the Sympetalae (EP), Gamopetalae (BH), Herbaceae (H). It contains one family, the Plantaginaceae.

PLANTAGINEAE (BH) = PLANTAGINACEAE.

PLAQUE. A clear area caused by a bacteriophage in a bacterial colony.

PLASM, PLASMA = PROTOPLASM.

PLASMAGEL. See *ectoplasm*.

PLASMAGENE. A self-reproducing particle in the cytoplasm of a cell, affecting the characteristics of the cell bearing it. Being outside the nucleus, it is usually transmitted only in the female gamete.

PLASMALEMMA. A thin membrane surrounding the protoplasm, or a cell organelle. It is about $1/100\ \mu$ thick and consists of fat and protein. It is responsible for the restricted penetration of many substances into the cell. Severe damage to it kills the cell. In plants it is in contact with the cell-wall, and around vacuoles.

PLASMA-MEMBRANE = PLASMALEMMA.

PLASMASOL = ENDOPLASM.

PLASMATOOGOSIS. Said of the Pythiaceae which produce a bud-like outgrowth in the host tissue.

PLASMATOPAROUS. Said of a spore which germinates to produce a naked protoplast, which then forms a wall and produces a germ-tube.

PLASMODERM. A fine thread of protoplasm passing through fine pores in the cell-wall, thus forming a connection between the cytoplasm of adjacent cells.

PLASMODIAL STAGE. A naked amoeboid stage in the life-history of the Colaciales. These bud off uninucleate portions which metamorphose into uniflagellate swarmers.

PLASMODIC GRANULE. A very small dark-coloured particle on the surface of the periderm and frequently on the spores of the Cribariaceae.

PLASMODIOCARP. A sporangium, formed by some myxomycetes, which is irregular or sinuous in form.

PLASMODIOPHORACEAE. The single family of the Plasmodiophorales.

PLASMODIOPHORALES. An order of the Mycetozoa. The single order of the Phytomyxinae. These are parasites producing plasmodia in the cells of roots and stems of higher plants and in a few algae and aquatic fungi. The swarm spores have two flagella anteriorly.

PLASMODIUM. A multinucleate, naked, amoeboid plant-body as found in the Myxothallophyta and some Chrysophyta.

PLASMGAMY, PLASTGAMY. (1) The fusion of cytoplasm, in contrast to the nucleoplasm.
(2) The fusion of 2 sexual cells.

PLASMOLYSIS. The collapsing of the cell-protoplasm from the cellulose cell-wall, due to the movement of water from the cell-sap into a solution of higher osmotic concentration surrounding the cell.

PLASMON. The cytoplasm of an individual cell, considered as a single hereditary unit. The sum of the plasmagenes.

PLASMOSOME = NUCLEOLUS. A small cytoplasmic granule. A type of nucleolus which stains with acidic dyes, and disappears during mitosis without mingling with the chromosomes.

PLASTICITY. The ability of a material to change shape continuously under applied stress, and retain the impression after the removal of the stress. The plasticity of a soil gives an indication of its moisture content.

PLASTIC MATERIAL. Any substance which is used up in growth processes.

PLASTID. A small, variously shaped, self-propagating body in the cytoplasm of plant cells. They are possibly special centres of chemical activity. They may be coloured (*chromoplast*) or colourless (*leucoplast*). Chloroplasts are specialized plastids.

PLASTID INHERITANCE. Inheritance determined by plastogenes.

PLASTID MUTATION. A change in a chloroplast affecting its ability to produce chlorophyll.

PLASTIDOME. The total compliment of plastid in a cell.

PLASTIN. An acidophillic substance occurring in masses in the nuclei of cells.

PLASTOCHRONE. The period of time that elapses between the formation of one leaf primodium and the next, on a shoot which has a stable spiral phyllotaxy.

PLASTOCONT = CHONDRIOSOME.

PLASTOGENE. A gene attached to a plastid, which determines the likeness of the daughter plastid to the parent.

PLASTONEMA. The deeply staining peripheral material in the sporogenous tissue of mosses.

PLASTOSOME. The lightly staining internal cytoplasm in the sporogenous tissue of mosses.

PLATANACEAE (EP, BH, H). A family of the Rosales (EP), Unisexuales (BH), Hamamelidales (H). These are woody with alternate 3-5-lobed stipulate leaves and pendulous heads of regular unisexual flowers. There are no sepals or petals. There are 8-3 free stamens. The superior ovary consists of 1 carpel with 2-1 ovules. A caryopsis and endosperm are present.

PLATY-. A Greek prefix meaning 'broad'.

PLATYCARPOUS. Broad fruited.

PLATYFORM. Flattened.

PLATYPHYLLOUS. (1) Broadly lobed.
(2) Having wide leaves.

PLATYSPERM. A seed which is flattened in transverse section.

PLACTASCALES = ASPERGILLALES.

PIECTENCHYMA. A thick tissue formed by hyphae becoming twisted and fused together.

PLECTOBASIDIALES = SCLERODERMATALES.

PLECTOMYCETAE. An old class of the Ascomycetes, including all the genera that do not form apothecia or perithecia.

PLECTOSTELE. An actinostele in which the xylem is divided into a number of longitudinal plates.

PLEIANDROUS. Having a large and indefinite number of stamens.

PLEIO-, PLEO-. A Greek prefix meaning 'several'.

PLEIOCHASIUM. A cymose inflorescence in which each branch bears more than 2 lateral branches.

PLEIOMEROUS. Having a large number of parts or organs.

PLEIOSPOROUS. Many-spored.

PLEIOTAXY. An increase in the number of whorls in a flower.

PLEIOTOMY. Multiple apical division with the production of multiplets.

PLEIOTROPY. The production of more than one physiologically uncorrelated effect by one gene. This is attributed to one gene initiating two or more chains of reactions.

PLEIOTROPISM. The condition when one factor has an effect simultaneously on more than one character in the offspring.

PLEISTOCENE. A geological period lasting approximately from 1 million to 10,000 years ago. During it the four major Ice Ages occurred.

PLEOMORPHIC. (1) Having more than one independent spore-stage or form in the life-cycle.
(2) Of dermatophytes, changes due to 'degeneration' in culture.

PLEOPHAGOUS. Said of a parasite which attacks several species of host plant.

PLEOSPORACEAE. A family of the Pseudosphaeriales. The stromata are small, perithecium-like, subepidermal, often eventually external. The asci are mostly parallel with pronounced masses of stromatic tissue between them.

PLEROME. The central core of tissue in the growing point of a stem or root of vascular plants. It gives rise to the vascular cylinder and pith.

PLEROTIC. Said of the oöspores of the Pythiaceae which fill the oögonium.

PLETHYSMOTHALLUS. A dwarf stage produced by some brown algae. It bears reproductive organs which produce either unilocular or neutral spores.

PLEUR-, PLEURO-. A Greek prefix meaning 'side-'.

PLEURACROGYNOUS. Produced at the tip and also at the sides.

PLEUROCARPOUS. (1) Said of a moss sporangium which is borne laterally on rudimentary branch.
(2) Having the fruit in a lateral position.

PLEUROGENOUS. Borne in a lateral position.

PLEUROMEIACEAE. A family of the Isoetales. Sporophytes had an upright unbranched stem, with leaves at the upper end and a massive rhizophore. The sporophylls which bore sporangia on the abaxial surface, were in definite stobili.

PLEURORHIZAL. Said of the embryo of a flowering plant, when the radical is placed against the edges of the cotyledons.

PLEUROSPOROUS. Bearing the spores on the sides.

PLEUROZIACEAE. A family of the Acrogynae. The lateral 'leaves' are unequally lobed and the ventral lobes are saccate.

PLICATE. Folded like a fan.

PLICATE AESTIVATION. A type of valvate aestivation in which the perianth segments are plicate.

PLIOCENE. A geological period which is a subdivision of the Tertiary. It lasted approximately from 15 until 1 million years ago.

PLOCOSPERMACEAE (H). A family of the Apocynales (H). The pollen is granular. The fruit is of 2 fused carpels and dehisces along both sides. There are 4 ovules in pairs. The lower pair are erect, and the upper pendulous. The seeds have a tuft of hairs at the apex, and the style is twice lobed in the upper part.

PLUMBAGINACEAE (EP, BH, H). A family of the Plumbaginales (EP), Primulales (BH, H). These are shrubs, undershrubs, or herbs, with simple leaves, often with water- or chalk-secreting glands, and a compound inflorescence or bisexual flowers. The corolla lobes are free or fused. The stamens are in 1 whorl. The superior ovary consists of 5 fused carpels with 5 stigmas. There is 1 loculus with 1 ovule. The endosperm is starchy.

PLUMBAGINALES (EP). An order of the Sympetalae (EP). It contains the single family—the Plumbaginaceae.

PLUMBEUS. Latin meaning 'lead-coloured'.

PLUME. A light, hairy or feathery appendage on a seed or fruit, aiding in wind-dispersal.

PLUMED DISSEMINULE. A fruit or seed bearing a plume.

PLUMOSE. (1) Like a feather.
(2) Hairy.

PLUMULE. The terminal bud of an embryo in seed plants. It is a rudimentary shoot.

PLURI-. A Latin prefix meaning 'many'.

PLURICELLULAR. Composed of two or more cells.

PLURILOCULAR. Said of a sporangium or ovary which is divided by septa into several compartments.

PLURISPOROUS. Having two or more spores.

PLURIVALENT. In certain types of cell-division, said of compound chromatin rods formed of more than two chromosomes.

PLURIVOROUS. Attacking a number of hosts, or living on a number of substrates.

PLUS STRAIN, + STRAIN. One of the two strains of a heterothallic mould, often distinguished from the corresponding (− strain) by its stronger growth.

P.M.C. = POLLEN MOTHER CELL.

PNEUMATHODE. An outlet of the ventilating system of a plant, it usually has some loosely packed cells on the surface of the plant; through it exchange of gases is facilitated.

PNEUMATOPHORE. A specialized root which grows vertically upwards into the air from roots embedded in the mud, and, being of loose construction make gaseous exchange possible for the submerged roots.

PNEUMOTAXIS. (1) The response of reaction of an organism to the stimulus of carbon dioxide in solution.
(2) The response to the stimulus of gases in general.

PNEUMOTROPISM = PNEUMOTAXIS.

Po. A flower class, containing flowers which offer pollen only to visitors.

POCULIFORM. Cup-shaped.

POD. A dry fruit formed from a single carpel, having a single loculus containing 1 (rarely) to many seeds, and usually opening at maturity by splitting along both ventral and dorsal sutures.

PODAXACEAE. A family of the Lycoperdales. A stipe is present. The basidia are in clusters on the glebal hyphae. The basidia and their supporting hyphae are not destroyed at maturity, but go to make up the powdery contents of the sporocarp.

PODETIUM. An upright secondary thallus of lichens. It is stalk-like, cup-shaped, or much branched.

PODO-. A Greek prefix meaning 'stalk'.

PODOACEAE (H). A family of the Sapindales (H). These are spineless trees or shrubs with alternate leaves and no stipules. There are more than two sepals. A disk is present. There are usually 8 stamens inserted within the disk or unilaterally. There are few axile ovule. The seeds are not arillate, and the embryo is not twisted.

PODOCARP. A stalk to a carpel.

PODOPHYLLACEAE (H). A family of the Ranales (H). There is no disk, and the torus is sometimes enlarged with enclosed carpels. There are as many, or twice as many stamens as there are petals. The anthers are introrse. The ovary consists of 1 carpel. The ovules are basal or axile and inserted on the adaxial suture of the carpels. The seeds are not arillate.

PODOSTEMACEAE (EP, BH, H). A family of the Rosales (EP), Multiovulatae aquaticae (BH), Podostemales (H). These are aquatic herbs found in rushing water, usually in the tropics. The flowers are regular or zygomorphic, bisexual, and without a perianth. There are many to 1 stamens which may be free or united. The ovary consists of 2 fused superior carpels, with 2-1 loculi and a thick central placenta. There are many or few anatropous ovules. The fruit is a capsule.

PODOSTEMALES (H). An order of the Herbaceae. These are submerged fresh-water herbs, looking rather like mosses. There are no petals. The 4-1 stamens are free or partly united. The ovary is syncarpous with parietal or central placentas. The seeds are minute with no endosperm.

PODSOL, PODZOL. A type of soil profile developed under good drainage, relatively cool climate, and high rainfall. These conditions lead to leaching so that this soil is typified by having a layer of humus on top, below which is a greyish, leached zone, below which is a darker layer where the leached salts have accumulated.

-POGON. A Greek suffix meaning 'a beard of hair'.

POINT MUTATION = MUTATION.

POISEUILLE'S LAW. A law describing the rate of mass flow: $dm = -(r^4/8n).(dp/dx).dt$, where dm = the quantity of liquid passing a given plane in a tube, r = radius of the tube, dp/dx = pressure gradient, dt = time, n = coefficient of viscosity of the liquid.

POIUM. A meadow formation.

POL. The pole of a resting nucleus which lies nearest to the centrosome.

POLAR. At the end, especially of spores, bacteria etc.

POLAR CAP. A group of fine protoplasmic strands formed early in division at the pole of a dividing nucleus, and contributing to the formation of the spindle.

POLAR CLEFT. A linear expansion of the outer tissue of the raphe of some Chrysophyta.

POLAR FUSION NUCLEUS. The nucleus formed in the embryo sac by the union of the two polar nuclei, later it fuses with a male nucleus to give rise to the first endosperm nucleus.

POLAR PYRENOID. A pyrenoid which is not wholly enveloped in a sheath of starch grains.

POLARIBILOCULAR, POLARILOCULAR = ORCULIFORM.

POLARILOCULAR SPORE = BLASTENIOSPORE. A two-celled spore with a very thick median septum traversed by a canal.

POLARINUCLEATE. Said of a spore which has an oil-droplet at each end.

POLARIZED LIGHT. Light in which the waves oscillate in one plane only.

POLARIZATION. (1) Of chromosomes at telophase of mitosis and later, the maintenance of their proximal parts on the polar side of the nucleus.

(2) Of chromosomes at zygotene, the movements of their ends towards the part or parts of the nuclear surface where the centrosomes lie.

(3) Of centromeres, the initiation of orientated division during mitotic metaphase.

POLAR NODULE. (1) A swelling at the end of the raphe of the Chrysophyta.

(2) Button-like thickening of wall material, filling the pores between heterocysts of the Cyanophyta.

POLAR NUCLEI. Two nuclei in the embryo-sac which unite to give the polar fusion nucleus.

POLE. (1) One end of an elongated spore.

(2) One end of the achromatic spindle, where the spindle fibres come together.

POLEMONIACEAE (EP, BH, H). A family of the Tubiflorae (EP), Polemoniales (BH, H). These are usually herbs with alternate or opposite exstipulate leaves. The flowers are bisexual, usually regular, with the flower parts in fives. The carpels are fused and superior. There may be 3, 2, or 5, each with many to 1 erect ovules. The fruit is a capsule, and the seeds contain endosperm.

POLEMONIALES (BH, H). An order of the Gamopetalae (BH), Herbaceae (H). These are herbs, rarely shrubs or climbers. The corolla is regular and sympetalous. The stamens are the same number as the corolla-lobes, alternating with them, and joined to the petals. The ovary is superior, with 1-5 loculi. There are many-to-few ovules on parietal or axile placentas.

POLIOPLASM. Granular protoplasm.

POLITUS. Latin meaning 'polished'.

POLLEN. The microspores of Gymnosperms and Angiosperms.

POLLEN CHAMBER. The cavity formed in the apex of the nucellus of Gymnosperms, in which pollen grains lodge after pollination has occurred. The pollen grains slowly develop there and ultimately bring about fertilization.

POLLEN FLOWER. A flower which produces no nectar, but liberates large amounts of pollen, which attracts insects.

POLLEN MOTHER CELL = MICROSPOROCYTE.

POLLEN SAC. A cavity in an anther where the pollen is formed.

POLLEN TUBE. A tube formed on germination of a pollen grain that carries male gametes to the egg, and one to the central fusion nucleus with which it fuses to form the endosperm nucleus from which arises the endosperm.

POLLINATE. To transfer pollen from the anther to the stigma.

POLLINIUM. A mass of pollen grains held together by a sticky substance, and transported as a whole during pollination.

POLLINODIUM = ANTHERIDIUM.

POLLUTION CARPET. In stagnant and polluted water, a slimy layer occurring on the bottom. It consists mainly of bacteria, detritus feeding protozoa, and fungi.

POLY-. A prefix meaning 'many'.

POLYADELPHOUS. Said of an andrecium when several of the stamens are joined by their filaments to form several bundles.

POLYANDRY. (1) Having a large and indefinite number of stamens.
(2) Of oöspores formed when more than one functional antheridium is present.

POLYARCH. Said of a stele which has many protoxylem strands.

POLYASCUS. Having many asci, especially when they are in one hymenium and not separated by sterile bands.

POLYCARPELLARY. Consisting of many carpels.

POLYCARPIC. Able to fruit many times in succession.

POLYCARPOUS. Having a gynecium made up of several free carpels.

POLYCENTRIC. (1) Having many centres of growth and development, and more than one reproductive organ.
(2) Of a chromosome or chromatid which has several centromeres.

POLYCEPHALOUS. Many headed.

POLYCHASIUM. A cymose inflorescence in which the branches arise in groups of 3 or more at each node.

POLYCHLAMYDEOUS CHIMAERA. A periclinal chimaera in which the skin is more than two layers of cells thick.

POLYCLINAL CHIMAERA. A chimaera which is made up of more than two components.

POLYCORMIC. Said of a woody plant which has several strong vertical trunks.

POLYCOTYLEDONOUS. Having more than 2 cotyledons.

POLYCYCLIC. Said of a stele in which there are 2 or more concentric rings of vascular tissue.

POLYEDER. A solitary resting cell which develops from a zoospore of some of the Hydrodictyaceae. They grow and form zoospores which remain within a vesicle, ultimately forming a vegetative colony.

POLYEMBRYONY. The production of more than 1 embryo within the testa of 1 seed of a flowering plant, either from 1 or several zygotes, the extra zygotes being sexual or parthenogenic, form reduced or unreduced eggs. The extra eggs may be derived from vegetative nucellar cells, sister mother-cells, sister spores, or sister nuclei within 1 embryo-sac.

POLYENERGID = COENOCYTE.

POLYENERGID NUCLEUS. A nucleus which contains several sets of chromosomes.

POLYGALACEAE (EP, BH, H). A family of the Geraniales (EP), Polygalinae (BH), Polygalales (H). These are herbs, shrubs or trees, with simple, entire usually alternate exstipulate leaves. The bisexual zygomorphic flowers are racemes, spikes or panicles. There are usually 5 free sepals, with 2 larger and petalloid. There are 3 free petals, with 1 often keel-like. There are usually 2 whorls of 4 stamens (or fewer) which are fused low down. The superior ovary usually is of 2 fused carpels and 2 loculi with 1 ovule in each. The fruit is a capsule, nut or drupe, and the seed may or may not have endosperm.

POLYGALALES (H). An order of the Lignosae (H). See *Polygalaceae*.

POLYGALEAE (BH). = POLYGALACEAE (EP).

POLYGALINAE (BH). An order of the Polypetalae (BH). These are herbs or shrubs wih exstipulate leaves. There are usually 5 sepals and petals (rarely 4 or 3). The stamens are as many, or twice as many as the petals. The ovary is bilocular (rarely one or more). The endosperm is fleshy (rarely absent).

POLYGAMOUS. (1) Having the antheridia and archegonia variously disposed in the same species.
   (2) Having male, female and bisexual flowers on the same plant.

POLYGENES. (1) Members of a polygenic system.
   (2) Genes whose differences or mutations are too slight to be identified by their individual effects in an individual, and are therefore presumed to have only a small effect on the selective advantages of the individual.

POLYGENIC BALANCE. A condition produced by the adjustment of the proportions of polygenes having opposite effects. A balanced set of polygenes within a chromosome is a *balanced polygenic combination*. Where the balance is achieved within a single representative of the chromosome, and is hence displayed by a homozygote, it is an *internal balance*, and where the balance is achieved by two homologous chromosomes acting together, and hence displayed by a heterozygote, it is a *relational balance*.

POLYGENETIC SYSTEM. Said of genes having effects similar and supplementary to each other, and small in comparison to the total variation. The members of such a system are replaceable in their effects.

POLYGONACEAE (EP, BH, H). A family of the Polygonales (EP, H), Curvembryae (BH). The leaves are usually ochreate. The flowers are haplo- to hetero-chlamydeous, bisexual and regular. The ovary is superior with 1 loculus, which usually contains 1 basal, erect ovule. The fruit is a nut, and the seeds contain endosperm.

POLYGONALES (EP, H). An order of the Archichlamydeae (EP), Herbaceae (H). There is one family, the Polygonaceae, (*q.v.*).

POLYGYNOUS. Said of a flower having many distinct styles.

POLYHAPLOID = HAPLOPOLYPLOID.

POLYMER. A compound composed of two or more smaller units of the same type. The combination is usually brought about by a simple condensation, and the polymer has the same empirical formula as the individual unit.

POLYMERIC GENES. Genes that are not allelomorphic, but have the same, and possibly a cumulative effect, on the phenotype.

POLYMERY. The condition when a whorl consists of many members.

POLYMITOSIS. The intercalation of rapid supernumerary mitosis in the life-cycle immediately after meiosis, with or without division of the chromosomes.

POLYMORPHISM. (1) The occurrence in the same habitat of 2 or more distinct forms of a species in such proportions that the rarest cannot be supposed to be maintained by recurrent mutation from any other.
(2) Of a fungus having more than one spore-form.

POLYNUCLEATE = MULTINUCLEATE.

POLYNUCLEOTIDE. A unit made up of several nucleotide units.

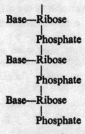

POLYOICIUS. Said of a bryophyte in which antheridia and archegonia may occur on the same plant, or on separate plants.

POLYOXYBIONTIC. Requiring copious supplies of oxygen.

POLYPEPTIDE. A condensation product of dipeptides.

POLYPETALAE (BH). A division of the Dicotyledons. The flowers are usually with 2 whorls of perianth, the inner polyphyllous.

POLYPETALOUS. Said of a flower which has the petals free from each other.

POLYPEPTIDE. A peptide formed of three or more amino-acid units, but never as many as make a peptone.

POLYPHAGY. Said of a parasitic fungus which has the thallus branching through and attacking several host cells at the same time.

POLYPHENOL. A compound made up of several phenol units.

POLYPHENOL OXIDASE. An enzyme bringing about the oxidation of polyphenols.

POLYPHYLETIC. A group of species classified together, is polyphyletic when some of its members have had quite distinct evolutionary histories, not being descended from a common ancestor which is also a member of the group: so that, if the classification is to correspond to phylogeny, the group should be broken-up into several distinct groups.

POLYPHYLOUS. Consisting of members which are separate from one another.

POLYPLANETISM. Said of zoospores, when they have resting and motile phases alternating.

POLYPLOID. An individual having 3 or more times the haploid number of chromosomes.

POLYPODIACEAE. A family of the Leptosporangiatae. The sorus is mixed. The sporangium has a vertical annulus and a definite stomium in the region of transverse dehiscence. The jacket-layer of the antheridium is composed of ring-shaped cells.

POLYPORALES. A class of the Eubasidiae. The hymenium is exposed from its inception, and is never on distinct gills. The basidiospores are perched obliquely on the basidium.

POLYPORACEAE. A family of the Polyporales. They are distinguished by the hymenium being distributed over deep or shallow pores.

POLYSACCHARIDE. A condensation product of monosaccharides. They are insoluble in water. The general formula is $(C_nH_{2n-2}O_{n-1})_x$, where n = 4-6 and x may be several thousands.

POLYSAPROBE. An organism able to live in heavily contaminated water.

POLYSEPALOUS. Said of a flower which has the sepals free from one another.

POLYSIPHONOUS. Said of an algal tallus consisting of a central row of elongated cells surrounded by one or more layers of peripheral cells.

POLYSOMIC. Said of an otherwise diploid individual which has one chromosome represented more than twice.

POLYSOMIC INHERITANCE. This arises where any chromosome in a polyploid or polysomic individual has more than one possible partner at meiosis.

POLYSPERMOUS. Containing many seeds.

POLYSPERMY. The entrance of more than 1 antherozoid into an egg.

POLYSPORE. (1) A multicellular spore.
(2) Of the Rhodophyta, an asexual spore, more than one of which is produced in a sporangium.

POLYSPOROUS. Many spored.

POLYSPORY. The formation of more than the normal number of spores.

POLYSTELIC. Having more than one stele.

POLYSTEMONOUS = POLYANDROUS.

POLYSTICHINEAE. A sub-class of the Heterogeneratae. There is a paren-chymatous thallus produced by vertical and transverse divisions of intercalary cells. The sporophyte may produce zoospores or neutral spores. The gametophyte is always macroscopic, isogamous, anisogamous or oögamous.

POLYSTICHOUS. Arranged in several rows.

POLYSTOMELLACEAE. A family of the Hemisphaeriales. The ascocarp has a more or less spherical structure, and are produced externally on strands of hyphae from the internal mycelium.

PLYTOMOUS. Having several branches arising at the same level.

POLYTRICHINALES. An order of the Eubrya. The capsules are angular in cross-section, and the peristome is several cells thick.

POLYURONIDES. These are gums, found in the soil, especially in the humus fraction. 10-30% of the organic carbon in soil is in the form of polyuronides.

POME. A false fruit, the greater part of which is developed from the receptacle of the flower, and not from the ovary.

POMIFORM. Apple-shaped.

POMOLOGY. The study of cultivated fruits and fruit trees.

POMONA. An account of fruits.

PONTEDERIACEAE (EP, BH, H). A family of the Farinosae (EP), Coronarieae (BH), Liliales (H). These are water plants often with two-ranked leaves, and spikes of bisexual, zygomorphic flowers. The perianth has 2 whorls of 3 free members, and a long tube. The 6 (3 or 1) stamens are free and attached to the tube. The superior ovary has 3 fused carpels and 3 loculi and several ovules, or 1 loculus with 1 ovule. The fruit is a capsule or nut, and the seeds are endospermous.

POPULATION. (1) A number of the same species living in a definite area.
(2) Genetically, a mating group, limited for special consideration, either by environment or breeding.
(3) Statistically, the hypothetical infinitely large series of potential observations or individuals, of which those observations or individuals actually obtained from a sample.

PORANDROUS. Said of anther which open by pores, and not slits.

PORE. (1) A small hole.
(2) The aperture of a stoma.
(3) The ostiole in the Pyrenomycetes.
(4) One of the tubular cavities lined by basidia in the pore-bearing fungi.
(5) A minute vertical canal running through the areola of some of the centric diatoms.

PORELLACEAE. A family of the Acrogynae. The lateral 'leaves' are unequally lobed and the capsule is incompletely dehiscent.

PORE PLATE. The single small asymmetrical plate in the epitheca of the Dinophysidales.

PORE SPACE. The spaces between soil-particles, which is filled with air or water.

PORICIDAL. Said of anthers which open by pores.

POROGAMAE. A division of the Angiospermae, including the members in which fertilization takes place through the micropyle.

POROGAMY. The entry of the pollen tube through the micropyle.

POROID. (1) Having more or less obvious pores.
(2) A fine incomplete canal running through the areola of some centric diatoms.

POROMETER. An instrument for measuring the rate at which air passes through a portion of a leaf. It is a means of measuring the degree to which the stomata are open.

POROSE. Said of cells which are pierced by pores.

POROUS DEHISCENCE. The liberation of pollen from anthers, or seeds from fruits through pores in the wall of the containing structure.

PORPHYROPECTIC. A type of soil aggregate in which the minerals are uncoated, and in a massive, or at least only finely cracked, ground mass, which separates easily. It occurs in laterites.

PORPHYROPEPTIC. A type of soil aggregate in whch the minerals are always covered by colloidal coatings or thick films, with the appearance of being embedded in a ground mass. They are characteristic of Mediterranean red earths.

PORRECT. Extending forward.

PORTULACACEAE (EP, BH, H). A family of the Centrospermae (EP), Caryophyllinae (BH), Caryophyllales (H). These are herbs or undershrubs with fleshy leaves, and often hair-like stipules. The regular bisexual flowers are in cymes. There are usually 2 free sepals and 5-4 free petals. The stamens are free, in 2 whorls of 5, 1 whorl of 5, or fewer of many. The ovary is superior, or semi-inferior, and consists of 5-3 fused carpels and 1 loculus with 2 to many ovules on a basal placenta. The fruit is a capsule, and the seeds have endosperm.

POSIDONIACEAE (H). A family of the Juncaginales (H). These are submerged marine perennials with the rhizome or stem densely covered with persistent leaf-bases. The leaves are sheathing at the base, and the sheaths are open and ligulate. The flowers are bisexual and in spikes. There are no bracts or perianth (may be 3 scales). There are 4-3 hypogynous stamens. The anthers are large, extrorse and sessile, with a thick connective produced beyond the loculi which are separate. The pollen is thread-like. The ovary is unilocular and superior. The fruit is fleshy and indehiscent. There is no endosperm.

POSITION EFFECT. The difference in effect of 2 or more genes according to their spatial relations in the chromosome.

POSITIVE REACTION. A taxis or tropism from a region where the stimulus is weaker to one where it is stronger.

POSTCINGULAR PLATE. One of a series of plates next to the girdle, in the hypotheca of the Peridiniales.

POSTERIOR. (1) Inserted on the back of another organ.
(2) The part of the flower nearest the axis.
(3) The rear.

POST-FERTILIZATION STAGES. The developmental processes which go on between the union of the gametic nuclei in the embryo-sac and the maturity of the seed.

POST-FLORAL MOVEMENT. A change in the position of the flower stalk or inflorescence stalk after fertilization has occurred, so as to bring the fruit into a more favourable position for development, or placing the seeds in good conditions for germination.

POSTICAL. Relating to or belonging to the back or lower part of a leaf or stem.

POSTICOUS. Outward or behind.

POSTMEIOTIC DIVISION. The first nuclear division after meiosis has been completed.

POST-REDUCTION. As opposed to *pre-reduction*, the segregation of differences between partners at the second, as opposed to the first meiotic division.

POSTVENTITIOUS. Delayed in development.

POTALIACEAE (H). A family of the Loganiales (H). There are 16-5 contorted corolla lobes. There are 4 or more stamens. The fruit is a berry with the seeds often embedded in the pulp. The seeds are never winged.

POTAMIUM. A river formation.

POTAMOGETONACEAE (H). A family of the Potamogetonales. These are fresh-water plants with a perianth of 4 free segments, which are clawed and valvate. The ovules are lateral and the fruiting carpels sessile.

POTAMOGETONALES (H). An order of the Calyciferae (H). These are fresh-water or marine perennials, with rhizomes. The leaves are alternate or opposite, sheathing at the base. The leaf-sheath is often ligule-like, and linear to nearly orbicular at the apex. The flowers are small single, in spikes or racemes. They may be bisexual or unisexual without bracts. There are 4-3 or no perianth segments and 4-1 stamens. The anthers are bilocular extrorse and usually on very short filaments. The gynecium consists of several free carpels or 1, with 1 pendulous or lateral ovule. The fruit is indehiscent, and the seeds have no endosperm.

POTAMOPLANKTON. The plankton of rivers and streams.

POTAMOUS. Living in rivers and streams.

POTENCE. The property of a group of polygenes, corresponding to the degree of dominance of the major gene.

POTOMETER. An instrument for measuring the rate of water-uptake of a shoot.

POTTIALES. An order of the Eubrya. The gametophores are generally erect and bear many-ranked leaves with mid-ribs. The sporophytes are acrocarpic. The capsule has a peristome, and is simple with 16 teeth.

POUSSIEROID. Said of a stage of meiotic division, prior to prophase in which the chromatin is distributed as fine granules.

PPM. Parts per million.

PRAECOX. Latin meaning 'appearing early'.

PRAEMORSE. As if bitten off.

PRAIRIE. The grass country east of the Rocky Mountains.

PRAIRIE SOILS. Soils developed under grassland. They develop under podsolic conditions, but they are far to the north so that the severity of the climate precludes tree-growth.

PRATAL. Growing in meadows.

PRATENSIS. Latin meaning 'of meadows'.

PRECINGULAR PLATE. One of the series of plates adjoining the girdle in the epitheca of the Peridinales.

PRECOCITY. The property in the nucleus at meiosis of beginning prophase before the chromosomes have divided. *Differential precocity* is when some chromosomes, or their parts, condense, divide, or pair before the rest of the chromosomes during prophase.

PRECOCITY THEORY. That meiosis is primarily distinguished from mitosis by showing a precocity of prophase, which successively determines pairing, crossing-over, chiasma formation and non-division of the centromeres. A second division is inserted together with the reduction of chromosome number and the segregation of differences.

PREFOLIATION = VERNATION.

PRELAMELLAR CHAMBER. A transverse chamber developed in the lower surface of the pileus of agarics. In this the gills and hymenium ultimately develop.

PREMEIOTIC MITOSIS. The nuclear division immediately preceding the organization of nuclei which will divide by meiosis.

PREMORSE. Looking as if the end has been bitten off.

PREPOTENCY. The ability of some pollen to bring about fertilization more readily than other pollen.

PREREDUCTION. Disjunction in the heterotype division.

PRESENCE AND ABSENCE THEORY. That the dominance of an allelomorph is due to the presence of something in it, which is absent from its alternative.

PRESENTATION TIME. The minimum time that an organism has to be exposed to a stimulus before there is a perceptible response.

PREVENTITIOUS BUD. A dormant bud which will produce an epicormic bud under suitable conditions.

PREVERNAL. Flowering early in the year.

PREVERNAL ASPECT. The condition of the vegetation of a community very early in the year.

PRICKLE. A hard pointed dermal appendage, that does not contain a vascular bundle.

PRIMARY. (1) First formed.
(2) Most important.

PRIMARY AXIS. (1) The main shoot of a plant.
(2) The main stalk of an inflorescence.

PRIMARY BODY. The part of the plant formed directly from cells cut off from the apical meristem.

PRIMARY CELL WALL. The cell-wall that surrounds the protoplast until it is nearly mature, and fully grown. It is thin, usually non-stratified and contains less cellulose and more pectin than the mature wall. It later persists as the middle lamella.

PRIMARY CONSTRICTION = CENTROMERE.

PRIMARY INCREASE. The increase in the size of a stem or root, brought about by the addition of cells from the cambium.

PRIMARY LAMELLA. The first-formed layer of the wall of a spore.

PRIMARY MEDULLARY RAY. A vascular ray passing radially from the pith to the cortex.

PRIMARY MERISTEM. The region of active cell division which has persisted from time of its origin in the embryo.

PRIMARY MYCELIUM. The haploid mycelium growing from a basidiospore.

PRIMARY NODE. The node at which the cotyledons are inserted.

PRIMARY PHLOEM. The phloem formed from a procambial strand and present in a primary vascular bundle. It consists of protophloem and metaphloem.

PRIMARY SERE. A plant succession beginning on land which has never borne vegetation in recent geological time.

PRIMARY STRUCTURE. The simplest structure of a protein.

PRIMARY SUCCESSION. A succession starting from bare soil.

PRIMARY THICKENING. The first layers of wall material to be laid down on the very young cell-wall, often rich in pectic material.

PRIMARY TISSUE. Tissue formed from cells derived from the primary meristems.

PRIMARY TRISOMIC. A plant which has the ordinary diploid chromosome complement, together with one extra chromosome.

PRIMARY UNIVERSAL VEIL = PROTOBLEM.

PRIMARY UREDO = UREDIUM, UREDINIUM.

PRIMARY XYLEM, PRIMARY WOOD. The xylem formed from a procambial strand and present in a primary vascular strand. It consists of protoxylem and metaxylem.

PRIME TYPE. In *Datura*, one of the homologous types of chromosome structure distinguished from other types by interchange, and used as a basis of reference.

PRIMINE. The outer integument of an ovule.

PRIMITIVE. Original; first-formed; of early origin.

PRIMODIUM. The earliest recognizable rudiment of an organ or structure in development.

PRIMOFILICES. A sub-class of the Filicinae. They are fossil forms of the Palaeozoic. The sporangia are generally single at the tip of an ultimate dichotomy of a leaf, and are elongate with a jacket-layer of more than 1 cell thick. They open by a longitudinal slit or terminal pore.

PRIMORDIAL. Primitive.

PRIMORDIAL CELL. A cell which has not yet formed a cell-wall.

PRIMORDIAL LEAF. (1) The next leaf formed after the cotyledons.
  (2) The small mass of tissue from which a leaf starts its development.

PRIMORDIAL MERISTEM = PROMERISTEM.

PRIMOSPORE. A spore very like a cell of a thallus.

PRIMULACEAE (EP, BH, H). A family of the Primulales (EP, BH, H). These are herbs with usually alternate, exstipulàte leaves and regular (rarely zygomorphic) bisexual flowers. The 5 sepals and petals are fused, and the 5 free stamens are joined to the petals alternating with the lobes. There are rarely 5 staminodes. The superior (rarely half-inferior) ovary is unilocular, with many ovules on a free-central placenta. The fruit is a capsule, and the seeds contain endosperm.

PRIMULALES (EP, BH, H). An order of the Sympetalae (EP), Gamopetalae (BH), Herbaceae (H). The flowers are bisexual or unisexual and regular (rarely zygomorphic). The flower parts are in fives (rarely many to 4), usually with 1 whorl of epipetalous stamens, and rarely 5 opposite the calyx lobes. The petals are usually fused. The ovary is superior to inferior, with 1 loculus and many to 1 ovules on basal or free-central placentas.

PRINCIPES (EP). An order of the Monocotyledoneae. These are tree-like or woody plants, sometimes climbing, with fan or feathery leaves, and regular usually unisexual flowers in spikes, or spike-like racemes, usually in a spathe. The perianth has 2 whorls of three free lobes, and 1, 2, or 3 whorls of 3 free stamens (or many). The ovary is of 3 free or fused carpels, usually with 1 ovule in each. The fruit is a berry or drupe, and the seeds are rich in endosperm.

PRISMATIC LAYER. A secondary tissue cut off by the cambium of the Isoetaceae. It lies internally to the cambium and has been variously interpreted as secondary xylem, or secondary phloem.

PROBABILITY. The proportion of an infinite (and hypothetical) series of cases in which an event, possible in any one of them, actually occurs.

PROBABILITY FUNCTION. The function relating observations to the parameter from whose value their frequencies may be predicted. A *simple probability function* relates to the occurrence of a single event, and a *compound probability function* relates to more than one event.

PORBAND = PROPOSITUS.

PROBASIDIUM. A teliospore, or the basidium from nuclear fusion to the start of the epibasidium, sterigma, or spore development in any basidiomycete.

PROCAMBIAL STRAND = PROCAMBIUM.

PROCAMBIUM. A tissue of elongated, narrow cells grouped into strands, differentiated in the plerome just behind the growing point in stem and root of vascular plants. It gives rise to the vascular system by further development.

PROCARP. The multicellular female organ of the Rhodophyta, consisting of an archcarp and trichogyne.

PROCERUS. Latin meaning 'lofty'.

PROCESS. A general term for an extension or projection.

PROCHROMOSOME. A body of heterochromatin seen in the resting stage of a nucleus.

PROCUMBENT. Said of a stem which lies on the ground for all, or most of its length.

PRODIPLOIDIZATION HYPHA. A hypha which may be diplodized.

PROEMBRYO. (1) = PROTONEMA.
 (2) The group of cells, few in number, formed as the zygote begins to divide, and from one or some of which the embryo-proper is organized.

PROENZYME = ZYMOGEN.

PROFILE POSITION. A position assumed by chloroplasts and by leaves when the edge of the structure is turned towards the position from which the brightest light is coming.

PROGAMETANGIUM. A short side-branch developed in the Mucorales, from which a gametangium is cut off by a transverse septum.

PROGENY TEST. The method of assessing the genetic character of an individual by the performance of its progeny.

PROGRESSIVE CLEAVAGE. The cleavage of the nucleus or cells one after the other to give 2, 4, 8 etc. daughter-cells one after the other.

PROHYBRID. A mycelium having additional nuclei from hyphal fusions.

PROLAMINE. A simple protein, insoluble in water and saline solution, but soluble in 70% ethanol.

PROLATE. Somewhat globular, but flattened equitorially.

PROLETARIAN. A plant having little or no reserves of food material.

PROLIFERATION, PROLIFICATION. (1) A renewal of growth in a mature organ after a period of inactivity.

(2) The production of vegetative shoots from a reproductive structure.

(3) The formation of a sporangium inside the empty walls of a previously discharged sporangium.

(4) The production of off-shoots which may become detached and established as new plants.

PROLIFEROUS. (1) Bearing off-shoots.

(2) Producing abnormal or supernumerary outgrowths.

(3) Producing progeny by means of off-shoots.

(4) Showing excessive development in some respect.

PROLIFICATION. The development of buds in the axils of sepals and petals.

PROLINE. An imino acid, a frequent product of protein hydrolysis.

$$H_2C\text{------}C\text{---}H_2$$
$$H_2C\qquad C\text{---}H\text{---}COOH$$
$$N$$
$$H$$

PROLONGED. Drawn out into a long point which is not hollowed along its sides.

POMERISTEM. The meristem in an embryo and at the growing point. It consists of actively dividing, uniformly shaped cells.

PROMETAPHASE. The stage between the dissolution of the nuclear membrane, and the aggregation of the chromosomes on the metaphase plate.

PROMITOSIS. The special type of nuclear division during the growth stage in the Plasmodiophoraceae.

PROMYCELIUM = EPIBASIDIUM. A short germ-tube, put out by some fungal spores, on which other spores of different types develop.

PROPAGATION. The increase in the number of plants by vegetative means.

PROPAGULUM. A small reproductive branch which becomes detached from the parent, and grows into a new plant. Referred especially to such structures found in the brown algae.

PROPER EXIPLE. A hyphal layer around the fructification of lichen. It contains no algal cells.

PROPHASE. The stage in mitosis or meiosis when the chromosomes appear within the nucleus, and, in meiosis undergo pairing.

PROPHYLL = BRACTEOLE.

PROPLASTID. A minute, self-reproducing inclusion in the cytoplasm from which a plastid may develop.

PROPOSITUS. The individual through which a pedigree is ascertained.

PROP ROOT. A root formed from a stem, usually close to the ground. It helps to support the stem.

PROSENCHYMA. (1) A tissue composed of cells with pointed ends. The cells are often empty, and are concerned with affording support and with conducting material.

(2) Plectenchyma in which the hyphae are evident and distinct.

PROPLECTENCHYMA. A false tissue of elongated fungal hyphae.

PROSORUS. In the Chytridiales, a cell giving rise to a group of sporangia.

PROSPORANGIUM. Of Phycomycetes, a sporangium-like body which puts out a vesicle (sporangium) in which zoospores may develop, and from which they are freed.

PROSPORY. The formation of sporangia on a very young plant.

PROSTHETIC GROUP. A non-protein group attached to a protein. If the protein is an enzyme, the prosthetic group is essential for its functioning.

PROSTRATE = PROCUMBENT.

PROTANDROUS. (1) The male gametes developing before the female gametes.

(2) The anthers ripening before the stigma(s) of the same flower is receptive.

PROTAMINE. A simple protein of relatively low molecular weight, and a high proportion of bases.

PROTEACEAE (EP, BH, H). A family of the Proteales (EP, H), Daphnales (BH). These are woody plants with alternate exstipulate leaves and spikes or racemes of cyclic, homochlamydeous (apparently haplochlamydeous) flowers. The flowers are bisexual or unisexual, regular or zygomorphic. the petaloid perianth is in 2 whorls of 2 free lobes. The stamens are anteposed, and usually attached to the perianth. The superior ovary consists of 1 carpel. The fruits are of various types, and the seeds contain no endosperm.

PROTEALES (EP) (H). An order of the Archichlamydeae, containing the single family, the Proteaceae (*q.v.*). (H) places it in the Lignosae.

PROTEASES. Enzymes attacking the peptide bond in protein molecules.

PROTECTIVE LAYER. A layer of suberized cells lying across the place where a leaf comes away at leaf-fall. It checks water-loss, and the entry of parasites.

PROTEIN. A compound containing carbon, hydrogen, oxygen, and nitrogen, and frequently sulphur and phosphorus. They are synthesized from amino-acids and are one of the main constituents of protoplasm. They are mainly α-amino-acid residues joined by peptide linkages.

PROTEINASES = PROTEASES.

PROTEOLYTIC ENZYME. Any enzyme taking part in the breaking down of proteins.

PROTEOSES. These are protein derivatives, soluble in water, and not coagulated by heat. They are precipitated by saturation with ammonium or zinc sulphate, and condense to form protein.

PROTERANDRY = PROTANDROUS.

PROTHALLIAL CELL. (1) A single cell cut off early in the division of a microspore of some heterothallic pteridophytes. It represents the vegetative tissue of the male gametophyte.

(2) A small cell in the pollen grain of the gymnosperms, representing the male prothallus.

PROTHALLUS, PROTHALLIUM. (1) = PROTONEM.

(2) The independent gametophyte plant of the Pteridophyta. It is a small, green parenchymatous thallus bearing antheridia and archegonia. It shows

little differentiation, and is usually prostrate on the soil-surface, to which it is attached by rhizoids. It may be subterranean and mycotrophic.

(3) The similar stages in the life-cycle of the gymnosperms.

(4) The earliest stages in the development of a lichen thallus.

PROTHECIUM. A primitive or rudimentary perithecium.

PROTISTA. A term sometimes used to include all unicellular organisms, both plant and animal.

PROTO-. A prefix meaning 'primitive' or 'primordial'.

PROTOAECIUM, PROTOPERITHECIUM, PROTOUREDIUM. A haploid structure which, after diploidization becomes a fruiting structure.

PHOTOARTICULATAE = HYENIALES.

PROTOASCOMYCETAE. A sub-class of the Ascomycetae. The ascus is formed directly from the zygote, and the asci are borne singly on a mycelium without any development of a sheath of sterile tissue.

PROTOBASIDIUM. A primitive basidium.

PROTOBLEM. A loose flocculent layer covering the universal veil.

PROTOCOCCACEAE. A family of the Ulotrichales. They may be unicellular or multicellular with the cells united in small groups. The chloroplasts are parietal and irregular in outline. There is no production of motile zooids.

PROTOCONIDIUM = HEMISPORE.

PROTOCORM. A tuber-like structure formed in the early stages of the development of club-mosses and some other plants which appear to live in close association with fungi when young. In the club-mosses, it develops from the distal 4 cells of the eight-celled embryo and produces rhizoids and conical *protophylls* which are leaf-like in function.

PROTODERM = DERMATOGEN.

PROTOGAMY. The union of gametes without the fusion of their nuclei.

PROTOGASTRACEAE. A family of the Protogastrales, in which there is no columella, and the hymenial cavity is more or less spherical.

PROTOGASTRALES. An order of the Gasteromyceteae. These are small, hypogeous or epigeous, with a single hymenial cavity lined by an even hymenium.

PROTOGONIDIUM. The first of a series of gonidia.

PROTOGYNOUS. (1) The development of the female gametes before the male.

(2) The stigma(s) of a flower being receptive before the anthers of the same flower discharge their pollen.

PROTOHYMENIAL. Having a primitive hymenium.

PROTOLYSIS. The decomposition of chlorophyll by light.

PROTOMYCETACEAE. The single family of the Protomycetales.

PROTOMYCETALES. An order of the Phycomyceteae, of doubtful validity. The walls of the mycelium respond to tests for cellulose. They are parasites, entirely within the tissues of the aerial parts of higher plants. The hyphae are coenocytic (with occasional cross-walls) and branched with large thick-walled intercalary or terminal resting spores, which produce numerous small spores within them on germination. The small spores escape and may unite in twos or threes and invade other host plants.

PROTONEMA. (1) One of the initial branches produced from a germinating spore of the Charales. One branch of the protonema ultimately produces the new plant.

(2) The filamentous or plate-like stage produced from the developing spore of the mosses. It produces the gametophore by vegetative means.

PROTOPECTIN. A substance similar to pectin, but consisting of longer chains, perhaps connected by metals. It is found in cell-walls, especially of algae.

PROTOPERITHECIUM. See *protoaecium*.

PROTOPHLOEM. The first phloem formed from the procambial strands. It may be generalized in structure with poorly formed sieve tubes and no companion cells.

PROTOPHYLL. A sterile leaf. See *protocorm*.

PROTOPHYTE. (1) The gametophyte in an alternation of generations when the two generations are unalike.

(2) A simple unicellular plant.

PROTOPLASM. The substance within and including the plasma-membrane of a cell or protoplast, but excluding vacuoles, cell-wall mass secretions. It is differentiated into the nucleus and cytoplasm.

PROTOPLASMIC CIRCULATION. The streaming motion that may occur in the protoplasm of living cells.

PROTOPLASMIC RESPIRATION. Respiration proceeding at the expense of protein materials in a starved plant.

PROTOPLAST. The protoplasm, as distinct from the cell-wall.

PROTOSIPHONACEAE. A family of the Chlorococcales. They have solitary, spherical to tubular multinucleate cells of which one side may be elongated into a colourless rhizoidal process. The cells may contain a single large reticulate parietal chloroplast or many small parietal chloroplasts. Some genera produce zoospores while others form biflagellate gametes only.

PROTOSOME. A hypothetical central body in a gene.

PROTOSPORE. Said of the Synchytriaceae when a uninucleate piece of protoplasm becomes the sporangium.

PROTOSTELE. The most simple and primitive form of stele. It consists of a central rod of xylem surrounded by a cylinder of phloem. It is present in the stems of some ferns and lycopods, and is found more or less universally in roots.

PROTOTHALLUS. The first stages in the formation of a lichen thallus, often before the fungus and alga have become associated. The term is sometimes applied to the fringe of hyphae growing out from the edges of a mature thallus.

PROTOUREDIUM. See *protoaecium*.

PROTOXYLEM. The xylem derived from the apical growing point. Its elements are extensible and become partly thickened before elongation is complete.

PROVIRUS. A plasmagene which may develop into a virus.

PROXIMAL. (1) Situated towards the point of attachment.

(2) The part of a chromosome arm which is nearer to the centromere than another part.

PRULAURASIN = MANDELONITRILE.

PRUINA. A powdery bloom or secretion on the surface of a plant.

PRUINOSE. (1) Covered with a waxy or powdery bloom.
 (2) Covered with minute points which give a frosted appearance to the surface.

PRUNASIN. A glucoside in the bark of wild cherry. It is d-mandelonitrile glucoside.

PRUNIFORM. Shaped like a plum.

PRURIENS. Latin meaning 'causing itching'.

PSAMMOPHILE. A plant which lives in sandy soil.

PSAMMOPHYTE. A plant which occurs only in sand.

PSEUDO-. A prefix meaning 'false'.

PSEUDOAETHALIUM. Of the myxomycetes, a group of separate sporangia looking like an aethalium.

PSEUDOAMITOSIS. An irregular nuclear division caused by treating the cells with poison.

PSEUDOAPOGAMY. The replacement of the normal fusion of sexual nuclei by the fusion of two female nuclei, a female nucleus with a vegetative nucleus, or two vegetative nuclei.

PSEUDOAPOSPORY. The formation of a spore without meiosis first taking place. The spore is then diploid.

PSEUDOAXIS = MONOCHASIUM.

PSEUDOBERRY. A fleshy fruit which looks like a berry, but in which some of the fleshy parts are derived from an enlarged perianth.

PSEUDOBULB. A swollen internode, formed by some orchids.

PSEUDOCAPILLITIUM. In the myxomycetes, a sterile structure in the fruit-body, which has no connection with the sporogenous protoplasm.

PSEUDOCILIUM. A long immobile cytoplasmic process produced by the cells near the edge of the colonies of the Tetrasporales.

PSEUDOCOLUMELLA. Of the Physaraceae, lime-knots massed like a columella in the centre of the sporangium.

PSEUDO-COMPATIBILITY. The occurrence, under exceptional circumstances of fertilization, such as would normally be excluded by incompatibility.

PSEUDOCYPHELLA. A cavity like a cyphella on the under side of the thallus of some lichens.

PSEUDOGAMY. (1) A union between two vegetative cells that are not closely related.
 (2) The development of an egg-cell into a new individual as the result of the stimulus of a male gamete, but without actual nuclear fusion.

PSEUDOIDIUM. A separated hyphal cell capable of germination.

PSEUDO-MARINE. Applied to freshwater forms bearing a superficial resemblance to marine types, but not necessarily closely related to them.

PSEUDOMIXIS. Fusion between two vegetative cells, or between cells which are not differentiated as gametes.

PSEUDOMORPH. A stroma made up of plant parts kept together by plectenchyma.

PSEUDOMYCELIUM. Said of yeasts etc. when the cells are loosely united in chains.

PSEUDOMYCORRHIZA. An association between a fungus and a higher plant in which the fungus is distinctly parasitic.

PSEUDONUCLEOLUS. A net knot.

PSEUDOPARAPHYSES = PARAPHYSOIDS.

PSEUDOPARENCHYMA. A mass of loosely interwoven filaments which looks like parenchyma in section.

PSEUDOPERIANTH. A cylindrical upgrowth, one cell thick, around a archegonium of the Marchantiaceae.

PSEUDOPERIDIUM. A sheath of sterile hyphae surrounding the aecidium of the Uredinales.

PSEUDOPERITHECIUM. (1) Of the Laboulbeniales, a perithecium-like structure in which asci and spores become free.
  (2) = PROTOPERITHECIUM.
  (3) The perithecium-like fruit-body characteristic of the Pseudosphaeriles.

PSEUDOPHYSIS. A paraphysoid structure in the Cyphellaceae, which is thin-walled, smooth, swollen at the end, and moniliform.

PSEUDOPLASMODIUM. A mass of closely associated uninucleate myxamoebae which retain their identity.

PSEUDOPODIUM. (1) A mobile and variable extension of the protoplasm.
  (2) A leafless stalk of the gametophyte of the Sphagnaceae. It bears the sporophyte, and remains short until the sporophyte is mature, then elongates.
  (3) A leafless branch formed by some mosses, and bearing gemmae.

PSEUDOPYCMIDIUM. A pycnidium-like structure of hyphal tissue, as in certain Fungi-Imperfecti.

PSEUDORAPHE. A longitudinal strip running the length of a valve of the Pennales. It lacks a longitudinal slit.

PSEUDORHIZA. A 'rooting' base, as in the fungus *Collybia radicata*.

PSEUDOSEPTUM. A septum which is perforated by one or more pores; found in the lower fungi.

PSEUDOSPERM. A small indehiscent fruit which looks like a seed.

PSEUDOSPHAERIACEAE. A family of the Pseudosphaeriales. The stromata are small and perithecium-like. They are internal or eventually external. The asci are more or less parallel, and are formed in monascal cavities which remain separated for a long time by the stromatal tissue which connect from the base to the top of the stroma.

PSEUDOSPHAERIALES. An order of the Ascomycetae. The ascocarps consist of perithecium-like stromata in which the asci are formed in separate monascal cavities, or the intervening tissues dissolve very early so that the asci form a parallel or diverging cluster in a single large cavity. There are no true paraphyses. The ostioles are formed by the breaking or dissolving of the apical parts of the stromata.

PSEUDOSPORE. (1) An encysted myxamoeba.
  (2) An obsolete term for a basidiospore.

PSEUDOSPOROCHNACEAE. A family of the Psilophytales. They were relatively large rootless shrubs with a conspicuous main trunk, ending in many dichotomously forked main branches, which bore many leafless branchlets with terminal sporangia.

PSEUDOSTIPULE. An appendage at the base of a leaf-stalk, which looks like a stipule but is really part of the lamina.

PSEUDOZYGOSPORE = AZYGOSPORE.

PSILIUM. A prairie formation.

PSILOPHYTACEAE. A family of the Psilophytales. The plant body is relatively small and differentiated into rhizome and aerial branches. The aerial branches are dichotomously forked, flattened or cylindrical and sometimes covered with spines. The sporangia have a jacket-layer several cells thick and are borne terminally on certain branches.

PSILOPHYTALES. An order of the Psilophytinae. These are fossil forms of the early to Mid-Devonian period. The rootless and dichotomously branched plant-body has subterranean and aerial portions. The sporangia are borne terminally on all or only some of the aerial branches which are leafless or have many small leaves.

PSILOPHYTINAE. A class of the Pteridophyta. The sporophytes are rootless, and the branches may or may not bear leaves. The sporangia are borne at the tips of elongated or greatly reduced branches.

PSILOTACEAE. A family of the Psilotales (*q.v.*).

PSILOTALES. An order of the Psilophytinae. The sporophyte is rootless and dichotomously branched, usually with a rhizome and aerial shoots which may be leafless, or have many small leaves. The sporangia are in twos or fours, directly on the stem or at the base of leaves. The gametophyte is subterranean and lacks chlorophyll. The antheridia lie partly embedded in the gametophyte.

PSYCHROPHILIC. Growing best at low temperatures.

PTERATE. Having wings.

PTERIDOPHYTA. A division of the Plant Kingdom. The sporophyte is vascular and independent of the gametophyte at maturity. Spores, or gametophytes formed from them are liberated from the sporophyte at maturity. Generally they have stems, leaves and roots. They are homosporous or heterosporous, and the gametophyte is smaller and simpler than the sporophyte.

PTERIDOSPERMAE. An ancient group of fossil plants. They were fern-like, but produced seeds, often apparently on the edges of the leaves.

PTERO-. A Greek prefix meaning 'wing'.

PTEROPSIDA. The vascular plants which have leaf-gaps in the stele. It includes the ferns, gymnosperms and angiosperms.

PTEROSTEMONACEAE (H). A family of the Cunoniales (H). The leaves are simple with small stipules which are never adnate to the petiole. Petals are present and they may be scale-like. There are 10 stamens, 5 of which are fertile. The filaments are toothed near the apex. The ovary is inferior, and the capsule is septicidal.

PTILIDIACEAE. A family of the Acrogynae. The lateral leaves have several apical teeth or lobes, and the ventral leaves, although smaller, are similar in form to the lateral leaves.

PTYXIS. The way in which an individual leaf is folded in the bud.

PUBERULENT, PUBERULOUS. Slightly pubescent.

PUBESCENT. Covered with fine hairs.

PUCCINIACEAE. A family of the Uredinales. The teliospores are usually stalked, simple or compound. If they are not stalked, they are produced successively as simple or compound teliospores which escape from the sorus dry, or embedded in slime. Aecia are only very exceptionally produced on the Pinaceae.

PUFFING. The simultaneous and violent discharge of ascospores from many asci at the same time.

PUGIONIFORM. Shaped like a dagger.

PULCHELLUS. Latin meaning 'beautiful'.

PULLULATION. Budding, sprouting.

PULSULE. A vacuole in the protoplast of the Dinophyceae. It looks superficially like a contractile vacuole, but has a distinct wall, and does not contract.

PULVERULENT. Looking as though covered by dust.

PULVINATE. Shaped like a cushion.

PULVINULE. The small pulvinus of a leaflet.

PULVINUS. A swelling at the base of a leaflet or leaf-stalk, concerned with their movement in response to a stimulus.

PUMILUS. Latin meaning 'low, small'.

PUNCTARIALES. An order of the Polystichineae. The sporophyte is medium-sized and parenchymatous, without marked internal differentiation of tissues. Growth is by intercalary cell-division. The reproductive organs may or may not be in definite sori. The sporophyte may produce zoospores or neutral spores. The gametophytes are microscopic and either isogamous or anisogamous.

PUNCTATE. Dotted, or marked with small hollows.

PUNCTIFORM. Of bacterial colonies which are very small, but can be seen without a lens. About a millimetre, or smaller in diameter.

PUNCTUM VEGETATIONIS. Latin meaning 'growing point'.

PUNGENS (Latin), PUNGENT. Ending in a sharp, hard point.

PUNICACEAE (EP, H). A family of the Myrtiflorae (EP), Myrtales (H). (BH) include it in the Lythraceae. These are woody plants with entire leaves and showy axillary bisexual regular flowers in top-shaped axis. There are 7-5 sepals and petals which are free, and many free stamens. The ovary consists of many fused carpels with superimposed whorls, with many ovules united to the axis. The fruit is berry-like, and the seeds have no endosperm.

PUNICEUS. Latin meaning 'bright carmine'.

PURE CULTURE. A culture of a pure stock of one species of plant. The term is used especially for bacteria, fungi, and bacteria.

PURE LINE. (1) The descendants obtained from self-fertilization of a single homozygous parent.

(2) An inbred homozygous strain.

PURINE GROUP. A group of cyclic diureides derived from one molecule of a dibasic hydroxyacid and two molecules of urea. The simplest is purine.

310

PURPUREUS. Latin meaning 'purple'.

PUSILLUS. Latin meaning 'small, weak, slender'.

PUSTULE. (1) A mass of fungal spores and the hyphae bearing them.
(2) A pimple or blister.

PUSULE. A small vacuole present in the protoplast of some lower plants, which is able to expand and contract.

PUTAMEN. The hard endosperm of a drupe, such as a stone of a plum.

pv. The negative logarithm of the settling velocity of a particle, in cm. per sec. The smaller the particle the lower it settles, given that the density is constant, so that the *pv* is an indication of the particle size.

PYCNIDIOPHORE. A fruit-body having pycnidia.

PYCNIDIOSPORE. A spore formed in a pycnidium.

PYCNIDIUM. The fertile layer of an Ascomycete when it lies in a cup- or flask-shaped cavity that is open from the beginning.

PYCNIOSPORE. Of Uredinales, a spore from a pycnium, a spermatium. Sometimes used for a pycnidiospore.

PYCNIUM = SPERMOGONIUM.

PYCNOCONIDIUM. A conidium formed inside a pycnidium.

PYCNOGONIDIUM = PYCNIDIOSPORE, PYCNIOSPORE, STYLO-SPORE.

PYCNOSIS. (1) Contraction of the nucleus as the cell dies. It forms a compact, strongly-staining mass.
(2) The formation of a perithecium under the cover of the tissue of a stroma.

PYCNOSPORE. (1) = SPERMATIUM.
(2) A spore formed inside a pycnidium.

PYCNOTHECIUM. A fruit-body formed by pycnosis.

PYRENOMYCETES. A group of ascomycetes including those which produce true perithecia with ostioles, or stromata containing cavities within which the asci are developed. The orders of this group are not necessarily related.

PYCNOXYLIC WOOD. A compact wood which contains little or no parenchyma.

PYGMAEUS. Latin meaning 'dwarf'.

PYRAN. The ring form of a hexose sugar.

PYRENE. A small hard body containing a single seed, rather like the stone of a drupe, but many pyrenes may occur in a single fruit.

PYRENIUM. An obsolete term for the sporocarp of the Sphaeriales.

PYRENOCARPEAE. A series of the Ascolichenes in which the fruit-body is a perithecium.

PYRENOCARP = PERITHECIUM.

PYRENOID. A small round protein granule which stores starch around it as a sheath. They are found singly or in numbers embedded in the chloroplasts of various algae and bryophytes.

PYRENOMYCETAE. A sub-class or order of the Ascomycetae. It includes all members whose fruit-body is a perithecium.

PYRENULALES. An order of the Pyrenomycetes. The ascocarps are true perithecia which are dark coloured. These fungi form a component of some lichens.

PYRIDINE. A heterocyclic compound of a ring of five carbon atoms with one nitrogen atom, with hydrogen atoms attached to the carbon atoms.

PYRIDOXAL. A naturally occurring chelator.

PYRIDOXINE = ADERMINE.

PYRIFORM. Pear-shaped.

PYRIMIDINE. A component of thiamine.

PYRMIDINE NUCLEUS. A heterocyclic ring.

PYROLACEAE (EP, H). A family of the Ericales (EP, H). (BH) include it in the Ericaceae and Monotropeae. These are evergreen or saprophytic herbs with alternate leaves and bisexual regular flowers which are solitary or in racemes. The sepals and petals are 5 in number, with the former fused or free. The stamens are hypogynous. The superior ovary consists of 5-4 fused carpels with many ovules in each. The fruits are lodicular capsules, and the seeds have a fleshy endosperm.

PYROPHYLLIC ACID. A possible degradation product in the breakdown of orthoclase.

PYROPHILOUS. Growing on ground which has recently been burnt over.

PYROPHYLLITE. A soil mineral related to pyrophyllic acid. No strong forces hold the sheets of the mineral together, as the outer layer has negative oxygen ions, but no hydroxyl ions.

PYROPHYTA. A division of the Algae. The photosynthetic pigment is in yellow-green to gold-brown chromatophores. The storage products are starch, starch-like substances, or oil. The cell-wall has cellulose. Most are unicellular, and biflagellate with the two flagella usually dissimilar in form, position, and motion. Some flagella have no flagella. The immobile genera may have non-motile or motile spores. There is sexual reproduction in a few genera.

PYROPHYTE. A plant which is protected by a thick bark from permanent damage by forest-fire.

PYROXYLOPHILOUS. Living on burnt wood.

PYRUVIC ACID. The final product of glycolysis of a hexose sugar in respiration. In the presence of oxygen it enters the Citric Acid cycle during which it is completely oxidized to carbon dioxide and water; without oxygen it is converted to ethyl alcohol.

One of the carbon dioxide acceptors in photosynthesis, being transformed to oxaloacetic acid.

$$CH_3.CO.COOH + CO_2 = COOH.CH_2CO.COOH.$$
$$\text{pyruvic acid} \qquad\qquad \text{oxaloacetic acid}$$

PITHIACEAE. A family of the Peronosporales. The well-developed mycelium has no constrictions. Asexual reproduction is by biflagellate zoospores that are formed in sporangia which remain attached or function as conidiosporangia. These may germinate directly into hyphae. Sexual reproduction is oögamous. A periplasm is cut off around a single egg in an oögonium.

PYXIDATE. Having a lid.

PYXIDIUM. A capsule which dehisces by means of a transverse circular split, so that the upper part of the pericarp forms a lid.

Q

Q_{10}. The rate of increase of a chemical process (expressed as a multiple of the initial rate) produced by raising the temperature 10°C. For biological processes, as with many chemical ones it is often between 2 and 3.

Q_{CO_2}. The number of cubic mm. of carbon dioxide, measured at N.T.P. (S.T.P.) exchanged per hour per milligram dry weight.

Q_{O_2}. The number of cubic mm. of oxygen, measured at N.T.P. (S.T.P.) exchanged per hour per milligram dry weight.

QUADRATE. A square of vegetation (usually 1 sq. m.) selected at random for the examination of the vegetation of a given area.

QUADRI-. A Latin prefix meaning 'four'.

QUADRIPARTITION. The division of a spore-mother-cell into four spores.

QUADRIPLEX. Having four dominant genes.

QUADRIPOLAR SPINDLE. An achromatic spindle with 4 poles. It is found in spore-mother-cells which are dividing.

QUADRIVALENT. (1) An association of 4 chromosomes held together between diplotene and metaphase of the first meiotic division, usually by chiasmata.

(2) A nucleus having 2 pairs of homologous chromosomes, or an individual containing such nuclei.

QUADRUPLEX. A polyploid in which a particular dominant allelomorph is present 4 times.

QUATERNATE. Arranged in fours.

QUANTITATIVE INHERITANCE. The inheritance of characters, where the variation is continuous.

QUANTUM. A 'packet' of energy. (See a reliable physics text-book for an elaboration of this definition.)

QUAQUAVERSAL. Bending every way.

QUARTZ. A mineral which makes up part of the sand and silt fraction of the soil. It has no nutritional value to plants.

QUATERNARY STRUCTURE. The three-dimensional structure of the large protein super-molecules.

QUARTET, QUARTETTE. The group of 4 related cells or nuclei formed as a result of meiosis.

QUERCITIN. A flavonal occurring in the free state or combined in glycosidal forms in a wide range of flowers, leaves etc.

QUIINQUACEAE (H). A family of the Guttiferales (H). These are non-resinous trees, shrubs, or climbers with interpetiolar stipules. The stamens are free. The seeds have no endosperm and are tomentose.

QUINARY. In fives.

QUINCUNCIAL AESTIVATION. A particular type of imbricate aestivation in a five-petalled corolla. Two petals overlap their neighbours by both edges, two are overlapped on both their edges, and one overlaps one neighbour, and is overlapped by the other.

QUINOL = HYDROQUINONE.

QUINQUE-. A Latin prefix meaning 'five'.

QUINQUEFOLIATE. Having 5 leaflets.

QUINTUPLINEVED. Having 5 main veins in a leaf.

R

RACE. A sub-species, which forms a genetically, and usually geographically, distinct mating group within a species.

RACEME. (1) A definite inflorescence, with the main axis bearing stalked flowers which are borne in acropetal succession.
(2) A group of sporangia borne in a similar manner.

RACEMULE. A small raceme.

RACHILLA. The axis in the centre of a grass spikelet.

RACHIS. (1) The axis of a pinnately compound leaf, to which the leaflets are attached.
(2) The main axis of an inflorescence.

RACKET CELL, RAQUET CELL. Of dermatophytes, a hyphal cell, having a swelling at one end.

RADIAL BUNDLE. A vascular bundle having the primary xylem and phloem lying on alternate radii. They are usually found in roots.

RADIAL DOT = CASPARIAN STRIP, when it is cut in transverse section, or certain longitudinal sections and appears as a dot on the wall of the endodermis.

RADIAL SYMMETRY. If an organ or organism can be split down any radius longitudinally to give two identical halves, it is radially symmetrical.

RADIAL VASCULAR BUNDLE = RADIAL BUNDLE.

RADIAL WALL. An anticlinal wall placed on or across the radius of an organ.

RADIANT = RADIATE.

RADIANT UMBEL. An umbel in which the outer flowers are larger than those in the middle.

RADIATE. (1) Said of a capitulum which has ray-florets.
(2) Said of a flower which spreads like a ray from the periphery of any densely packed inflorescence.
(3) Said of a stigma in which the receptive surface radiate outwards from the centre.
(4) Spreading from the centre.

RADICAL. Said of leaves which arise from the base of the plant at ground-level.

RADICANT, RADICANS (Latin). Rooting.

RADICATE. Rooted.

RADICATING. (1) Rooting.
(2) Said of a stipe which is like a root.

RADICATION. The general characteristic of the root system of a plant.

RADICELLOSE. Bearing rhizoids.

RADICICOLOUS. Said of a parasite which attacks roots.

RADICIFEROUS. Bearing roots.

RADICIFORM. Shaped like a root.

RADICLE. (1) The embryonic root of seed-plants.
(2) Any very small root.
(3) A rhizoid of a moss.

RADICULAR. Belonging to, or related to the radicle.

RADICULOSE. Said of the 'stem' of a moss which bears many rhizoids at the base.

RADIOSPERM. (1) A seed which is approximately circular in cross-section.
(2) A plant, especially fossil forms, bearing such seeds.

RADIUS. The group of ray florets in a capitulum.

RADULACEAE. A family of the Acrogynae. They are distinguished by the unequally lobed lateral leaves, and rhizoids on the ventral lobes.

RADULA SPORE. A spore on a sterigma which has no relation to the growing point of the hypha.

RAFFLESIACEAE (EP). A family of the Aristolochiales (EP). (BH) and (H) include it in the Cytinaceae (EP). These are thalloid parasites with very short shoots with a single flower or raceme terminally. They are usually unisexual, regular, with the perianth of similar members. The 5–4 perianth lobes are fused, and the many free stamens are on a column. The inferior ovary consists of 8-6-4 fused carpels with parietal placentas, or any twisted loculi. The fruit is a berry with many seeds, and the seeds contain endosperm.

RAFFINOSE. A trisaccharide sugar ($C_{18}H_{32}O_{16}$), made up of a galatose residue, a glucose residue and a fructose residue.

RAIN FACTOR. An expression of the humidity as a function of rainfall and temperature. The ratio of the mean annual rainfall in mm. to the mean annual temperature in °C.

RAMAL, RAMEAL. Relating to a branch.

RAMBLER. A weak-stemmed plant that leans on and scrambles over the surrounding vegetation.

RAMENTACEOUS. Covered with ramenta.

RAMENTUM. A thin brown scale, one cell thick, occurring on the stems, petioles and leaves of ferns. *plu.* Ramenta.

RAMICOLE, RAMICOLOUS. Living on twigs.

RAMIFICATION. A branch.

RAMO-CONIDIUM. A spore derived from a part or branch of conidiophore.

RAMOSE. Much branched.

RAMULAR. Relating to a branch.

RAMULAR TRACE = BRANCH TRACE.

RAMULUS. A very small branch of a stem or leaf.

RANALES (EP, BH, H). An order of the Archichlamydeae (EP), Polypetalae (BH), Herbaceae (H). The leaves are simple or divided, alternate (rarely opposite) and rarely with stipules. They are herbaceous, often with scattered vascular bundles in the stem, or softly woody with broad medullary rays. The flowers are bisexual and hypogynous (rarely perigynous). The parts are hemicyclic (rarely completely cyclic). Petals are usually present, there are many free stamens, and the gynecium consists of free carpels (rarely 1 carpel). The seeds have copious endosperm and a minute embryo. EP, and BH include woody types.

RANDOM MATING. The situation within a population where each individual has an equal chance of mating with any other individual, including itself; subject to any known restrictions, *e.g.* dioecy or incompatibility.

RANDOMIZATION. The process of arriving at a random combination of types of event.

RANDOMIZED BLOCK. An experimental design involving one restraint.

RANDOMNESS. That combination of types of event in which the distribution of classes of one has no causal relation with the distribution in classes of others, *i.e.* arrived at by chance.

RANGE. The area over which a species is spread in the wild.

RANGIFEROID. Branched like a reindeer's horn.

RANUNCULACEAE (EP, BH, H). A family of the Ranales (EP, BH, H.) These are usually herbs often with divided leaves. The flowers are usually regular and bisexual, but they may be zygomorphic or fully cyclic. The perianth is rarely differentiated into sepals and petals. There are usually many free stamens. The gynecium is superior and of many to 1 free carpels (rarely united), each with many to 1 ovules. The fruit is a follicle or capsule (rarely a berry), and the endosperm is oily.

RANUNCULACEOUS. Having the characters of a buttercup.

RAPACEOUS. Latin meaning 'turnip-shaped'.

RAPATEACEAE (EP, BH, H). A family of the Farinosae (EP), Coronarieae (BH), Xyridales (H). These are perennial herbs with two-ranked narrow leaves. The inflorescence is terminal with 2 large spathes enclosing a head of spikelets, each of many bracts and a terminal bisexual regular flower. There are 3 sepals and petals, and each whorl is fused. There are 2 whorls of 3 free stamens. The superior ovary consists of 3 fused carpels and 3 loculi, with many to 1 ovules in each. The fruit is a capsule, and the seeds contain endosperm.

RAPHE. (1) An elongated mass of tissue, containing a vascular bundle, and lying on the side of an anatropous ovule, between the chalaza and the attachment to the placenta.

(2) A slit-like line running longitudinally on the valve of a diatom, indicating the position of a narrow slit in the wall; it bears a nodule at each end and one in the middle.

RAPHIDE. A needle-shaped crystal, usually of calcium oxalate, occurring in bundles, rounded masses, or singly in certain plant cells.

RAVENELIN. A metabolic product of *Helminthosporium ravenelii* and *H. turcicum*.

RAW HUMUS = MOR.

RAY. The non-vascular tissue developed in a stele. That between the primary vascular bundles is the *interfascicular ray*, and that developed in the secondary vascular tissue, by division of the cambium, is a *vascular ray*.

RAY FLORET. One of the small flowers radiating out form the margin of a capitulum or other dense inflorescence.

RAY INITIAL. One of the cells of the cambium which takes part in the formation of a ray.

RAY TRACHEID(E). A thick-walled cell which occurs in the vascular rays of pines. It has bordered pits and conducts aqueous solution horizontally.

RDP = RIBULOSE DIPHOSPHATE.

REACTION. Any change in the activity of an organism in response to a stimulus.

REACTION TIME. The time interval between the application of a stimulus and the appropriate reaction.

RECAPITULATION. The theory that the ancestral characters are reflected in the development of an individual.

RECEDENT. A term applied either to the genom or the plasmon when they are of subsidiary importance in heredity.

RECEPTACLE. (1) In fungi, a spore-bearing structure, especially if it is more or less concave.
(2) In algae, the swollen end of a branch bearing reproductive organs.
(3) In liverworts, a cup containing gemmae.
(4) In ferns, the cushion of tissue bearing the sporangia.
(5) In flowering plants, (a) The more or less enlarged end of the flower-stalk bearing the flower-parts; (b) The enlarged end of a peduncle bearing the flowers of a crowded inflorescence.
(6) In fungi, an axis having one or more organs.

RECEPTIVE BODY. A small branched or unbranched process from the stroma, capable of being 'spermatized' by micro-condia.

RECEPTIVE HYPHA = FLEXUOUS HYPHA = TRICHOGYNE.

RECEPTIVE PAPILLA. In some phycomycetes, a small outgrowth from the oögonium into the antheridium, to which the antheridium become attached.

RECEPTIVE SPOT. The clear area in the female gametes of some fungi and algae, through which the male gamete enters.

RECEPTIVENESS. The condition of a stigma when effective pollination, and fertilization is possible.

RECEPTIVITY = SUSCEPTIBILITY.

RECESSIVE. The relationship of two allelomorphs where the single gene heterozygote does not resemble one of the two homozygous parents because its effect is masked by the dominant allelomorph.

RECIPROCAL CROSS. A cross where the sources of male and female gametes are reversed.

RECIPROCAL TRANSLOCATION = INTERCHANGE. A mutual interchange of particles between two chromosomes.

RECOLONIZATION. The re-establishment of vegetation on an area which has been stripped of plants.

RECOMBINATION. The formation in the offspring, of combinations of genes not present in either parent. This is brought about by crossing-over or segregation at meiosis, and/or the random union of different sorts of gametes at fertilization.

RECTINERVED. Having straight or parallel veins.

RECTIPETALY. The tendency of plant members to grow in a straight line.

RECTISERAL. Arranged in straight rows.

RECURRENT. Said of the small veins of a leaf, when they bend back towards the mid-rib.

RECURVED. Bent or curved backwards.

RED ALGAE = RHODOPHYTA.

REDOX CHAIN. An oxidation-reduction system in which one substance is oxidized and another reduced.

RED SNOW. Snow stained by a surface growth of unicellular algae, which are rich in haematochrome.

REDUCED. Simplified in structure as compared with some probable ancestral form.

REDUCED APOGAMY = MEIOTIC EUAPOGAMY. The development of a sporophyte from a cell or cells of a gametophyte without the fusion of gametes. This gives a plant which has the gametic number of chromosomes.

REDUCED FERTILIZATION. The substitution for a normal sexual fusion between male and female gametes or nuclei of some other nuclei, *e.g.* the fusion of two female nuclei.

REDUCING CENTRE. The $-CH.OH$ group in the C_1 position of a monosaccharide or disaccharide sugar. This causes the positive reaction with the Fehling's test.

REDUCTASES. Flavoproteins which are enzymes that bring about reductions, *e.g.* of nitrate (NO_3^-) ions to ammonium (NH_4^+) ions. They are also important in the oxidation of NAD and NADP.

REDUCTION. The halving of the number of chromosomes at meiosis.

REDUCTION DIVISION = MEIOSIS.

REDUCTION SEPARATION. The separation of homologous parts of non-sister chromatids at anaphase of one or other meiotic division.

REDUCTIVE AMINATION. The reduction of a substance by the addition of an NH_3^- radical, *e.g.* α ketoglutaric acid is transformed to glutamic acid.

$$\begin{array}{ll}
\begin{array}{c}
COOH \\
| \\
CH_2 \\
| \\
CH_2 \\
| \\
C=O \\
| \\
COOH \\
\text{α-ketoglutaric acid}
\end{array}
& + DPNH + NH_3 \rightleftharpoons
\begin{array}{c}
COOH \\
| \\
CH_2 \\
| \\
CH_2 \\
| \\
HC-NH_3 \\
| \\
COOH \\
\text{glutamic acid}
\end{array}
+ DPN + H_2O
\end{array}$$

REDUPLICATION. The occurrence of a segment of a chromosome twice in a haploid set.

REFLEXED. Turned-back abruptly.

REFLORESCENCE = DOUBLE FLOWERING.

REFRACTED. Bent backwards abruptly from the base.

REFRINGENT = REFRACTED.

REGENERATION. (1) The growing again of a part or organ lost by injury.
(2) In bryophytes, the growth of a diploid gametophyte direct from a sporophyte after injury.

REGMA. A fruit, which breaks up into rounded one-seeded portions when ripe.

REGOLITH, RHEGOLITH. The fragmental unconsolidated debris mantling the rocks of the earth's crust.

REGRESSION. (1) See *Galton's Law*.

(2) The dependence of one variate, termed the dependent, on another, termed the independent variate.

(3) A tendency to return from an extreme to an average condition, as when a tall parent gives rise to plants of average stature.

REGRESSION COEFFICIENT. The rate of change of the dependent variate on the independent variate.

REGRESSION LINE. A straight line or curve showing a regression in a co-ordinate representation.

REGULAR = ACTINOMORPHIC.

REGUR. A black cotton-soil in India. The natural vegetation is grass steppe. It is usually clayey, cracking on drying, with a low organic matter content.

REJUVENESCENCE. (1) The conversion of the contents of a cell into one or more cells of a different and usually more active character.

(2) The renewal of growth from old or injured parts.

RELATIONAL COILING. The loose coiling of two chromatids during despiraling.

RELATIVE SEXUALITY. The occurrence in a species of strains giving gametes able to fuse with those produced by either of the normal strains.

REMOTE. Said of the gills of agarics which do not reach the stipe but leave a free space around it.

RENDZINA. A humus-carbonate soil. The form is generally determined by the parent material which is always calcareous. There is no B-horizon, the humus-containing A-horizon rests directly on the calcareous parent rock.

RENIFORM. Kidney-shaped, either solid or flat.

RENNER EFFECT. The competition between 4 genetically different spores formed at one meiosis in regard to which shall form the embryo-sac.

R-ENZYME. One of the enzymes which breaks down starch. It breaks down the $1:6$ linkages.

REPAND. Having a slightly wavy edge.

REPEATED EMERGENCE. A condition in fungi in which the zoospores after swimming for some time, encyst and then emerge from the cysts without any change in morphology.

REPENS, REPTANS (Latin). Prostrate and rooting.

REPENT = REPENS.

REPETITION. A method of spore germination where a new spore is produced like the first.

REPLICATE. Folded back as when the edge of an apothecium is turned outwards and downwards.

REPLICATE SEPTUM. A septum in some algae which bears a collar-like appendage projecting into the cavity of the cell.

REPLICATION. The equal incorporation of all combinations two or more times in an experimental design, which is then said to be replicated.

REPLUM. A thin wall dividing the fruit into two chambers, formed by an ingrowth from the placentas, and not a true part of the carpel-walls.

REPRODUCTION. The process of generating new individuals whereby the species is perpetuated.

REPRODUCTIVE PHASE. The stage in the growth of a plant when it changes from purely vegetative growth to producing reproductive bodies.

REPRODUCTOCENTRE. Of the Chytridiales, having the development of one or more reproductive bodies at the centre of gravity of the thallus.

REPTANT = REPENT.

REPULSION. (1) The presence of two given genes in different homologous chromosomes.

(2) The tendency shown by dominant characters to separate.

RESEDACEAE (EP, BH, H). A family of the Rhoeadales (EP), Parietales (BH), Resedales (H). These are herbs with alternate stipulate leaves, and racemes of bisexual zygomorphic flowers which have a posterior disk. There are 8-4 free sepals, 0-8 free petals, and 10-3 free stamens. The superior ovary consists of 6-2 fused carpels, which are free above, and contains 1 loculus with many to 1 ovules. The fruit is a capsule, and there is no endosperm.

RESEDALES (H). An order of the Herbaceae (H). It contains the single family—the Resedaceae.

RESERVE CELLULOSE. Cellulose present in endosperm or other storage tissue, and subsequently used in the nutrition of the plant.

RESIDUAL SHRINKAGE. The shrinkage that occurs when a soil dries, after the loss of volume is directly proportional to the loss of water. During residual shrinkage the loss of volume of the soil becomes less than the loss of volume of the water, due to the emptying of the interstitial spaces.

RESIN. An acidic substance, either a phenolic derivative or an oxidation product of terpenes. The resins are insoluble in water, but soluble in alcohol, ether, and carbon disulphide, and burn with a sooty flame. They are products of secretion or disintegration which are usually found in special cavities or passages.

RESIN CANAL. An intercellular space, often bordered by secreting cells, containing resin or turpentine.

RESINOGENIC, RESINOGENETIC. Giving rise to resin.

RESISTANCE. (1) The ability of an organism to withstand more than the normally lethal dose of a toxic substance.

(2) The ability of a plant to remain productive if attacked by a parasite which would normally kill it, or reduce its productivity below an economic level.

RESPIRATION. (1) The chemical reactions which break-down complex molecules to simple ones, releasing energy, sometimes called *cell-*, *internal-*, or *tissue-respiration*.

(2) The taking in of oxygen and the releasing of carbon dioxide. This is better called *gaseous exchange*.

RESPIRATORY CHROMOGEN. A colourless substance which gives rise to a coloured one on oxidation or reduction. It may play a part in respiration.

RESPIRATORY ENZYME. An enzyme which catalyses oxidation-reduction reactions.

RESPIRATORY INDEX. The number of milligrams of carbon-dioxide released from 1 gm. of plant material (dry-weight) at 10°C, when the amount of respirable material is unlimited, and when the oxygen is in the same proportion as it is in the ordinary atmosphere.

RESPIRATORY QUOTIENT. The ratio of the volume of carbon-dioxide expired to the volume of oxygen consumed during the same time. It might give a rough idea of the food material being oxidized.

RESPONSE. The change in an organism, or part of it, usually adaptive, produced by a stimulus.

RESPONSE CURVE. A graph showing the relationship between the amount of a stimulus or substance applied, and the response. The term is used particularly in relation to experiments such as fertilizer trials etc.

RESTANS. Latin meaning 'persistent'.

RESTIACEAE (BH) = RESTIONACEAE (EP).

RESTING CELL. A cell which is not undergoing division.

RESTING NUCLEUS = METABOLIC NUCLEUS. A nucleus which is not dividing.

RESTING PERIOD, RESTING STAGE. Any time in the life of a plant or plant organ when no growth or activity appears to be in progress.

RESTING SPORE. (1) A spore which germinates after a resting period.
   (2) A thick-walled spore which can survive adverse conditions, remaining dormant for some time before it germinates.

RESTING STAGE. Said of a nucleus, when the chromosomes are not clearly visible as individuals.

RESTIONACEAE. A family of the Farinosae (EP), Glumaceae (BH), Juncales (H). These are rush-like xerophytic or marsh herbs with a creeping rhizome and two-ranked bracts or scale-leaves on a stem. The flowers are in spikes in the axils of bracts. They are usually unisexual and regular. The perianth is sepaloid, in two whorls of 3-2 members in each. There are 3-2 free stamens. The superior ovary consists of 3-1 fused carpels, with 3-1 styles and 3-1 loculi, with 1 ovule in each. The fruit is a capsule or nut, and the seeds contain endosperm.

RESTITUTION. The formation of the original chromosome after breakage, especially by X-ray.

RESTITUTION NUCLEUS. The single nucleus found instead of two due to the failure of the first or second division of meiosis.

RESTRAINT. A limitation of the random arrangement of combinations in an experimental design, so that error variation, while still capable of being estimated without bias, is reduced or potentially reduced.

RESUPINATE. Reversed in position, usually through 180°, *e.g.* the hymenium of some fungi being on the upper surface, or the twisting of the flower as in orchids.

RETICULATE. (1) Lattice-like.
   (2) Having the surface marked by a network of fine upstanding ridges.

RETIFORM. Having the appearance of being netted.

RETINERVED. Net-veined.

RETROCULTURE. The re-isolation of a pathogen from a host into which it has been introduced experimentally.

RETROCURVED, RETROFLEXED. Bent backwards.

RETRORSE. (1) Pointing backwards.
   (2) Pointing in a direction contrary to normal.

RETROSERRATE. Having marginal teeth strongly directed backwards.

RETROVERSE = RETRORSE.

RETUSE. Having a bluntly rounded apex with a central notch.

REVERSE. The underside.

REVERSION = ATAVISM.

REVOLUTE. Rolled backwards and usually downwards.

RHACHILLA = RACHILLA.

RHACHIS = RACHIS.

RHAGADIOSE. Deeply marked with cracks or fissures.

RHAMNACEAE (EP, BH, H). A family of the Rhamnales (EP, H), Celastrales (BH). These are woody plants (rarely herbs), often climbing with simple stipulate leaves and small greenish or yellowish flowers often in axillary cymes. There are 5-4 free sepals and petals (the latter may be absent), and 5-4 free stamens. The ovary is superior to inferior and of 5-2 fused carpels with 1 ovule in each. The fruit is dry or a drupe, and the seeds have little or no endosperm.

RHAMNALES (EP, H). A family of the Archichlamydeae (EP), Lignosae (H). The flowers are cyclic, with the perianth in 2 whorls (the petals are sometimes absent), and regular. There is 1 whorl of stamens before the petals. The ovary consists of 5-2 fused carpels each with 1-2 ascending ovules, which have a dorsal, lateral, or ventral raphe with 2 integuments.

RHAMNEAE (BH) = RHAMNACEAE (EP).

RHAMNOSE. A methyl pentose $C_5H_{10}O_6$.

RHAPE = RAPHE.

RHEOTROPISM. A tropism in response to water.

RHEXIGENOUS = LYSIGENIC.

RHIOPTELEACEAE (H). A family of the Juglandales (H). The leaves are pinnate and stipulate. The bisexual flowers are fertile, and the female ones sterile. The ovary is superior, and bilocular, with 1 loculus empty. The ovule is axile, and the fruit two-winged.

RHIPIDIACEAE. A family of the Saprolegniales. The mycelium has a well-developed holdfast, with a more or less thickened basal segment, from which arise the slender, often constricted hyphae, which end in zoosporangia or sex-organs, or both may arise on the basal segment. Only zoospores of the secondary type are produced. The oögonium contains a single gamete surrounded by periplasm.

RHIPIDIUM. A monochasial cyme in which the branches lie in one plane, so that the whole is fan-shaped.

RHIZANTHOUS. Apparently forming flowers from the root.

RHIZIDIACEAE. A family of the Chytridiales. They are endobiotic, epibiotic or interbiotic and have rhizoids or haustoria. Most have the fertile portion outside the host. Reproduction is by uniflagellate zoospores or gametes. The zoospore cyst enlarges into a zoosporangium or prosporangium.

RHIZIDIOMYCETACEAE. A family of the Hypochytriales. It includes the eucarpic, monocentric, epibiotic members of the order.

RHIZINE. A rhizoid-like outgrowth from the underside of a lichen thallus. It may be single, or a strand of rhizoids.

RHIZOBLAST. Of zoospores, the thread joining the belepharoplast and the nucleus.

RHIZOBOLAEA = CARYOCARACEAE.

RHIZOCALINE = CALINE.

RHIZOCARPAE = SALVINIALES.

RHIZOCARIC, RHIZOCARPOUS. Producing flowers underground, as well as in the normal position.

RHIZOCHLORIDALES. An order of the Xanthophyceae. The protoplasts are amoeboid with pseudopodia. They are uninucleate or multinucleate, with one or many chromatophores.

RHIZOCHRYSIDALES. An order of the Chrysophyceae. The protoplasts are amoeboid, and if flagellate stages are formed, they are only temporary.

RHIZOCORM. A stout fleshy rhizome.

RHIZODERMIS = PILIFEROUS LAYER.

RHIZODINIALES. An order of the Dinophyceae. The vegetative cells are naked and amoeboid.

RHIZOFLORA. The microflora growing in the rhizosphere.

RHIZOGENIC, RHIZOGENETIC. Producing roots.

RHIZOID. (1) A single- or several-celled, hair-like structure at the base of a moss 'stem', and on the underside of liverworts and fern prothalli. They serve for anchorage, and in the moss, hold water by capillarity.
(2) A short hypha which attaches a fungus to the substrate and collects nutrients.

RHIZOIDAL CELL = BASAL CELL.

RHIZOME. An elongated underground stem bearing buds in the axils of reduced scale-leaves. It is usually horizontal, and serves for perennation and vegetative propagation.

RHIZOMORPH. A compact strand of fungal hyphae. It elongates by apical growth and transports food material from one part of the thallus to another. It also helps in the spread of the fungus over or through the substratum.

RHIZOMYCELIUM. In some Chytridiales, a gradually attenuated system of rhizoidal branches in which fertile areas develop at various points on the branching system.

RHIZOPHILOUS. Growing on roots.

RHIZOPHORACEAE (EP, BH, H). A family of the Myrtiflorae (EP), Myrtales (BH, H). These are woody plants usually with opposite, stipulate leaves and usually bisexual flowers which are solitary or in racemes. There are 16-3 (usually 8-4) free sepals, and as many (or no) free petals. There are many to 8 free stamens. The ovary is inferior to semi-inferior and consists of 5-2 (rarely 6) fused carpels, each with 2-4-many pendulous axile ovules. The fruit usually has 1 seed in each loculus. The seed may or may not contain endosperm.

RHIZOPHORE. In the Selaginellales, leafless downward-growing branches which bear adventitious roots at their apices.

RHIZOPIN. A growth substance increasing the carbon-dioxide production by yeasts.

RHIZOPLANE. The interface between the roots and the soil.

RHIZOPLAST = RHIZOBLAST. A delicate fibre connecting the parades-mone of the neuromotor apparatus to a small, intranuclear centrosome.

RHIZOPDAL STAGE. An amoeboid stage formed by some algal macrozoo-spores before they develop into filaments.

RHIZOSPHERE. The region in the soil surrounding the roots of a plant, and affected by its secretions.

RHODO-. A Greek prefix meaning 'rose-red'.

RHODOPHYCEAE. The single class of the Rhodophyta.

RHODOPHYTA. A division of the Thallophyta. The red algae. They are distinguished by their sexual reproduction, in which non-flagellate male gametes are transported to lodge against the female sex-organ (*carpogonium*). In some the zygote develops directly into spores, but in most the development of the spores is indirect. There are no flagellate asexual spores. The plastids contain red phycoerythrin as well as chlorophyll.

RHODOPLAST. The chromatophore of the red algae.

RHODOSPOROUS. Having pink spores.

RHODOTORULACEAE. A family of the Saccharomycetales. The yeast-like cells contain carotinoid pigments, which give the colonies a red to orange colour.

RHODYMENIALES. An order of the Floridaea. These are the tetrasporic Floridaea in which the auxillary cell is a specialized cell differentiated before fertilization. It is the terminal member of a two-celled filament borne upon the supporting cell of the carpogonial filament.

RHOEDALES (EP, H). An order of the Archichlamydeae (EP), Herbaceae (H). These are usually herbs, with the flowers in racemes. The flower-parts (except sometimes for the stamens) are in whorls. The perianth is usually differentiated into sepals and petals, or it is apetalous. The flowers are regular or zygomorphic and hypogynous. The ovary consists of many to 2 fused carpels, and the ovules have 2 integuments.

RHOIUM. A creek formation.

RHOMBOIDAL. Having sides of equal length, but not a square, and being attached by one of the acute angles.

RHYNIACEAE. A family of the Psilophytales. These were small with the plant body differentiated into subterranean and aerial portions. The subterranean part was rhizome- or corm-like, with unicellular rhizoids. The aerial part was leafless with dichotomously branched, cylindrical branches. The sporangia were borne singly at the ends of the branches. They were indehiscent and had a jacket-layer several layers thick.

RHYNCHOSPOROUS. Having beaked spores.

RHYNCHOSPOROUS FRUIT. A fruit which ends in a beak.

RHYTIDOME. A tissue cut-off outside a periderm. The cells die leaving a crust made up of alternate layers of cork and dead phloem or cortex.

RIB. One of the larger veins of a leaf.

RIBITOL. A sugar alcohol found in riboflavin and flavin-adenine-nucleotide, and perhaps in all plant cells.

RIBOFLAVIN. Vitamin B_2. It is found in all plant cells, and forms a co-enzyme which is concerned in cellular oxidations.

RIBONUCLEASE. The enzyme which breaks down RNA.

RIBONUCLEIC ACID. It is found on the nucleolus, ribosomes, endoplasmic reticulum and free in the cytoplasm. That on the ribosomes and endoplasmic reticulum is 'messenger' RNA—the template for protein synthesis, and that in the cytoplasm is 'transfer' RNA which conveys amino-acids to the correct site on the 'messenger' RNA.

RIBOSE.   A pentose sugar, which exists naturally in the furanose form.

D-ribofuranose

RIBOSE-9-ADENINE = ADENOSINE.   A nucleoside consisting of ribose joined to the prurine adenine.

1-β-Dribose-9-adenine

RIBOSE-3-URACIL.   A nucleoside consisting of ribose combined with the pyrimidine uracil.

326

RIBOSOME. A macromolecule containing protein and RNA. It is seen as a dense particle (150-200 Å) in electron micrographs. They are found in all types of cells in which protein is being synthesized.

RIBULOSE DIPHOSPHATE. Ribulose 1,5-diphosphate is a carbon dioxide receptor during photosynthesis. Each molecule is formed into two molecules of phosphoglyceric acid.

$$
\begin{array}{l}
CH_2OP \\
| \\
C=O \\
| \\
CHOH + CO_2 + H_2O \longrightarrow \\
| \\
CHOH \\
| \\
CH_2OP \\
\text{Ribulose diphosphate}
\end{array}
\qquad
\begin{array}{l}
CH_2OP \\
| \\
2 \ CHOH \\
| \\
COOH \\
\\
\text{Phosphoroglyceric acid}
\end{array}
$$

RICCARDIACEAE. A family of the Anacrogynae. There are no clear distinguishing features. Most genera have a thallose gametophyte and sporophyte with an apical elaterophore, but both these features may be absent.

RIELLACEAE. A family of the Sphaerocarpales, distinguished by the asymmetrical gametophyte.

RIGENS. Latin meaning 'rigid'.

RIGOR. An inert condition assumed by a plant when conditions for growth are unfavourable.

RIMOSE. Having the surface marked by a network of intersecting cracks.

RIMULOSE. Having small cracks.

RIND. (1) The outer layers of the bark of a tree.
(2) The outer layers of a fruit-body or sclerotium of a fungus.

RING. (1) = ANNULUS.
(2) Of bacteria, growth at the surface of a liquid culture which sticks to the container.
(3) In mitosis, chromosomes which have no ends.
(4) In meiosis, chromosomes associated in rings, usually by terminal chiasmata, and in twos, fours, sixes etc.

RING CELL. One of two cells lying one above the other, cut off during the early stages of development of the most advanced Leptosporangiatae antheridia. Each cell completely surrounds the antheridium, and with the apical cap-cell, completely enclose the large central cell.

RINGENS. Latin meaning 'gaping'.

RING BARK. Bark which splits-off in more or less complete rings.

RING POROUS. Said of a wood which contains more vessels, or larger vessels in the spring wood than elsewhere, so that it is marked in cross section by rings or portions of rings of small holes.

RINGENT. Said of a corolla consisting of two distinct, widely gaping lips.

RIPARIAL, RIPARIAN, RIPARIOUS, RIPARIUS (Latin). Living on the banks of rivers and streams.

RIVALIS. Latin meaning 'growing by brooks'.

RIVULOSE. Marked with lines, giving the appearance of rivers as shown on a map.

RNA = RIBONUCLEIC ACID.

ROD ORGAN = PHARANGEAL ROD.

ROESTELOID. Of an aecium which is long and tube-like.

ROESTELIA. A type of aecium with an elongated cylindrical periderm, split into segments.

ROGUE. (1) A variant from the standard type of variety.
(2) To remove such variants and plants infected by disease etc. from a crop.

ROOT. The lower portion of the axis of a higher plant. It is usually branching and does not bear leaves or buds. It anchors the plant in the soil, and by means of the root-hairs absorbs water and dissolved materials.

ROOT CAP. A cap of loosely arranged cells which covers the apex of the growing point of a root, and protects it as it grows through the soil. It is formed from the promeristem, dermatogen or from the calyptrogen.

ROOT HAIR. A tubular outgrowth from the cells of the piliferous layer of a root. They have thin walls, and are in close contact with the soil particles, so that they can absorb water and dissolved materials. They occur in greatest abundance just behind the root tip.

ROOT NODULE. Small swelling on the roots of various plants (especially legumes) produced as a result of invasion by nitrogen-fixing bacteria.

ROOT PRESSURE. The pressure under which aqueous solutions from the living cells in the root are transported to the xylem.

ROOTSTOCK = RHIZOME.

ROOT TUBER. A swollen root containing reserve food material.

ROOT TUBULE = ROOT NODULE.

RORIDULACEAE (H). See *Byblidaceae* (EP).

ROSACEAE (EP, BH, H). A family of the Rosales (EP, BH, H). These are herbs, shrubs or trees, with usually alternate stipulate leaves and regular (rarely zygomorphic) flowers. The flower parts are in fives (may be 3-8 or more). The axis is flat or hollowed. There are usually 5 free sepals and petals (or no petals), and 4-2, or more times as many free stamens which are bent inwards in the bud. There are as many carpels as calyx lobes (or 2-3 times as many), or many, or rarely 1. These are free, or united to the hollow axis. The ovary is usually unilocular, with 2 ovules in each carpel. The fruit is a follicle, achene, drupe, or pome, and the endosperm is thin or absent.

ROSALES (EP, BH, H). An order of the Archichlamydeae (EP), Polypetalae (BH), Lignosae (H). The flowers are cyclic, rarely spirally arranged, with sepals and petals (rarely the petals are absent), and regular or zygomorphic. The carpels are free or fused, and superior to inferior. They sometimes have a thick placenta and many ovules.

ROSETTE. (1) A group of leaves arising from a short stem, and so lying close together on or near the ground.
(2) The 4 cells in the embryo of a pine, which lie just above the suspensor.

ROSTELLATE. Beaked.

ROSTELLUM. A beak-like outgrowth from the column in the flower of an orchid.

ROSTRATE. (1) Ending in a long, and usually hard point.
(2) Of the lid of a moss, when elongated.

ROSTRUM. Any beak-like process.

ROSULE. A rosette of leaves.

ROSULATE. Forming a small rosette.

ROTACEOUS. Latin meaning 'rotate', 'wheel-shaped'.

ROTATION. The movement of the protoplasm in a cell, in a constant direction.

ROTUND. Approximately circular.

ROTUNDATE = ORBICULAR.

ROXBURGHIACEAE (H). A family of the Dioscoreales (H). The flowers are bisexual. There are 4 stamens. The ovary is inferior to semi-inferior, and unilocular.

RQ = RESPIRATORY QUOTIENT.

RUBENS. Latin meaning 'blush-red'.

RUBER. Latin meaning 'red'.

RUBESCENT. Turning pink or red.

RUBIACEAE (EP, BH, H). A family of the Rubiales (EP, BH, H). These are herbs or woody plants with decussate entire leaves and interpetiolar stipules, and regular flowers in cymes, which are often condensed into heads. The flower-parts are 5-4 (rarely more). The calyx is usually open, and the petals valvate or convient. The inferior ovary is of 2 fused carpels, each with 1 to many ovules. The fruits are of various forms, and the seeds may or may not have endosperm.

RUBIALES (EP, BH, H). An order of the Sympetalae (EP), Gamopetalae (BH), Lignosae (H). These are woody plants or herbs with opposite, usually simple leaves. The flowers are usually regular with 5-4 sepals and petals. The inferior ovary is of 1 (or more fused) carpel, each with many to 1 anatropous ovules.

RUBIGINOUS. Rust-coloured.

RUBROFUSARIN. A red pigment produced by *Fusarium culmorum*.

RUBROGLAUCIN. A red pigment produced by *Aspergillus glaucus* series.

RUDERAL, RUDERALIS (Latin). A plant which grows on rubbish heaps or waste places.

RUDIMENTARY. Refers to an organ or part of an organ which is imperfectly developed.

RUFESCENT. Becoming reddish-brown.

RUFFLING. A faint crumpling of a leaf attacked by a virus disease.

RUGOSE. Having a wrinkled surface.

RUGULOSE. Delicately wrinkled.

RUMINATE. (1) Said of an endosperm into which the inner layer of the testa protrudes.

(2) Said of an endosperm which is mottled in two or more different colours.

RUNCINATE. Said of a leaf having a lamina composed of lobes with their points pointing backwards.

RUNNER. A prostrate shoot which roots at the end, and here gives rise to a new plant.

RUN-OFF. The free water which runs away from the surface of a soil. It is one of the major factors in soil-erosion.

RUPESTRAL, RUPESTRINE. Living on walls or rocks.

RUPICOLOUS. Living or growing among rocks.

RUPPIACEAE (H). A family of the Potamogetonales (H). These plants which inhabit salt marshes have no perianth. The ovule is apical, and the fruiting carpels are stipitate.

RURALIS. Latin meaning 'living in rustic places'.

RUSCACEAE (H). A family of the Liliales (H). The leaves are reduced to scales, and the flowers are borne on leaf-like cladodes. The filaments are joined to form a column. The fruit is a berry.

RUST FUNGI = UREDINALES.

RUTACEAE (EP, BH, H). A family of the Geraniales (EP, BH), Rutales (H). These are usually woody with alternate or opposite, simple or compound, exstipulate leaves. The flowers are regular or zygomorphic, usually bisexual, with 5-4 sepals and petals, and a disk. There are 2 whorls of stamens with the outer opposite to the petals, or 5, 4, 3, 2, (rarely many). The superior ovary has 5-4 (rarely 3-1 or many) fused ovules with many to 1 ovules. The fruits are of various forms, and the seeds may or may not have endosperm.

RUTALES (H). An order of the Lignosae (H). These are trees, shrubs or climbers (rarely herbs). The leaves are simple or compound, and often dotted with glands. There are rarely stipules. The flowers are hypogynous to slightly perigynous and usually bisexual. The sepals are mostly imbricate, and the petals are contorted to valvate, free or joined near the base. The disk is mostly conspicuous. The ovary is superior, and contains 1-2-to many seeds. Endosperm may or may not be present.

RUTILANT, RUTILANS (Latin). Brightly coloured in red, orange or yellow.

S

SABIACEAE (EP, BH, H). A family of the Sapindales (EP, BH, H). These are woody plants, often climbers with alternate, exstipulate leaves. The flowers are bisexual or unisexual in racemes or cymose racemes. The 5-2 sepals are fused, and the 5-4 petals are free. The 5 free stamens are antepetalous. The ovary consists of 3-2 fused carpels, each with 2 ovules. The fruit is unilocular and one-seeded. The seeds have no endosperm.

SABULOSE, SABULINE. Growing in sandy places.

SAC. Any pouched or bag-like structure.

SACCATE. Like a sac or bag.

SACCATE FRUIT. A fruit having a bag-like envelope around it.

SACCHARASE = SUCRASE.

SACCHARINE. Covered with sparkling grains, like grains of sugar.

SACCHAROBIOSE = SUCROSE.

SACCHAROLYTIC. Said of bacteria which use simple carbohydrates and starches as sources of energy.

SACCHAROMYCETACEAE. A family of the Saccharomycetales. This family contains most of the industrial strains of yeasts. The ascus-forming yeasts.

SACCHAROMYCETALES. An order of the Ascomyceteae. The yeasts. They are filamentous, or unicellular. The fundamental distinguishing character is the production of a single asci instead of a cluster of asci from branched ascogenous hyphae.

SACCIFEROUS, SACCIFORM. Like a sac.

SACCULIFORM. Shaped like a little sac.

SACK-PUSTULE = PUSTULE.

SAGITTATE. Shaped like an arrow-head, with two barbs pointing backwards.

SAJONG SOIL. A black prairie-soil found in the central plains of China.

SALICACEAE (EP, BH, H). A family of the Salicales (EP, H), Incompletae (BH). These are woody with simple alternate stipulate leaves and spikes of dioecious flowers which lack a perianth. The disk is cup-like or reduced to scales. There are many to 2 free stamens, and a superior ovary of 2 fused carpels, and one loculus. The placentas are parietal and bear many anatropous ovules. The fruit is a capsule with many seeds which are small with a basal tuft of hairs, and no endosperm.

SALICALES (EP, H). An order of the Archichlamydeae (EP), Lignosae (H). It contains one family—the Salicaceae.

SALINE SOIL. A soil developed in an area of waterlogging and high evaporation, so that there is a high accumulation of salts, especially calcium carbonate, gypsum, and sodium salts.

SALSUGINOSUS (Latin), SALSUGINOUS. Growing in a salt-marsh.

SALTANT, SALTATION. A mutation of a fungus or bacterium in culture.

SALTATION. The movement of medium-sized soil-particles by wind. Sand particles are picked-up by eddy winds into the horizontal air-streams. The vertical lift is dissipated and the particle falls back to the ground, throwing up more particles.

SALT GLAND. A hydathode which exudes a saline solution, which evaporates, leaving a salt deposit on the leaf.

SALVADORACEAE (EP, BH, H). A family of the Sapindales (EP), Gentianales (BH), Celastrales (H). These are woody plants with opposite simple leaves and sometimes bristle-like stipules. The bisexual or unisexual flowers are regular and in panicles. The 4-2 sepals are fused, and the 5-4 petals may be fused or free. The 5-4 stamens are free. The superior ovary is of 2 fused carpels, with 2-1 basal ovules in each. The fruit is a berry or drupe, usually one-seeded. The seed has no endosperm.

SALVER-SHAPED. Said of a corolla which has the lower part long and tubular, and the upper part spreading horizontally.

SALVINIACEAE. A family of the Leptosporangiatae. They are heterosporous. The sporocarp wall is formed from the indusium, and contains only microsporangia *or* macrospores.

SALVINIALES. An order of the Leptosporangiatae, including the Salviniaceae and Marsiliaceae.

SAMARA. A single-seeded, dry indehiscent fruit, having a wing-like extension of the pericarp.

SAMARIFORM. Shaped like a samara.

SAMPLE. A finite series of observations or individuals taken at random from the hypothetical infinitely large population of potential observations or individuals.

SAMPLING ERROR. The variance of a statistic arising from, and a function of, the limited size of samples.

SAMYDACEAE (BH, H). A family of the Passiflorales (BH), Bixales (H). (EP) place it in the Flacourtiaceae (*q.v.*).

SAND BINDER. A plant which holds sand-dunes together by forming a mat of rhizomes and roots.

SAND CULTURE. An experimental method for determining the elements required for the healthy growth of a plant. The plants are grown in purified sand, and supplied with solutions of known constitution.

SANGUINEOUS, SANGUINEUS (Latin). Blood-red.

SANIO'S BAND, SANIO'S BEAM = CRASSULA.

SANTALACEAE (EP, BH, H). A family of the Santalales (EP, H), Achlamydosporeae (BH). (BH) include the Grubbiaceae, Myzodendraceae. These are semi-parasitic herbs, shrubs and trees with opposite or alternate leaves and small bisexual or unisexual flowers, with homochlamydeous perianths. The disk is perigynous or epigynous. The perianth lobes are free and in 2 whorls of 2, or of 2 and 3, with as many free stamens inserted on the perianth lobes. The inferior ovary is unilocular with axile placentas and 3-1 ovules. The fruit is a one-seeded nut or drupe. The seeds contain endosperm.

SANTALALES (EP, H). A family of the Archichlamydeae (EP), Lignosae (H). These are herbs, shrubs, or trees which are often parasitic. The flowers are cyclic, and usually with the perianth lobes alike. The free stamens are anteposed in 1 or 2 whorls. The ovary consists of 3-2 fused carpels (rarely 1), which is inferior (rarely superior), each with 1 pendulous ovule.

SAP. An aqueous solution of mineral salts, sugars and other organic substances, present in the xylem.

SAP CAVITY. A large fluid-filled vacuole in the middle of an adult cell.

SAPIDUS. Latin meaning 'with a pleasant taste'.

SAPINDACEAE (EP, BH, H). A family of the Sapindales (EP, BH, H). (BH. include the Aceraceae, Hippocastanaceae, Melianthaceae, and Staphyleaceae) These are woody plants with alternate leaves, and usually zygomorphic, bisexual or unisexual flowers. The disk is extrastaminal. The 5 sepals are free. The 5-3 petals are free (or absent), often with scales. There are usually 8 free stamens (rarely 10, 5, or many). The superior ovary consists of 3-2 fused carpels, each with 1 ovule (usually). The fruit is a capsule, drupe, nut, or schizocarp. The seeds have no endosperm.

SAPINDALES (EP, BH, H). An order of the Archichlamydeae (EP), Polypetalae (BH), Lignosae (H). These are usually woody, and similar to the Geraniales, but the ovules are pendulous with a dorsal raphe and the micropyle uppermost, or erect with a ventral raphe and the micropyle down.

SAPLING. A young tree.

SAPONACEOUS. Slippery as if covered with sap.

SAPONITE. A clay mineral in which all the octahedral holes are filled with magnesium ions, and 15-20% of the tetrahedral holes are filled with aluminium ions, the rest being filled with silicon.

SAPOTACEAE (EP, BH, H). A family of the Ebenales (EP, BH, H). These are woody plants with simple alternate leaves, secretory passages, and usually bisexual flowers. There are 8-4 free calyx lobes in 2 whorls, and as many fused petals in 1 whorl, or twice as many in 2 whorls. The petals sometimes have dorsal or lateral appendages. The stamens are free in 2 or 3 whorls, the outer ones are sometimes staminodes. The ovary consists of fused carpels and is superior. There are as many, or twice as many carpels as there are stamens in 1 whorl, each with one basal or axile ovule. The fruit is a berry. The seeds may or may not contain endosperm.

SAPROBE. (1) = SAPROPHYTE.
(2) A plant growing in foul water.

SAPROBIOTIC. Feeding on dead or decaying organic matter.

SAPROGENOUS. Growing on decaying matter.

SAPROLEGNIACEAE. A family of the Saprolegniales. The mycelium is not definitely constricted at intervals, nor do they have cellulin plugs. There is no enlarged basal portion and slender branches. The zoospores are mostly dimorphic, and the oögone has one or more eggs and lacks periplasm.

SAPROLEGNIALES. An order of the Phycomyceteae. These are eucarpic fungi, usually with well-developed branches and a tapering holdfast system. Most are saprophytic in soil and fresh-water, but some are parasitic in roots, algae, fungi, or fresh-water animals. The external hyphae are branched and coenocytic, nearly uniform in thickness, or with large basal segments and slender branches. Asexual reproduction is by biflagellate zoospores, often dimorphic, rarely of the primary type only, often of the secondary type only. The oögones have one or more eggs, with or without periplasm. Fertilization is by antherids to form thick-walled oöspores.

SAPROPHYLOUS = SAPROGEONOUS.

SAPROPHYTE. A plant living on, and deriving its food from dead organic matter.

SAPROPLANKTON. Plankton growing in polluted water.

SAPROZOIC. Feeding on dead or decaying organic material.

SAPWOOD. The outer region of the xylem in trees. It contains living cells and functional water-conducting and food-storage tissue.

SARCINIFORM. Formed in small packets.

SARCOCARP. The fleshy part of the pericarp of a drupe.

SARCODY. Conversion, so as to have a fleshy texture.

SARCOLAENACEAE (H). A family of the Ochnales (H). These are erect trees and shrubs. The leaves have extrapetiolar stipules. There are 3 sepals, which are usually involucrate at the base. The stamens are free among themselves, with the anther-connective not produced at the apex. The ovules are axile.

SARCOSPERMACEAE (H). A family of the Ebenales (H). The leaves are opposite, or sub-opposite with the flowers in axillary racemes or panicles. The flowers are bisexual, with epipetalous stamens, the fertile ones being opposite the corolla-lobes. The ovules ascend from the inner angles of the carpels. The seeds have a broad basal, or lateral hilum.

SARCOTESTA. A fleshy layer in the testa.

SARGENTODOXACEAE (H). A family of the Berberidales (H). These are shrubs or climbers, and the wood has broad primary medullary rays. The leaves are compound. The flowers are unisexual with many carpels spirally arranged.

SARMENTOSE. Having a stem which arises as a small arch from the roots, and then becoming prostrate.

SARRACENIACEAE (EP, BH, H). A family of the Sarraceniales (EP, H), Parietales (BH). These are herbs with pitcher-leaves. The bisexual flowers are in scapes or racemes and are regular. There are 8-5 free petals, 5 free petals, and many free stamens. The superior ovary consists of 5-3 fused carpels, with 5 or 3 loculi, which contain many ovules. The fruit is a capsule and the seeds contain endosperm.

SARRACENIALES (EP, H). An order of the Archichlamydeae (EP), Herbaceae (H). These are herbs with usually alternate, insectivorous leaves and regular flowers. They are spirocyclic to cyclic and hypogynous. The perianth lobes are all similar, or differentiated into sepals and petals. The superior ovary consists of 3-5 fused carpels with parietal or axile placentas, and many to 3 ovules. The seeds contain endosperm.

SATELLITE. A short segment of a chromosome separated from the rest by one long constriction, if terminal, or two if intercalary.

SATIVUS. Latin meaning 'cultivated or planted'.

SATURNINE. Said of acrospores which have a flat edge around the middle.

SAUCONITE. A clay mineral in which all the octahedral holes are filled with zinc ions.

SAURAUIACEAE (H). A family of the Theales (H). The leaves are simple and alternate. The bracts are not present. There are many stamens, and the anthers are not inflexed in the bud. The plants are not epiphytes.

SAURUACEAE (EP, H). A family of the Piperales (EP, H). (BH) include it in the Piperaceae. These are herbs with alternate leaves and spikes of achlamydeous, bisexual flowers. The stamens are free, and are 6 or less in number. The superior ovary consists of 4-3 free or fused carpels. The placentas are parietal, with many to 2 ovules. The seeds contain endosperm and perisperm.

SAVANNAH. Grass country broken by patches of forest or copse.

SAXATILIS (Latin), SAXICOLE, SAXICOLOUS. Growing on rocks or stones.

SAXIFRAGACEAE (EP, BH, H). A family of the Rosales (EP, BH), Saxifragales (H). (BH) include the Cephalotaceae, and Cunoniaceae. These are herbs, shrubs or trees with alternate leaves and various inflorescences of usually many bisexual, regular (rarely zygomorphic) flowers, with convex, flat or concave axis. The stamens are free and as many or twice as many as the petals. The number of carpels is equal to, or less than the number of petals, usually with free styles. The ovary has 2-1 (rarely 5) loculi, with swollen placentas and many ovules in several ranks. The ovary is superior or inferior. The fruit is a capsule or berry, and the seeds contain endosperm.

SAXIFRAGALES (H). An order of the Herbaceae (H). The leaves are radical, alternate or opposite, and lack stipules. The flowers are regular, more or less perigynous (rarely epigynous). There are petals. The stamens are free and definite. The carpels are free of fused with axile placentas. The seeds have copious endosperm and small straight embryos.

SAXIFRAGEAE (BH) = SAXIFRAGACEAE (EP).

SCABERULOUS. Slightly scabrous.

SCABRID. Having a roughed, file-like surface.

SCABROUS. (1) Having a surface covered with small wart-like projections. (2) Scurfy.

SCALARIFORM. Like a ladder.

SCALARIFORM THICKENING. Internal lignification of the wall of a xylem vessel or tracheid, laid down more or less transversely like the rungs of a ladder.

SCALE. (1) A thin, flat plant member, which may be green when very young, and is usually small.
(2) The hardened, usually non-green bract of a catkin.

SCALE BARK, SCALY BARK. (1) = RHYTIDOME.
(2) Bark which becomes detached in irregular patches.

SCALE HAIR. A multicellular flattened hair.

SCALE LEAF. A membranous tough leaf, which is usually smaller than the normal leaf, and is usually protective.

SCALLOPED. Said of a margin bearing rounded teeth.

SCALPELLIFORM. Shaped like the blade of a pen-knife.

SCALY BULB. A bulb in which the swollen leaf-bases do not form a complete circle in cross-section.

SCANDENS (Latin), SCANDENT. Climbing.

SCAPE. A flower-stalk which is leafless or nearly so, arising from the middle of a rosette of leaves. It bears a flower, several flowers, or a crowded inflorescence.

SCAPHOID. Boat-shaped.

SCAPIGEROUS. Having a scape or scapes.

SCAPINACEAE. A family of the Acrogynae. A family in which the ventral lobes of the lateral leaves are larger than the dorsal lobes.

SCARIOSE, SCARIOUS. Dry, thin, and with a dried-up appearance, usually at the tips and edges.

SCATTERED. Occurring singly, spaced-out, and without any apparent order.

SCENODESMACEAE. A family of the Chlorococcales. They reproduce only by autospores which become apposed to one another at the time of liberation to form a coenobium.

SCHEUCHZERIACEAE (EP, H). A family of the Helobieae (EP), Alismatales (H). (BH) include it in the Naiadeae. These are marsh herbs with narrow leaves and racemes or spikes of bisexual or unisexual regular flowers. The perianth segments are free and in 2 whorls of like segments, each 3 in number. They are bract-like. There are 6 free stamens in 2 whorls of 3. The superior ovary consists of 2 whorls of 3 carpels which are free or united. The outer whorl is sometimes absent. Each carpel contains 1 or 2 anatropous ovules.

SCHISANDRACEAE (H). A family of the Magnoliales (H). These are climbing or trailing shrubs with simple, alternate leaves, which are pellucid-dotted. The flowers are unisexual and solitary. There are 15-9 sepals and petals which are imbricate and scarcely indistinguishable. The many stamens are united to form a globular mass. There are many carpels with 2-3 ovules which are in a globose mass, or spread on an elongated axis in the fruit. The fruit is baccate. The seeds have an oily endosperm and are embedded in pulp. The embryo is small.

SCHIZAEACEAE. A family of the Leptosporangiatae. All the sporangia develop simultaneously and the sporangia are borne singly. The sporangia are generally protected by an indusial outgrowth from the leaf-margin. The sporangia dehisce longitudinally by a transverse, ring-shaped, annulus at the distal end of the sporangial jacket.

SCHIZOCARP. A dry fruit formed from a syncarpous ovary that splits at maturity into its constituent carpels, which are usually one-seeded.

SCHIZCOTYLY. A forking of the cotyledons.

SCHIZOGENOUS. (1) Formed by cracking or splitting.
(2) Said of a secretory cavity which is formed by the separation of the cells.

SCHIZOGONIACEAE. The single family of the Schizogoniales.

SCHIZOGONIALES. An order of the Chlorophyta. These are filamentous, plate-like or solid cylinders. The cells are uninucleate with a single stellate chloroplast. There are no zoospores or gametes formed. Reproduction is by aplanospores and akinetes only.

SCHIZOGONY. The division of a schizont.

SCHIZOMERIDACEAE. A family of the Ulvales. The thalli are solid cylinders several cells in diameter.

SCHIZYMYCETE A.

SCHIZONT. A vegetative thallus, which has no wall and undergoes simple or multiple division.

SCHIZPHYCEAE = MYXOPHYCEAE.

SCHIZOPHYTA. An artificial group of plants containing the Bacteria and Myxophyceae.

SCHIZOSACCHAROMYCETACEAE. A family of the Saccharomycetales, distinguished by the fact that conjugation of two vegetative cells initiates ascus-formation.

SCHIZOSTEGIALES. An order of the Eubrya. These grow in dimly illuminated caves, and appear to be luminous. The protonema are persistent and has lens-shaped cells. The gametophytes have 'leaves' in two rows, and the small capsule lacks a peristome.

SCION. (1) A portion of a plant, usually a piece of young stem which is inserted into a rooted stock in grafting.
(2) A young plant formed at the end of, or along the course of, a runner.
(3) A stolon.

SCIOPHYLLOUS. Having leaves which can tolerate shade.

SCIOPHYTE. A plant which grows in shady situations.

SCIOPHYTIUM. A shade formation.

SCISSILE. (1) Capable of being split.
(2) Said of the flesh of a pileus which can be separated into horizontal layers.

SCITAMINEAE (EP, BH). An order of the Monocotyledoneae (EP), a family of the Epgynae (BH). These are tropical herbs, sometimes large or woody, with cyclic flowers, with the whorls alike or unalike. They are usually zygomorphic with the parts in threes. There are typically 6 stamens in 2 whorls of 3, but there may be great reduction. The inferior ovary usually has 3 loculi, with large ovules. The seeds are usually arillate, with perisperm and endosperm.

SCLERANTHACEAE (BH) = ILLECEBRACEAE (BH).

SCLERANTHIUM. A dry one-seeded fruit enclosed in the hardened remains of the calyx.

SCLEREIDE, SCLERIDE. (1) See *sclerenchyma*.
(2) A thick-walled cell mixed with the photosynthetic cells of a leaf, giving them mechanical support.
(3) A stone cell.

SCLERENCHYMA. Mechanical tissue consisting of cells with thick lignified walls and small lumens. If the cells are elongated they are called *fibres*, and have pointed ends, usually occurring in bundles. When the cells are oval or rounded, they are called *sclerides*. These occur singly or in groups.

SCLEROCAULOUS. Having a hard, dry, stem.

SCLERODERMATACEAE. The single family of the Sclerodermatales.

SCLERODERMATALES. An order of the Gasteromyceteae. These are medium to large, with the fruit-bodies mostly above the surface at maturity. The hymenial cavity is obliterated or replaced by nests or clusters of basidia. At maturity the whole gleba or portions of it become a powdery mass of spores with only rudimentary or lacking capillitium.

SCLEROPHYLL. A hard, stiff leaf which is heavily cutinized.

SCLEROPHYLLOUS VEGETATION. Woody plants with hard, tough, and generally small leaves. It is characteristic of dry places.

SCLEROPHYLLOUS. Having sclerophylls.

SCLEROSED, SCLEROTIC. Having hard, usually lignified walls.

SCLEROSIS. The hardening of cell-walls, or tissues, usually by thickening and lignification.

SCLEROTESTA. A hard layer in the testa of a seed in which a sarcotesta is also present.

SCLEROTIC = SCLEROSED.

SCLEROTIC CELL = SCLERIDE, STONE CELL.

SCLEROTINIACEAE. A family of the Inoperculatae. The apothecia arise from a definite sclerotium or stromatized portion of the sub-stratum, and are stipitate. The asci are cylindrical to club-shaped with 8 spores. The ascospores are ellipsoid, and usually colourless, smooth and one-celled. Most are parasitic.

SCLEROTIUM. (1) A compact mass of fungal hyphae often with a thickened rind. It varies in size from a pin-head to a man's head. They are organs of perennation, and may give rise to fruit-bodies.
(2) Of Myxomycetes, the firm, resting condition of a plasmodium.

SCOBICULAR, SCOBICULATE, SCOBIFORM. Looking like saw-dust.

SCOBINATE. Said of a surface which is roughened.

SCOLECITE. (1) = ARCHICARP.
(2) A loosely coiled hyphae at the centre of a perithecium which later gives rise to ascogenous hyphae.

SCOLECOSPORE. A long, thread- or worm-like spore.

SCOLECOSPOROUS. Having thread-shaped or worm-shaped spores.

SCORPIOD CYME. A cyme in which the branches develop alternately to left and right, but do not lie all in one plane. The bud, the axis of the inflorescence is coiled.

SCOTOPHYTE. A plant which lives in the dark.

SCRAMBLER. A plant which has long, weak shoots, and grows over and above other plants.

SCROBICULATE, SCROBICULATUS (Latin). Having the surface dotted all over with small, rounded depressions.

SCROBICULUS. A small pit, or rounded depression.

SCROPHULARIACEAE (EP, BH, H). A family of the Tubiflorae (EP), Personales (BH, H). These are herbs or shrubs, rarely trees, with alternate, opposite or whorled leaves, and variously arranged flowers, but they are never terminal. The flowers are bisexual, more or less zygomorphic, with 5 sepals and petals. There are 4 or 2 stamens which are free. The superior ovary consists of 2 fused carpels, each with many or few ovules on axile placentas. The fruit is a capsule or berry. The seeds have endosperm.

SCROPHULARINAEA (BH) = SCROPHULARIACEAE (EP).

SCROTIFORM. Like a bladder.

SCRUPOSE. Roughened with very small hard points.

SCURFY. Having the surface covered with small bran-like scales.

SCUTATE = SCUTIFORM.

SCUTELLAR EPITHELIUM. A layer of elongated cells covering the scutellum, lying against the endosperm, and producing enzymes which assist in the utilization of the substances stored therein.

SCUTELLATE. Saucer-shaped.

SCUTELLUM. (1) The cotyledon of a grass embryo. It is flattened and in intimate contact with the endosperm, acting as an absorptive organ.
(2) A shield-like cover, as in some ascoma.

SCUTIFORM. Shield-shaped.

SCYPHIFEROUS. Having scyphi.

SCYPHIPHOROUS. Cup-bearing.

SCYPHOSTEGIACEAE (H). A family having affinities with the Celastrales and Bixales. The leaves are alternate and lack stipules. The plants are dioecious, and the flowers unisexual. The petals are united. The 3 stamens are connate. The carpels are free and enclosed in an enlarged disk, which ultimately forms a false fruit.

SCYPHUS. The cup-shaped widening at the distal end of the potedium of some lichens.

SCYTOPETALACEAE (EP, H). A family of the Malvales (EP), Tiliales (H). These are woody plants with leathery leaves which are alternate. The long-stalked flowers are in bunches or racemes. The calyx is dish-like. The 7-3 petals are free and valvate. There are many free stamens. The 6-4 carpels are fused each containing 6-2 pendulous ovules. The fruit is woody or a drupe, and is one-seeded.

SEAWEEDS. Algae living in, or in close proximity to the sea.

SEBACEOUS. Looking like lumps of tallow.

SECEDING. Of gills which are:
(1) Separating from the stipe, or
(2) At first adnate, then free.

SECONDARY ASSOCIATION. The coming together of bivalent chromosomes during meiosis.

SECONDARY CELL-WALL. The layers of wall-material deposited on the primary wall as it ages. The amount of cellulose in usually higher than in the primary wall, and the amount of pectin less.

SECONDARY CORTEX = PHELLODERM.

SECONDARY GROWTH = SECONDARY THICKENING.

SECONDARY MERISTEM. A region of active cell-division that has arisen from permanent tissue.

SECONDARY MYCELIUM. (1) The mycelium of the binucleate segments, bearing clamp-connections, found in many Basidiomycetes.
(2) Hyphae which grow down from the developing fruit-body of a fungus. They absorb food-material from the substratum.

SECONDARY NUCLEUS. The nucleus formed in the embryo-sac by the union of two polar nuclei.

SECONDARY PETIOLE. The petiole of a leaflet of a compound leaf.

SECONDARY PHLOEM. The phloem formed by the activity of a cambium.

SECONDARY SEGREGATION. The segregation in an allopolyploid of differences between its ultimate diploid parents.

SECONDARY SPORE. Spores, other than basidiospores, formed by basidiomycetes.

SECONDARY SUCCESSION. A succession arising after the ground has been cleared of its original vegetation.

SECONDARY THICKENING. The formation of additional vascular tissue by the activity of a cambium.

SECONDARY TISSUE. Tissue formed by secondary thickening.

SECONDARY WALL LAYER = SECONDARY CELL WALL.

SECONDARY CONTRACTION. Shortening and thickening of the chromatid threads in diplonema stage of meiosis, as diakinesis comes on.

SECOND DIVISION. The second of two divisions at meiosis.

SECOTIACEAE. A family of the Hymenogastrales. The columella is produced downwards below the pileus to form a distinct stipe. The gleba is free from the stipe, at maturity, at least below.

SECRETION. (1) A substance produced within a cell and passed out through the plasma membrane, and having a special function in the organism.
(2) The passage of the substance from the cell.
(3) In contrast to simple diffusion, the cell does work against the forces of diffusion, and possibly in the case of water, osmosis.

SECRETORY CELL. A cell which carries out secretion. The wide range of substances produced include oils, resins, nectar etc. These substances may be retained within the cell or exuded from it.

SECRETORY DUCT, SECRETORY PASSAGE. An intercellular space, usually elongated, in which secretions accumulate.

SECRETORY TISSUE. A group of secretory cells.

SECTION. (1) A division of a genus, containing a number of closely related species.
(2) A thin piece of material cut so as to be capable of being examined by a microscope.

SECTORIAL CHIMAERA. A chimaera in which a plant consists of two of more types of tissue, which are arranged in cross-section as sectors of a circle.

SECTORING. Mutation in plate-cultures of bacteria and fungi resulting in one or more sectors of the culture having a changed form of growth.

SECUND. Having the lateral members all turned to one side.

SECUNDINE. The inner integument, where two are present.

SEDOHEPULOSE. A seven-carbon sugar playing a part in photosynthesis.

SEED. A product of a fertilized ovule, consisting of an embryo, enclosed in the testa which is derived from the integument(s). Food reserves may or may not be contained in the endosperm.

SEED LEAF, SEED LOBE. A cotyledon of a flowering plant.

SEED STALK = FUNICLE.

SEED VESSEL. A dry fruit.

SEEDLING. The young plant developing from a germinating seed.

SEGETALIS. Latin meaning 'growing in grain fields'.

SEGMENT. (1) A portion of the lamina of a leaf, when it is deeply lobed, but not divided into leaflets.
    (2) A daughter-cell, cut off from a single apical cell.
    (3) The free part, when a calyx or corolla consists of partially fused lobes.
    (4) A multinucleate portion of a filament or hypha, which is cut-off by a cross-wall.

SEGMENTAL INTERCHANGE. The exchange of pieces, between two chromosomes which are not homologous.

SEGMENT HALF-CELL. One of the two smaller cells cut off from the pialac cells of most Acrogynae. They ultimately give rise to a 'leaf'.

SEGREGATE. (1) Keep separate.
    (2) In taxonomy a group which is based on part of an earlier group.

SEGREGATION. The separation at meiosis of the members of a pair of allelomorphs.

SEISMONASTY. The response to a non-directional mechanical stimulus.

SELAGINACEAE (EH, H). A family of the Lamiales (BH, H). These are shrublets or perennial herbs, with opposite, rarely sub-opposite leaves. The flowers are in terminal spikes, corymbs or heads, and the inflorescence has no common involucre. The anthers are unilocular. The bilocular ovary is entire or slightly lobed, with 1 ovule in each loculus. The style is terminal.

SELAGINEAE (BH) = SELAGINACEAE (EP).

SELLAGINELLACEAE. The single family of the Selaginellales.

SELAGINELLALES. An order of the Lycopodinae. The sporophytes are herbaceous and usually without any secondary thickening of the stem. The leaves are microphyllous and ligulate. The roots are borne on rhizophores. The sporophytes are heterosporous, with the sporophylls in strobili. The micro-gametophytes and macrogametophytes are markedly different in size, and the antherozoids are biflagellate.

SELECTION. (1) Statistical, discrimination in sampling or arrangement, as opposed to randomness.
    (2) Biological, any non-random process which will lead to individuals of different genotype being represented unequally by their progeny in later generations of a population of self-propagating units. Selection can be natural or artificial.

SELECTION PRESSURE. The measure of the action of selection in tending to alter the frequency of a gene in a given population.

SELECTIVE ABSORPTION. The ability of a cell, or the cells of an organ, to accumulate a particular ion in a higher concentration in its vacuole, than it is in the surrounding medium; or alternatively not to accumulate a substance which is present in the surrounding medium, and would normally pass into it by diffusion.

SELECTIVE ADVANTAGE. The genotypic condition of a cell, individual, or genetic class of individuals which increases its chances relative to others, of representation in later generations of cells or individuals.

SELECTIVE FERTILIZATION. The fusion of gametes of different sexes in a non-random manner.

SELECTIVE TRANSPORT. The preferential movement of a particular ion from one part of a plant to another, so that the ion appears in higher concentration in some parts than others.

SELF-COLOURED. Having the same colour all over.

SELF-COMPATIBLE = SELF-FERTILE.

SELF-FERTILIZATION. The fusion of male and female gametes produced by the same individual.

SELF-INCOMPATIBLE = SELF-STERILE.

SELF-POLLINATION. The transfer of pollen from the anther to the stigma of the same flower. More generally it may mean a similar transfer to stigmas on the same plant, or within the same clone.

SELF-PROPAGATION = REPRODUCTION.

SELF-STERILE. The inability of gametes of opposite sex, but produced by the same individual, to fuse.

SELFING = SELF-POLLINATION.

SEMI-. Latin meaning 'half', but also means 'somewhat', 'more or less'.

SEMI-AMPLEXICAUL. Said of a leaf-base which half clasps the stem.

SEMI-APOGAMY. The fertilization of one cell by another, when the cells are not of opposite sexes, but at least one of them is semi-gametic, *i.e.* can function as a gamete.

SEMI-CELL. One of the two halves of a cell of a desmid.

SEMI-HETEROTYPIC DIVISION. The first division of meiosis, when the chromosomes fail to pair properly. It gives rise to a restitution nucleus.

SEMI-LUNAR. Shaped like a half-moon.

SEMI-NATURAL VEGETATION. Vegetation which is influenced by constant interference by man, although not actually planted by him.

SEMINIFEROUS. Seed-bearing.

SEMIPERMEABLE MEMBRANE. A membrane which permits the passage of a solvent, but is impermeable to stated dissolved substances.

SEMISTERILE. Said of a gene heterozygote of structural hybrid in which approximately half the male and female gametes are inviable.

SEMITERETE. Half-cylindric.

SEMPERVIRENT, SEMPERVIRENS (Latin). Evergreen.

SENESCENT. Said of that stage in the life-history of an individual when the rate of metabolic activities declines, and there is a change in the physiology prior to death.

SENILITY. Senescence in an individual or race.

SENSE ORGAN. Any organ or cell adapted for the reception of stimuli.

SENSE ORGANELLE. A specialized part of a cell functioning as a sense organ.

SENSE TRIPLET. Three adjacent nucleotides on RNA, necessary for the production of an amino-acid. This is theoretically the smallest number which will code the production of an amino-acid.

SENSIBILITY. The degree of reaction of a plant attacked by a parasite.

SENSITIVE PERIOD. The period of development during which the action of a gene is sensitive to the influence of external conditions.

SENSITIVE VOLUME. The hypothetical volume in which an ionization must produce a given gene mutation.

SENSITIVITY = SENSIBILITY.

SEPAL. One of the parts (lobes) forming the calyx of a flower. It is usually green, and protects the rest of the flower in the bud.

SEPALINE. Like a sepal.

SEPALODY. An abnormal condition when some members of a flower are converted to sepals..

SEPALOID = SEPALINE.

SEPARABLE, SEPARATING. (1) Becoming detached normally.
(2) = SECEDING.

SEPARATION. The liberation of a reproductive body from the parent plant.

SEPARATION DISK. A bi-concave disk of intercellular material found occasionally between the cells of a filament of the blue-green algae. They aid in the break-up of the filament into hormogones.

SEPARATION LAYER = ABSCISSION LAYER.

SEPTAL. Of hedgerows.

SEPTATE. (1) Divided into cells or segments by walls.
(2) Said of the ovary of flowering plants when it is divided into chambers by partitions.

SEPTATE FIBRE. A fibre in which the lumen is divided by transverse walls.

SEPTEMFID, SEPTEMIFID. Deeply divided into seven parts.

SEPTENATE. Having the parts in sevens.

SEPTICIDAL. Said of a fruit which opens by the loculi splitting through the middle, and so forming several compartments.

SEPTIFRAGAL. Said of the dehiscence of a fruit, when the outer wall breaks away from the walls of the loculi.

SEPTOBASIDIACEAE. A family of the Auriculariales. They are parasitic on or symbiotic with scale-insects. The basidia usually have well-developed hypobasidia, and 1-4-celled epibasidia. Hypobasidia sometimes lacking. The basidiospores are on distinct sterigmata. Conidia are often produced, and clamp-connections are apparently lacking.

SEPTUM. (1) A wall between two adjoining cells.
(2) A transverse wall in a fungal hypha, an algal filament, or a spore.
(3) The partition between two loculi of an ovary.

SEQUENCE HYPOTHESIS. A theory how RNA codes the production of a protein. It is that a particular sequence of nucleotides on the RNA molecule is responsible for the manufacture of a particular amino-acid.

SERAL COMMUNITY. A stage in the succession of the development of a plant community towards its climax.

SERE. A particular example of a plant succession.

SERIAL BUD. A supernumerary bud beside an axillary bud.

SERIATE. Arranged in a row.

SERICEROUS. (1) Like silk.
(2) Having a silken sheen.

SERINE. A water soluble amino-acid found in protein.

$$CH_2OH$$
$$|$$
$$CH—NH_2$$
$$|$$
$$COOH$$

SEROLOGY. A method of identifying microorganisms and viruses, their chemical components, and their relations to one another. One acts as an antigen in blood serum, against which the activities of another can be examined.

SEROTINAL, SEROTINOUS, SEROTINUS (Latin). Appearing late in the year.

SERRATE. Said of a toothed margin, when the teeth are pointed, and project forwards.

SERRATULATE, SERRULATE. Serrate, but with very small teeth.

SESAMOID. Granular.

SESQUI-. A Latin prefix meaning 'one-and-a-half'.

SESQUIPEDALIS. Latin meaning 'one-and-a-half feet'.

SESSILE. (1) Lacking a stalk.
(2) Fixed and stationary.

SESSILE BENTHOS. Plants growing attached to the bottom of water.

SESTON. Very small planktonic organisms which are retained only by the finest net.

SETA. (1) A bristle.
(2) A slender, straight prickle.
(3) A long hollow outgrowth from a cell-wall.
(4) A single elongated cell, or row of cells with scanty colourless contents, found in some algae.
(5) A thick unicellular structure found among the asci of some ascomycetes.
(6) The multicellular stalk which bears the capsule of liverworts and mosses.

SETACEOUS. Shaped like a bristle.

SET OF CHROMOSOMES. A group of chromosomes consisting of one of each kind contained in the nucleus of a gamete.

SETOSE. Bristly.

SEX, SEXUAL DIFFERENTIATION. The production by an individual or group of individuals, of gametes of two types, differing in size and mobility (male and female gametes) such that the one type can only fuse with the other type.

SEXANGULAR. Having six angles.

SEX CELL = GAMETE.

SEX CHROMOSOMES. Chromosomes of which there is a homologous pair in the nucleus of one sex (the homogametic sex) and a dissimilar pair or only one chromosome, in those of the other (the heterogametic sex).

SEX DETERMINATION. The process by which a haploid spore or egg, or diploid zygote comes to develop the properties of one or other sex.

SEXFARIOUS. In six rows.

SEX-INTERGRADE. A plant bearing staminate and pistilate flowers, but belonging to a species which is normally dioecious.

SEX-LIMITED. Said of the inheritance of differences which are expressed in one sex only, or if in the two sexes, then differently.

SEX LINKAGE. A gene having special distribution with reference to sex, as a result of being carried on the X chromosome.

SEXPARTITE. Divided deeply into six lobes or segments.

SEX RATIO. The ratio of the males to females in a population at a given time. It is usually expressed as the number of males per 100 females.

SEXUAL DIMORPHISM. Structural differences between the males and females of a species, especially if the differences are superficial.

SEXUAL ISOLATION. Genetic isolation by a sexual mechanism.

SEXUAL REPRODUCTION. The fusion of male and female gametes, or their nuclei, to form a zygote which develops into a new individual.

SHADE CHROMOPHORE. A plant inhabiting a shaded rock-crevice.

SHADE PLANT. (1) A plant which grows best in the shade.
(2) A plant grown to shade a crop-plant.

SHAGGY. Covered with long, weak hairs.

SHEATH. (1) A leaf-base when it forms a tubular casing around the stem.
(2) A cup of tubular cells without protoplasts formed around the base of the ligule of the Lycopodinae.

SHEET EROSION. The removal of the surface soil over a wide area by water, in contrast to gullying.

SHELL BARK = RHYTIDOME.

SHIKIMIC ACID. A cyclic compound formed from an aromatic unit, and intermediary in the synthesis of lignin.

SHOOT. The part of a plant which develops from the plumule.

SHOOT-TENSION = TRANSPIRATION PULL.

SHORT-CYCLIC. Said of a rust which lacks various stage(s) in the life-cycle.

SHORT DAY PLANT. Plants which produce flowers only if there is less than twelve hours daylight per day.

SHORT SHOOT. A small shoot, borne in the axile of a scale leaf, and bearing the foliage leaves.

SHRUB. A woody plant in which the side shoots are well-developed, so that there is no trunk. They are less than thirty feet high.

SIALLITIC SOIL. A soil in which the clay fraction has a medium to high silica-alumina ratio.

SICCUS. Latin meaning 'dry, juiceless'.

SIEROZEM. A grey semi-desert soil, with a fairly low organic matter content, and with calcareous and gypseous horizons near the surface.

SKELETAL SOIL. A soil derived from physically weathered material.

SIEVE AREA. A limited area on the longitudinal wall of a sieve tube, perforated by many small pores, through which material may pass.

SIEVE FIELD. One of the perforated areas into which a sieve-plate may be divided by a net-work of thick strands of wall material.

SIEVE PLATE. See *Sieve tube*.

SIEVE TUBE. A characteristic element of phloem. It translocates food materials synthesized into plant. The cells are living, thin-walled and in longitudinal rows. They are connected by perforations in their transverse walls, through which pass strands of cytoplasm. The perforated walls are called *sieve-plates*. Sieve areas may be present in the longitudinal walls.

SIGILLARIACEAE. A family of the Lepidodendrales. The trunk is massive and sparingly branched. The leaf scars are in vertical series, often on vertical ribs.

SIGMOID. Curved like the letter S.

SIGNIFICANCE. The measure of reliability of a difference between observation and expectation. A difference is *significant* if, on the hypothesis being tested, the probability of obtaining one as large or larger, is lower than some chosen level, termed the *level of significance*.

SIKYOTIC. Parasitic by the fusion of the protoplasm of the host and parasite.

SILICLE, SILICULA, SILICULE. See *Siliqua*.

SILICOLOUS. Growing on rocks which contain much silica.

SILIQUA, SILIQUE. A dry elongated fruit, formed from a superior ovary of two united carpels, and divided by a septum between the carpels into two loculi. Dehiscence is by the separation of the carpels from below upwards, leaving the seeds exposed on the septum. A *silicle* is similar, but short and broad, never more than four times as long as broad.

SILKY. Said of a surface covered with fine hairs which are flattened and gleam like silk.

SILURIAN. A geological period from 340-310 million years ago.

SIMARUBACEAE (EP, BH, H). A family of the Geraniales (EP, BH), Rutales (H). (BH) include the Brunelliaceae, Cneoraceae, and Koeberliniaceae. These are woody plants with a bitter bark. The leaves are alternate or opposite, usually pinnate and without stipules. The flowers are regular and usually unisexual. There are 5-4 free sepals and petals and a disk. There are many, 10 or 5 free stamens. The superior ovary consists of 5 or less fused carpels. The fruits are of various kinds. The seeds have little or no endosperm.

SIMARUBEAE (BH) = SIMARUBACEAE (EP).

SIMPLE. Not divided into several like parts.

SIMPLE FRUIT. A fruit formed from one ovary.

SIMPLE LEAF. A leaf with the lamina in one piece. It may be lobed, but the lobes never reach the mid-rib.

SIMPLE SORUS. A sorus containing one sporangium.

SIMPLE TISSUE. A tissue made up of cells all of the same kind.

SIMPLE UMBEL. An umbel in which the flower-stalks arise directly from the top of the main stalk.

SIMPLE VENATION. A type of venation in which only the mid-rib is clearly visible.

SIMPLEX. The condition of a polyploid in which all the chromosomes of one homologous type carry the dominant allelomorph of a particular gene once.

SIMPLEX GROUP. The haploid complement of chromosomes.

SINGLE FLOWER. A flower which has one set of petals, with no sign of doubling. It is referred loosely to the flower-head of the Compositae, when it is like a daisy.

SINISTRORSE. Turned to the left.

SINKER. A haustorium formed by mistletoe.

SINUATE. (1) Having the margin divided into wide irregular teeth or lobes, which are separated by shallow notches.
(2) Said of the gills of agarics which curve suddenly on reaching the stipe.

SINUOUS. Waved from side-to-side.

SINUS. A depression or notch in a margin between two lobes.

SIPHON. An elongated cell which extends the whole length of a joint in some red algae.

SIPHONALES. An order of the Chlorophyceae. The plant-body is a multinucleate coenocyte, which is branched. A few produce zoospores or aplanospores, and nearly all produce gametes, which are isogamous, anisogamous, or oögamous.

SIPHONEOUS. Tubular.

SIPHONOCLADIALES. An order of the Chlorophyceae. These are siphonaceous when young, like the siphonales, but develop many multinucleate segments when older. Vegetave multiplication is frequent, but rarely by zoospores. Sexual reproduction is by biflagellate isogametes.

SIPHONOGAM. A plant in which the contents of the pollen-grain pass into the embryo-sac through a pollen-tube.

SIPHONOSTELE. A stele in which the xylem and phloem form concentric cylinders around a central pith. It is *ectophloic* if the xylem is nearest the pith, and *amphiphloic* if the phloem is nearest the pith. An amphiphloic siphonostele = *solenostele*.

SIROBASIDIACEAE. A family of the Tremellales. The hymenium is external, and the basidia are formed in chains, by successive transformation of the basidiogenous hyphae into basidia, beginning at the apex. The basidia are four- or two-celled, the septum in the latter case being oblique. The basidiospores are sessile.

SIROLPIDIACEAE. A family of the Lagenidiales. These are parasitic on marine algae. The fungus lies free within the host-cell and produces a cellulose cell-wall. Usually this cell elongates to form several zoosporangia by cross-walls. Laterally biflagellate zoospores are produced which escape fully-formed through an inoperculate exit-tube. Resting spores are not known.

SISTER CELL. One of two cells formed by the division of a pre-existing cell.

SISTER NUCLEUS. One of the two nuclei formed by the division of a pre-existing nucleus.

SKEIN. The nuclear reticulum.

SKEWNESS. Asymmetry is a frequency distribution.

SKIN. (1) = EPIDERMIS.

(2) The outer tissue of the periclinal chimaera.

SKIOPHYTE = SCIOPHYTE.

SKYLIGHT. Diffuse sunlight, in contrast to the direct glare of the sun.

SLASHED. Cut deeply by tapering incisions.

SLEEP MOVEMENT. The folding of the leaflets of a compound leaf at night time. This brings the stomatal surfaces together.

SLIDING CARRIER. A hypothetical carrier of ions across membranes. It is supposed to slide along the membrane-surface, pick up the ion, and slide through a pore in the membrane, depositing the ion on the other side. The carrier stays in contact with the membrane all the while.

SLIDING GROWTH. The movement of developing tracheids and vessels along their mutually touching longitudinal walls as the cells elongate.

SLIME FLUX. An exudation of a watery solution of sugars and other substances from trees which are wounded, or attacked by parasites.

SLIME-FUNGI, SLIME-MOULDS = MYXOMYCETAE.

SLIME PLUG. A mass of slimy material which blocks the pores in a sieve-plate.

SLIME STRING. A sticky mass of food material passing through a pore in a sieve-tube.

SLING FRUIT. A fruit from which the seeds are ejected by elastic tissue.

SLIPPING ZONE. A shiny surface a little below the rim of a pitcher on which insects slip and fall to the bottom of the pitcher.

SLOOTING = GULLYING.

SMARAGDINE. Emerald, or dark, bluish-green.

SMILACACEAE (H). A family of the Liliales (H). These are terrestrial, rarely swamp plants, with climbing or straggling stems which are often prickly. The leaves are usually well-developed, or there are cladodes with axillary flowers. The inflorescence is not subtended by a spathe-like leaf-sheath. The flowers are usually dioecious. The stamens have free filaments, and the anthers are unilocular. The ovary is superior.

SMOOTH. Said of a surface which bears no appendages.

SMUT = USTILAGINALES.

SNAIL PLANT. A plant which is pollinated by snails.

SOBOLE. A creeping underground stem which develops leaf-buds and roots at the nodes.

SOBOLIFEROUS. Bearing soboles.

SOCIAL PLANTS. Species which grow in large groups, and occupy wide areas.

SOCIETY. A minor community within a consociation, arising as a result of local variation in the environment, and dominated by species other than the consociation dominant.

SOFT BAST. The sieve-tubes and phloem parenchyma.

SOIL FLORA. The plants, chiefly fungi, living in the soil.

SOIL SOLUTION, SOIL WATER. The aqueous solutions of salts and other materials, occupying the spaces between the soil-particles.

SOL. A colloidal solution.

SOLANACEAE (EP, BH, H). A family of the Tubiflorae (EP), Polemoniales (BH), Solanales (H). (BH) include the Nolanaceae. These are herbs or shrubs with alternate leaves, and usually bisexual, regular flowers, with 5 sepals and petals. The flowers are terminal, solitary, or in cymose umbels. There are 5 free stamens. The ovary consists of 2 fused carpels and is superior. Each carpel has many to 1 ovules on axile placentas. The fruit is a berry or capsule, and the seeds contain endosperm.

SOLANALES (H). An order of the Herbaceae (H). These are herbs or climbers with alternate, exstipulate leaves. The corolla is regular, and there are as many stamens as corolla-lobes, and alternating with them. The ovary is superior with 4-1 loculi, (usually 2) or of almost free carpels. There are many to 1 ovules on axile placentas. The seeds have some endosperm and the embryo is often curved.

SOLARIZATION. The temporary stopping of photosynthesis in a leaf which is exposed to strong light for a long time.

SOLDERED. Joined.

SOLENOSTELE. The vascular tissue in plant when the phloem lies on both sides of the xylem.

SOLIFLUCATION. A downflow, or slip of surface soil material.

SOLITARY. (1) Single.
  (2) Of flowers occurring one in each axil.

SOLOD. A later stage in the development of a saline soil, after leaching.

SOLONCHAK. A saline soil developed under hot, dry conditions, where the salts accumulate near the surface.

SOLONETZ. A solonchak which has been leached of its salt, so that the surface exchangable ions are not sodium, but calcium or magnesium.

SOLOPATHOGENIC. Of a smut, a pathogenic monosporidial strain.

SOLOTI SOIL. A degraded alkaline soil, in which the sesquioxides have been washed out, leaving a bleached, eluviated horizon, relatively rich in silica.

SOLUTION. The abnormal separation of parts which are normally united.

SOMATIC. Referring to all parts of a plant, except the germ mother-cells and gametes.

SOMATIC APOGAMY. The development of the sporophyte from the tissues of a gametophyte, without the fusion of nuclei. It therefore has the same chromosome number as the gametophyte.

SOMATIC DOUBLING. The doubling of the number of chromosomes in the nuclei of somatic cells.

SOMATIC MITOSIS. The division of a metabolic nucleus.

SOMATIC MUTATION. A mutation in a somatic cell, rather than in a germ-cell.

SOMATIC SEGREGATION. The formation, by mitosis of cells differing from one another, either through mutation, or by somatic crossing-over in the nucleus, or by an unequal assortment of cytoplasmic determinants.

SOMATOTROPISM. A directed growth movement in a plant so that the members come to be placed in a definite position in relation to the substratum.

SONNERATIACEAE (EP, H). A family of the Myrtiflorae (EP), Myrtales (H). (BH) include it in the Lythraceae. These are woody plants with opposite exstipulate leaves and bisexual or unisexual regular flowers with bell-shaped axes. There are 8-4 free sepals, 8-4 or no free petals, and many free stamens. The inferior ovary consist of 15-4 fused carpels which form a hollow axis. There are 15-4 loculi with many ovules. The fruit is a capsule or berry-like, and there is no endosperm.

SORALIUM. A group of soredia surrounded by a distinct margin formed from the thallus of a lichen.

SORAL MEMBRANE. The wall surrounding the sorus in some lower fungi.

SORBITOL. The sugar alcohol of glucose.

$$
\begin{array}{c}
CH_2OH \\
| \\
H-C-OH \\
| \\
HO-C-H \\
| \\
H-C-OH \\
| \\
H-C-OH \\
| \\
CH_2OH \qquad \textit{D-Sorbitol}
\end{array}
$$

SORDID, SORDIDUS (Latin). Dirty-coloured.

SOREDIAL BRANCH. A branch of a lichen thallus formed by a sorideum beginning to develop before it has become detached from the parent thallus.

SOREDIATE. Having small patches on the surface.

SOREDIUM. An organ of vegetative reproduction of lichens. One or more algal cells become enclosed in hyphae. This small body becomes detached and gives rise to a new thallus.

SORIFEROUS. Bearing sori.

SOROCARP. The fruiting structure of the Acrasiales.

SOROPHORE. The sorus-stalk.

SOROSIS. A fleshy fruit formed from a number of crowded flowers, *e.g.* a pineapple.

SOROSPHERE. A hollow ball of spores formed by some of the lower plants.

SORUS. (1) A group of fern sporangia.
(2) A powdery mass of soredia on the surface of a lichen thallus.
(3) A fruiting structure in certain fungi, especially the spore-mass of rusts and smuts.
(4) A group of fruit-bodies in the Synchytriaceae.

SPACE PARASITE. A plant which inhabits the intercellular spaces in another plant, obtaining shelter, but possibly nothing else.

SPADICEOUS, SPADICEUS (Latin). (1) Chestnut coloured.
(2) Like a spadix.
(3) Bearing a spadix.

SPADIX. A spike of flowers with a fleshy axis, enclosed in a spathe.

SPARGANIACEAE (EP, H). A family of the Pandanales (EP), Typhales (H). (BH) include it in the Typhaceae. These are herbs with rhizomes and two-ranked leaves. The flowers are unisexual, but are in heads with the males above and the female below. The perianth consists of 6-3 free sepaloid lobes. There are 6-3 free stamens. The ovary consists of 2-1 fused carpels, each with 1 pendulous ovule. It is superior. The fruit is drupaceous, and the endosperm floury.

SPATHE. A large bract, often coloured or membraneous, enclosing a spadix.

SPATHIFLORAE. An order of the Monocotyledonae. These are herbs or woody, sometimes climbing rarely forming an erect stem. The flowers are cyclic with the perianth lobes all similar or distinguished into sepals and petals, or absent. The flowers are bisexual or unisexual, with the parts in twos or threes (often reduced to one stamen or carpel), and are borne on a spadix, which may or may not be enclosed in a spathe.

SPATHULATE, SPATULATE. Spoon-shaped.

SPAWN. The mycellium of a mushroom.

SPECIALIZED. (1) Having special adaptations to a particular habitat or mode of life which result in wide divergences of characters from ancestral organisms.
(2) The tendency of a parasite to attack only one species or variety of a host plant.

SPECIES. The smallest unit of classification commonly used. In sexually reproducing organisms it is a maximum interbreeding, or potentially interbreeding group, breeding true within its own limits in nature.

SPECIFIC CHARACTERS. The constant characteristic by which a species is distinguished.

SPECIFIC CONDUCTIVITY OF WOOD. The rate at which water flows through a piece of wood of standard cross-sectional area in a stated length of time.

SPECIFIC NAME = EPITHET.

SPECIOSUS. Latin meaning 'handsome'.

SPECTABILIS. Latin meaning 'remarkable'.

SPECTAUS. Latin meaning 'opposite'.

SPERMAGONE = SPERMOGONIUM.

SPERMAGONIUM = SPERMOGONE.

SPERMAPHYTA = PHANEROGAMAE. The seed-bearing plants *i.e.* the Gymnospermae and Angiospermae.

SPERMAPHYTIC. Seed-bearing.

SPERMATIOPHORE. A hypha bearing a spermatium.

SPERMATIUM. A non-motile male sex-cell, present in red algae, in some ascomycetes and basidiomycetes, and some lichens.

SPERMATPHYTA. Seed plants.

SPERMATOPLAST. A male gamete.

SPERMATOZOID = ANTHEROZOID.

SPERMOGONE, SPERMOGONIUM. Of fungi, a flask-shaped, or flattened hollow, structure in which spermatia are formed.

SPERMOPHTHORACEAE. A family of the Saccharomycetales. The only species *Spermophthora gossypii* causes stigmatomycosis of cotton and tomato.

SPHACELARIALES. An order of the Isogeneratae. The growth of the thallus is by a single large apical cell, and its cells are in transverse tiers. The sporophyte produces haploid zoospores, or diploid neutral spores. The gametes are isogamous or anisogamous.

SPHACELATE. Dark and shrunken.

SPHAERIACEAE. A family of the Sphaeriales. The perithecial walls are firm and dark, and they are superficial or in a subiculum. The ostiole is simple or on a low papilla. The asci do not dissolve at maturity.

SPHAERIALES. An order of the Ascomycetae. The perithecia are dark-coloured, usually with firm walls, and without a marked stromatic outer layer. They are not lichen forming.

SPHAEROBOLACEAE. The single family of the Sphaerobolales.

SPHAERBOLALES. An order of the Gasteromycetae. These are small, growing on decayed wood or manure. The whole gleba is expelled as a single ball by the eversion of the inner layer of the peridium, which splits stellately at the top. The gleba has many distinct cavities, or these are obliterated by the ingrowing basidia.

SPHAROCARPACEAE. A family of the Sphaerocarpales. The gametophyte is bilaterally symmetrical and each sex-organ is surrounded by an involucre.

SPHAEROCARPALES. An order of the Hepaticae. Each archegonium has an individual involucre.

SPHAEROCARPOUS. Having a globular fruit.

SPHAEROCEPHALOUS. Having the flower crowded into a rounded head.

SPHAEROCRYSTAL. A rounded crystalline mass of calcium oxalate found in the cells of some plants.

SPHAEROCYST. Globose cells in the tissues of fungi.

SPHAEROPSIDACEAE. A family of the Sphaeropsidales. The pycnidia resemble typical perithecia, and are tough, leathery to brittle, and dark-coloured.

SPHAEROPSIDALES. An order of the Fungi Imperfecti. The conidia are produced within pycnidia or modifications of such structures.

SPHAEROSEPALACEAE (H). A family of the Ochnales (H). These are erect trees or shrubs, with the leaves having extrapetiolar stipules. The stamens are irregularly connate at the base, and the anther-connectives are not produced at the apex. They are erect from the base of the loculi.

SPHAERRAPHIDE. A rounded spiky mass of calcium oxalate found, usually singly, in the cells of many plants.

SPHAGNACEAE. The single family of the Sphagnales.

SPHAGNALES. The single order of the Sphagnobrya.

SPHAGNOBRYA. A subclass of the Musci. The protonema is broadly thallose. The sporophyte is lifted above the gametophyte on a stalk of gamophytic tissue. The sporogenous tissue of the sporophyte develops from the amphithecium of an embryo.

SPHENOID, SPHENOIDAL. Wedge-shaped.

SPHENOPHYLLALES. An order of the Equisetinae. The sporophytes had stems with longitudinal ridged internodes and whorls of sessile leaves at the nodes. The primary xylem is an actinostele and the protoxylem exarch. There is a secondary xylem. The sporangia were monosporous or primitively heterosporous.

SPHENOPSIDS = ARTHROPHYTES = EQUISETALES.

SPHERE-CRYSTAL = SPHAERRAPHIDE.

SPHERIDIUM = CAPITULUM.

SPHEROME. A cell-inclusion which gives rise to globules of fats and oils.

SPHEROPLASTS = MITOCHONDRIA.

SPHERULE. A chlamydospore-like structure.

SPICATE. (1) Spike-like.
(2) In spikes.

SPICULA. A small spike.

SPICULAR CELL. A hard, thick-walled cell, spindle-shaped or branched, occurring among the thin-walled soft tissue.

SPICULATE. Said of a surface covered by fine points.

SPICULE. An obsolete term for a sterigma.

SPICULISPORIC ACID. A metabolic product of *Penicillium spiculisporum*.

SPICULUM. A little spine.

SPIGELIACEAE (H). A family of the Loganiales (H). These are herbs, rarely shrublets. There may or may not be an epicalyx. The corolla lobes are valvate. There are 4 or more stamens inserted in the corolla tube. There are usually many ovules in each ovary or loculus. The fruit is a capsule which is usually septocidally dehiscent. The seeds are not winged.

SPIKE. A raceme of sessile flowers.

SPIKELET. One of the units of the inflorescence of a grass. It consists of a central rachis bearing one or more sterile glumes at the base, followed by one or more flowers, each enclosed between a flowering glume and palea. All the parts are crowded together.

SPINDLE. (1) = FUSEAU.
(2) The axially differentiated crystalline ('fibrous') part of the cytoplasm within which the centromeres of the chromosomes are orientated during metaphase and anaphase. It is probably composed of longitudinally orientated protein molecules. Movement of the chromatids apart during anaphase may be due to pulling by contractile fibres in the spindle.

SPINDLE ATTACHMENT. The position of the centromere.

SPINDLE FIBRE. The orientated protein molecules of the spindle.

SPINDLE ORGAN. See *Cladochytriaceae*.

SPINE. The end of a branch or leaf which has become rounded in section, and is hard and sharply pointed.

SPINESCENT. (1) Spiny.
(2) Tapering.

SPINICARPOUS. Having spiny fruit.

SPINIFEROUS. Bearing thorns.

SPINIFORM. Like a thorn.

SPINIGER, SPINIGEROUS. Producing thorns.

SPINOSE. Spiny.

SPINULE. A very small spine or prickle.

SPINULOSIN. An antibiotic active against bacteria, but not fungi. It is produced by *Penicillium spinulorum*, *P. cinerascens*, and *Aspergillus fumigatus*.

SPIRAL. A coil of the chromosome, chromatid, or chromosome-thread at mitosis or meiosis.

SPIRAL CELL, SPIRAL TRACHEID(E), SPIRAL VESSEL. The cell, tracheid(e), or vessel in which the secondary wall is laid down in the form of spirally arranged thickened.

SPIRAL FLOWER. A flower having its members arranged in spirals.

SPIRAL HYPHA. A hypha ending in a flat or helical coil.

SPIRAL TRACHEID(E). See *Spiral cell*.

SPIRAL VESSEL. See *Spiral cell*.

SPIREME. The tangle of thread-like chromosomes at the beginning of prophase of mitosis.

SPIRILLIUM. Cork-screw shaped bacterium, some of which are motile with terminal flagella.

SPIROCHAETAE. These are elongated, spirally twisted organisms, which move by flexing the body, and not by flagella. Some are free-living and others are parasitic in animals and man. They are usually classified with the Bacteria.

SPIROLOBOUS. Said of an embryo having spirally rolled cotyledons.

SPLASH CUP, SPLASHING CUP. An open cup-like structure in some fungi and lichens, from which the reproductive bodies are discharged by falling drops of water.

SPLENDENS. Latin meaning 'glittering'.

SPLINT WOOD. Wood in which the living cells are scattered throughout.

SPODOGRAM. A preparation of the ash of a portion of a plant, especially a woody portion, used in investigation of structure.

SPONGIOPLASM. The more viscid constituents of cytoplasm, forming a threadwork.

SPONGY LAYER, SPONGY MESOPHYLL, SPONGY PARENCHYMA, SPONGY TISSUE. A loosely constructed layer of irregularly shaped cells separated by large intercellular spaces. It lies just above the lower epidermis of a dorsiventral leaf and contains chloroplasts.

SPONTANEOUS GENERATION = ABIOGENESIS.

SPONTANEOUS MOVEMENT. The movements of a plant, which do not depend on an external stimulus.

SPORADIC. Scattered over a wide area.

SPORANGIAL SAC, SPORANGIAL VESICLE. A very thin-walled and often evanescent outgrowth from the sporangium of many lower fungi, in which the zoospores complete their development, and from which they are set free.

SPORANGIOCYST. A thick-walled sporangium which is able to remain alive, but inert under unfavourable conditions.

SPORANGIOGENIC BAND. A strip of cells 2-3 cells broad and several cells tall in the fertile spike of *Osmundaceae*. It ultimately becomes differentiated into blocks of archesporial cells and sterile cells.

SPORANGIOLE, SPORANGIOLUM. A sporangium which contains one, or a very few, spores.

SPORANGIOLIFEROUS HEAD. A rounded group of sporangia.

SPORANGIOPHORE. Of fungi, hyphae bearing one or more sporangia, sometimes morphologically distinct from vegetative hyphae.

SPORANGIOSPORE. A spore, especially a non-motile spore formed within a sporangium.

SPORANGIUM. The structure containing asexual spores.

SPORE. An asexual reproductive body. It may be unicellular or multicellular, and produced by haploid or diploid generations. Spore production is a phase in a life-cycle inserted between two sexual phases (if present).

SPORE BALL. A globular mass of spores, which is either solid or hollow.

SPORE GROUP. A multicellular spore, each cell of which is capable of independent germination.

SPORELING. (1) The early stages of a plant developing from a spore.
(2) A young fern plant.

SPORE, MEMBRANE, SPORE WALL. The firm membrane surrounding contents of a spore.

SPORE MOTHER CELL. A diploid cell which gives rise to four haploid spores by meiosis.

SPORE PRINT. The pattern formed by the spores of an agaric, when the cap is placed on a sheet of paper, so that the spores fall on it.

SPORE SAC. Part of a moss sporangium which contains the spores.

SPORE TETRAD. A group of 4 spores formed from a spore-mother cell.

SPORI-, SPORO-. A Greek prefix meaning 'seed'.

SPORIDESM, SPORODESM = SPORE GROUP. An obsolete term for a compound spore or spore ball, the components of which are merispores.

SPORIDIIFEROUS. Bearing sporidia.

SPORIDIOLE. A small spore.

SPORIDIUM. (1) A spore formed from a promycelium.
(2) An obsolete term for an ascospore.

SPOROBLAST. A one cell segment of a spore group.

SPOROBOLA. The curve made by a basidiospore after discharge from its sterigma, as it falls under the influence of gravity.

SPOROBOLOMYCETACEAE. A family of the Saccharomycetales. Ascus formation is unknown. Aerial conidia are formed and are violently discharged. Budding also takes place.

SPOROCARP. (1) A fruit-body of the fungi.
(2) A multi-cellular structure in which spores are formed.
(3) The spore containing structure of the Marsiliaceae and Salviniaceae.

SPOROCHNALES. An order of the Haplostichineae. Each branch of the sporophyte terminates in a tuft of hairs. Growth of the branches is trichothallic due to inter-calary cell division at the base of each hair. The unilocular sporangia are usually terminal in dense clusters. The gametophytes are microscopic and oögamous.

SPOROCLADIA. The special sporogenous branches in the Kickxellaceae.

SPOROCYST. A cyst producing asexual spores.

SPOROCYTE. A spore mother-cell.

SPORODOCHIUM. (1) = ACERVULUS.
(2) A mass of conidiophores tightly packed together on a stroma or mass of hyphae.

SPOROGENOUS. Spore-forming.

SPOROGENOUS CELL = SPORE MOTHER CELL.

SPOROGENOUS LAYER = HYMENIUM.

SPOROGENOUS TISSUE. A layer or group of cells from which spore mother cells are formed.

SPOROGONIUM. (1) The spore-producing structure of the liverworts and mosses.
(2) The sporophyte generation of these plants.

SPORONT. A thallus on which spores will be produced.

SPOROPHORE. Of fungi, a general term for any structure producing and bearing spores.

SPOROPHYLL. A leaf that bears sporangia. It may resemble an ordinary leaf, except for the presence of sporangia, *e.g.* in the ferns, or it may be highly modified, *e.g.* the stamens and carpels of flowering plants.

SPOROPHYTA = CRYPTOGAMIA.

SPOROPHYTE. An individual of the diploid generation of plants, usually producing spores, and is formed by the union of sexual cells produced by the gametophyte.

SPOROPLASM. The spore-producing protoplasm within the epiplasm of a sporangium or ascus.

SPORT. A dissimilar form of an individual, or part of an individual, arising directly or indirectly from gene, plastid, chromosome, or nuclear (polyploid) mutation. The indirect origins are due to segregation of a recessive gene, cellular sorting-out of plastids, and tissue sorting-out of chimaeras. The term is usually confined to new, unforeseen forms.

SPORULATION. The production of spores.

SPREAD. The establishment of a species in a new area.

SPREADING. (1) Diverging gradually outwards.
(2) Of bacterial or fungal growth in culture, stretching out from the line of inoculation, sometimes over the whole surface of the medium.

SPRING WOOD. Secondary xylem formed during the spring and early summer. It is distinguished by the large, thin-walled elements of which it is composed.

SPROUT CELL. A cell formed as a bud from a mother cell.

SPROUTING, SPROUT GERMINATION. (1) The elongation of axillary or adventitious buds.
(2) The production of daughter cells as rounded outgrowths as in yeasts.

SPUR. (1) = SHORT SHOOT.
(2) A short branch on which flowers and fruits are borne.
(3) The extension of the base of a leaf beyond its point of attachment.
(4) A hollow horn-like extension of a petal, or gamopetalous corolla.

SPURIOUS DISSEPIMENT = FALSE SEPTUM. A partition in a fruit which is not an ingrowth from the edges of the carpels, nor an upgrowth from the receptacle.

SPURIOUS FRUIT. A group of fruit having the appearance of a single fruit.

SPUR PELORY. An abnormal condition in which all the petals of an irregular flower develop spurs, so that the flower becomes regular.

SQUAMA. A scale or scale-like structure.

SQUAMATE, SQUAMOSE, SQUAMOUS. Scaly.

SQUAMULAE INTRAVAGINALIS.   Small scales within a sheating leaf-base.

SQUAMULA.   A small scale.

SQUAMULOSE.   Covered with small scales.

SQUARE.   The unopened flower bud of cotton.

SQUARROSE, SQUARROSUS (Latin).   (1) Spreading in all directions and bent backwards.
(2) Rough, with many scales or hairs standing out at right-angles.

STABILIZATION.   The establishment of an equilibrium between the vegetation of an area and its environment.

STABLE COMMUNITY.   A plant community which remains unaltered in its general characters for a long time.

STACHYURACEAE (EP, H).   A family of the Parietales (EP), Hamamelidales (H). (BH) include it in the Ternstroemiaceae. These are small shrubs, with alternate leaves, without stipules, and bisexual flowers in axillary racemes. There are 4 free sepals and petals, and 2 whorls of 4 free stamens. The superior ovary consists of 4 fused carpels with many ovules. The fruit is a berry with many seeds, which are arillate and contain endosperm.

STACKHOUSIACEAE (EP, BH, H).   A family of the Sapindales (EP), Celastrales (BH, H). These are herbs with alternate exstipulate leaves and the flowers in spikes or cymes. There are 5 free sepals, petals, and stamens. The superior ovary consists of 5-2 fused carpels, with the same number of loculi, each with 1 erect ovule. The fruit is a schizocarp, and the seeds contain endosperm.

STAG-HEADED.   Said of a tree which, owing to disease or attack by insects lacks twigs at the top, and has dead main branches standing up like antlers.

STALAGMOID.   Said of spores shaped like a long tear or drop.

STALING.   The accumulation in the substratum of waste metabolic products from bacteria or fungi rendering the substratum unfit for the further growth of the organisms.

STAMEN.   The microsporophyll of a flower; made up of the anther and filament.

STAMINATE.   Having stamens, but no carpels.

STAMINODE.   An abortive stamen. It may be highly modified or reduced.

STAMINOSE.   Said of a flower in which the stamens are very obvious.

STANDARD.   The large petal which stands up at the back of the flower of the Papilionaceae.

STANDARD DEVIATION.   The distance of the abscissa of the point of inflection (maximum slope) from the mean in a normal curve. It is generalized as the square root of the variance, and called the *standard error* when the deviation can be regarded as an error, *i.e.* outside the estimates of the parameters.

STANDARD ERROR.   See *Standard deviation*.

STAPHYLEACEAE (EP, H).   A family of the Sapindales (EP, H). (BH) include it in the Sapindaceae. These are woody plants with opposite lobed stipulate leaves and the flowers in panicles or racemes. There are 5 free sepals, petals and stamens. The superior ovary consists of 3-2 fused carpels which are free above, and contain many-to-few pendulous ovules. The fruit is a capsule, and the seeds contain endosperm.

STAR-LIKE. Said of a flower with rather narrow, pointed radiating petals.

STARCH. A polysaccharide with the general formula $(C_6H_{10}O_5)_n$, where n is large. It is formed by the condensation of glucose units and is a storage substance in many plants. It does not reduce Fehling's solution, but forms a blue-black compound with iodine solution.

STARCH CRESCENT. A strand of cell, crescentric in cross-section, containing starch grains.

STARCH GRAIN. A rounded or irregular-shaped inclusion in a cell, made up of a series of layers of starch, giving a stratified appearance, and surrounding a central hilum.

STARCH PLANT. A plant in which the carbohydrate formed in excess of immediate requirements is stored in the leaf as temporary starch.

STARCH SHEATH. (1) A one-layered cylinder of cells lying on the inner boundary of the cortex of a young stem, with prominent starch grains in the cells. It is homologous with an endodermis.
  (2) A layer of starch grains around the pyrenoid in an algal cell.

STARTER. The pure culture, or mixture, of microorganisms used for starting a fermentation process.

STASIS. The stoppage of growth.

STATE, STAGE. A phase in a life-cycle.

STATENCHYMA. A tissue composed of cells containing statoliths.

STATION. The place where a plant grows.

STATISTIC. The estimate of a parameter arrived at from an observed sample. It bears the same relation to the sample as the parameter does to the population.

STATOCYST, STATOCYTE. A cell which contains statolith(s).

STATOLITH. A solid inclusion in a plant cell, often a starch grain, free to move under the influence of gravity. It may be the means whereby position in respect to gravity influences a plant organ.

STEADY STATE. The state reached in a reversible reaction in equilibrium under a particular set of conditions.

STEGMA. A small elongated cell nearly filled with silica.

STEGNOSPERMACEAE (H). A family of the Pittosporales (H). The sepals and petals are imbricate, with the petals shorter than the sepals. There are 10 stamens and the anthers open lengthwise.

STEGOCARPOUS. Said of the capsule of a moss when it has a lid.

STELE. The vascular cylinder. The cylinder or core of vascular tissue in the centre of stems and roots. It consists of xylem, phloem, and pericycle, and in some cases pith and medullary rays. It is surrounded by an endodermis. The detailed structure differs in different groups of plants.

STELLATE. Star-like.

STELLULATE. Resembling a small star.

STEM. The axis of a plant bearing leaves with buds in their axils. It may be above or below ground, and the leaves may be functional or scales. The vascular bundles are arranged in a circle forming a hollow cylinder, or are scattered.

STEM BODY. The part of a spindle between two groups of chromosomes separating at anaphase.

STEMONACEAE (EP) = **ROXBURGHIACEAE (BH, H).** A family of the Liliiflorae (EP). These are perennial herbs with rhizomes, often climbing stems, and axillary inflorescences. The flowers are bisexual, regular with the whorls each of two members. The perianth is sepaloid. The superior ovary consists of 2 fused carpels with 1 loculus. The fruit is a capsule, and the seeds contain endosperm.

STEM SUCCULENT. A plant with a succulent stem and very small leaves which are often reduced to spines.

STEM TENDRIL. A tendril which is a modified stem.

STEM XEROPHYTE. A plant characteristic of very dry places, with ephemeral or very reduced leaves. The photosynthetic tissue is in the peripheral layers of the stem.

STENOCARPUS. Latin meaning 'narrow-fruited'.

STENOMERIDACEAE (H). A family of the Dioscoreales (H). The flowers are bisexual. There are 6 stamens with the connectives produced beyond the anther loculi. The ovary is inferior with many ovules in each loculus, superimposed in two series. The fruits are elongate with many winged seeds.

STENOPETALOUS. Having narrow petals.

STENOPHYLLOUS. Having narrow leaves.

STENOTHERY. Tolerance of only a very narrow range of temperature.

STEP ALLELOMORPHISM. The occurrence of a series of multiple allelomorphs with overlapping effects, which can be supposedly related to the linear order in the distribution of units of change within the gene.

STEPPE. A dry grassy plain.

STEPHANOKONTAN. Bearing a crown of cilia.

STERCULIACEAE (EP, BH, H). A family of the Malvales (EP, BH), Tiliales (H). These are trees, shrubs, and herbs with alternate, simple or compound, stipulate leaves, and bisexual or unisexual flowers in complex inflorescences. The sepals are fused. The petals are convolute or absent. The stamens are in 2 whorls, the outer being staminodes, and the inner are often branched, all are more or less united. The anthers are bilocular, often forming an androgynophore. The superior ovary is usually of 5 fused carpels which are antepetalous, each having many to 2 ovules. The fruit is a schizocarp, and the seeds contain endosperm.

STEREIDE. A stone cell.

STEREOISOMER. Having isomers consisting of the same groups of atoms, differing in their arrangement in space.

STEREOME. A general term for the mechanical tissue in a plant.

STEREOME CYLINDER. A cylinder of strengthening tissue in a stem, usually just outside the phloem.

STEREOTAXIS, STEREOTROPISM. The response or reaction of an organism to the stimulus of contact with a solid object.

STERIGMA. Of fungi, a minute stalk, bearing a spore or chain of spores.

STERILE. (1) Unable to reproduce sexually.
(2) Free from living microorganisms.

STERILE CELL. The terminal cell in a chain of aecidiospores.

STERILE FLOWER. A staminate flower.

STERILE GLUME. One of the glumes at the base of a grass spikelet which does not subtend a flower.

STERILE VEIN. A strand or sheet of interwoven hyphae occurring with the spore-bearing hyphae in the fruit-bodies of some fungi.

STERILIZATION. To make sterile.

STERNOTRIBE. A flower which dusts pollen on the underside of a visitor.

STEROIDS. Saturated hydrocarbons containing 17 carbon atoms in a system of rings, 3 six-membered and 1 five-membered, condensed together (six atoms being shared between rings).

STEROLS. These are compounds with the general chemical ring-structure of a steroid, but with a long side-chain and an alcohol group. Those occurring in plants (*phytosterols*) are oxidation products of cholesterol ($C_{27}H_{45}OH$).

STERRHIUM. A moor formation.

STICHIDIUM. A special branch of the thallus in red algae on, or within which tetraspores are formed.

STICHOBASIDIUM. A basidium, usually elongated and cylindrical, in which the spindle of the dividing nuclei lie obliquely or longitudinally.

STICKY POINT. The soil-water content of a soil when it can just be worked with the hands without being adhesive.

STIGMA. (1) The receptive part of the stigma.
    (2) = EYESPOT.

STIGMATEACEAE. A family of the Hemisphaeriales. The ascocarps have a more or less radial structure, and arise subcutanularly, but emerge, at least at maturity. The vegetative mycelium is scanty or lacking.

STIGMATOCYST = HYPHOBASIDIUM.

STILBACEAE. A family of the Moniliales. The rather long conidiophores are in a more or less compact column or synnema. The conidiophores spread at the top and have the spores at the top or down the sides of the tip.

STILBEACEAE (H). A family of the Verbenales (H). These are shrubs with verticillate leaves. If an indumentum is present, it is not stellate. The flowers are in dense terminal spikes. The calyx is actinomorphic and the corolla is more or less zygomorphic. The stamens are usually one less than the corolla-lobes. The ovary has 9-2 loculi with 1 erect ovule in each.

STILBELLACEAE. A family of the Moniliales. The conidiophores are more or less separate, not united into synnemata nor on sporodochia. The conidia and conidiophores are hyaline or bright-coloured, never brown or black.

STILBOID. Said of a fungus having a stalked head of spores or other reproductive structures.

STILT ROOT. An adventitious root formed by a stem from a point above ground level passing downwards into the soil and affording support to the plant.

STIMULOSE. Bearing stinging hairs.

STIMULUS. Any change in the environment of an organism or of part of it which is intense enough to produce a change in the activities of the living material, without itself providing energy for the new activities.

STINGING HAIR. A multicellular hair with a brittle tip, which breaks off on contact with an animal, leaving a sharp edge which penetrates the skin and injects an irritant fluid.

STIPATE. Crowded.

STIPE. (1) The stalk of a carpel or pistil.
(2) The stalk of fruit-bodies of certain higher fungi.
(3) The stalk of the thallus of sea-weeds.
(4) A leaf-stalk.

STIPEL. One of the two small leaf-like appendages present at the base of a leaflet in some compound leaves.

STIPELLATE. Having stipels.

STIPITATE. (1) Having a stalk.
(2) On a special stalk.

STIPITATIC ACID. A metabolic product of *Penicillium stipitatum*.

STIPULAR TRACE. The vascular tissue running to a stipule.

STIPULATE. Having stipules.

STIPULES. Basal appendages of a leaf or petiole. They may photosynthesize, or be scales, and may protect the axillary buds.

STIPULOSE. Bearing conspicuous stipules.

STIRPS. A well-established variety that keeps its characters in cultivation.

STOCK. (1) An artificial mating group.
(2) A rooted stem into which a scion is inserted during grafting.
(3) A perennial portion of a herbaceous perennial.
(4) A race.

STOLON. (1) A horizontally growing stem that bears adventitious roots at the nodes, and scale leaves. New plants or tubers may develop terminally.
(2) A long hypha lying on the substratum and producing tufts of rhizoids and sporangiophores at intervals.

STOLONIFEROUS. Producing stolons.

STOMA. (1) A pore in the epidermis of plants, present in large numbers, particularly on leaves, through which gaseous exchange takes place. Each stoma is surrounded by the sausage-shaped guard cells which open and close the stoma by changes in their turgidity.
(2) It may be used to include the pore and guard cells.

STOMATOPODIUM, STOMOPODIUM. A hyphal branch or 'plug', above or in a stoma.

STOMIUM. The place in the wall of a fern sporangium whose rupture occurs at maturity, releasing the spores. It consists of thin-walled cells.

STONE. The hard endocarp of a drupe.

STONE-CELL. A thick-walled cell, not much longer than broad, with lignified walls.

STONE FRUIT. A drupe.

STOOL. A plant from which off-sets may be taken, or with several stems arising together.

STORAGE PITH. A pith in which starch or water is stored.

STORAGE TRACHEID(E). A thick-walled cell, resembling a tracheid(e) without living contents, in which water is stored.

STOREY. A layer of vegetation.

STORIED CORK. A type of cork composed of short cells arranged in somewhat irregular radial groups.

STRAIN. (1) A natural or artificial mating group, uniform in some particular.
(2) A variety of species, with distinct morpholical and/or physiological characters.

STRAMINEOUS. Straw-coloured.

STRAND PLANT. A plant growing by the sea, where it is not submerged at high-tide, but is splashed by spray.

STRASBURGERIACEAE (H). A family of the Ochnales (H). These are erect trees and shrubs. The stipules are intrapetiolar. The stamens are free with the anthers sub-sagittate and versatile, and the connectives not produced at the apex. There are no staminodes. There are 2 ovules in each loculus on axile placentas.

STRATIFICATION = LAYERING (1) and (2).

STRATIFIED CAMBIUM. A cambium in which the cells, seen in tangential section, appear arranged in fairly regular horizontal rows.

STRATIFIED THALLUS. A lichen thallus composed of a layer of algal cells between layers of fungal hyphae.

STRATOSE. Made up of well-defined layers.

STRATOSE THALLUS. A lichen thallus having the tissues in horizontal layers.

STRATUM SOCIETY. A plant society occurring as a well-defined layer in a plant community, *e.g.* shrubs in a wood.

STREAK. (1) A layer of tissue differing in colour or structure from the tissues on each side of it.
(2) A furrow.
(3) A kind of virus disease in which necrotic streaks develop.

STRELITZIACEAE (H). A family of the Zingiberales (H). The leaves and bracts are distichous. The sepals are free or almost adnate to the corolla. The median petal does not form a labellum. There are 6-5 stamens with bilocular anthers. The fruit is a capsule.

STREPSMEMA, STREPSISTINE. A stage in meiotic prophase, when crossing over takes place.

STREPTOCOCCUS. A coccus in which the individuals tend to be grouped in chains.

STREPTOMYCIN. An antibiotic, active against Gram+ and some Gram− bacteria, produced by *Streptomyces griseus*.

STREPTOTHRICIN. An antibiotic active against bacteria and fungi, produced by *Streptomyces lavendulae*.

STRIATE. Marked with parallel, longitudinal lines, furrows, ridges, or streaks of colour.

STRICT, STRICTUS (Latin). Stiff and rigid.

STRIGOSE. Bearing hairs which are usually rough, and all pointing in the same general direction.

STRIGILLOSE. Slightly strigose.

STRIOLATE. Finely striate.

STRIPE = STREAK (3).

STRIPED. Having longitudinal stripes of colour.

STROBILACEOUS. Of, or resembling a cone.

STROBILATE, STROBILIFORM. Of the nature of a cone.

STROBILE, STROBILUS. A cone; a group of sporophylls with their sporangia, more or less tightly packed around a central axis, forming a well-defined group.

STROMA. (1) A tissue-like mass of fungal hyphae in, or from which fruit-bodies are produced.
(2) The zone between the intergranum lamellae of a chloroplast.

STROMA STARCH. Starch formed in the stroma of a chloroplast when photosynthesis is active.

STROMBULIFORM. Said of a spirally twisted fruit.

STROMBUS. A spirally coiled pod.

STROPHULAR CLEFT. A small opening in the seed coat through which water will enter before germination can take place. This will only occur after the removal of the strophular plug.

STROPHULAR PLUG. A mass of suberin blocking the strophular cleft.

STROPHIOLE = CARUNCLE.

STRUCTURAL CHANGE. Change in the genetic structure of one or more chromosomes.

STRUCTURAL CONTROL = GENOTYPIC CONTROL.

STRUCTURAL DEVIATION. Any departure from the usual structure of a plant.

STRUCTURAL HYBRID. A hybrid whose parental gametes differed in respect of the arrangement of their genes.

STRUCTURAL MUTATION. An intergenic change in the linear arrangement of genes.

STRUCTURE. The potentially permanent linear order of the particles, chromomeres or genes in the chromosome.

STRUMA. A swelling on one side at the base of a moss capsule.

STRUMOSE. (1) Having a swelling at one side of the base.
(2) Having cushion-like swellings.

STRYCHNACEAE (H). A family of the Loganiales (H). The epicalyx is lacking. There are 4 or more stamens, and usually many ovules in each ovary or loculus. The fruit is a drupe or berry and the seeds are not winged.

STUFFED. Said of the stipe of an agaric when it is cottony or spongy inside, with a tougher outer layer.

STUPOSE. (1) Tow-like.
(2) Said of a tissue formed from hyphae which are not gelatinized.

STYLAR. Referring to the style.

STYLAR CANAL. A tube or mass of loose tissue running through the centre of the style. It is through this that the pollen tube grows.

STYLE. The narrow part of the gynecium bearing the stigma.

STYLIDIACEAE (EP, BH, H). A family of the Campanulatae (EP), Campanales (BH), Goodeniales (H). These are herbs with simple exstipulate leaves and bisexual or unisexual usually zygomorphic flowers. There are 5 sepals and petals. The petals are usually united. There are 3-2 stamens, which are free or united to the style. The anthers are extrorse. The inferior ovary consists of 2 fused carpels, with 2 or 1 loculi. The fruit is septicidal or indehiscent, and the seeds contain endosperm.

STYLIDIEAE (BH). = STYLIDIACEAE (EP).

STYLIDIUM. The upper part of an archegonium.

STYLOPODIUM. The swelling at the base of the style.

STYLOSPORE. An obsolete term for a spore on a pedicel or hypha.

STRYACACEAE (EP, BH, H). A family of the Ebenales (EP, BH), Styacales (H). (BH) include the Symplocaceae. These are woody plants with simple alternate leaves, without stipules, and stellate or scaly hairs. The bisexual flowers are small. There are 5-4 sepals and petals which are fused in 2 whorls. The 10-8 stamens are united at the base or rarely in a tube. The superior (rarely half-inferior) ovary consists of 5-3 fused carpels, each with 1 or a few ovules. There are 3-5 loculi below and one above. The fruit is a drupe, indehiscent, or a capsule, with one or a few seeds. The seeds contain endosperm.

STYRACALES (H). An order of the Lignosae (H). These are trees or shrubs, often with a stellate indumentum. The leaves are simple and alternate with no stipules. The flowers are regular with valvate sepals, and imbricate, valvate or rarely contorted petals which are free or united. The stamens are free or joined to the corolla lobes. They are few and alternate with the lobes or are more numerous, and the anthers open lengthwise. The ovary is superior to inferior with axile placentas. The seeds have copious endosperm.

SUAVEOLENS, SUAVEOLET. Latin meaning 'fragrant'.

SUBCENTRIC OÖSPHERE. An oösphere with the protoplasm surrounded by one layer of fatty globules and with 2-3 additional layers on one side only.

SUB-CLIMAX. A community which has not attained the full development possible under the prevailing conditions because of some limitations imposed by an edaphic or biotic factor.

SUB-CULTURE. A culture of bacteria or fungi prepared from a pre-existing culture.

SUB-ENTIRE. Said of a margin which is very slightly indented.

SUB-EPIDERMAL TISSUE = HYPODERMIS.

SUBERECT. Upright below, nodding at the top.

SUBERIFICATION = SUBERIZATION.

SUBERIN. A complex mixture of oxidation and condensation products of fatty acids, present in the walls of cork-cells, rendering them impervious to water.

SUBERIN LAMELLA. A layer of wall material impregnated with suberin.

SUBERIZATION. The deposition of suberin.

SUBERIZED. Transformed into cork.

SUB-EROSE. Appearing as if somewhat gnawed and eroded.

SUBEROSE, SUBEROUS. Of a corky texture.

SUB-GENUS. A subdivision of a genus, higher than a species.

SUB-GLOBOSE. Not quite spherical.

SUB-HYMENIAL LAYER. A layer of hyphae immediately below the hymenium.

SUBICLE, SUBICULUM. A felted or cottony mass of hyphae underlying the fruit-body of some fungi.

SUB-LITTORAL. Growing near the sea, but not on the shore.

SUBMERGED, SUBMERSED. Growing under water.

SUB-PETIOLATE. Said of a bud which is concealed by the petiole.

SUBRAMOSE. (1) Not branching freely.
(2) Having few branches.

SUBSIDIARY CELL = ACCESSORY CELL.

SUB-SPECIES. A sub division of a species, larger than a race.

SUB-SPONTANEOUS. Said of a plant which has been introduced, but maintained itself fairly successfully by its ordinary means of reproduction.

SUBSPORANGIAL VESICLE. A swelling on the sporangiophore immediately beneath the terminal sporangium.

SUBSTANTIVE VARIATION. Variation in the constitution of an organ or organism, as opposed to the variation in the number of parts.

SUBSTOMATIC CHAMBER. A large intercellular chamber below a stoma.

SUBSTRATE. (1) The substance on which an enzyme acts.
(2) The medium on which a fungus and bacteria (or other plant) grow, especially in culture.
(3) The solid object to which a plant is attached, *e.g.* a seaweed.

SUBSTRATUM = SUBSTRATE (2) and (3).

SUBTENDING. Having a bud, or developed from a bud, or a sporangium, in its axil.

SUBTERRANEAN. Growing beneath the surface of the soil.

SUBTRACTION. The loss of hereditary factor.

SUBULA. A delicate, sharp-pointed prolongation of an organ.

SUBULATE. Awl-shaped.

SUB-UNIVERSAL VEIL = PROTOBLEM.

SUCCESSION. The progressive change in the composition of a plant population during development of vegetation, from initial colonization to the attainment of the climax.

SUCCINEUS. Latin meaning 'amber-coloured'.

SUCCINIC ACID. The acid produced from α-ketoglutaric acid, in the citric acid cycle, by oxidation of CO to CO_2.

$$
\begin{array}{ccc}
\text{COOH} & & \text{COOH} \\
| & & | \\
\text{CO} & & \text{CH}_2 + \text{CO}_2 \\
| & & | \\
\text{CH}_2 + [\text{O}] & \longrightarrow & \text{CH}_2 \\
| & & | \\
\text{CH}_2 & & \text{COOH} \\
| & & \\
\text{COOH} & & \textit{succinic acid}
\end{array}
$$

SUCCINIC DEHYDROGENASE. An enzyme which catalyses the oxidation of succinic acid.

SUCCISE, SUCCISUS (Latin). Ending abruptly, as if cut off.

SUCCUBOUS. Having the lower edge of the leaf in front of the stem and overlapping the upper edge of the next leaf below it, on the same side of the stem.

SUCCULENT (1) Juicy, soft and thick.

(2) A xerophyte storing water in the tissue, giving it a fleshy appearance.

SUCKER. (1) A shoot arising below ground.

(2) A new shoot on an old stem.

(3) The modified root of a parasite, by which it absorbs materials from the host.

SUCRASE. An enzyme which catalyses the hydrolysis of sucrose.

SUCROSE. Cane sugar. A disaccharide compound of glucose and fructose.

SUCTION PRESSURE. When referred to a cell, the force which is available for taking in water. It is equivalent to the osmotic pressure of the cell-sap (OP), less the inward pressure of the cell-wall (WP) and the osmotic pressure of the surrounding medium (OPM), which is zero if the cell is in pure water. SP = OP − WP − OPM.

SUFFRUTESCENT, SUFFRUITICOSE. Said of a plant in which many of the branches die after flowering, leaving a persistent woody base.

SUFFUSE. Spread out on the substratum.

SULCATE. Marked by distinct longitudinal parallel furrows.

SULPHUR BACTERIA. Bacteria which live in situations where oxygen is scarce or absent, and which act on sulphur compounds releasing the element.

SULPHUREOUS. Pale, clear yellow.

SUMMER ANNUAL. A plant which lives for a short time in the summer, setting seed at the end of its growth, then dying.

SUMMER SPORE. A spore germinating without a resting period, living only for a short time.

SUMMER WOOD = AUTUMN WOOD. Secondary wood formed in the summer as secondary thickening comes to an end for the season; the elements are often thick-walled and smaller than those of spring wood.

SUM OF SQUARES. The sum of the squared deviations of observations from their mean.

SUN PLANT. A plant tolerant of exposure to much bright light, or needing much light.

SUPERAXILLARY. Developed from above the axil.

SUPERFAMILY. A group of related families.

SUPERFICIAL. On the surface of the substratum.

SUPERFILICALES. A series of the Filicinae in which the sporangia are borne on the abaxial surface of the leaf blade.

SUPER GENE. A group of genes acting as a mechanical unit in particular allelomorphic combinations.

SUPERIOR. (1) = HYPOGYNOUS.

(2) Said of an annulus near the top of the stipe.

SUPERIOR PALEA. See *Palea*.

SUPERIOR RADICLE. A radicle which points towards the apex of a fruit.

SUPER-ORDER. A division of a phylum between a class and an order.

SUPER-PARASITE. A parasite on another parasite.

SUPERPOSED. Placed one above the other.

SUPINUS. Latin meaning 'lying face upwards'.

SUPPRESSION. Failure to develop, referred to a member which is normally present.

SUPPRESSIVE. Said of a plasmagene which suppresses the expression of an alternative condition in a particular respect in a hybrid individual and its descendants.

SUPPRESSOR. A gene, one of whose allelomorphs renders indetectable the difference determined by the allelomorphs of another gene.

SUPRA-AXILLARY. Arising above an axil.

SURCULOSE. Bearing suckers.

SURCULUS. A sucker.

SUSCEPTIBILITY. The whole properties of a plant, which dispose it to be attacked by a parasite.

SUSCEPT. A living organism which is susceptible to a given disease or pathogen.

SUSPENDED. Said of an ovule which hangs from the top of an ovary.

SUSPENSOR. (1) A cell supporting the gametangium of the Mucorales.
(2) A structure developing with the embryo of seed plants and some pteridophytes, elongating and pushing the embryo into the nutritive endosperm or prothallus.

SUTURE. (1) A line of junction.
(2) In flowering plants, the line of fusion of edges of a carpel is the *ventral suture*. The mid-rib of the carpel is the *dorsal suture*, not implying any fusion of parts to form it, but to distinguish it from the ventral suture.

SWARM-CELL. Of myxomycetes, and some Chytridiales, a motile cell acting, before and after division, as an isogamete.

SWARM SPORANGIUM = ZOOSPORANGIUM.

SWARM SPORE = ZOOSPORE.

SWITCH PLANT. A plant with small leaves, often reduced to non-functional scales, and long thin stems and branches, with photosynthetic tissue in their cortical regions.

SYCONIUM, SYCONUS. The fruit of a fig, consisting mainly of the much-enlarged receptacle of the inflorescence.

SYLVESTRAL. Growing in woods and shady hedges.

SYMBIOSIS. The living of two (or more) organisms in close association to their mutual benefit.

SYMMETRICAL = REGULAR.

SYMPETALAE (EP). A division of the dicotyledons in which the petals are united into a gamopetalous corolla.

SYMPETALOUS = GAMOPETALOUS.

SYMPHOGENOUS. Of pycnidia etc. formed by the growth and division of several hyphae.

SYMPLAST. A uninucleate cell formed by fragmentation of the nucleus within a single energid.

SYMPLOCACEAE (EP, H). A family of the Ebenales (EP), Styracales (H). (BH) include it in the Styracaceae. These are woody plants with alternate exstipulate leaves and bisexual flowers with the parts in fives. There are the same number or twice the number of petals as sepals, and they are fused. The stamens are free in 3-1 whorls and joined to the corolla. The ovary consists of 5-2 carpels which are fused and inferior (or half-inferior). Each carpel has 4-2 pendulous ovules. The fruit is a drupe and the seeds contain endosperm.

SYMPODIUM. A composite axis produced and increased in length by successive development of lateral buds, just behind the apex. The main axis stops growing.

SYNANDRIUM. A mass of united anthers.

SYNANDROUS. Having several united stamens.

SYNANGIUM. A compound structure formed by the lateral fusion of sporangia. Found in certain ferns.

SYNANTHAE (EP). An order of the Monocotyledonae (EP). These are often palm-like plants, climbers or large herbs with unisexual flowers alternating over the surface of a spike. The male flowers are naked or with a thick perianth with many to 6 stamens. The female flower is naked or with 4 fleshy, or scaly perianth segments with a long thread-like staminode in front of each. It consists of 2 or 4 fused carpels, with 2 or 4 placentas and many ovules. The unilocular ovaries are sunk in the spike and united, thus forming a multiple fruit with many seeds. The seeds have endosperm.

SYNAPSIS = SYNDESIS.

SYNAPTENE = ZYGOTENE.

SYNCARP. A multiple fleshy fruit.

SYNCARPOUS. Composed of united carpels.

SYNCARYON = SYNKARYON.

SYNCHYTRIACEAE. A family of the Chytridiales. These are holocarpic, without rhizoids, and the discharge tubes are inoperculate. The vegetative cell divides internally into numerous zoosporangia, or the cell acts as a prosorus, and the zoosporangia develop in a cell developing from it.

SYNCLADOUS. Said of branches growing in groups from the same point.

SYNCOTYLOUS. Having united cotyledons.

SYNCOTYLY. The union of cotyledons.

SYNDESIS. In meiotic nuclear division, the fusion of homologous chromosomes.

SYNDIPLOIDY. Doubling; the fusion of sister nuclei to give a doubled chromosome number, particularly in the divisions immediately preceding meiosis.

SYNDROME. A group of symptoms characteristic of the same infection or abnormal genetic condition, but not necessarily all appearing together.

SYNECOLOGY. The study of plant communities.

SYNEMMA. A group of united filaments, when the flower has monadelphous stamens.

SYNERGIDAE, SYNERGIDS. The two cells, which, together with the egg nucleus make up the egg apparatus of the embryo sac of flowering plants.

SYNERGISM. (1) The combined activities of agencies such as drugs, hormones etc. which separately influence a reaction in the same direction, such that the effect produced is greater than the sum of effects of each agent acting alone.
    (2) A similar effect brought about by two organisms.

SYNGAMY. The fusion of gametes.

SYNGENESIOUS. Said of stamens which are united by their anthers to form a tube around the style.

SYNGYNOUS = EPIGYNOUS.

SYNIZESIS. In meiotic prophase, contraction of the chromatin to one side of the nucleus, obscuring the individual loops.

SYNKARYON, SYNKARYON. A pair of nuclei in close association in a hyphal segment, and dividing at the same time to give daughter nuclei which associate in the same way.

SYNNEMA. An erect bunch of hyphae, bunched tightly together and producing spores.

SYNOECIOUS, SYNOICOUS. In mosses, having antheridia and archegonia in the same head surrounded by a common involucre.

SYNOECY. The condition where male and female gametes are produced by the same individual.

SYNONYM. A systematic or proper name which has become superseded, or is, by the rules of nomenclature, not tenable.

SYNUSIA. An ecological unit based on the life-forms of plants growing in company.

SYRTIDIUM. A dry sand-bar formation.

SYSTEMATICS. The study of the classification of living things, with emphasis on their evolutionary relationships.

SYSTEMATIC. Generally distributed throughout the organism.

SYSTROPHE. The clumping of chloroplasts when exposed to very bright light.

SYSTYLOUS. Said of the lid of mosses when it adheres to the columella.

SYNTHETASES = LIGASES.

SZIK SOILS. Hungarian alkali soils containing very small proportions of soluble salts.

T

t. The ratio of an observed deviation from its estimated standard deviation.

TABASHEER, TABASHIR. A mass of silica found in stems of bamboos.

TABESCENT. Shrivelling.

TABULAR. Horizontally flattened.

TACCACEAE (EP, BH, H). A family of the Liliiflorae (EP), Epigynae (BH), Haemodorales (H). These are perennial herbs with tubers and large, entire or cymosely branched leaves. The bisexual flowers with long thread-like bracts, which are in cymose umbels. The perianth segments are free, petaloid, and in 2 whorls of 3. There are 2 whorls of 3 free stamens. The inferior ovary consists of 3 fused carpels and 1 loculus with parietal placentas and many ovules. There are 6 petaloid stigmas. The fruit is a capsule or berry, and the seeds contain endosperm.

TACHYSPOROUS. Said of a plant which liberates its seeds quickly.

TACTILE BRISTLE. A stiff hair which transports a contact stimulus.

TACTILE PIT. A sharply defined area of a thin cell-wall in an epidermal cell of a tendril, which appears to be concerned in the perception of pressure.

TAIGA. A wet woodland soil of Siberia.

TAILED. Bearing a long slender point.

TAMARICACEAE (EP, BH, H). A family of the Parietales (EP), Caryophyllinae (BH), Tamaricales (H). (BH) include the Fouquieriaceae. These are shrubs or herbs with small alternate exstipulate leaves and regular bisexual flowers with 6–4 sepals and petals. There are as many or twice as many stamens as petals, or many in bundles. The superior ovary consists of 5–2 fused carpels with many ascending ovules on basal placentas. The fruit is a capsule with hairy seeds which may or may not have endosperm.

TAMARICALES (H). An order of the Lignosae (H). These are trees or shrubs. The flowers are hypogynous, regular and usually bisexual. The leaves are alternate or opposite, often very small and without stipules. The sepals are imbricate or valvate, and the petals are free or fused. The stamens are mostly definite. The carpels are fused and the placentas are parietal or basal. The seeds may or may not have endosperm and are often hairy.

TANGENTIAL LONGITUDINAL SECTION. A section through a plant member parallel to a tangent to its surface.

TANNINS. A group of complex compounds containing phenols, hydroxy-acids or glucosides.

TANNIN SAC. A cell containing much tannin.

TAPERING. Said of a leaf base, which becomes gradually narrowed towards the petiole.

TAPETAL. Relating to the tapetum.

TAPETAL PLASMODIUM. A multinucleate mass formed by the breakdown of the cell walls between the cells of the tapetum.

TAPETUM. A food rich layer of cells around a group of spore-mother-cells in vascular plants. They disintegrate to liberate the contents which is absorbed by the developing spores.

TAPHRINACEAE. A family of the Taphrinales. The asci form an indeterminate subcuticular hymenium, or forming tufts of asci emerging from between epidermal cells. They always live in green tissues of fruits. The mycelium sometimes is perennial in the host plant.

TAPHRINALES. An order of the Ascomycetae. They are not parasitic of algae. The apothecia have no definite limiting border. The asci form a hymenium without paraphyses, usually with very limited hypothecial tissues.

TAPHRIUM. A ditch formation.

TAP ROOT. A root system with a prominent main root, directed vertically downwards bearing smaller lateral roots. It may be swollen to store food.

TARGIONIACEAE. A family of the Marchantiales, in which the female receptacles are enclosed by a sheath developed from the tissues lateral to it.

TARTAROUS. Said of the surface of a lichen when it is rough and crumbly.

TASSEL. The staminate inflorescence of maize.

TAUTONYM. A name in which the specific epithet is a repetition of the generic name.

TAWNY. Dark, brownish-yellow.

TAXACEAE. A family of the Coniferae. These are mostly dioecious with the cone-formation imperfect. There are usually few or even one terminal carpel, with 2-1 ovules each. The seeds project beyond the carpels or they may be naked, or with a fleshy aril, or a drupaceous testa.

TAXIS. The movement of a whole organism towards or away from a stimulus.

TAXONOMIC SERIES. The range of extant living organisms, ranging from the simplest to the most complex forms.

TAXONOMY. The science of classifying living things.

T.D.P. = THERMAL DEATH POINT.

TECOPHILAEACAE (H). A family of the Liliales (H). These are terrestrial, rarely swamp plants, with linear alternate leaves (or cladodes with axillary flowers) never in a whorl at the top of the stem. The inflorescence is not subtended by a spathe and the flowers are usually bisexual. The filaments are usually free with bilocular (usually) anthers which open by terminal pores or pore-like slits. The ovary is semi-inferior.

TEEN SUDA. A pedocalic soil occurring in the transition zone between the desert soils and the equatorial humid belt.

TEETH. Small marginal teeth.

TEGEM, TEGMEN. The inner coat of a testa.

TELEBLEM, TELEOBLEMA = UNIVERSAL VEIL.

TELEOLOGY. The interpretation of structures in terms of purpose and utility.

TELEMORPHOSIS. The growth of a gametophyte into the clear area between two strains of a heterothallic fungus grown in solid culture.

TELEUTOSORUS. A group of teleutospores with their supporting hyphae, forming a pustule on the surface of the host.

TELEUTOGONIDIUM, TELEUTOSPORE. A thick-walled spore, consisting of two or more cells, formed by a rust fungus towards the end of the growing season. They can remain dormant for some time and then germinate to give one or more promycelia on which the basidiospores are formed.

TELEUTOSTAGE. The stage in the life-history of a rust fungus when teleutospores are formed.

TELIAL STAGE = TELEIOSTAGE.

TELIOSPORE = TELEUTOGONIUM.

TELIOSPOREAE. A sub-class of the Basidiomyceteae. These are parasitic on all parts of the Pteridophyta, Coniferae, and Angiospermae. The mycelium produces primary and secondary phases. The latter produces diploid teliospores, which germinate to produce a haploid promycelium, on which there are borne 4 sporidia. The mycelia from these fuse, if they are compatible, and the diploid mycelium invades a new host. There are various other forms of asexual reproduction.

TELIUM = TELEUTOSORUS.

TELMATIUM. A wet meadow formation.

TELOCENTRIC. Said of a chromosome or chromatid having a terminal centromere.

TELOKINESIS = TELOPHASE.

TELOMERE. The rounded end of a chromatid.

TELOMITIC. Having chromosomes attached to the fibres of the spindle by their ends.

TELOPHASE. The terminal stage of mitosis or meiosis during which nuclei revert to the resting stage.

TELOSYNAPSIS, TELOSYNDESIS = METASYNDESIS. The end-to-end association of chromosomes in zygotene and pachytene.

TEMPERATE PHAGE = LYSOGENIC PHAGE. A bacteriophage that does not kill its host.

TEMPERATURE COEFFICIENT. The ratio of the rate of progress of any reaction or process in a plant, at a given temperature, to the rate at a temperature 10°C lower.

TEMPLATE RNA. See *RNA*.

TEMPORARY COLLENCHYMA. Collenchyma present in a young organ, and disappearing as secondary thickening progresses.

TEMPORARY STARCH. Starch which is stored for a time in the chloroplasts when the plant is forming carbohydrates more rapidly than they are being used or removed from the leaf.

TENDRIL. A stem, leaf, or part of a leaf modified as a branched or unbranched filamentous structure, used by many climbers for attachment to a support by twining, or by adhesive terminal disks.

TENELLUS. Latin meaning 'very tender or dainty'.

TENTACLE. One of the hairs on the leaf of an insectivorous plant, which helps in capturing insects and produces enzymes which digest the prey.

TERMIFOLIUS. Latin meaning 'thin-leaved'.

TEPAL. A perianth segment, not differentiated into a calyx and corolla.

TERATOLOGY. The study of monstrosities.

TEREBRATE. Having scattered perforations.

TEREBRATOR. The trichogyne of a lichen.

TERETE. (1) Cylindrical.
(2) Elongated, cylindrical-conical, tapering to a point.

TERFEZIACEAE. A family with affinities in the Aspergillales and Tuberales.

TERMINAL. (1) Situated at the tip.
(2) Said of secondary wood parenchyma when this develops only at the end of the growing season, and therefore at the limit between one annual ring and the next.

TERMINAL AFFINITY. The property by which chromosomes are held together end-to-end from diplotene till first metaphase of meiosis, or brought together in this way at metaphase.

TERMINALIZATION. The expansion of the association of the two pairs of chromatids on one side of a chiasma at the expanse of that on the other side. So called because the resulting 'movement' of the chiasma is towards the ends of the chromosomes.

TERMINALS. The fine end-branches of the veins of a leaf.

TERMINUS PHIALOSPORE, TERMINUS SPORE. A phialospore of a one-spored phialide, terminating its growth.

TERNARY. Having three parts.

TERNATE. (1) Said of a compound leaf with three leaflets.
(2) Arranged in threes, as branches arising at about the same point from a stem.

TERNSTROEMIACEAE (BH) (H). A family of the Guttiferales (BH). See *Theaceae (H)*. (EP) distribute this family, but most of the genera are in the Theaceae (EP).

TERPENES. These are unsaturated hydrocarbons ($C_{10}H_{16}$), essential oils with straight-chain or ring-structure derived from isoprene.

$$CH_3 \qquad\qquad H$$
$$C - C$$
$$CH_2 \qquad\qquad CH_3$$
Isoprene

TERRA ROSA. A red soil, associated with limestone, and occurring in countries bordering the Mediterranean Sea.

TERRA ROXA. A red earth of Brazil.

TERREIN. A metabolic product of *Aspergillus terreus*.

TERRESTRIAL, TERRICOLOUS. (1) Living in the soil.
(2) Living in the ground.

TERRESTRIC ACID, ETHYLCAROLIC ACID. A metabolic product of *Penicillium terrestre*.

TERTIARY. A geological period from 70-1 million years ago.

TERTIARY WALL, TERTIARY LAYER, TERTIARY THICKENING. The deposit of wall thickening on the inner surface of a secondary cell-wall, trecheid, or vessel, usually in the form of rings or a loose spiral.

TERTIARY STRUCTURE. The structure of a protein where the polypeptide chain is bent sharply to form a roughly spherical, three-dimensional mass.

TESSELLATE. Said of a surface marked in squarish areas, like a pavement.

TEST = TESTA.

TESTA. The seed coat. It is derived from the integuments and is several layers thick. It is protective in function.

TESTACEOUS. The colour of old red brick.

TEST CROSS. A cross of a double or multiple heterozygote to the corresponding double or multiple recessive. It is used to estimate linkage relationships to behaviour.

TETRACENTRACEAE (H). A family of the Hamamelidales (H). The stipules are single or adnate to the petiole. The indumentum is often stellate. A calyx and usually petals are present, and the ovary is superior.

TETRACYCLIC. Said of a flower with 4 whorls or members.

TETRACYTE. One of the 4 cells formed after a meiotic division.

TETRAD. (1) A group of 4 spores formed by meiosis within a spore mother cell.
(2) The paired chromosomes of meiosis, after each chromosome has duplicated itself, and the pair is visibly four-stranded.

TETRAD DIVISION. The nuclear and cell divisions occurring when a spore mother cell divides to give 4 spores.

TETRADIDYMOUS. Four-fold.

TETRADYNAMOUS. Having 4 long, and 2 short stamens.

TETRAGONOUS. Having 4 angles and 4 convex faces.

TETRAHEDRAL. Having 4 triangular faces.

TETRAKONTON. Having 4 flagella.

TETRAMERISTACEAE (H). A family of the Theales (H). They are not epiphytes. The leaves are alternate and simple. The petals are imbricate. There are 4 stamens, and the ovary is unilocular, with 1 basal ovule.

TETRAMEROUS. (1) Having 4 parts.
(2) Arranged in fours.
(3) Arranged in multiples of 4.

TETRAPHIDALES. An order of the Eubrya. The peristome of the capsule has only four teeth.

TETRAPLOID. Having 4 times the haploid number of chromosomes.

TETRAPOLAR. Said of basidiomycetes which have the sex factors in four groups, AB, Ab, aB, ab.

TETRARCH. Having 4 strands of xylem.

TETRASACCHARIDE. A carbohydrate formed by the condensation of four hexose units.

TETRASOMIC. Said of a nucleus, or organism which has the diploid number of chromosomes, except of one which is present 4 times.

TETRASPORALES. An order of the Chlorophyceae. The immobile vegetative cells may be temporarily metamorphosed into a flagellated motile stage. Most genera have the cells united to form non-filamentous colonies. Asexual reproduction is by zoospores, aplanospores, or akinetes. Sexual reproduction is by biflagellate gametes.

TETRASPORANGIUM. The sporangium of the tetrasporophyte of the red algae. Initially it is unicellular, but the nucleus divides meiotically to give 4 tetraspores.

TETRASPORE. (1) See *Tetrasporangium*.
(2) A diploid spore produced by some Phaeophyceae.

TETRASPOROPHYTE. The diploid phase in the life-cycle of the red algae. It produces tetraspores.

TETRASTER. A complex mitotic figure formed in an ovum after polyspermy.

TETROSE. A monosaccharide sugar with four carbon atoms, $C_4H_8O_4$.

TETRASTICHOUS. Arranged in 4 rows.

THALAMIFLORAL. Said of a flower which has all its members inserted separately on the receptacle, with the gynecium superior.

THALAMIUM. The hymenium of an apothecium.

THALAMUS. The receptacle of a flower.

THALASSIUM. A sea formation.

THALASSOPHYTE. A sea-weed.

THALLIFORM. Like a thallus.

THALLINE EXCIPLE. An exciple which contains algal cells.

THALLINE MARGIN. The margin of an apothecium of a lichen, when it has the same structure as the thallus, usually coloured like the thallus.

THALLOPHYTA. The most primitive division of the plant kingdom. It includes the algae, fungi, lichens, and possibly the bacteria and myxomycetes. The plant body (*thallus*) is simple, being unicellular, multicellular, and even large, but never differentiated into stem, leaf, and root. Asexual reproduction is by spores, and sexual reproduction by the fusion of gametes produced in sex-organs which consist essentially of single cells.
Some authors do away with this division, and elevate each class to the status of a phylum.

THALLOSPORE. An asexual spore of the fungi, having no conidiophore, or one which is not separated from the hypha or conidiophore producing it.

THALLUS. See *Thallophyta*.

THAMNIDIACEAE. A family of the Mucorales. The terminal primary sporangium is many-spored with a well-developed columella. The secondary sporangia are in the form of few-celled or one-celled sporangioles which are usually indehiscent. The sporangioles are in more or less dichotomous branches formed laterally along the main sporangiophore.

THEACEAE (EP, H). A family of the Parietales (EP), Theales (H). (BH) include it in the Ternstroemiaceae. These are woody plants with simple, usually alternate, exstipulate leaves, and bisexual regular flowers. There are 7-5 free sepals and 9-5 free petals which may be united below. The many to 5 stamens are free or may be united in bundles. The superior ovary consists of 5-3 or many-to-2 fused carpels, with many-to-1 ovules in each, on axile placentas. The fruit is a capsule, and the seeds may or may not contain endosperm.

THEALES (H). An order of the Lignosae (H). These are trees or shrubs, rarely woody climbers, sometimes epiphytic. The leaves are simple, alternate, and exstipulate. The flowers are hypogynous to subperigynous, and mostly bisexual. The sepals are imbricate (rarely contorted) and the petals are contorted or imbricate. The many-to-few stamens are free or slightly joined. The ovary is superior with axile placentation. The seeds have little or no endosperm.

THECA.   (1) An ascus.
  (2) The capsule of a moss.
  (3) A pollen sac.
  (4) An anther.

THECASPORE. An ascospore.

THECIFEROUS. Containing asci.

THECIUM. The hymenium of an ascomycete or lichen.

THEINE = CAFFEINE.

THELOPHORACEAE. A family of the Polyporales. The hymenium is smooth and entirely resupinate, or reflexed, or partially stipitate, with the hymenial surface on the underside. The spore fruit is a thin weft of hyphae, or may have a more definite structure, being papery, leathery, corky or even slightly woody. Both stichobasidial and chiastobasidial types of basidia are present

THEOPHRASTACEAE (EP, H). A family of the Primulales (EP), Myrsinales (H). (BH) include it in the Myrsinaceae. These are woody plants with alternate exstipulate leaves, often crowded at the ends of stems and branches. The flowers are bisexual or unisexual, and regular (rarely zygomorphic). There are 5 free sepals and 5 fused petals. There is 1 whorl of 5 free stamens and 1 of 5 staminodes. The superior ovary has 1 loculus with many ovules on free-central or basal placentas. The fruit is a drupe with many to 2 seeds, which contain endosperm.

THERMAL DEATH POINT. The temperature at which an organism is killed or a plant virus inactivated, after a given time, usually ten minutes.

THERMAL EMISSIVITY. The loss of heat from a leaf by radiation, conduction and convection.

THERIUM. A hot-spring formation.

THERMOGENIC SOIL. A soil developed in sub-tropical or tropical regions where the dominant factor in its formation is the high temperature.

TERMONASTY. The response to general non-directional temperature stimulus.

THERMOPHILE, THERMOPHILLOUS, THERMOPHILIC. Said of a plant which requires a high temperature for growth, or which can tolerate exposure to high temperatures.

THERMOPHILIC BACTERIA. Bacteria which require a temperature of from 45°C to 65°C for their development.

THERMOPHYTE. A plant growing in warm situations.

THERMO-STAGE. A stage in the life-history of a flowering plant, when at the onset of development from the embryo in the seed, low temperatures are needed to ensure further normal development.

THERMOTAXIS. The response or reaction of an organism to the stimulus of heat.

THERMOTOLERANT. Able to endure high temperatures, but not growing well under such conditions.

THERMOTROPISM = THERMOTAXIS.

THEROPHYLLOUS. (1) Having leaves only in the warmer part of the year. (2) Deciduous.

THEROPHYTE. An annual plant passing the winter, or dry season as a seed.

THIAMIN = ANEURIN.

THICKENING FIBRE. One of the spiral bands of thickening on the wall of a cell, tracheid or vessel.

THIGMOTAXIS. The response of an organism to the stimulus of touch or contact.

THIGMOTROPISM = THIGMOTAXIS.

THIOCTIC ACID = LIPOIC ACID.

THIOL. An —SH grouping commonly present in proteins.

THIOUREA.

Capable of breaking the dormancy in seeds.

THISMIACEAE (H). A family of the Burmanniales (H). The perianth is regular, inflated or bell-shaped, with the lobes more or less filiform or variously appendaged. There are usually six stamens. The ovary is unilocular and the fruit is not winged.

THIXOTROPHY. The conversion of a gel to a sol by shaking.

THORN. A leaf, part of a leaf, or shoot, with a vascular bundle, and ending in a hard, sharp point.

THRAUSTOCHYTRIACEAE. A family of the Lagenidiales. The zoosporangium is epibiotic with rhizoids.

THRECONINE. An amino-acid found in proteins.

THRESHOLD. That intensity of stimulus below which there is no response by a given irritable tissue.

THROAT. The opening of a gamopetalous corolla or gamosepalous calyx.

THRUM-EYED. Having the throat of the corolla more or less closed by the anthers.

THRYPTGEN, THRYPTOPHYTE. An organism increasing the sensitivity of a suspect to outside factors.

THURIACEAE (EP, H). A family of the Farinosae (EP), Juncales (H). (BH) include it in the Juncaceae. These are perennial herbs with narrow leaves and heads of regular bisexual regular flowers. The perianth lobes are similar, free and in 2 whorls of 3. There are 6 free stamens. The superior ovary consists of 3-fused carpels with 3 loculi, each with many to 1 ovules. The fruit is a capsule and the seeds contain endosperm.

THYMELAEACEAE (EP, BH, H). A family of the Myrtiflorae (EP), Daphnales (BH), Thymelaeales (H). These are shrubs or trees (rarely herbs) with entire opposite or alternate exstipulate leaves. The bisexual flowers are solitary, in racemes, or in spikes, with a cup-like or tubular axis. There are 5-4 free sepals, 5-4 or no free petals, and 5-4 or 10-8 free stamens. The superior ovary consists of 5-2 fused carpels, or 1, each with 1 pendulous ovule. The seeds may or may not contain endosperm.

THYMELAEALES (H). An order of the Lignosae (H). (EP) originally included it as an order of the Archichlamydeae. These are mostly woody with alternate or opposite leaves which have no stipules. If stipules are present they are small and glandular. The flowers are often in heads surrounded by an involucre of leafy bracts. Usually there are no petals and the calyx is petalloid. There are few-to-1 ovules and the seeds may or may not have endosperm.

THYRIOTHECIUM. An inverted perithecium in which the asci hang down.

THYROID. Shield-shaped.

THYRSE, THYRSUS. (1) A densely branched inflorescence with the main branching racemose, but the lateral branching cymose.

(2) Any closely branched inflorescence with many small-stalked flowers.

TIGELLUM = PLUMULE.

TILIACEAE (EP, H, BH). A family of the Malvales (EP) (BH), Tiliales (H). (BH) include the Elaeocarpaceae. These are usually woody plants with alternate stipulate leaves and regular bisexual flowers. There are 5 free sepals and 5, or no free petals. There are many to 10 free stamens, or they are in bundles. The anther are bilocular. The superior ovary consists of α-2-fused carpels each with α-1 ovules. The seeds contain endosperm.

TILIALES (H). An order of the Lignosae (H). These are trees or shrubs with an indumentum which is mostly stellate. The leaves are simple to compound, mostly alternate and usually with stipules. The flowers are regular, hypogynous, and bisexual or unisexual. The calyx is usually valvate. The stamens are free to monadelphous with the anthers bi- or uni-locular. The ovary is superior with axile placentation. The seeds have copious endosperm.

TILL = BOULDER CLAY.

TILLER. A grass shoot produced at the base of a stem.

TILLETIACEAE. A family of the Ustilaginales. The promycelium is not septate and has 4 to many sporidia at its blunt apex. The teliospores mostly arise as lateral outgrowths from hyphal cells, or are intercalary. The sporidia are expelled from the promycelium.

TILOPTERIDALES. An order of the Isogeneratae. The thallus is freely branched and trichothallic. The upper part is monosiphonous and the lower polysiphonous.

TIPHIUM. A pool formation.

TIRIUM. A bad-land formation.

TISSUE. A region consisting of cells of the same type and performing the same function. They are associated in large numbers and bound together by cell-walls.

TISSUE CULTURE. The growth of detached pieces of tissue in nutrient solutions under sterile conditions.

TISSUE SYSTEM. The whole tissues in a plant having the same function, whether or not they are in continuity, and whatever their position in the plant.

TISSUE TENSION. The mutual compressions and stretchings exerted by the tissues of a living plant.

TOADSTOOL. The fruit body of the agarics, other than mushrooms.

TOLERANCE. (1) The ability of a plant to withstand adverse environmental conditions.
(2) The ability of a plant to withstand the development of a parasite within it, without showing serious symptoms of disease.

TOMENTOSE. Covered with a felt or cottony hairs; downy.

TOMENTUM. A covering of cottony hairs.

TONOPLAST. The inner plasma membrane, bordering the vacuoles.

TOOTH. (1) Any small irregularity on the margin of a leaf.
(2) The free tip of a petal of a gamopetalous corolla.

TOPIARY. Formal ornamental gardening.

TOPOTAXIS. The response or reaction of an organism to a stimulus when the organism orientates itself in relation to the stimulus, and then moves towards or away from it.

TOP YEAST. A yeast which vegetates at the surface of the fluid in which it is living.

TORF = ACID HUMUS.

TORFSCHLAMM = DY.

TORMENTOSE = COTTONY.

TORSION. Twisting without marked displacement.

TORSION PAIRING. The non-homologous association of chromosomes at pachytene which releases a torsion without satisfying an attraction, when it occurs in continuance of homologous association.

TORTUOSE, TORTUOUS. Elongated and cylindrical, with evenly spaced swellings; necklace-like.

TORULOPSIDACEAE. A family of the Saccharomycetales. They are similar to the Saccharomycetaceae, but differ in the complete absence of ascus-formation.

TORULOSE, TORULOSE. Cylindrical, with slight contractions.

TORUS. (1) = RECEPTACLE (of a flower).

(2) The small thickening in the middle of the closing membrane of a bordered pit.

TOTIPOTENT. Bisexual.

TOUCHWOOD. Wood, much decayed by fungal attack. It crumbles readily, and is easily ignited by a spark when dry.

TOVIACEAE (EP, H). A family of the Rhoeadales (EP), Capparidales (H). (BH) include it in the Capparidaceae. These are herbs with the flowers in terminal raceme. The flowers are bisexual and regular with 8 free sepals, petals, and stamens. The superior ovary consists of 8-6 fused carpels, with placentas which reach the centre. There are many ovules. The fruit is a berry and the seeds contain endosperm.

TOXIC. Poisonous.

TPN = TRIPHOSPHOROPYRIDINE NUCLEOTIDE.

TRABANT = SATELLITE.

TRABECULA. (1) = CRASSULA.

(2) A rod-like structure or a rod-like cell running across a cavity.

(3) A gill primodium.

(4) Plates of undifferentiated primodial tissue in a developing gleba, forming a branch of a dendroid columella.

(5) Plates of sterile tissue running transversely across the sporangia in some Lycopodinae.

(6) Longitudinal plates of tissue across the lumen of a stem in *Selaginella*, supporting the stele.

TRABECULATE. (1) Said of peristome teeth marked by transverse bars.

(2) Having trabeculae.

TRACE ELEMENT. An element needed in very small amounts for healthy growth and development. Frequently it is a metal involved in the structure of an enzyme complex.

TRACER. An isotope incorporated into a compound of biological importance, so that the passage of the isotope atoms can be followed through the physiological processes of the plant.

TRACHEA. A conducting element of the xylem. See *vessel*.

TRACHEID(E). A non-living element of xylem formed from a single cell. It is elongated with tapering ends and thick, lignified and pitted, walls. It runs parallel to the long axis of an organ, overlapping and communicating with adjacent tracheids by means of pits. It conducts water and mineral salts in solution, and gives support.

TRACHEIDAL. Of the nature of a tracheid.

TRACHEOPHYTA = PTERIDOPHYTA and SPERMOPHYTA.

TRACHYSPERMOUS. Having seeds with rough surfaces.

TRACTILE FIBRE. The spindle fibre which begins to develop from an attachment to a chromosome and extend to the pole of the spindle.

TRAIT. A character.

TRAMA = CONTEXT. The somewhat loosely packed hyphae which occupy the middle of the gill of an agaric.

TRANSAMINATION. The transfer of ammonia during the formation of amino acids from keto acids.

TRANSAMINASE. An enzyme which carries out transamination.

TRANSDUCTION. The carrying of genetic material from an invaded and lysed bacterium to another bacterium by a bacteriophage.

TRANSECT. A line or belt of vegetation selected for charting plants. It is designed to study changes in the composition of the vegetation across a particular area.

TRANSEPTATE. Having all the septa placed transversely.

TRANSFER RNA. See *RNA*

TRANSFORMATION. The introduction of genetic material from dead bacteria in suspension to living bacteria in the same suspension.

TRANSFUSION CELL = PASSAGE CELL.

TRANSFUSION TISSUE. A tissue lying on either side of the vascular bundle in the leaves of most gymnosperms. It is thought to represent an extension of the vascular bundle performing the function of lateral veins. It is made up of empty cells with pitted and occasionally internally thickened walls, and parenchymatous cells which contain protein.

TRANSGRESSIVE SEGREGATION. The appearance in a segregating generation (F_2, back-cross etc.) of an individual or individuals falling outside the limits of variation set by the parents and F_1 of the cross in one or more characters.

TRANSITION CELL. A thin-walled cell at the end of a leaf, representing the last of the phloem.

TRANSITION REGION. The portion of the axis of a young plant in which the change from root structure to stem structure occurs.

TRANSITIONAL. Said of an inflorescence which has some racemose and some cymose characters.

TRANSITORY STARCH. Starch found temporarily in a leaf in which photosynthesis is proceeding more quickly than the removal or consumption of carbohydrates.

TRANSLATOR. A mechanism standing between two stamens, receiving pollen, half from each. Found in the Asclepiadaceae.

TRANSLOCASE = PERMEASE.

TRANSLOCATION. (1) The movement of soluble organic food material etc. through tissues.
    (2) The transfer of part of a chromosome into a different part of a homologous chromosome, or into a non-homologous chromosome.

TRANSPIRATION. The loss of water vapour from a plant, especially through the stomata.

TRANSPIRATION CURRENT, TRANSPIRATION STREAM. The stream of water which passes through a plant from the roots to the leaves, whence it escapes, chiefly as water vapour.

TRANSVERSE, TRANSVERSAL. (1) Broader than long.
    (2) Lying across the long axis of a body or organ.

(3) Lying crosswise between two structures.

(4) Connecting two structures in a crosswise fashion.

(5) Attached by the longer side.

TRAPACEAE (H). A family of the Lythrales (H). These are floating herbs with two kinds of leaves. They are similar to the Onagraceae, but the cotyledons are unequal.

TRAPEZIFORM. Of unsymmetrical four-sided shape.

TRAUMATIC. Relating to wounds.

TRAUMATIC RESPONSE. A reaction to wounding.

TRAUMATONASTY. A nastic movement following wounding.

TRAUMOTAXIS. The movement of protoplasts and nuclei after wounding.

TRAUMOTROPISM. The development of curvature following wounding.

TREE. A tall, woody, perennial plant, having a well-marked trunk with few or no branches persisting from the base.

TREE FERN = CYATHEACEAE.

TREHALOSE. A carbohydrate found in some fungi.

TREMANDRACEAE (EP, BH, H). A family of the Geraniales (EP), Polygalinae (BH), Pittosporales (H). These are shrubs with entire or toothed leaves and solitary axillary flowers. The flowers are solitary, bisexual with the parts in whorls of 5, 4, or rarely 3 parts. The sepals are free, and the petals valvate. The free stamens are in 2 whorls. The superior ovary consists of 2 fused carpels with 2-1 ovules in each. The fruit is a capsule, and the seeds contain endosperm.

TREMANDEAE (BH) = TREMANDRACEAE (EP).

TREMELLACEAE. A family of the Tremellales. The fruit-body is flat, cushion-shaped, lobed, or pileate, and is gelatinous, waxy, or somewhat dry. The hymenial surface is exposed. The basidia are single and terminal on the supporting hyphae with a more or less elongated extension from each wall, ended by a sterigma. The basidia are normally cruciately four-celled, but sometimes have 3 or even 2 cells. The primary septum is vertical or oblique. Clamp connections are present in the mycelium of many species.

TREMELLALES. An order of the Heterobasidiae. The basidium is rounded and divided cruciately by 4 septa which are vertical. The nuclear divisions are stichobasal.

TREMELLOID. Gelatinous.

TRENTEPOHLIACEAE. A family of the Ulotrichales. They have irregularly branched filaments with loose or compact branching. The protoplasts have many to 1 chloroplasts but are uninucleate at least when young. The zoospores develop in sporangia which differ in shape from the vegetative cells.

TRIARCH. Having 3 strands of xylem in the stele.

TRIASTER. A complex mitotic figure resulting from triple mitosis, as in an ovum after polyspermy.

TRIASSIC. A geological period from 190 to 170 million years ago.

TRIBE. A section of a family consisting of a number of related genera.

TRICARBOXYLIC ACID CYCLE = CITRIC ACID CYCLE.

TRICARPELLARY. Consisting of 3 carpels.

TRICHIDIUM = STERIGMA.

TRICHITE. A hypothetical crystal, which is very thin and elongated, and presumed to be present in very large numbers in a starch grain.

TRICHOCOMACEAE. A family of the Aspergillales. The ascocarps are not subterranean, and are stromatic at the base. The mass of asci and ascospores are pushed out at the top as a columnar structure. Conidia are not known.

TRICHOBLAST. (1) A colourless rod-like or granular particle. They surround the gullet in the Cryptomonadales.
(2) A branched multicellular filament produced from a small trichoblast initial at the apex of the filament of the Ceramiales. The trichoblast may divide to ultimately form a surface layer of spermatogonia.

TRICHOGYNE. A unicellular or multicellular organ which projects from a female sex-organ. It receives the male gamete or male nucleus before fertilization. It occurs in some of the green and red algae, ascomycetes and lichens.

TRICHOME. (1) A hair.
(2) In the Cyanophyta, a single row of cells which with their sheath make up the filament.

TRICHOME HYDATHODE. A multicellular hair which secretes water.

TRICHOPELTACEAE. A family of the Hemisphaeriales. The mycelium is conspicuous, external and radial, or forming parallel ribbons of closely united hyphae. The cover of the ascocarp arises as a thickening of the vegetative mycelium and is radial at least towards the margin.

TRICHOPHYLLOUS. Said of a plant of dry places which has the young stems and leaves protected from desiccation by a thick coating of hairs.

TRICHOPODACEAE (H). A family of the Dioscoreales (H). The flowers are bisexual. There are 6 stamens with the connectives produced beyond the anther-loculi. The ovary is inferior with 2 ovules in each loculus. The fruits are short and 1-seeded.

TRICHOTHALLIC GROWTH. A type of growth of an algal filament, in which cell division is confined to a few (or 1) cells near or at the base of the filament.

TRICHOTHECIN. An anti-fungal metabolic product of *Trichothecium roseum*.

TRICHOTHYRIACEAE. A family of the Erysiphales. The aerial mycelium is dark and the outer periderm not brittle. They are parasitic on epiphyllous Meliolaceae and similar fungi. The perithecium is inverted at maturity, with the morphological base upwards.

TRICHOTOMOUS. Having 3 equal or nearly equal branches arising from the same part of a stem.

TRICOTYLOUS. Having 3 cotyledons.

TRIENNIAL. Lasting for 3 years.

TRIFARIOUS. Arranged in 3 rows.

TRIFID. Divide about half-way down into 3 parts.

TRIFOLIATE. Said of a compound leaf having 3 leaflets.

TRIFURCATE. Bearing 3 prongs.

TRIGENIC. Said of a hereditary difference determined by 3 gene differences.

TRIGONAL. Triangular in cross-section.

TRIGONE. A thickened angle of a cell.

TRIGONIACEAE (EP, H). A family of the Geraniales (EP), Polygalales (H). (BH) include it in the Vochyiaceae. These are woody plants, often climbing, with alternate or opposite leaves with or without stipules. The bisexual flowers are obliquely zygomorphic. The 5 sepals are fused, the 5 petals are free and the 5, 6, 10-12 stamens are more or less fused below. The superior ovary consists of 3 fused carpels, each with many to 2 ovules. The fruit is a capsule, and the seeds contain endosperm.

TRIHYBRID. A trigenic hybrid.

TRILLIACEAE (H). A family of the Liliales (H). These are terrestrial, rarely swamp plants, with the inflorescence not subtended by a spathe. The leaves are opposite or whorled at the top of the stem. The flowers are mostly bisexual. The stamens have the filaments usually free from each other, and bilocular anthers.

TRIMENIACEAE (H). A family of the Laurales (H). The stamens are free with the anthers opening by longitudinal slits. The carpels are fused and the ovary is superior (rarely inferior). The fruit is berry-like or a drupe.

TRIMEROUS. Arranged in threes or multiples of 3.

TRIMITIC. Having 3 kinds of hyphae.

TRIMONOECIOUS. Having bisexual and unisexual flowers (both male and female) on the same plant.

TRIMORPHIC. Said of a species which has 3 kinds of flowers, differing in the relative lengths and positions of the filaments, anthers and stigmas.

TRINACRIFORM. Having 3 prongs.

TRIOECIOUS. Having bisexual, male, and female flowers on distinct plants of the same species.

TRIOLEIN. A fat formed from oleic acid (only) and glycerol.

TRIOSE. A monosaccharide sugar with 3 carbon atoms, $C_3H_6O_3$.

TRIOSE PHOSPHATE. The compound formed from a triose sugar and phosphoric acid.

TRIOSE PHOSPHATE DEHYDROGENASE. An enzyme which catalyses the oxidation of triose phosphate.

TRIPARTITE. Divided nearly to the base into 3 parts.

TRIPHOSPHOPYRIDINE NUCLEOTIDE. A substance closely related to DPN, and concerned with biological oxidations, and reductions, *e.g.* glyceraldehyde-3-phosphate is oxidized by the removal of H^+ to form TPNH. In photosynthesis it cooperates with TPNH to reduce PGA to triose, and in respiration triose is converted to PGA in the opposite way.

TRIPINNATE. Said of a pinnately compound leaf, with pinnately divided leaflets, which themselves are pinnately divided.

TRIPLE FUSION. The union in the embryo-sac of the two polar nuclei and a male nucleus. It is the starting point for the development of the endosperm.

TRIPLET. (1) The three adjacent nucleotides necessary to code a single amino acid. It is the minimum number.
(2) One of three individuals resulting from the division of an ovum.

TRIPLEX. A tetraploid zygote which has three doses of any given dominant.

TRIPLEX STRAND. A hypothetical three-stranded structure formed by two polynucleotide strands of DNA and a strand of coded RNA bases, the position of the latter being determined by the base-pairs on the DNA strands.

TRIPLINERVED. Having 3 main veins in a leaf.

TRIPLOID. Having three times the haploid number of chromosomes in the nucleus.

TRIPLO-POLYPLOID. A relatively triploid individual derived from an allopolyploid.

TRIQUETROUS. Having 3 angles and 3 concave surfaces.

TRISACCHARIDE. A carbohydrate formed by the condensation of three glucose units.

TRISOMIC. An otherwise diploid individual having one chromosome represented three times.

TRISTICHOUS. In 3 rows.

TRISTIS. Latin meaning 'dull coloured'.

TRISPORIN. A metabolic product of *Helminthosporium tritivulgaris*.

TRIURIDACEAE (EP, BH, H). A family of the Triuridales (EP, H), Apocarpae (BH). These are small saprophytes with scale leaves and small bisexual or unisexual flowers on long stalks. There are 8-3 free perianth lobes which are petal-like. There are 3, 4, or 6 free stamens, and many carpels each with 1 basal ovule. The pericarp is thick and there is much endosperm.

TRIURIDALES (EP, BH, H). An order of the Monocotyledonae (EP, BH), Calyciferae (H). It contains one family—the Triuridaceae.

TRIVALENT. Said of a chromosome which is threefold.

TRIVIAL NAME = SPECIFIC NAME.

TROCHLEAR. Pulley-shaped.

TROACHODENDRACEAE (EP, H). A family of the Ranales (EP), Magnoliales (H). (BH) include it in the Magnoliaceae. These are trees or shrubs with alternate, exstipulate leaves. The flowers are bisexual or unisexual and lack a perianth. They are solitary or in racemes. There are many free stamens, and many to 5 free superior carpels. The fruits are capsules or achenes.

TROPAEOLACEAE (EP, H). A family of the Geraniales (EP, H). (BH) include it in the Geraniaceae. These are usually climbers with a sensitive petiole. There may or may not be stipules. The bisexual flowers are zygomorphic, with 5 sepals and petals. The dorsal sepal forms a spur. There are 8 free stamens, and the superior ovary is trilocular with 1 ovule in each loculus. The fruit is schizocarp, and the Pseeds have no endosperm.

TROPHIC. Pertaining to nutrition.

TROPHOCHONDRIOMA. A mitochondria concerned with nutrition.

TROPHOCHROMATIN. A substance within the nucleus which controls the nutrition of the cell.

TROPHOCHROMIDIA. Chromidia concerned with nutrition.

TROPHOCYST. A hyphal swelling from which a sporangium is produced.

TROPHOGONE, TROPHOGONIUM. Of ascomycetes, an atheridium whose only function is to supply food.

TROPHOPHYLL. A vegetative leaf.

TROPHOPHYTE. A plant which is xerophytic at one period of the year, and hygrophytic at another.

TROPHOPLASM. Protoplasm which is mainly concerned with nutrition.

TROPHOPLAST. A plastid.

TROPHOTROPISM. A reaction in a growing organ induced by the chemical nature of the environment.

TROPIC CURVATURE. A curvature of a plant organ caused by one-sided growth under the influence of a stimulus falling on the plant from one side.

TROPISM. A response to a stimulus by a growth curvature, the direction of the curvature being determined by the direction from which the stimulus originates.

TRUCATED. Said of a soil profile in which the top layers have been removed by erosion.

TRUE FRUIT. A fruit formed only from the ovary.

TRUFFLE. The subterranean fruit-body of the fungi of the Tuberales.

TRUMPET HYPHA. A filament inside the thallus of a brown alga, which is markedly enlarged at each transverse septum.

TRUNCATE. Blunt-ended, as if cut off abruptly.

TRUNK. The upright massive main stem of a tree.

TRYPTOPHANE. An amino acid in protein.

TYROSINE. An amino acid found in protein.

TUBE GERMINATION. Germination of a spore by the formation of a hypha (germ tube).

TUBE NUCLEUS. The non-generative nucleus in a pollen-tube. It probably plays a part in regulating the development and behaviour of that organ.

TUBER. A swollen underground stem or root, containing stored food, and acting as an organ of perennation and vegetative propagation.

TUBERALES. An order of the Ascomycetae in which the ascocarp is closed, and completely subterranean.

TUBERCLE. (1) A rounded protuberance.

(2) A small swelling on the roots of legumes and other plants, inhabited by symbiotic bacteria.

TUBERCULARIACEAE. A family of the Moniliales. An artificial group, based on the fact that the conidiophores arise from a sporodochium.

TUBERCULAROID. Having a warted surface.

TUBERCULATE. Provided with tubercles.

TUBERCULIFORM. Wart-like.

TUBERIFORM, TUBEROUS. Having the form of a tuber.

TUBEROUS. Thickening and forming tubers.

TUBIFLORAE (EP). An order of the Sympetalae (EP). These are usually herbs, typically with the flowers having 4 whorls of like parts. There are usually a few carpels in the ovary, and if the flower is zygomorphous, a few stamens, which are fused to the petals. The ovary has one integument.

TUBULE, TUBULUS. (1) The neck of a perithecium.
(2) A pore lined by a hymenium bearing basidia.

TUFTED. Having many short crowded branches, all arising at about the same level.

TULASNELLACEAE. The single family of the Tulasnellales.

TULASNELLALES. An order of the Heterobasidae. The basidia are rounded, bearing large rounded epibasidia, usually separated from the hypobasidium by septa, and often falling free. The nuclear divisions are chiastobasidial.

TULSTOMATACEAE. A family of the Lycoperdales. The fruit-body has a stipe, and the basidia line definite cavities, or these may be more or less obliterated. At maturity the whole gleba except the capillitium and spores, dissolves into a powdery mass.

TUMERACEAE (H). A family of the Loasales (H). The petals are contorted in the bud. The stamens have free filaments. The ovary is superior with 3 styles, and the seeds are arillate.

TUMESCENT. Somewhat tumid.

TUMID. Swollen; inflated.

TUNDRA. A cold desert characterized by the scanty, xerophytic vegetation in which mosses and lichens are dominant.

TUNIC = EXOCARP.

TUNICA. A coat, especially a thin, white membrane round the peridiole in most species of the Nidulariaceae.

TUNICATE, TUNICATED. Having a coat or covering.

TUNICATE BULB = LAMINATED BULB.

TURBINATE. Shaped like a top and attached at the point.

TURBINATE CELL, TURBINATE ORGAN. A swelling on a vegetative thallus of the Cladochytriaceae.

TURF. An association in which grasses predominate.

TURGESCENCE. The condition of cells or tissues which are distended by water.

TURGID. (1) Said of a cell which is distended, being well-supplied with water.
(2) Said of a young or soft plant member which is stiff or rigid owing to the internal pressure arising from a plentiful supply of water.

TURGOR. The state of a cell when the cell-wall is stretched by an increase in the volume of the vacuole by the intake of water.

TURGOR PRESSURE. The hydrostatic pressure set up within a cell by the water present acting against the elasticity of the wall.

TURION. (1) A swollen perennating bud, containing much stored food, formed by a number of water plants. It comes away from the parent, remains inert through the winter and gives rise to a new plant in the following spring.
(2) A scaly sucker or shoot arising from underground.

TURNERACEAE (EP, BH). A family of the Parietales (EP), Passiflorales (BH). These are herbs, trees, or shrubs with alternate leaves with or without stipules. The flowers are axillary, or in racemes or cymes. They are bisexual, regular with a tubular axis. There are 5 free sepals, petals, and stamens. The ovary consists of 3 fused carpels, and is superior. There are many to 3 ovules in each carpel, on parietal placentas. The fruit is a capsule. The seeds are arrilate, and contain endosperm.

TWINS. Individuals arising from the division of a fertilized egg into two; each half developing into an individual.

TWISTED AESTIVATION = CONTORTED AESTIVATION.

TWO-EDGED. Flattened and having two sharp edges.

TWO-LIPPED = BILABIATE.

TYLOSE, TYLOSIS. A balloon-like enlargement of the membrane of a pit in the wall of a vessel or tracheid, and a xylem parenchyma cell lying next to it. It protrudes and blocks the cavity of the wood element.

TYMPANIFORM, TYMPANOID. Like a drum-head.

TYNDELL EFFECT. In a sol whose particles are too fine to be seen with the naked eye, a milkiness shown in the path of a lateral beam of light.

TYPE SPECIMEN. The original specimen(s) on which a description of a new species etc. is made.

TYPHACEAE (EP, BH, H). A family of the Pandanales (EP), Nudiflorae (BH), Typhales (H). (BH) include the Sparganiaceae. These are herbs with rhizomes and linear two-ranked leaves. The naked flowers are in spikes with female ones below and male above. There are 5-2 free stamens, and a superior ovary of 1 carpel on a hairy axis. The ovary contains 1 ovule. The fruit is a nutlet, and the seed contains endosperm.

TYPHALES (H). An order of the Corolliferae (H). These are aquatic or marsh herbs with rhizomes and elongated leaves which are sheathing at the base. The flowers are unisexual and very small, being crowded into clusters or dense spikes. The perianth is much modified and reduced. There are more than 2 stamens and the ovary is unilocular with 1 pendulous ovule.

U

UDP = URIDINE DIPHOSPHATE.

UDPG = URIDINE DIPHOSPHGLUCOSE.

UFERFLUCHT. A concentration of limnetic plankton towards the centre of a lake.

ULGINOSE, ULGINOUS. Growing in wet places.

ULMACEAE (EP, H). A family of the Urticales (EP, H). (BH) include it in the Urticaceae. These are trees or shrubs with two-ranked simple, stipulate leaves. The bisexual, or unisexual flowers are in cymes and the perianth whorls are similar, and sepaloid. There are 5-4 free perianth lobes and 5-4 or 10-8 free stamens. The superior ovary consists of 2 fused carpels, and is usually unilocular with 1 pendulous ovule. The fruit is a nut or drupe, and the seeds usually have no endosperm.

ULMIC ACID. An acid occurring in humus.

ULMIN. A substance, related to ulmic acid, found in humus.

ULOTRICHACEAE. A family of the Ulotrichales. The filaments are unbranched and the cells are uninucleate. There is a single parietal girdle-shaped chloroplast, and the cell-walls are not of overlapping pieces. Asexual reproduction is by biflagellate or quadriflagellate zoospores, aplanospores, or akinetes. Sexual reproduction is isogamous by motile gametes.

ULOTRICHALES. An order of the Chlorophyceae. The cells are uninucleate (except for the old cells in some genera), and usually have 1 parietal, laminate chloroplast. The simple or branched filaments may form a pseudoparenchymatous mass. Asexual reproduction is by biflagellate or quadriflagellate zoospores, aplanospores, or akinetes. Sexual reproduction is isogamous, anisogamous or oögamous.

ULTRAMICROBE. An agent of obscure nature, able to cause disease in organisms, but too small to be seen with a light-microscope.

ULVACEAE. A family of the Ulvales. The thallus is an expanded sheet 1-2 cells thick.

ULVALES. An order of the Chlorophyceae. The uninucleate cells divide in three planes to give a parenchymatous thallus, which may be a sheet, tube, or solid cylinder. Asexual reproduction is isogamous or anisogamous, by biflagellate gametes.

UMBEL. A raceme in which the axis has not elongated, so that the flowers stalks arise at the same point. Thus the flowers are in a head, with the oldest at the outside.

UMBELLALES (BH) = UMBELLIFLORAE (EP); UMBELLALES (H) = UMBELLIFERAE (EP).

UMBELLATE. (1) Having the characters of an umbel.
(2) Producing umbels.

UMBELLIFERAE (EP, BH, H). A family of the Umbelliflorae (EP, BH), Umbellales (H). These are herbs with tap roots or rhizomes and hollow stems, with alternate, usually much-divided, sheathing exstipulate leaves. The flowers are usually bisexual, regular, small and with the parts in fives. They are in umbels, which are compound or simple. The calyx is often indistinguishable. There are as many stamens as petals. The inferior ovary consists of 2 fused carpels, with 2 styles on a swollen base. The fruit is a schizocarp with mericarps on a carpophore, each usually with 5 ribs with vittae between. The endosperm is oily.

UMBELLIFLORAE (EP). An order of the Archichlamydeae. The flowers are bisexual and regular, with the parts in fours or fives (rarely many in a whorl). They are in umbels, cyclic, with sepals and petals unalike, usually with one whorl of stamens which are epigynous. The inferior ovary has 5-1 or many fused carpels, each with 1 (rarely 2) pendulous ovules with 1 integument. There is a rich endosperm.

UMBILICATE. Peltate and depressed in the centre.

UMBONATE. With a central boss.

UMBRACULIFEROUS. Like an expanded umbrella.

UMBRINUS. Latin meaning 'umber-coloured'.

UMBROSUS. Latin meaning 'of shady places'.

UNAVAILABLE NUTRIENTS. Minerals which for some reason, *e.g.* insolubility, cannot be absorbed by the plant from the soil.

UNBALANCED POLYPLOID = ANEUPLOID.

UNCATE, UNICINATE. Hooked.

UNDERSHRUB. A low-growing woody plant.

UNDULATE. Wavy.

UNFIXABLE. Eversporting from seed.

UNFREE WATER. Water which is held closely by the soil colloids so that it is not readily available to plants. It freezes well below 0°C.

UNGULATE. Like a horse's hoof in form.

UNICELLULAR. Consisting of one cell.

UNIJUGATE. Said of a compound leaf having one pair of leaflets.

UNILATERAL. Said (1) of members which are inserted on one side of an axis, or are all turned to one side of the axis,
(2) of a raceme with all the flowers turned to one side,
(3) of a stimulus falling on the plant from one side.

UNILOCULAR. Consisting of one compartment.

UNINUCLEATE. Having one nucleus.

UNIPOLAR. At one end only, especially of a bacterial cell.

UNISERIATE. (1) Arranged in a single row, series, or layer.
(2) Said of a vascular ray which is one cell wide in cross-section.

UNISEXUAL = DIOECIOUS. Of one sex.

UNISEXUALES (BH). A series of the Incompletae (BH). The flowers are unisexual, with the perianth sepaloid, or much reduced, or absent. The carpels are fused to form the ovary, or it is of one carpel. There are 1 or 2 ovules in each carpel. Endosperm may or may not be present.

UNISTRATOSE. Forming a single layer.

UNIT CHARACTER. One whose differences from an alternative is transmitted as a unit in heredity. These are independent characters which are traceable in each generation, and assorted and distributed by the laws of chance.

UNIT OF VEGETATION = COMMUNITY.

UNIVALENT. One of the single chromosomes which separate in the first meiotic division.

UNIVERSAL VEIL. A coating of hyphae which completely surrounds the fruit-body of a fungus.

UNREDUCED GAMETE. A gamete having the diploid number of chromosomes.

UNSEPTATE. Lacking cross-walls.

UNSTABLE COMMUNITY. A plant community which does not remain constant over a period of years.

UNSTABLE GENE. A gene liable to frequent mutations.

UNSYMMETRICAL. Said of a flower in which all the parts are not regular.

UPPER = POSTERIOR.

URACIL. A pyrimidine base.

URCEOLATE. Urn-shaped.

URCEOLUS. An urn-shaped, or pitcher-shaped structure.

UREASE. An enzyme which catalyses the splitting of urea into ammonia and water.

UREDICOLE. Growing as a parasite on rust fungi.

UREDINALES. An order of the Teliosporeae. The Rusts. They are obligate parasites of pteridophytes, conifers, and flowering plants. The teliospores are single or united into crusts or columns, or are several together in compound spores, remaining within the host tissue or bursting through the epidermis or cuticle. Spermogonia are normally produced, the sperm cells diploidizing special receptive hyphae. Typically three types of spore are produced, aeciospores the product of the diploidization of the monocaryon mycelium which arises from sporidia; urediospores (repeating spores) and teliospores from which arise the promycelia and sporidia. The sporidia are always expelled violently.

UREDINIAL STAGE = UREDO-STAGE.

UREDINIOSPORE = UREDOSPORE.

UREDINIUM, UREDIUM = UREDOSORUS.

UREDINIOSPORE, UREDOGONIUM = UREDOSPORE.

UREDOSORUS. A pustule consisting of uredospores, with their supporting hyphae, and some sterile hyphae.

UREDOSPORE. An orange or brownish spore formed by rust fungi when growth is vigorous. It serves as a means of rapid propagation and gives rise to a mycelium, which may produce more uredospores or later in the year, teleutospores.

UREDOSTAGE. The phase in the life-history of a rust fungus when uredospores are formed.

URIDINE = RIBOSE-3-URACIL. A nucleoside formed from ribose and uracil.

URIDINE DIPHOSPHATE (UDP). See *uridine diphosphoglucose.*

URIDINE DIPHOSPHOGLUCOSE (UDPG). Glucose-1-phosphate is converted to UDPG in the presence of UDP during carbohydrate metabolism.

URN. The capsule of a moss.

URN-SHAPED. Having the shape of a rounded vase, swollen in the middle.

URTICACEAE (EP, BH, H). A family of the Urticales (EP, BH, H). (BH) include the Moraceae, Ulmaceae, and Cynocrambaceae. These are usually herbs with opposite or alternate, stipulate leaves. There is no latex. The flowers are usually unisexual, in cymes, with the perianth lobes alike. The 5-4 perianth lobes are usually free, and there are 5-4 free stamens opposite the perianth lobes. The superior ovary is of 1 carpel with 1 basal ovule. The fruit is a nut or drupe and the seed contains endosperm.

URTICALES (EP, BH, H). An order of the Archichlamydeae (EP), Unisexuales (BH), Lignosae (H). These are herbs, shrubs or trees, with alternate or opposite stipulate leaves and cymose inflorescences. The flowers are cyclic and the perianth whorls are alike (rarely different) or absent. The flowers are regular and bisexual or unisexual, usually with the 2 whorls of the perianth of 2 lobes each (rarely 2 and 3). The stamens are in front of the perianth lobes. The superior ovary consists of 2 fused carpels, or 1, with 1 ovule. The fruit is a drupe or nut.

URTICLE. A small fruit with the pericarp free from the seed.

USTILAGINACEAE. A family of the Ustilaginales. The promycelium is transversely septate into several, usually 4 cells. The teliopores arise in the tissues of the host from transformed hyphal cells, and are mostly distributed by air currents. The sporidia are not expelled from the promycelium.

USTILAGINALES. An order of the Teliosporeae. They are obligate parasites of flowering plants, or in many cases facultative saprophytes. The teliospores are single or united in columns or balls, remaining within, or bursting out of, the host tissue. They are mostly distributed by air-current. There are no spermogonia or special receptive hyphae. Diploidization is by means of union of compatible spores, hyphae etc. Typically only teliospores and often hyaline thin-walled conidia are produced. Sporidia are expelled violently in 1 family, not so in the two others.

UTERUS. The peridium of some fungi, especially the Gasteromycetes.

UTERICLE. A more or less inflated, membranous bladder-like envelope, surrounding the fruits of some plants, and covering certain fungi.

UTRICULAR, UTRICULIFORM. (1) Like a bladder.
(2) Pertaining to the utriculus.

UTRIFORM. Bag-like.

UVA. A berry formed from a superior ovary.

UVARIOUS. Latin meaning 'like a bunch of grapes'.

V

VACCINIACEAE (BH, H). A family of the Ericales (H). Part of the Ericaceae (EP).

VACUOLAR. Resembling or pertaining to the vacuole.

VACUOLAR MEMBRANE. The protoplasmic membrane which bounds a vacuole, separating it from the surrounding protoplasm.

VACUOLATE. (1) Vesicular.
(2) Having a vacuole(s).

VACUOLE. A fluid-filled space in a cell. A single vacuole, taking up most of the volume of the cell is present in many plant cells, and contains a cell-sap which is isotonic with the protoplasm.

VACUOLIZATION. The formation of vacuoles.

VAGINA. A sheathing leaf-base.

VAGINATE. Sheathed.

VAGINULE. A minute sheath, surrounding the base of the seta in the bryophytes.

VAHLIACEAE (H). A family of the Saxifragales (H). The leaves are alternate, and the flowers are rarely solitary. The anthers are bilocular and there are no staminodes. The petals are free or absent. The carpels are united and the ovary is superior or inferior, with the stigma dorsal to the carpels. The style is usually free or absent. The placenta hangs from the top of the ovary, and the fruit is a capsule.

VALERIANACEAE (EP, BH, H). A family of the Rubiales (EP), Asterales (BH), Valerianales (H). These are herbs, often shrubby, with opposite exstipulate leaves and cymose umbels or heads or bisexual or unisexual flowers. The calyx is indistinct in the flower, but later enlarges to form a pappus. The 5-4-3 petals are fused, often forming a spur at the base. There are 4-1 free stamens. The inferior ovary consists of 3 fused carpels, but only 1 develops and contains 1 pendulous ovule. There is no endosperm.

VALERIANALES (H). An order of the Herbaceae. These are perennial or annual herbs, with alternate or opposite exstipulate leaves, which may be reduced. The flowers are zygomorphic and sometimes unisexual, in cymes, heads or on verticillate branches, sometimes in an involucre of bracts. The perianth is epigynous and the petals fused. The 4 (usually) stamens alternate with the corolla lobes. The ovary is inferior with 3-1 loculi, only 1 of which is fertile. There is 1 pendulous ovule, and the fruit is indehiscent.

VALID. Of names, in accord with the International Rules of Botanical Nomenclature.

VALINE. An amino acid found in protein.

VALLECULA. Latin for 'grooves in flowers'.

VALLECULAR CANAL. A large intercellular space running in the cortex of *Equisetum*, outside the carinal canal.

VALONIACEAE. A family of the Siphonocladiales. All the cells except the rhizoids are similar. Reproduction is by segmentation of the vegetative thallus or by biflagellate swarmers. Sometimes the zooids germinate directly, sometimes they unite in pairs.

VALSOID. (1) Having the perithecia in a circle in the stroma.
(2) Having groups of perithecia with their beaks pointing inwards, or even parallel to the surface.

VALVATE. Having the margins not overlapping.

VALVATE DEHISCENCE. The liberation of pollen from anthers, or seeds from dry fruits by means of little flaps of upraised wall-material.

VALVES. (1) A part of a fruit-wall which separates at dehiscence.
(2) One of the two halves of the cell-wall of a diatom.

VALVULAR. Opening by means of valves.

VAN DER WALL'S FORCE. A weak form of linkage between two groups.

$$HC-CH_3 - - - - - - CH_3-C-H$$

VARIABILITY. The capacity of an individual or group of individuals to produce gametes having genotypic variation.

VARIENCE. The mean square deviation of a variate from its mean. Estimated as the Mean Square. The square of the Standard Deviation.

VARIANCE RATIO. The ratio of two estimated variances. Twice the natural antilogarithm of z. Sometimes denoted by F.

VARIANT. (1) A variable quantity whose measurements or frequencies form all or part of the data for analysis.
(2) A specimen differing slightly in its characteristics from the type.

VARIATION. (1) The occurrence of differences in the permanent structure of cells.
(2) The occurrences of differences between individuals due to differences in the permanent structure of their cells, or to differences in the environment.
(3) The differences between the offspring of a single mating.

VARICOSE. Dilated.

VARIEGATION. Irregular variation in colour of a plant organ, *e.g.* leaves or flowers due to the suppression of normal pigment development. This may be due to the action of a marginal genotype, somatic mutation, or infection.

VARIETY. A subdivision of a species, owing to its uniformity, either to genetic isolation in nature, or to artificial propagation in cultivation.

VARNISHED. Said of a surface which appears to be covered by a thin shining film.

VASCULAR. Pertaining to, or having vessels which convey fluids.

VASCULAR ANASTOMOSIS. A small transverse vascular bundle, acting as a link between the main vascular bundles of a stem or root.

VASCULAR BUNDLE, VASCULAR STRAND. The longitudinal strand of conducting tissue, consisting essentially of xylem and phloem.

VASCULAR CRYPTOGAM. A non-flowering plant which has vascular tissue.

VASCULAR CYLINDER = STELE.

VASCULAR PLANT. A plant having a vascular system, *i.e.* the Pteridophyta and Spermophyta.

VASCULAR RAY = MEDULLARY RAY.

VASCULAR RAY INITIAL. A cell of the cambium which divides to give daughter cells which are converted into cells which form a medullary ray.

VASCULAR SYSTEM. The xylem and phloem forming a continuous system throughout all parts of vascular plants.

VASCULAR TISSUE = VASCULAR BUNDLE.

VASCULIFORM. Shaped like a little pot.

VASICENTRIC = PARATRACHEAL.

VASIFORM TRACHEID(E). A wide tracheid capable of conducting water.

VAUCHERIACEAE. A family of the Siphonales. The thallus is sparingly branched and tubular. Asexual reproduction is by zoospores or aplanospores both of which are produced singly in a sporangium. Sexual reproduction is oögamous. The oögonium contains one egg which remains in it after fertilization.

VECTOR. An animal which transmits a pathogen.

VEGETATION. All the plants in a given area.

VEGETATIVE CORE. The apical meristem of a shoot.

VEGETATIVE REPRODUCTION, VEGETATIVE MULTIPLICATION, VEGETATIVE PROPAGATION. Asexual reproduction by detachment of part of the plant which then develops into a complete plant.

VEIL, VELUM. (1) An evanescent membrane over an apothecium.
(2) A sheath of hyphae forming a complete membrane over the fruit-body of an agaric.

VEIN. One of the vascular bundles in a leaf.

VEIN ISLET. A very small patch of photosynthetic tissue in a leaf, more or less surrounded by a small vein.

VELAMEN. Water absorbing tissue occurring on the outside of aerial roots in certain plants, *e.g.* epiphytic orchids, consisting of several layers of dead cells often with spirally thickened and perforated walls, which soak up water running over it.

VELLOZIACEAE (EP, H). A family of the Liliiflorae (EP), Haemodorales (H). (BH) include it in the Amaryllidaceae. These are herbs or shrubs with linear crowded leaves and terminal solitary, bisexual flowers, with the parts in threes. The perianth is petalloid. The stamens are free, and 6 in number, or 6 bundles. The ovary is inferior consisting of 3 fused carpels and 3 loculi with many ovules on lamella placentas. The fruit is a capsule and the seeds contain endosperm.

VELUM. (1) = VEIL.

(2) A membranous outgrowth from the base of a leaf of *Isoetes*. It covers or partly covers the sporangium.

VELUM PARTIALE = PARTIAL VEIL.

VELUM UNIVERSALE = UNIVERSAL VEIL.

VELUTINATE, VELUTINOUS. Having a velvety surface.

VENATION. The arrangement of the vein in a leaf.

VENENATUS. Latin meaning 'poisonous'.

VENOSE. With veins.

VENTER. The swollen basal region of an archegonium, containing the egg-cell.

VENTILATION TISSUE. The sum total of the intercellular spaces in a plant, by means of which gases can circulate through the plant body.

VENTRAL (1) Anterior or in front.

(2) Uppermost.

(3) Nearest to the axis.

VENTRAL CANAL CELL. An unwalled cell which lies in the venter of an archegonium, above the egg of which is a sister cell.

VENTRAL SUTURE. The presumed line of junction of the edges of the unfolded carpel.

VENTRAL TRACE. One of the two laterally placed vascular strands often present in the wall of a carpel.

VENTRICOSE.(1) Swollen in the middle.

(2) Having an inflated bulge at one side.

VERBENACEAE (EP, BH, H). A family of the Tubiflorae (EP), Laminales (BH), Verbenales (H). (BH) include the Phrymaceae. These are herbs or woody plants with usually opposite or whorled, entire or divided, leaves. The flowers, which are in cymose umbels, are bisexual, usually zygomorphic, with the flower parts in fives or fours (rarely more). The lobes of the calyx and corolla are fused, the latter often being two-lipped. There are 4 didynamous, or 2, stamens. The superior ovary consists of 2 (rarely more) fused carpels, each with 2 ovules, and 4 loculi. The fruit is a drupe or schizocarp, and there is usually no endosperm.

VERBENALES (H). An order of the Lignosae (H). These are trees or shrubs (rarely herbs) with the branches often quadrangular. The leaves are alternate, opposite or whorled, and lack stipules. The flowers are bisexual with a persistent calyx. The corolla is regular or zygomorphic with 5-4 imbricate lobes. There are as many stamens as corolla lobes or one fewer (rarely 5 or 2), inserted on the corolla tube. The anthers are bilocular. The ovary is superior with 9-1 loculi. The style is terminal and simple. There are 2-1 ovules in each loculus, and they are erect, axile or rarely pendulous. The fruit is a drupe or berry and the seeds have little or no endosperm.

VERMICULAR. Shaped like a worm.

VERNAL, VERNALIS (Latin). Of or belonging to spring.

VERNAL ASPECT. The condition of a plant community in spring, characterized by some species being particularly active at this time.

VERNALIZATION. Treatment of plant with low temperatures in the initial stages of seedling development, inducing a rapid development towards a physiologically older condition and resulting in a shortening of the interval between sowing and flowering.

VERNATION. The manner in which the leaves are arranged in the bud.

VERRUCA. A granular or wart-like outgrowth on a thallus.

VERRUCIFORM. Resembling a wart.

VERRUCOSE, VERRUCOUS. Said of a surface covered with wart-like upgrowths.

VERRUCULOSE. Slightly verrucose.

VERSATILE. Said of an anther which is attached at the tip of the filament, by a small area on its dorsal side, so that it turns freely in the wind, helping the dispersal of the pollen.

VERSICOLOROUS. (1) Not all the same colour.
(2) Changing colour with age.

VERSIFORM. (1) Said of organs of the same kind which are not all the same shape.
(2) Changing form with age.

VERTICEL = WHORL.

VERTICILLASTER. A kind of inflorescence which looks like a dense whorl of flowers, but is really a combination of two crowded dichiasal cymes, one at each side of the stem.

VERTICILLATAE (EP). An order of the Archichlamydeae. See *Casuarinales*.

VERTICILLATE. Arranged in whorls.

VESICLE. A thin-walled globular swelling usually at the end of a hypha.

VESICULAR. (1) Like, or pertaining to, a vesicle.
(2) Like a bladder.

VESICULAR BODIES. Thin-walled vesicles in the sub-hymenium of certain Hymenomycetes.

VESICULAR WATER. Water which is held in the gel structure of soil colloids.

VESICULOSE. (1) Swollen like a bladder.
(2) Appearing as if made up of small bladders.
(3) Made up from, or full of vesicles.

VESSEL. A non-living element of the xylem consisting of a tube-like series of cells arranged end-to-end, running parallel to the long axis of the organ in which it lies, and in communication with adjacent elements by means of numerous pits in the side-walls. It functions in the conduction of water and mineral salts, and acts in mechanical support.

VESTIGIAL ORGAN. Functionless rudiments.

VESTITURE. A covering, especially of hairs.

VEXILLAR. Relating to the standard.

VEXILLUM = STANDARD.

VIABLE. Able to live, said of seeds, spores etc. which are capable of germination.

VIABILITY. The measure of the number of individuals surviving in one class, relative to another standard class.

VICILLIN. A globulin occurring in pea seeds.

VICINISM. An unexpected outcrossing.

VILLOSE, VILLOUS. With long weak hairs.

VILLUS. A thin branching outgrowth from the 'stem' of a moss.

VINOUS. Of the colour of red wine.

VIOLACEAE (EP, BH, H). A family of the Parietales (EP, BH), Violales (H). These are herbs or woody, with alternate, stipulate leaves and bisexual regular, or zygomorphic flowers. There are 5 free sepals, petals and stamens. The superior ovary consists of 3 fused carpels, each with α-1 ovules on parietal placentas. The fruit is a capsule or berry and the seeds contain endosperm.

VIOLALES (H). an order of the Lignosae (H) with one family, the Violaceae (H).

VIOLAREAE (BH) = VIOLACEAE (EP) and part of the Ochnaceae (EP).

VIRENS. Latin meaning 'green'.

VIRESCENT. An abnormal green condition, sometimes accompanied by the development of small, crowded leaf-like structures, due to the attack of a parasite or other disease.

VIRGATE. (1) Long, slender and stiff, much branched.
(2) Streaked.

VIRGATUS. Latin meaning 'twiggy'.

VIROLOGY. The study of viruses.

VIROSE. (1) Poisonous.
(2) Having a strong unpleasant smell.

VIRULENCE. (1) The capacity of a parasite to cause disease.
(2 The degree or measurement of pathogenicity.

VIRUS. An ultra-microscopic particle whose reproduction in the cell and transmission by natural infection gives characteristic reactions of cells and individuals.

VISCID, VISCIDOSUS (Latin). Said of a surface which is sticky or clammy.

VISCID DISSEMINULE. A seed or spore which has a sticky surface, or sticky hairs, and is dispersed by becoming attached to an animal.

VISCIN. The sticky substance produced in the fruits of mistletoe.

VITACEAE (EP, H) = AMPELIDACEAE (BH). A family of the Rhamnales (EP, H), Celastrales (BH). These are climbing shrubs, often with tendrils opposite the leaves. It is like the Rhamnaceae, but the fruit is a berry. The petals are valvate, often united above, and falling as a whole. The superior ovary consists of 8-2 fused carpels. The seeds contain endosperm.

VITAL STAINING. The staining of cells or their parts while they are living, and without killing them.

VITELLINE. Egg-yellow.

VITTA. (1) A stripe.
(2) A resin or oil canal in the pericarps of some fruit.

VITTATE. Having longitudinal ridges or stripes.

VIVIANACEAE (H). A family of the Pittsporales (H). The leaves are opposite. The sepals are valvate, and the petals contorted. The embryo is curved or coiled.

VIVIPAROUS. (1) Producing bulbils or young plants, instead of and in place of flowers.

(2) Said of a seed which germinates before it is detached from the parent plant.

VLEI SOILS. African soils, rather like gley soils, occurring in depressions subject to seasonal wetness.

VOCHYSIACEAE (EP, BH, H). A family of the Geraniales (EP), Polygalinae (BH), Polygalales (H). (BH) include the Trigoniaceae. These are woody, rarely herbs, with opposite or whorled simple leaves, with or without stipules, and bisexual obliquely zygomorphic flowers. There are 5 fused sepals, 1 is often spurred, and 3-1 corolla lobes which are free. There is 1 stamen, and staminodes. The ovary, which is superior or inferior consists of 3 fused carpels with many to 2 ovules. The fruit is indehiscent or a capsule. The seeds have no endosperm.

VOLCANIC SOIL. A soil derived from volcanic ash under tropical conditions. If basic they develop rapidly into fertile soils, but take much longer if they are acid.

VOLUBLE, VOLUBILIS (Latin). Twining.

VOLUNTEER. A crop growing from unplanted seeds, *e.g.* seed shed by the previous crop.

VOLUTIN GRANULES. (1) Granular cytoplasmic inclusions which stain intensely with basic dyes. They are believed to contribute to the formation of chromatin.

(2) Stored food substances in fungi, especially yeasts.

VOLVA. A sheath of hyphae enclosing the whole of the fruit-body of some agaric, becoming ruptured as the fruit-body enlarges and sometimes remaining as a cup or pouch around the base of the stipe.

VOLVATE. Possessing a volva.

VOLVOCACEAE. A family of the Volvocales. It contains all the motile genera in which the cells lie in a disk or hollow sphere, not in superimposed tiers.

VOLVOCALES. An order of the Chlorophyceae. It contains the genera in which the vegetative cells are flagellate and motile. They may be solitary or in colonies.

VOLUME SPECIFIC HEAT. Used for soils; the amount of heat required to heat 1 cc. of soil through 1°C.

VULGARIS. Latin meaning 'common'.

W

W. A class of flower pollinated by wind.

WART. A small blunt-tipped rounded upgrowth.

WARTED. Bearing warts.

WATER BLOOM, WATER FLOWER. Large masses of algae, mostly Myxophyceae which sometimes develop suddenly on bodies of fresh water.

WATER CALYX. A calyx in the form of a closed sac, into which hydathodes secrete much water, so that the other flower parts continue their development without risk of damage from dryness.

WATER CULTURE. An experimental means of determining the mineral requirement of a plant. The plant is grown with its roots dipping in solutions of known composition.

WATER FREE SPACE. The space in a tissue in which salts move by diffusion so that their concentration equals that of the external medium.

WATER PORE = HYDATHODE.

WATER STORAGE TISSUE. A group of large, often thin-walled cells, inside a plant, in which water is stored, and from which it is withdrawn in times of drought.

WATER VESICLE. A much enlarged epidermal cell which stores water.

W CHROMOSOME. Sometimes used for the X chromosome where the female is the hetergamous sex.

WEATHERING. The various ways in which the parent material is broken down to form the mature soil.

WEED. A plant growing where it is not wanted by man.

WEIGHT. A differential value assigned to an estimate of a quantity, relative to other estimates of the same quantity, for the purpose of combining the estimates. Usually its invariance or amount of information is taken as the weight for each estimate of a series.

WEIGHTED MEAN. A mean obtained when different classes of observations or quantities are given different weights in calculation.

WHITE ALKALINE SOIL = SALINE SOIL.

WHORL. The arrangement of organs in such a way that they arise from another organ at a common level.

WHORLED. Arranged in a whorl.

WIESENBODEN = GLEY SOIL.

WILD TYPE. Said of an organism or gene of the type predominant in the wild population.

WILT. The collapse of leaves due to unfavourable water relations. It may be due to excessive transpiration, blocking of the xylem elements, pathogens or parasites.

WILTING COEFFICIENT. The percentage water in a soil when a plant begins to wilt.

WILTING POINT. The pF value when the soil cannot supply water rapidly enough to make up the losses by transpiration.

WIND DISPERSAL. The dispersal of spores, seeds and fruits by wind.

WIND POLLINATION = ANEMOPHILY. The conveyance of pollen from anthers to stigmas by wind.

WING. (1) One of the lateral petals in the Papilionaceae.

(2) A flattened outgrowth from a fruit or seed increasing the area without increasing the weight, serving in wind-dispersal.

(3) The downwardly continuing base of a decurrent leaf.

IN TERA CEAE (H) = WINTERANACEAE (EP, BH) = CANEL-LAC EAE . A family of the Parietales (EP, BH), Magnoliales (H).

WINTER ANNUAL. A plant which lives for a short time, grows, and sets its seeds in the colder parts of the year, dies, and survives the rest of the year as seed.

WINTER-GREEN PLANTS. Small plants which retain their green leaves throughout the winter.

WITCHES BROOM. A dense tuft of weak branches formed on a woody plant attacked by a parasite.

WOOD = XYLEM.

WOOD FIBRE. A thick-walled elongated dead element found in wood. It is developed by the elongation and lignification of the wall of a single cell, but differs from a tracheid in its thicker wall, and general inability to conduct water.

WOOD PARENCHYMA = XYLEM PARENCHYMA.

WOOD RAY = XYLEM RAY.

WOOD RAY PARENCHYMA. The parenchymatous cells in a wood ray.

WOOD VESSEL = VESSEL.

WOODY TISSUE. Tissue which has become hard because of the presence of lignin in the walls.

WOOL. A tangled mass of long, soft, whitish hairs.

WOOLLY = LANATE.

WORKING DEPTH. The average distance reached by the general root system of a plant in the soil.

WORKING MEAN. A value approximating to the value of the mean, used for the purpose of lightening the calculations of the mean, sum of squares etc.

WORONINACEAE. A family of the Lagenidiales. The fungus remain naked and amoeboid for a considerable time, forming a 'plasmodium'. Eventually separating into several segments which produce 'cell walls' and become zoosporagia, or the whole 'plasmodium' enlarges to form a single zoosporangium in close contact with the host wall. Clusters of angular, or single round, resting spores may be formed. They are parasitic in algae and fungi.

WOUND CORK. A layer of cork cambium and cork formed below and around wounds, if not too large. It heals the damage, and prevents the entry of parasites into the plant.

WOUND HORMONE. A substance produced in wounded tissue, which is able to influence the subsequent development of parts of the plant.

WOUND PARASITE. A parasite which can only gain entry into the body of a plant through a wound.

WOUND TISSUE. A pad of parenchymatous cells formed by a cambium after wounding. It may give rise to groups of meristematic cells from roots and buds form.

X

x. The basic or haploid number of chromosomes.

XANTH-, XANTHO-. A Greek prefix meaning 'yellow'.

XANTHEIN. A yellow pigment sometimes present in cell sap.

XANTHOPHYLL. $C_{40}H_{56}O_2$. A yellow pigment located in plastids. It occurs with chlorophyll in chloroplasts and occurs in other chromoplasts.

XANTHORRHOEACEAE (H). A family of the Agavales (H). The stem is simple or little branched, and the perianth is dry and more or less glumaceous. It has 6 free, or almost free, segments.

X BODY. An inclusion in a plant cell suffering from a virus disease.

X CHROMOSOME. A heterochromosome associated with the determination of sex, and sometimes occurring alone.

XENIA. The effect of more distantly related pollen, as contrasted with more closely related pollen on the maternal tissue of a fruit.

XENOGAMY. Pollination of a flower from a flower of the same species, but another plant.

XERARCH SUCCESSION. A succession starting on land where conditions are very dry.

XERIC = XEROPHYTE.

XERIC ENVIRONMENT. An environment in which the soil contains very little water, and the atmospheric conditions favour the rapid loss of water from plants.

XEROMORPHIC. Said of plants which are protected from loss of water by unusual morphological characteristics, *e.g.* thick cuticle.

XEROMORPHY. Having morphological characters which aid in the reduction of water-loss.

XEROPHILE. A plant which tolerates a dry habitat.

XEROPHYTE. A plant which can live where the water supply is scanty or there is physiological drought.

XEROSERE = XERARCH SUCCESSION.

X-GENERATION. The gametes.

XANTHOPHYCEAE. A class of the Chrysophyta. The yellow-green chromatophores have more carotinoids than chlorophyll. There is no starch, food being stored as fats or leucosin. The cell-walls are usually of two overlapping halves, and rarely contain cellulose. They are unicellular or multicellular. The motile vegetative and reproductive cells are biflagellate, with the flagella of unequal length and inserted anteriorly. Asexual reproduction is by zoospores or aplanospores. Sexual reproduction is rare, and isogamous by zoogametes.

XYLARIACEAE. A family of the Sphaeriales. The stroma is well-developed and entirely fungal, it is almost always external, at least eventually, and covered at first by a conidial layer. The asci are long and cylindrical. The ascospores are unicellular, inequilaterally ellipsoid, and dark-brown. The paraphyses are filiform.

XYLEM. Wood. The vascular tissue which conducts water and mineral salts throughout the plant and provides mechanical support. It consists of vessels, and/or tracheids, fibres, and some parenchyma.

XYLEM CORE. A solid strand of xylem in the centre of a stele.

XYLEM MOTHER CELL. A daughter cell cut off from a cell of the cambium which is later converted into a component of the xylem.

XYLEM PARENCHYMA. Parenchymatous cells occurring in the xylem, apart from those occurring in the vascular rays.

XYLEM RAY. That part of the vascular ray which traverses the xylem.

XYLIC GAP. A gap in the xylem opposite a leaf base.

XYLOCHROME. A mixture of substances to which the colour of the heart-wood is due. It contains tannins, gums, and resins.

XYLOGENOUS. Living in, or on wood.

XYLOMA. A sclerotium-like body which forms spores internally, and does not put out branches which develop into sporangiophores.

XYLAN A hemicellulose made of xylose units.

XYLOSE. A pentose sugar. 1-xylose is found widely in plants.

$$
\begin{array}{c}
\text{CHO} \\
\mid \\
\text{HO—C—H} \\
\mid \\
\text{H—C—OH} \\
\mid \\
\text{HO—C—H} \\
\mid \\
\text{C—H}_2\text{OH}
\end{array}
$$

XYRIDACEAE (EP, BH, H). A family of the Farinosae (EP), Coronarieae (BH), Xyridales (H). These are perennial herbs with long linear leaves. The bisexual flowers have the parts in threes, are heterchlamydeous, and are in axillary spikes. The calyx is zygomorphic with two small lateral lobes. The 3 petals are fused, with a tube. The 3 free stamens are on the petals, with sometimes 3 outer staminodes. The superior ovary consists of 3 fused carpels, with 1 loculus and many ovules. The fruit is a trivalved capsule, and the seeds have endosperm.

XYRIDALES (H). An order of the Calyciferae. See *Xyridaceae*.

XYRIDEAE (BH) = XYRIDACEAE.

Y

Y CHROMOSOME. With diploid sex-differentiation, the sex chromosome which is present and pairs with the X chromosome in the sex heterozygote. With haploid sex differentiation, the sex chromosome of the male.

YEAST = SACCHAROMYCETACEAE.

YELLOWS. A general term applied to plant diseases characterized by yellowing of the leaves and stunting. It may be caused by a virus, fungus, insect, or toxin.

YELLOW EARTHS. A group of soils in the tropics, sub-tropics, and warm-temperate regions. The yellow colour is probably connected with a high degree of hydration of the ferric oxide.

YELLOW SNOW. Snow coloured yellow by the growth on it of certain algae; sometimes seen in the Alps and Antarctic regions.

YOUTH FORM = JUVENILE FORM.

Z

z. The natural logarithm of the ratio of the two estimated standard deviations.

ZANNICHELLIACEAE (H). A family of the Najadales. They are perennials. There are 1 or more carpels, and the ovules are apical and pendulous.

Z CHROMOSOME. Sometimes used for the Y chromosome, when the female in the heterogametic sex.

ZEAXANTHOL = LUTEIN.

ZECANIN. A glutelin protein.

ZEIN. A prolamine in maize seed.

ZETA POTENTIAL. The difference in potential between the fixed layer of solvent held to the surface of the solid in a hydrophilic gel, and the freely mobile layer which extends into the bulk of the solvent.

ZEUGITE. A cell in which nuclear fusion occurs.

ZINGIBERACEAE (EP, H). A family of the Scitamineae (EP), Zingiberales (H). (BH) include it in the Scitamineae. These are perennial herbs with tuberous rhizomes and lanceolate petiolate leaves, with a ligule. The bisexual usually zygomorphic flowers are in simple or compound inflorescences. The 3 sepals and petals are fused to form tubes. There is 1 stamen, the remainder are staminodes. The inferior ovary consists of 3 fused carpels, with 3 loculi and many ovules. The fruit is a capsule. The seeds have endosperm and perisperm, and are usually arillate.

ZINGIBERALES (H). An order of the Calyciferae (H). The characters as for the Zingiberaceae, but there may be 6-5 stamens.

ZOID = ZOOSPORE.

ZOIDIOPHILOUS. Pollinated by animals.

ZONAL SOIL. A broad group of soils occurring in one of the main climatic regions of the world.

ZONATE. Said of tetraspores formed in a row of 4 and not in a tetrahedral group.

ZONATION. (1) The formation of bands of different colour on the surface of a plant.

(2) The formation, by fungi in culture, of a concentric band of colour, texture, abundance of spores etc.

(3) The occurrence of vegetation in bands each having its characteristic dominant species.

(4) A stage in the development of an oögonium when the contents are arranged in two or more well-marked zones.

ZONE. A band of colour, hairs etc. on the surface.

ZONE LINES Narrow dark-brown or black lines in decaying wood, generally caused by fungi.

ZOOBIOTIC. A parasite living in association with an animal.

ZOOCHOROUS. Said of spores or seeds dispersed by animals.

ZOOGAMITE. A motile gamete.

ZOOGLOEA. A mucilaginous mass of bacteria embedded in a slimy material derived from the swollen cell-walls.

ZOOGONIUM = ZOOSPORE.

ZOOPAGALES. An order of the Phycomycetae. The mycelium is very slender, nonseptate, or septate when older. It is attached to the host by more or less complicated haustoria. Asexual reproducton is by various forms of conidia. Sexual reproduction is by the fusion of gametangia to form zygospores of various shapes. They are parasitic on amoebae, nematodes or insect larvae.

ZOOPHILY. Pollinated by animals.

ZOOSPORANGIUM. A sporangium producing zoospores.

ZOOSPORE. An asexual spore which is motile by 1 to many flagella.

ZOSTERACEAE (H). A family of the Aponogetonales (H). They are marine. There is 1 stamen which is unilocular. There are 2 stigmas and the ovary has 1 pendulous ovule.

ZYGONEMATACEAE. A family of the Zygnematales. The cylindrical cells are in unbranched filaments. The cells have 1 or more peripheral chloroplasts which are spiral and single, or 2, axial and stellate. The amoeboid gametes pass through a tubular connection and are never released.

ZYGNEMATALES = CONJUGALES. An order of the Chlorophyceae. They lack flagellate reproductive cells, and sexual reproduction is by amoeboid gametes.

ZYGOMORPHIC. Divisible in half by one longitudinal plane only.

ZYGONEMA = ZYGOTENE.

ZYGOPHASE. The diploid phase in a life-history.

ZYGOPHORE. A mycelial branch bearing a gametangium.

ZYGOPHYLLACEAE (EP, BH, H). A family of the Geraniales (EP, BH), Malpighiales (H). These are usually shrubby with opposite, often pinnate, stipulate leaves. The regular bisexual flowers are in cymes or compound inflorescences and have the parts in fours or fives. There is a disk or gynophore. There are 10-8 (rarely 15) free stamens often with united basal appendages. The superior ovary consists of 5-4 (or more) fused carpels with many to 1 ovules. The fruit is usually a capsule or schizocarp, and there may or may not be endosperm.

ZYGOPTERIDIDALES = COENOPTERIALES.

ZYGOPTERIDINEAE. A sub-order of the Coenopteridales. The pinna were in 2 or 4 series inserted at an angle to the plane of the leaf blade. The stems were erect, prostrate or clambering with a more or less complicated actinostele. The sporangia were in pedicellate clusters in the ultimate segments of the leaf. The jacket layer was more than one cell thick, with a definite or indefinite annulus.

ZYGOSPORE. A thick-walled resting spore formed by the fusion of the contents of two gametangia.

ZYGOSPOROPHORE. The suspensor of the Zygomycetes.

ZYGOTE. A fertilized ovum, before further differentiation.

ZYGOTENE. The second stage in meiotic prophase in which the chromatin threads approximate in pairs and become loops.

ZYGOTIC. Relating to, or belonging to a zygote.

ZYGOTIC MEIOSIS. A meiosis occurring at the first two divisions of the nucleus resulting from gametic union.

ZYGOTIC NUMBER. The diploid chromosome number.

ZYGOTROPISM. The curvature of the gametophore of one strain of a fungus towards one of an opposite strain.

ZYMASE. An enzyme system in yeasts which breaks down hexose sugars to alcohol and carbon dioxide.

ZYMOGEN. A non-catalytic substance formed by plants and animals as a stage in the development of an enzyme. It is convertible into an enzyme and a protein by the action of a kinase or zymoexcitor.

ZYMOGENUS. Producing enzymes.

ZYMOHEXOSE. The enzyme which catalyses the splitting of hexose diphosphate to triose phosphate.

ZYTHIACEAE = NECTRIOACEAE.

ZWITTER ION. An ion with two active charges such as when the two active groups of an amphoteric substance are ionized together.